Generalized power rule Let u be some function of x. Let $f(x) = u^n$. Then
$$f'(x) = n \cdot u^{n-1} \cdot u'.$$

Chain rule Let u be some function of x. Let y be a function of u. Then
$$\frac{dy}{dx} = \frac{dy}{du} \cdot \frac{du}{dx}.$$

Logarithmic functions If $f(x) = \ln |x|$, then $f'(x) = \frac{1}{x}$.

If $f(x) = \ln |g(x)|$, then $f'(x) = \frac{g'(x)}{g(x)}$.

Exponential functions If $f(x) = e^x$, then $f'(x) = e^x$.
If $f(x) = e^{g(x)}$, then $f'(x) = g'(x) \cdot e^{g(x)}$.

Trigonometric functions

$f(x)$	$f'(x)$
$\sin x$	$\cos x$
$\cos x$	$-\sin x$
$\tan x$	$\sec^2 x$
$\cot x$	$-\csc^2 x$
$\text{Arcsin } x$	$\dfrac{1}{\sqrt{1-x^2}}$
$\text{Arccos } x$	$\dfrac{-1}{\sqrt{1-x^2}}$
$\text{Arctan } x$	$\dfrac{1}{1+x^2}$

Essential Calculus with Applications

Second Edition

The Ruck-A-Chucky bridge, shown here and on the cover, is designed to cross the reservoir of the proposed Auburn Dam on the north fork of the American River in northern California. The unusual design of the bridge was chosen because of the difficult terrain in the area. Conventional design would have required concrete piers; these piers would have been extremely costly because of the 450-foot water depth and the 40° slope of the canyon walls. In the accepted design, the bridge will be suspended by cables from the canyon walls. The cables form a *hyperbolic paraboloid*—cross sections in one direction are hyperbolas with equations of the form $x^2 - y^2 = a^2$, while cross sections in another direction are parabolas with equations of the form $y = ax^2$.

Scott, Foresman and Company

Glenview, Illinois   Dallas, Tex.   Oakland, N.J.
Palo Alto, Cal.   Tucker, Ga.   London, England

Essential Calculus with Applications

Second Edition

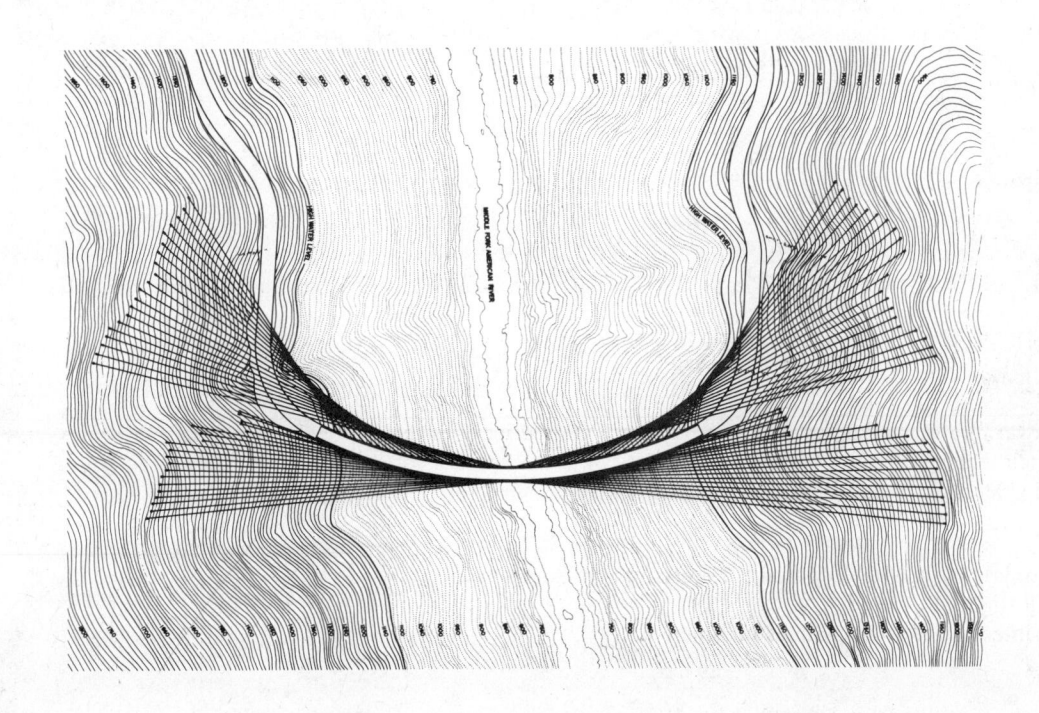

Margaret L. Lial/Charles D. Miller

American River College

Cover photo and title page drawing courtesy of Hedrich Blessing
Architects: Skidmore, Owings & Merrill
Engineers: T. Y. Lin International and Hanson Engineers

Copyright © 1980, 1975 Scott, Foresman and Company.
All Rights Reserved.
Printed in the United States of America.

Library of Congress Cataloging in Publication Data

Lial, Margaret L
    Essential calculus.

    Bibliography: p. 485.
    Includes index.
    1. Calculus. I. Miller, Charles David, 1942–
joint author. II. Title.
QA303.L482   1980   515   79-20317
ISBN 0-673-15248-0

   5 6 7-VHS-85 84 83

Preface

The Second Edition of *Essential Calculus with Applications* provides a thorough treatment of the ideas of calculus needed for further work in management, natural, or social sciences. The only prerequisite assumed is a course in algebra, although knowledge of trigonometry is needed for the optional chapter on trigonometry.

While the book contains an abundance of practical examples of the topics presented, the emphasis is on the mathematics, not the details of applications.

The Second Edition features many improvements:

- *Examples and applications* have been greatly increased in both number and quality. This edition has 40% more drill problems as well as more than twice as many applied problems. Calculator problems have also been added.
- *Applications are labeled in the margin* by discipline, so that those of particular interest to a given class may be easily identified.
- *Exercises have been added to the margin* of the text. By using these marginal exercises, students can test their understanding of the material as they go and immediately identify any difficulties. The answers to the marginal exercises are given on the same page.
- *Mathematical models* are discussed very early in the text and are emphasized in applications throughout the book.
- *Actual cases* follow most of the chapters of the book. These cases, including seven new to this edition, apply chapter topics to a variety of real-world situations. We have tried to make the cases useful to the instructor by keeping them short and simple.
- *New topics* include

    average rates of change   differential approximation
    average cost   optimal lot size
    probability density functions   volume
    trigonometry   logistic curves.
    improper integrals

- *Algebra* used in the exercises has been simplified. The exercises test the students' understanding of calculus, not sophisticated factoring or algebraic techniques.

Supplements available include the following:

- A *Study Guide with Computer Programs* by Margaret Lial features computer programs for many of the sections of this book. This *Study Guide* can be assigned for student purchase or used as a source for instructor-prepared materials.
- An *Instructor's Guide* features answers to even-numbered exercises, two alternate forms of a test for each chapter, and solutions to many of the exercises in the book. These solutions may be made available to students as desired.

We wish to thank the many users of the first edition of this text who filled out questionnaires which were useful in revising it. In addition, we received specific advice on this revision from Louis F. Bush, San Diego City College; Philip Crooke, Vanderbilt University; Herb Gindler, San Diego State University; Ernest P. Lane, Appalachian State University; Lakshmi N. Nigam, Quinnipac College; James H. Reed, University of South Florida; J. Bryan Sperry, Pittsburg State University; Jerry L. Young, Boise State University.

At Scott, Foresman, Dorothy Raymond, Sharon Noble, Greg Odjakjian, Charles Dawkins, John Gibbs, and Earl Karn were very helpful in identifying improvements needed in this edition. Pam Carlson and Margaret Prullage helped turn these ideas into the finished book.

Margaret Lial
Charles Miller

Contents

Algebra Pretest

1 Review of Algebra 1

1.1 The Real Numbers 1
1.2 Linear Equations and Inequalities 8
1.3 Polynomials 15
1.4 Quadratic Equations and Inequalities 21
1.5 Rational Expressions 27
1.6 Exponents and Roots 32
      Chapter Summary 40
      Chapter Test 41

2 Functions and Models 43

2.1 Functions 43
2.2 Linear Functions 51
2.3 Slope and the Equation of a Line 59
2.4 Linear Function Models 65
2.5 Polynomial Function Models 72
2.6 Rational Function Models 80
      Chapter Summary 88
      Chapter Test 88
      Case 1: Estimating Seed Demands—The Upjohn Company 90
      Case 2: Marginal Cost—Booz, Allen and Hamilton 91

3 The Derivative *93*

- 3.1 Limits *93*
- 3.2 Continuity *106*
- 3.3 Rates of Change *113*
- 3.4 Definition of the Derivative *121*
- 3.5 Some Shortcuts for Finding Derivatives *130*
- 3.6 Derivatives of Products and Quotients *138*
- 3.7 The Chain Rule *144*
- Chapter Summary *149*
- Chapter Test *150*

4 Applications of the Derivative *153*

- 4.1 Relative Maximums and Minimums *153*
- 4.2 Absolute Maximums and Minimums *164*
- 4.3 The Second Derivative Test and Curve Sketching *170*
- 4.4 Applications of Maximums and Minimums *181*
- 4.5 Economic Lot Size (Optional) *190*
- 4.6 Implicit Differentiation (Optional) *194*
- 4.7 Newton's Method (Optional) *201*
- 4.8 Approximation by Differentials *205*
- Chapter Summary *209*
- Chapter Test *210*
- Case 3: Compression of the Trachea and Bronchi Due to Coughing *212*
- Case 4: Pricing an Airliner—The Boeing Company *212*
- Case 5: A Total Cost Model for a Training Program *215*

5 Exponential and Logarithmic Functions   *218*

- 5.1 **Exponential Functions**   *218*
- 5.2 **Logarithmic Functions**   *226*
- 5.3 **Applications of Exponential and Logarithmic Functions**   *233*
- 5.4 **Derivatives of Exponential and Logarithmic Functions**   *241*
- **Chapter Summary**   *248*
- **Chapter Test**   *249*
- **Case 6: The Van Meegeren Art Forgeries**   *251*

6 Integration   *254*

- 6.1 **The Antiderivative**   *254*
- 6.2 **Area and the Definite Integral**   *261*
- 6.3 **The Fundamental Theorem of Calculus**   *269*
- 6.4 **Some Applications of the Definite Integral**   *273*
- 6.5 **The Area Between Two Curves**   *279*
- 6.6 **Integration by Substitution**   *289*
- 6.7 **Tables of Integrals**   *295*
- **Chapter Summary**   *298*
- **Chapter Test**   *299*
- **Case 7: Estimating Depletion Dates for Minerals**   *301*

7 Further Techniques and Applications of Integration   *304*

- 7.1 **Integration by Parts**   *304*
- 7.2 **Numerical Integration**   *308*
- 7.3 **Finding Volumes by Integration**   *313*
- 7.4 **Improper Integrals**   *318*
- 7.5 **Probability Density Functions**   *323*
- **Chapter Summary**   *332*
- **Chapter Test**   *333*
- **Case 8: A Crop-Planting Model**   *334*

8 Differential Equations *337*

- 8.1 General and Particular Solutions *337*
- 8.2 Separation of Variables *343*
- 8.3 Applications of Differential Equations *347*
- Chapter Summary *355*
- Chapter Test *356*
- Case 9: A Marine Food Chain *357*
- Case 10: Differential Equations in Ecology *359*

9 Functions of Several Variables *362*

- 9.1 Functions of Several Variables *362*
- 9.2 Graphing Functions of Two Variables *366*
- 9.3 Partial Derivatives *372*
- 9.4 Maximums and Minimums *379*
- 9.5 Lagrange Multipliers *387*
- 9.6 An Application—The Least Squares Line *393*
- Chapter Summary *400*
- Chapter Test *401*

10 The Trigonometric Functions *403*

- 10.1 Trigonometric Functions *403*
- 10.2 Finding Values of Trigonometric Functions *412*
- 10.3 Trigonometric Identities *423*
- 10.4 Inverse Trigonometric Functions *430*
- 10.5 Derivatives of Trigonometric Functions *434*
- 10.6 Antiderivatives *440*
- 10.7 Applications *444*
- Chapter Summary *449*
- Chapter Test *451*
- Case 11: The Mathematics of a Honeycomb *452*

Appendix *455*

Tables *456*
   1 Selected Powers of Numbers   *456*
   2 Squares and Square Roots   *457*
   3 Common Logarithms   *458*
   4 Powers of e *460*
   5 Natural Logarithms   *461*
   6 Integrals Involving Trigonometric Functions   *462*
   7 Trigonometric Functions   *464*

List of Symbols   *465*

Answers to Selected Exercises   *467*

Bibliography   *485*

Glossary   *487*

Index   *491*

ALGEBRA PRETEST

The answers to the test problems can be found on the reverse side of the next page.

[1.1]   *Answer* true *or* false *for each of the following.*

   1. -6 is an integer and a rational number
   2. $-\sqrt{5}$ is irrational, but not a real number

   Graph each of the following on a number line.

   3. $x > 4$
   4. $-2 \leq x < 3$
   5. Evaluate $-|-3 + 5| + |-2 - (-1)|$

[1.2]   *Solve the following equations.*

   6. $2x + 3(2x - 1) + 5 = 6x - 2$
   7. $\dfrac{3p}{2} + \dfrac{p - 2}{3} = 5p$

   Solve the following inequalities.

   8. $-6m > 42$
   9. $2z - 7 \leq 8z + 6$

[1.3]   *Factor as completely as possible.*

   10. $6k^2 - 9k^3$
   11. $r^2 - 4r - 12$
   12. $8y^2 - 10y - 3$
   13. $9k^2 + 12k + 4$
   14. $25g^2 - 16h^2$

[1.4]   *Solve the following quadratic equations.*

   15. $2p^2 - 11p = 6$
   16. $x^2 + 2 = 4x$
   17. $3m^2 - 2m - 1 = 0$
   18. Solve $2m^2 - 3m - 20 \leq 0$.

[1.5] Simplify as much as possible.

19. $\dfrac{3k^2}{2k} \cdot \dfrac{5k^3}{9}$

20. $\dfrac{6y^7}{9y^5} \div \dfrac{15y^9}{20y^8}$

21. $\dfrac{3}{k} - \dfrac{1}{4}$

22. $\dfrac{5}{z+3} - \dfrac{2}{z}$

23. $\dfrac{6}{a} + \dfrac{5}{3a} - \dfrac{2}{7a}$

[1.6] Simplify each of the following. Write all answers without exponents.

24. $3^2 \cdot 3^5 \cdot 3^{-8}$

25. $2^{-1} + 3^{-1}$

26. $\dfrac{9^{-1}}{9^{-5}}$

27. $100^{3/2}$

28. $16^{-5/4}$

ANSWERS TO ALGEBRA PRETEST

1. True
2. False
3. ────○─────────→
          4
4. ────●────○────→
       −2   3
5. -1
6. -2
7. $-4/19$
8. $m < -7$
9. $z \geq -13/6$
10. $3k^2(2 - 3k)$
11. $(r - 6)(r + 2)$
12. $(4y + 1)(2y - 3)$
13. $(3k + 2)^2$
14. $(5g + 4h)(5g - 4h)$
15. $6, -1/2$
16. $2 + \sqrt{2}, 2 - \sqrt{2}$
17. $1, -1/3$
18. $-5/2 \leq m \leq 4$
19. $5k^4/6$
20. $8y/9$
21. $(12 - k)/(4k)$
22. $(3z - 6)/[z(z + 3)]$
23. $155/(21a)$
24. $1/3$
25. $5/6$
26. 6561
27. 1000
28. $1/32$

Essential Calculus with Applications

Second Edition

1
Review of Algebra

This book is about the application of calculus to various subjects, mainly business, social science, and biology. In almost all of these applications, we begin with some real-world problem that we want to solve. We then try to create a mathematical model of that problem. We discuss mathematical models in more detail later on, but basically a **mathematical model** is an equation (or other mathematical expression) which represents the given problem.

For example, suppose our real-world problem is to find the area of a floor, 12 feet by 18 feet. Here we would construct a mathematical model by using the formula for the area of a rectangle, Area = Length × Width, or $A = LW$. By substituting our numbers, $L = 12$ and $W = 18$, into this formula, we can find the area $A = 12 \times 18 = 216$ square feet.

Most mathematical models involve algebra. We must first use algebra to set up a model and then to simplify the resulting equations. Since algebra is so vital to a study of the applications of mathematics, we begin the book with a review of some of the fundamental ideas of algebra that we will need. Algebra is a sort of advanced arithmetic, so we begin by looking at the real numbers, the basis of arithmetic.

1.1 THE REAL NUMBERS

A *set* is a collection of objects; in mathematics the objects in a set are usually numbers. Sets are written with *set braces:* {1, 2, 3} is the set of the three numbers 1, 2, and 3.

The numbers used in this book can be illustrated on a *number line*. To draw a number line, choose any point on a horizontal line and label it 0. Then choose any point to the right of 0 and label it 1. The distance between 0 and 1 gives a unit of measure that can be used repeatedly to locate points to the right of 1, which we label 2, 3, 4, and so on, and points to the left of 0, labeled -1, -2, -3, -4, and so on. A number line with several sample numbers located (or **graphed**) on it is shown in Figure 1.1.

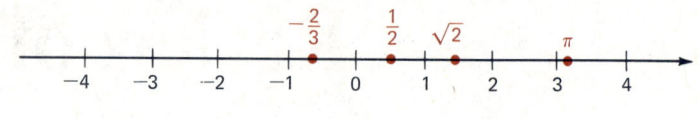

Figure 1.1

1. Draw a number line and graph the numbers $-4, -1, 0, 1, 2.5, 13/4$ on it.

Answer:

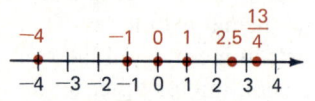

Work Problem 1 at the side.

Any number that can be associated with a point on the number line belongs to the set of **real numbers.*** All the numbers used in this book are real numbers. However, sometimes only a partial set of real numbers is used. The most common of these sets are as follows.

Natural or counting numbers	$\{1, 2, 3, 4, \cdots\}$
Whole numbers	$\{0, 1, 2, 3, 4, \cdots\}$
Integers	$\{\cdots, -3, -2, -1, 0, 1, 2, 3, \cdots\}$
Rational numbers	$\{$All numbers of the form p/q, where p and q are integers, with $q \neq 0\}$
Irrational numbers	$\{$All real numbers that are not rational$\}$

The three dots in the above sets show that the numbers continue indefinitely in the same way. Examples of these sets of numbers, as well as the relationships among them, are shown in Figure 1.2. Notice, for example, that the integers are also rational numbers and real numbers, but the integers are not irrational numbers.

One example of an irrational number is π, the ratio of the circumference of a circle to its diameter. The number π can be approximated by writing $\pi \approx 3.14159$ or $\pi \approx 22/7$ ($\approx$ means "is approximately equal to"), but there is no rational number that is exactly equal to π. Another irrational number can be found by constructing a triangle having a 90° angle, with the two shortest sides each 1 unit long, as shown in Figure 1.3. The third side can be shown to have a length which is irrational (the length is $\sqrt{2}$ units).

Many whole numbers have square roots which are irrational numbers; in fact, if a whole number is not the square of an integer, then its square root is irrational.

Figure 1.3

---

* Not all numbers are real numbers. An example of a number that is not a real number is $\sqrt{-1}$.

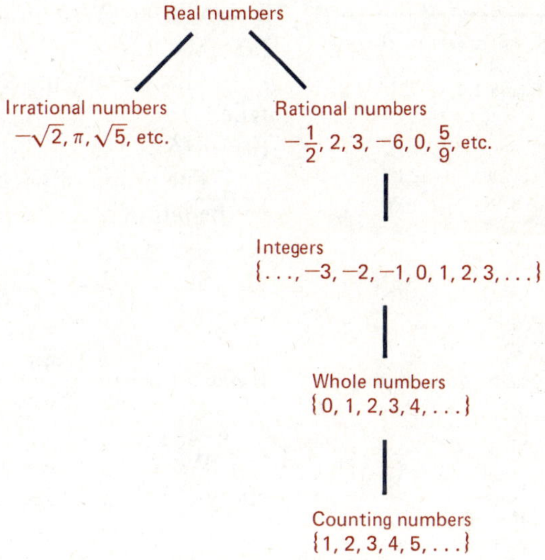

1.1 The Real Numbers

```
                        Real numbers
                       /            \
           Irrational numbers    Rational numbers
           −√2, π, √5, etc.      −1/2, 2, 3, −6, 0, 5/9, etc.
                                      |
                                   Integers
                              {..., −3, −2, −1, 0, 1, 2, 3, ...}
                                      |
                                Whole numbers
                                {0, 1, 2, 3, 4, ...}
                                      |
                               Counting numbers
                                {1, 2, 3, 4, 5, ...}
```

Figure 1.2

Example 1 Name all the sets of numbers that apply to the following.
(a) 6

Consult Figure 1.2 to see that 6 is a counting number, whole number, integer, rational number, and real number.

(b) 3/4

This number is rational and real.

(c) $\sqrt{8}$

Since 8 is not the square of an integer, $\sqrt{8}$ is irrational and real. ∎

Work Problem 2 at the side.

It is often necessary to write that one number is greater than or less than another number. To do so, we can use the following symbols.

$<$ means *less than* $\leq$ means *less than or equal to*
$>$ means *greater than* $\geq$ means *greater than or equal to*

The following definitions tell us how to use the number line to decide which of two given numbers is either greater or less.

If *a* is to the left of *b* on a number line, then $a < b$.

If *a* is to the right of *b* on a number line, then $a > b$.

---

2. Name all the sets of numbers that apply to the following.
(a) −2
(b) −5/8
(c) π

Answer:
(a) integer, rational, real
(b) rational, real
(c) irrational, real

4 Review of Algebra

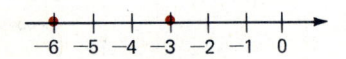

Figure 1.4

Example 2 Write *true* or *false* for the following.
(a) $8 < 12$
This statement says that 8 is less than 12, which we know is true.
(b) $-6 > -3$
The graph of Figure 1.4 shows both -6 and -3. Since -6 is to the *left* of -3, $-6 < -3$. Thus, the original statement is false.
(c) $-2 \leq -2$
Since $-2 = -2$, this statement is true. ∎

3. Write *true* or *false*.
(a) $-9 \leq -2$
(b) $8 > -3$
(c) $-14 \leq -20$

Answer:
(a) true
(b) true
(c) false

Work Problem 3 at the side.

We can use a number line to draw graphs of numbers, as shown in the next few examples.

Example 3 Graph all integers x such that $1 < x < 5$.
The only integers between 1 and 5 are 2, 3, and 4. These integers are graphed on the number line of Figure 1.5. ∎

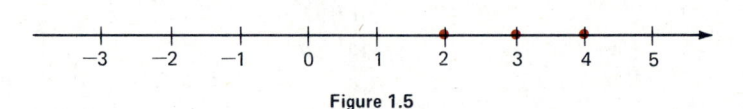

Figure 1.5

4. Graph all integers x such that
(a) $-3 < x < 5$
(b) $1 \leq x \leq 5$.

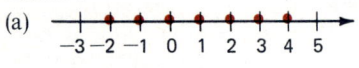

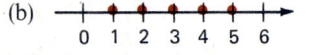

Answer:
(a)
(b)

Work Problem 4 at the side.

Example 4 Graph all real numbers x such that $-2 < x < 3$.
Here we need all the real numbers between -2 and 3 and not just the integers. To graph these numbers, draw a heavy line from -2 to 3 on the number line, as in Figure 1.6. Open dots at -2 and 3 show that neither of these points belongs to the graph. ∎

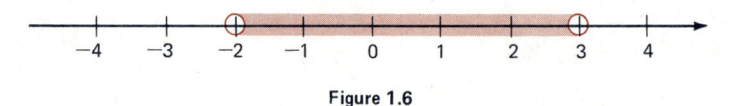

Figure 1.6

5. Graph all real numbers x such that
(a) $-5 < x < 1$
(b) $4 < x < 7$.

Answer:
(a)
(b)

Work Problem 5 at the side.

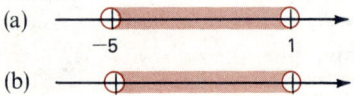

Example 5 Graph all real numbers x such that $x \geq -2$.
Start at -2 and draw a heavy line to the right, as in Figure 1.7. Put a heavy dot at -2 to show that -2 itself is part of the graph. ∎

Figure 1.7

6. Graph all real numbers x such that
(a) $x \geq 4$
(b) $-2 \leq x \leq 1$.

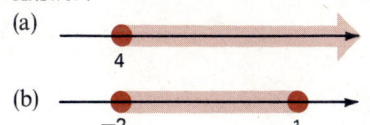

Answer:
(a)
(b)

Work Problem 6 at the side.

Distance is always given as a nonnegative number. For example, the distance from 0 to -2 on a number line is 2, the same as the distance from 0 to 2. We say that 2 is the absolute value of both numbers, 2 and -2. In general, the **absolute value** of a number a is the distance on the number line from a to 0. We write the absolute value of a as $|a|$. We can state the definition of absolute value more formally as

$$|a| = a \quad \text{if } a \geq 0$$
$$|a| = -a \quad \text{if } a < 0.$$

The second part of this definition requires a little care. If a is a *negative* number, then $-a$ is a *positive* number. Thus, for any value of a, we have $|a| \geq 0$.

Example 6 Find the following.
(a) $|5|$
Since $5 > 0$, then $|5| = 5$.
(b) $|-5| = 5$
(c) $-|-5| = -(5) = -5$
(d) $|0| = 0$
(e) $|8 - 9|$
First simplify the expression inside the absolute value bars.
$$|8 - 9| = |-1| = 1$$
(f) $|-4 - 7| = |-11| = 11$ ∎

7. Find the following.
(a) $|-6|$
(b) $-|7|$
(c) $-|-2|$
(d) $|-3 - 4|$
(e) $|2 - 7|$

Answer:
(a) 6
(b) -7
(c) -2
(d) 7
(e) 5

Work Problem 7 at the side.

1.1 EXERCISES

Name all the sets of numbers that apply to the following. See Example 1.

1. 6
2. -9
3. -7
4. 0
5. $1/2$
6. $-5/11$
7. $\sqrt{7}$
8. $-\sqrt{11}$
9. π
10. $1/\pi$

Label the following as true *or* false.

11. Every integer is a rational number.
12. Every integer is a whole number.
13. Every whole number is an integer.
14. Some whole numbers are not natural numbers.
15. There is a natural number that is not a whole number.
16. Every rational number is a natural number.
17. Every natural number is a rational number.
18. No whole numbers are rational.

Graph each of the following on a number line. See Example 3.

19. All integers x such that $-5 < x < 5$
20. All integers x such that $-4 < x < 2$
21. All whole numbers x such that $x \leq 3$
22. All whole numbers x such that $1 \leq x \leq 8$
23. All natural numbers x such that $-1 < x < 5$
24. All natural numbers x such that $x \leq 2$

Graph all real numbers x satisfying the following conditions. See Examples 4 and 5.

25. $x < 4$
26. $x > 5$
27. $6 \leq x$
28. $x \geq 8$
29. $x > -2$
30. $-4 < x$
31. $-5 < x < -3$
32. $3 < x < 5$
33. $-3 \leq x \leq 6$
34. $8 \leq x \leq 14$
35. $1 < x \leq 6$
36. $-4 \leq x < 3$

Evaluate the following. See Example 6.

37. $|-8|$
38. $|9|$
39. $-|-4|$
40. $-|-2|$
41. $|6 - 4|$
42. $|3 - 17|$
43. $-|12 + (-8)|$
44. $|-6 + (-15)|$
45. $|8 - (-9)|$
46. $|-3 - (-2)|$
47. $|8| - |-4|$
48. $|-9| - |-12|$
49. $-|-4| - |-1 - 14|$
50. $-|6| - |-12 - 4|$

In each of the following problems, fill in the blanks with either $=$, $<$, or $>$.

51. $|5|$ _____ $|-5|$ **52.** $|3|$ _____ $|-3|$

53. $-|7|$ _____ $|7|$ **54.** $-|-4|$ _____ $|4|$

55. $|10 - 3|$ _____ $|3 - 10|$ **56.** $|6 - (-4)|$ _____ $|-4 - 6|$

57. $|1 - 4|$ _____ $|4 - 1|$ **58.** $|10 - 8|$ _____ $|8 - 10|$

59. $|-2 + 8|$ _____ $|2 - 8|$ **60.** $|3 + 1|$ _____ $|-3 - 1|$

61. $|3| \cdot |-5|$ _____ $|3(-5)|$ **62.** $|3| \cdot |2|$ _____ $|3(2)|$

63. $|3 - 2|$ _____ $|3| - |2|$ **64.** $|5 - 1|$ _____ $|5| - |1|$

65. In general, if a and b are any real numbers having the same sign (both negative or both positive), is it always true that

$$|a + b| = |a| + |b|?$$

66. If a and b are *any* real numbers, is it always true that

$$|a + b| = |a| + |b|?$$

APPLIED PROBLEMS

Use inequality symbols to rewrite the following statements which are based on an article in Business Week *magazine. Use x to represent the unknown in each exercise.*

Example To rewrite "No more than 40% of the employees of Delmar Industries are men," we let x represent the percent of men. Then $x \leq 40$. ∎

67. Shaklee Products pays its distributors at least 35% of the money received from the sale of its products.

68. Over the last two years, company sales have varied between 102 million dollars and 179 million dollars.

69. The best 5300 of the Shaklee distributors make at least $12,600 per year. (Hint: Let x represent the annual income of the best 5300 distributors.)

70. No distributor made more than $346,000 last year.

71. Manufacturing costs for Shaklee Products are at least 15% of the retail price.

72. In the last few years the percent of the firm's sales in California has varied between 18% and 57% of total sales.

Status measurement

Sociologists measure the status of an individual within a society by evaluating for that individual the number x, which gives the percentage of the population with less income than the given person, and the number y, the percentage of the population with less education. The average status is defined as $(x + y)/2$, while the individual's status incongruity is defined by $|(x - y)/2|$. People with high status incongruities would include unemployed Ph.D's (low x, high y) and millionaires who didn't make it past the second grade (high x, low y).

73. What is the highest possible average status for an individual? The lowest?

74. What is the highest possible status incongruity for an individual? The lowest?

75. Jolene Rizzo makes more money than 56% of the population and has more education than 78%. Find her average status and status incongruity.

76. A popular movie star makes more money than 97% of the population and is better educated than 12%. Find the average status and status incongruity for this individual.

1.2 LINEAR EQUATIONS AND INEQUALITIES

One of the main uses of algebra is to solve equations. An **equation** states that two expressions are equal. Examples of equations include $x + 6 = 9$, $4y + 8 = 12$, and $9z = -36$. The letter in each equation, the unknown, is called the **variable.**

A **solution** of an equation is a number which, when substituted for the variable in the equation, produces a true statement. For example, if the number 9 is substituted for x in the equation $2x + 1 = 19$, we get

$$2x + 1 = 19$$
$$2(9) + 1 = 19$$
$$18 + 1 = 19,$$

a true statement. Thus, 9 is a solution of the equation $2x + 1 = 19$.

1. Is -4 a solution of the following equations?
(a) $3x + 5 = -7$
(b) $2x - 3 = 5$

Answer:
(a) yes
(b) no

Work Problem 1 at the side.

Equations which can be written in the form $ax + b = c$, where a, b, and c are real numbers, with $a \neq 0$, are called **linear equations.** Examples of linear equations include $5y + 9 = 16$, $8x = 4$, and $-3p + 5 = -8$. Equations that are *not* linear include absolute value equations, such as $|x| = 4$.

The following properties are used to help solve linear equations. For any real numbers a, b, and c, the following statements are true.

1.2 Linear Equations and Inequalities

$a(b + c) = ab + ac$ **Distributive property**

If $a = b$, then $a + c = b + c$. **Addition property of equality**
(The same number may be added to both sides of an equation.)

If $a = b$, then $ac = bc$. **Multiplication property of equality**
(The same number may be multiplied on both sides of an equation.)

Example 1 Solve the linear equation $5x - 3 = 12$.

First, use the addition property of equality and add 3 to both sides. We do this to isolate the term containing the variable on one side of the equals sign.

$$5x - 3 = 12$$
$$5x - 3 + 3 = 12 + 3 \quad \text{Add 3 to both sides}$$
$$5x = 15$$

On the left we have $5x$, but we want just $1x$, or x. To get $1x$, use the fact that $(1/5) \cdot 5 = 1$, and multiply both sides of the equation by $1/5$.

$$5x = 15$$
$$\frac{1}{5}(5x) = \frac{1}{5}(15) \quad \text{Multiply both sides by } \frac{1}{5}$$
$$1x = 3$$
$$x = 3$$

The solution of the original equation, $5x - 3 = 12$, is thus 3, which can be checked by substituting 3 for x in the original equation. ∎

2. Solve the following.
(a) $3p - 5 = 19$
(b) $4y + 3 = -5$
(c) $-2k + 6 = 2$

Answer:
(a) 8
(b) -2
(c) 2

Work Problem 2 at the side.

Example 2 Solve $2k + 3(k - 4) = 2(k - 3)$.

We must first simplify this equation using the distributive property. By this property, $3(k - 4)$ is $3k - 3 \cdot 4$, or $3k - 12$. Also, $2(k - 3)$ is $2k - 2 \cdot 3$, or $2k - 6$. The equation can now be written as

$$2k + 3(k - 4) = 2(k - 3)$$
$$2k + 3k - 12 = 2k - 6.$$

On the left, $2k + 3k = (2 + 3)k = 5k$, again by the distributive property. We have

$$5k - 12 = 2k - 6.$$

One way to proceed is to add $-2k$ to both sides.

$$5k - 12 + (-2k) = 2k - 6 + (-2k)$$
$$3k - 12 = -6$$

Now add 12 to both sides.

$$3k - 12 + 12 = -6 + 12$$
$$3k = 6$$

Finally, multiply both sides by 1/3.

$$\frac{1}{3}(3k) = \frac{1}{3}(6)$$
$$k = 2$$

The solution is 2. ∎

3. Solve the following.
(a) $3(m - 6) + 2(m + 4) = 4m - 2$
(b) $-2(y + 3) + 4y = 3(y + 1) - 6$

Answer:
(a) 8
(b) -3

Work Problem 3 at the side.

We can also solve equations involving fractions, as shown in the following example.

Example 3 Solve $\dfrac{r}{10} - \dfrac{2}{15} = \dfrac{3r}{20} - \dfrac{1}{5}.$

To solve this equation, first eliminate all denominators by multiplying both sides of the equation by a **common denominator,** a number which can be divided (with no remainder) by each denominator in the equation. Here the denominators are 10, 15, 20, and 5. Each of these numbers can be divided into 60, so that 60 is our common denominator. Multiply both sides of the equation by 60 and use the distributive property.

$$\frac{r}{10} - \frac{2}{15} = \frac{3r}{20} - \frac{1}{5}$$

$$60\left(\frac{r}{10} - \frac{2}{15}\right) = 60\left(\frac{3r}{20} - \frac{1}{5}\right)$$

$$60\left(\frac{r}{10}\right) - 60\left(\frac{2}{15}\right) = 60\left(\frac{3r}{20}\right) - 60\left(\frac{1}{5}\right)$$

$$6r - 8 = 9r - 12$$

Add $-6r$ and 12 to both sides.

$$6r - 8 + (-6r) + 12 = 9r - 12 + (-6r) + 12$$
$$4 = 3r$$

Multiply both sides by 1/3, ending up with

$$r = \frac{4}{3}. \quad \blacksquare$$

1.2 Linear Equations and Inequalities   11

4. Solve the following.

(a) $\dfrac{x}{2} - \dfrac{x}{4} = 6$

(b) $\dfrac{2x}{3} + \dfrac{1}{2} = \dfrac{x}{4} - \dfrac{9}{2}$

Answer:
(a) 24
(b) −12

Work Problem 4 at the side.

Example 4 Solve $\dfrac{4}{3(k+2)} - \dfrac{k}{3(k+2)} = \dfrac{5}{3}$.

Multiply both sides of the equation by the common denominator $3(k+2)$.

$$3(k+2) \cdot \dfrac{4}{3(k+2)} - 3(k+2) \cdot \dfrac{k}{3(k+2)} = 3(k+2) \cdot \dfrac{5}{3}$$

$4 - k = 5(k + 2)$
$4 - k = 5k + 10$ Distributive property
$4 - k + k = 5k + 10 + k$ Add k on both sides
$4 = 6k + 10$
$4 + (-10) = 6k + 10 + (-10)$ Add -10 on both sides
$-6 = 6k$
$-1 = k$ Multiply by 1/6

The solution is -1. ∎

5. Solve the equation

$\dfrac{5p+1}{3(p+1)} = \dfrac{3p-3}{3(p+1)} + \dfrac{9p-3}{3(p+1)}$.

Answer: 1

Work Problem 5 at the side.

An equation states that two expressions are equal; an **inequality** states that they are unequal. Examples of inequalities include $3x + 1 < 6$, $5y - 8 \geq 4$, and $-3x > 6$. We can solve inequalities using properties very similar to those used for solving equations. In the following properties, a, b, and c represent real numbers.*

If $a < b$, then $a + c < b + c$ **Addition property of inequality**

If $a < b$, and if $c > 0$, then $ac < bc$.   **Multiplication property of inequality, $c > 0$**
(Both sides of an inequality may be multiplied by the same *positive* number without changing the direction of the inequality symbol.)

If $a < b$, and if $c < 0$, then $ac > bc$.   **Multiplication property of inequality, $c < 0$**
(Both sides of an inequality may be multiplied by a *negative* number, *if the direction of the inequality symbol is reversed.*)

---

* The following properties can also be stated for $>$, $\geq$, or $\leq$.

Be careful with the second form of the multiplication property of inequality; if you multiply both sides of an inequality by a negative number, then the direction of the inequality must be reversed. For example, to solve $-3x < 12$, we would multiply both sides by $-1/3$, getting

$$-\frac{1}{3}(-3x) > -\frac{1}{3}(12) \qquad \text{Here we changed } < \text{ to } >$$

$$x > -4.$$

Example 5 Solve $3x + 5 \geq 11$.

First, add -5 to both sides.

$$3x + 5 + (-5) \geq 11 + (-5)$$
$$3x \geq 6$$

Now multiply both sides by $1/3$.

$$\frac{1}{3}(3x) \geq \frac{1}{3}(6)$$
$$x \geq 2$$

Why did we *not* change the direction of the inequality symbol? ∎

Work Problem 6 at the side.

Example 6 Solve $4 - 3y \leq 7 + 2y$.

Add -4 to both sides.

$$4 - 3y + (-4) \leq 7 + 2y + (-4)$$
$$-3y \leq 3 + 2y$$

Add $-2y$ to both sides. Remember that *adding* to both sides never changes the direction of the inequality symbol.

$$-3y + (-2y) \leq 3 + 2y + (-2y)$$
$$-5y \leq 3$$

Multiply both sides by $-1/5$. Since $-1/5$ is negative, we must change the direction of the inequality symbol.

$$-\frac{1}{5}(-5y) \geq -\frac{1}{5}(3)$$

$$y \geq -\frac{3}{5} \quad ∎$$

6. Solve the following.
  (a) $5x - 11 < 14$
  (b) $-3x \leq -12$
  (c) $-8x \geq 32$

Answer:
  (a) $x < 5$
  (b) $x \geq 4$
  (c) $x \leq -4$

7. Solve the following.
  (a) $8 - 6t \geq 2t + 24$
  (b) $-4r + 3(r + 1) < 2r$

Answer:
  (a) $t \leq -2$
  (b) $r > 1$

Work Problem 7 at the side.

1.2 EXERCISES

Solve the following equations. See Examples 1–4.

1. $4x - 1 = 11$
2. $3x + 5 = 23$
3. $-2k + 1 = 19$
4. $-5y - 6 = 14$
5. $4p - 11 + 3p = 2p - 1$
6. $8y - 5y + 4 = 2y - 9$
7. $9x - 2(x - 6) = 10x + 3$
8. $5r - 6(r + 4) = -5r - 4$
9. $2(k - 5) + 4k - 6 = 2k$
10. $2(z - 4) = 7z + 2 - 2z$
11. $\dfrac{2f}{5} - \dfrac{f - 3}{5} = 2$
12. $\dfrac{3w}{4} - \dfrac{2w}{3} = \dfrac{w - 6}{3}$
13. $\dfrac{m + 1}{8} - \dfrac{m - 2}{3} = 1$
14. $\dfrac{y - 6}{3} - \dfrac{y + 2}{5} = 2$
15. $\dfrac{2}{r} + \dfrac{3}{2r} = \dfrac{7}{6}$
16. $\dfrac{2}{q} - \dfrac{3}{4q} = -\dfrac{5}{12}$
17. $\dfrac{1}{m - 1} + \dfrac{2}{3(m - 1)} = -\dfrac{5}{12}$
18. $\dfrac{4}{a + 2} - \dfrac{1}{3(a + 2)} = \dfrac{11}{9}$
19. $\dfrac{3}{4(z - 2)} - \dfrac{2}{3(z - 2)} = \dfrac{1}{36}$
20. $\dfrac{2}{3x + 7} - \dfrac{5}{2(3x + 7)} = -\dfrac{1}{56}$

Solve the following inequalities. See Examples 5 and 6.

21. $6x \leq -18$
22. $4m > -32$
23. $-3p < 18$
24. $-5z \leq 40$
25. $-9a < 0$
26. $-4k \geq 0$
27. $2x + 1 \leq 9$
28. $3y - 2 < 10$
29. $6k - 4 < 3k - 1$
30. $2a - 2 > 4a + 2$
31. $m - (4 + 2m) + 3 < 2m + 2$
32. $2p - (3 - p) \leq -7p - 2$
33. $-2(3y - 8) \geq 5(4y - 2)$
34. $5r - (r + 2) \geq 3(r - 1) + 5$

APPLIED PROBLEMS

In the metric system of weights and measures, temperature is measured in degrees Celsius (°C) instead of degrees Fahrenheit (°F). To convert back and forth between the two systems, use the equations

$$C = \dfrac{5(F - 32)}{9} \quad \text{and} \quad F = \dfrac{9}{5}C + 32.$$

Review of Algebra

In each of the following exercises, convert to the other system. Round answers to the nearest tenth of a degree if necessary.

35. 20°C   36. 100°C   37. 59°F   38. 86°F

39. 100°F   40. 350°F   41. 40°C   42. 85°C

Loan interest

When a consumer borrows money, the lender must tell the consumer the true annual interest rate of the loan. The method of finding the exact true annual interest rate is given in Section 4.7. However, a quick approximate rate can be found by using the equation

$$A = \frac{2pf}{b(q+1)},$$

where p is the number of payments made in one year, f is the finance charge, b is the balance owed on the loan, and q is the total number of payments. Find the value of the variables not given in each of the following. Round A to the nearest percent and round other variables to the nearest whole numbers.

43. $p = 12$, $f = \$800$, $b = \$4000$, $q = 36$; find A
44. $p = 12$, $f = \$60$, $b = \$740$, $q = 12$; find A
45. $A = 14\%$ (or .14), $p = 12$, $b = \$2000$, $q = 36$; find f
46. $A = 11\%$, $p = 12$, $b = \$1500$, $q = 24$; find f
47. $A = 16\%$, $p = 12$, $f = \$370$, $q = 36$; find b
48. $A = 10\%$, $p = 12$, $f = \$490$, $q = 48$; find b

Unearned interest

When a loan is paid off early, a portion of the finance charge must be returned to the borrower. By one method of calculating finance charge (called the rule of 78), the amount of unearned interest (finance charge to be returned) is given by

$$u = f \cdot \frac{n(n+1)}{q(q+1)}$$

where u represents unearned interest, f is the original finance charge, n is the number of payments remaining when the loan is paid off, and q is the original number of payments. Find the amount of the unearned interest in each of the following.

49. Original finance charge = $800, loan scheduled to run 36 months, paid off with 18 payments remaining.

50. Original finance charge = $1400, loan scheduled to run 48 months, paid off with 12 payments remaining.

51. Original finance charge = $950, loan scheduled to run 24 months, paid off with 6 payments remaining.

52. Original finance charge = $175, loan scheduled to run 12 months, paid off with 3 payments remaining.

Management

According to a recent article in The Wall Street Journal, *the company that makes the Godzilla movies puts out about* $1.2 *million to make each movie. It then receives back about* $5 *million from worldwide showings of the film.*

53. Let x represent the number of Godzilla movies that the company makes. Write an equation for C, the cost in millions to make x films.

54. Write an equation for the total income, I, in millions from x films.

55. Use the formula Profit = Income − Cost to write an equation for the profit in millions from x films.

56. Find the company's total profit in millions if 5 films are made.

Car rentals

Bill and Cheryl Bradkin went to Portland, Maine, for a week. They needed to rent a car, so they checked out two rental firms. Avis wanted $14 *per day, with no mileage fee. Downtown Toyota wanted* $54 *per week and 7¢ per mile.*

57. Let x represent the number of miles that the Bradkins would drive in one week. Set up an inequality expressing the rates of the two firms. Then decide how many miles they would have to drive before the Avis car was the better deal.

1.3 POLYNOMIALS

A **polynomial,** one of the key ideas of algebra, is an expression of the form

$$a_n x^n + a_{n-1} x^{n-1} + \cdots + a_1 x + a_0,$$

where $a_0, a_1, a_2, \ldots, a_n$ are real numbers, n is a whole number, and $a_n \neq 0$. Examples of polynomials include

$$5x^4 + 2x^3 + 6x, \quad 8m^3 + 9m^2 - 6m + 3, \quad 10p, \quad \text{and} \quad -9.$$

Expressions that are *not* polynomials include

$$8x^3 + \frac{6}{x}, \quad \frac{9+x}{2-x}, \quad \text{and} \quad \frac{-p^2 + 5p + 3}{2p - 1}.$$

When we write $9p^4$, the entire expression is called a **term,** the number 9 is called the **coefficient,** p is the **variable,** and 4 is the **exponent.** The expression p^4 means $p \cdot p \cdot p \cdot p$, while p^2 means $p \cdot p$. We discuss exponents in more detail later in this chapter.

Polynomials can be added or subtracted by the distributive property. For example,

$$12y^4 + 6y^4 = (12 + 6)y^4 = 18y^4$$

and

$$-2m^2 + 8m^2 = (-2 + 8)m^2 = 6m^2.$$

16 Review of Algebra

Note that the polynomial $8y^4 + 2y^5$ cannot be further simplified. Two terms having the same variable and the same exponent are called **like terms;** other terms are called **unlike terms.** Only like terms may be added or subtracted. To subtract polynomials, we need to know that $-(a + b) = -a - b$. The next example shows how to add and subtract polynomials.

Example 1 Add or subtract as indicated.
(a) $(8x^3 - 4x^2 + 6x) + (3x^3 + 5x^2 - 9x + 8)$
Combine like terms.

$$(8x^3 - 4x^2 + 6x) + (3x^3 + 5x^2 - 9x + 8)$$
$$= (8x^3 + 3x^3) + (-4x^2 + 5x^2) + (6x - 9x) + 8$$
$$= 11x^3 + x^2 - 3x + 8$$

(b) $(-4x^4 + 6x^3 - 9x^2 - 12) + (-3x^3 + 8x^2 - 11x + 7)$
$$= -4x^4 + 3x^3 - x^2 - 11x - 5$$

(c) $(2x^2 - 11x + 8) - (7x^2 - 6x + 2)$
$$= (2x^2 - 11x + 8) + (-7x^2 + 6x - 2)$$
$$= -5x^2 - 5x + 6 \quad \blacksquare$$

1. Add or subtract.
(a) $(-2x^2 + 7x + 9)$
  $+ (3x^2 + 2x - 7)$
(b) $(4x + 6) - (13x - 9)$
(c) $(9x^3 - 8x^2 + 2x)$
  $- (9x^3 - 2x^2 - 10)$

Answer:
(a) $x^2 + 9x + 2$
(b) $-9x + 15$
(c) $-6x^2 + 2x + 10$

Work Problem 1 at the side.

Polynomials may be multiplied using the distributive property. For example, the product of $8x$ and $6x - 4$ is found as follows.

$$8x(6x - 4) = 8x(6x) - 8x(4) \qquad \text{Distributive property}$$
$$= 48x^2 - 32x \qquad x \cdot x = x^2$$

The product of $3p - 2$ and $5p + 1$ can be found by using the distributive property twice:

$$(3p - 2)(5p + 1) = (3p - 2)(5p) + (3p - 2)(1)$$
$$= 15p^2 - 10p + 3p - 2$$
$$= 15p^2 - 7p - 2.$$

Work Problem 2 at the side.

2. Find the following products.
(a) $-6r(2r - 5)$
(b) $11m(8m + 3)$
(c) $(5k - 1)(2k + 3)$
(d) $(7z - 3)(2z + 5)$

Answer:
(a) $-12r^2 + 30r$
(b) $88m^2 + 33m$
(c) $10k^2 + 13k - 3$
(d) $14z^2 + 29z - 15$

Multiplication of polynomials can be performed using the distributive property. The reverse process, where we write a polynomial as a product of other polynomials, is called **factoring.** For example, one way to factor the number 18 is to write it as the product $9 \cdot 2$. When 18 is written as $9 \cdot 2$, both 9 and 2 are called **factors** of 18. It is true that $18 = 36 \cdot \frac{1}{2}$, but we do not call 36 and $\frac{1}{2}$

factors of 18; we restrict our attention only to integer factors. Integer factors of 18 are 2, 9; −2, −9; 6, 3; −6, −3; 18, 1; −18, −1.

We will usually be concerned with factoring an algebraic expression such as $15m + 45$. This expression is made up of two terms, $15m$ and 45. Each of these terms can be divided by 15. In fact, $15m = 15 \cdot m$ and $45 = 15 \cdot 3$. We can use the distributive property to write

$$15m + 45 = 15 \cdot m + 15 \cdot 3 = 15(m + 3).$$

Both 15 and $m + 3$ are factors of $15m + 45$. Since 15 divides into all terms of $15m + 45$ (and is the largest number that will do so), it is called the **greatest common factor** for the polynomial $15m + 45$. The process of writing $15m + 45$ as $15(m + 3)$ is called **factoring out** the greatest common factor.

Example 2 Factor out the greatest common factor.
(a) $12p - 18q$
Both $12p$ and $18q$ are divisible by 6. Therefore,

$$12p - 18q = 6 \cdot 2p - 6 \cdot 3q = 6(2p - 3q).$$

(b) $8x^3 - 9x^2 + 15x$
Each of these terms is divisible by x.

$$8x^3 - 9x^2 + 15x = (8x^2) \cdot x - (9x) \cdot x + 15 \cdot x$$
$$= x(8x^2 - 9x + 15). \blacksquare$$

Work Problem 3 at the side.

A polynomial may not have a greatest common factor (other than 1), and yet may still be factorable. For example, the polynomial $x^2 + 5x + 6$ can be factored as $(x + 2)(x + 3)$. To see that this is correct, work out the product $(x + 2)(x + 3)$; you should get $x^2 + 5x + 6$.

If we are given a polynomial such as $x^2 + 5x + 6$, how do we know that it is the product $(x + 2)(x + 3)$? There are two different ways to factor a polynomial of three terms such as $x^2 + 5x + 6$, depending on whether the coefficient of x^2 is 1, or a number other than 1. If the coefficient is 1, proceed as shown in the following example.

Example 3 Factor $y^2 + 8y + 15$.
Since the coefficient of y^2 is understood to be 1, factor by finding two numbers whose *product* is 15, and whose *sum* is 8. Use trial and error to find these numbers. Begin by listing all pairs of integers having a product of 15. As you do this, also form the sum of the numbers.

3. Factor out the greatest common factor.
(a) $12r + 9k$
(b) $75m^2 + 100n^2$
(c) $6m^4 - 9m^3 + 12m^2$

Answer:
(a) $3(4r + 3k)$
(b) $25(3m^2 + 4n^2)$
(c) $3m^2(2m^2 - 3m + 4)$

18   Review of Algebra

$$\begin{array}{cc} \text{Products} & \text{Sums} \\ 15 \cdot 1 = 15 & 15 + 1 = 16 \\ 5 \cdot 3 = 15 & 5 + 3 = 8 \\ (-1) \cdot (-15) = 15 & -1 + (-15) = -16 \\ (-5) \cdot (-3) = 15 & -5 + (-3) = -8 \end{array}$$

The numbers 3 and 5 have a product of 15 and a sum of 8. Thus, $y^2 + 8y + 15$ factors as

$$y^2 + 8y + 15 = (y + 3)(y + 5).$$

We can also write the answer as $(y + 5)(y + 3)$. ∎

4. Factor the following.
(a) $m^2 + 11m + 30$
(b) $h^2 + 10h + 9$

Answer:
(a) $(m + 5)(m + 6)$
(b) $(h + 9)(h + 1)$

Work Problem 4 at the side.

Example 4   Factor $p^2 - 8p - 20$.
We need two numbers whose product is -20 and whose sum is -8. Make a list of all pairs of integers whose product is -20. Choose from this list that pair whose sum is -8. Doing so, you should find the pair -10 and 2; the product of these numbers is -20 and their sum is -8. Therefore,

$$p^2 - 8p - 20 = (p - 10)(p + 2). \quad \blacksquare$$

5. Factor the following.
(a) $a^2 - 7a + 10$
(b) $r^2 - 5r - 14$
(c) $m^2 + 3m - 40$

Answer:
(a) $(a - 5)(a - 2)$
(b) $(r - 7)(r + 2)$
(c) $(m + 8)(m - 5)$

Work Problem 5 at the side.

If the coefficient of the squared term is *not* 1, we must use trial and error, as shown in the next example.

Example 5   Factor $2x^2 + 9x - 5$.
The factors of $2x^2$ are $2x$ and x; the possible factors of -5 are -5 and 1, or 5 and -1. We use trial and error, trying various combinations of these factors until we find the one that works (if, indeed, any work). Let's try the product $(2x + 5)(x - 1)$. Multiply it out.

$$(2x + 5)(x - 1) = (2x + 5)(x) + (2x + 5)(-1)$$
$$= 2x^2 + 5x - 2x - 5$$
$$= 2x^2 + 3x - 5$$

This product is not the one we want. So we try another combination.

$$(2x - 1)(x + 5) = (2x - 1)(x) + (2x - 1)(5)$$
$$= 2x^2 - x + 10x - 5$$
$$= 2x^2 + 9x - 5$$

This combination leads to the correct polynomial; thus

$$2x^2 + 9x - 5 = (2x - 1)(x + 5). \quad \blacksquare$$

1.3 Polynomials 19

6. Factor the following.
(a) $3k^2 + k - 2$
(b) $3m^2 + 5m - 2$
(c) $6p^2 + 13p - 5$

Answer:
(a) $(3k - 2)(k + 1)$
(b) $(3m - 1)(m + 2)$
(c) $(2p + 5)(3p - 1)$

Work Problem 6 at the side.

There are two special types of factorizations that occur so often that we list them for future reference.

$x^2 - y^2 = (x + y)(x - y)$ **Difference of two squares**

$x^2 + 2xy + y^2 = (x + y)^2$ **Perfect square**

Example 6 Factor each of the following.
(a) $x^2 - 25 = x^2 - 5^2 = (x + 5)(x - 5)$
(b) $64p^2 - 49q^2 = (8p)^2 - (7q)^2 = (8p + 7q)(8p - 7q)$
(c) $x^2 + 36$ cannot be factored
(d) $x^2 + 12x + 36 = (x + 6)^2$
(e) $9y^2 - 24yz + 16z^2 = (3y - 4z)^2$ ■

7. Factor the following.
(a) $r^2 - 81$
(b) $9p^2 - 49$
(c) $y^2 + 100$
(d) $m^2 - 8m + 16$
(e) $100k^2 - 60k + 9$

Answer:
(a) $(r + 9)(r - 9)$
(b) $(3p + 7)(3p - 7)$
(c) cannot be factored
(d) $(m - 4)^2$
(e) $(10k - 3)^2$

Work Problem 7 at the side.

Finally, we list two more special types of factorizations that occur from time to time.

$x^3 - y^3 = (x - y)(x^2 + xy + y^2)$ **Difference of two cubes**

$x^3 + y^3 = (x + y)(x^2 - xy + y^2)$ **Sum of two cubes**

Example 7 Factor each of the following.
(a) $y^3 - 8 = y^3 - 2^3 = (y - 2)(y^2 + 2y + 4)$
(b) $m^3 + 125 = m^3 + 5^3 = (m + 5)(m^2 - 5m + 25)$
(c) $8k^3 - 27z^3 = (2k)^3 - (3z)^3 = (2k - 3z)(4k^2 + 6kz + 9z^2)$ ■

8. Factor the following.
(a) $a^3 + 1000$
(b) $z^3 - 64$
(c) $1000m^3 - 27z^3$

Answer:
(a) $(a + 10)(a^2 - 10a + 100)$
(b) $(z - 4)(z^2 + 4z + 16)$
(c) $(10m - 3z)(100m^2 + 30mz + 9z^2)$

Work Problem 8 at the side.

1.3 EXERCISES

Add or subtract as indicated. See Example 1.

1. $(8m + 9) + (6m - 3)$
2. $(-7p - 11) + (8p + 5)$
3. $(-2k - 3) - (7k - 8)$
4. $(12z + 10) - (3z + 9)$
5. $(2x^2 - 6x + 11) + (-3x^2 + 7x - 2)$
6. $(-3a^2 + 2a - 5) + (7a^2 + 2a + 9)$

7. $(-4y^2 - 3y + 8) - (2y^2 - 6y - 2)$

8. $(7b^2 + 2b - 5) - (3b^2 + 2b - 6)$

9. $(2x^3 - 2x^2 + 4x - 3) - (2x^3 + 8x^2 - 1)$

10. $(3y^3 + 9y^2 - 11y + 8) - (-4y^2 + 10y - 6)$

Find each of the following products.

11. $3p(2p - 5)$
12. $4y(8y + 1)$
13. $-9m(2m^2 + 3m - 1)$
14. $2a(4a^2 - 6a + 3)$
15. $(6k - 1)(2k - 3)$
16. $(8r + 3)(r - 1)$
17. $(3y + 5)(2y - 1)$
18. $(2a - 5)(4a + 3)$
19. $(6y - 2)(6y + 2)$
20. $(8p + 3)(8p - 3)$

Factor out the greatest common factor in each of the following. See Example 2.

21. $25k + 30$
22. $6m - 12$
23. $4z + 4$
24. $9y - 9$
25. $8x + 6y + 4z$
26. $15p + 9q + 12s$
27. $m^3 - 9m^2 + 6m$
28. $y^3 + 6y^2 + 8y$
29. $8a^3 - 16a^2 + 24a$
30. $3y^3 + 24y^2 + 9y$

Factor each of the following. If a polynomial cannot be factored, write "cannot be factored." See Examples 3 and 4.

31. $m^2 + 9m + 14$
32. $p^2 - 2p - 15$
33. $x^2 + 4x - 5$
34. $y^2 + y - 72$
35. $z^2 + 9z + 20$
36. $k^2 + 8k + 15$
37. $b^2 - 8b + 7$
38. $r^2 + r - 20$
39. $a^2 + 4a + 5$
40. $y^2 - 6y + 8$
41. $y^2 - 4y - 21$
42. $r^2 + r - 42$

Factor each of the following. If a polynomial cannot be factored, write "cannot be factored." Factor out the greatest common factor as necessary. See Example 5.

43. $6a^2 - 48a - 120$
44. $8h^2 - 24h - 320$
45. $3m^3 + 12m^2 + 9m$
46. $3y^4 - 18y^3 + 15y^2$
47. $2x^2 - 5x - 3$
48. $3r^2 - r - 2$
49. $3a^2 + 10a + 7$
50. $4y^2 + y - 3$
51. $15y^2 + y - 2$
52. $6x^2 + x - 1$
53. $3p^2 - 7p + 10$
54. $8r^2 + r + 6$

55. $5a^2 - 7a - 6$
56. $12s^2 + 11s - 5$
57. $24a^4 + 10a^3 - 4a^2$
58. $18x^5 + 15x^4 - 75x^3$
59. $32z^5 - 20z^4 - 12z^3$
60. $15x^4 - 7x^3 - 4x^2$

Factor each of the following. See Examples 6 and 7.

61. $x^2 - 64$
62. $y^2 - 144$
63. $9m^2 - 25$
64. $4p^2 - 9$
65. $9x^2 + 64$
66. $100a^2 + 9$
67. $z^2 + 14z + 49$
68. $y^2 + 20y + 100$
69. $m^2 - 6m + 9$
70. $a^2 - 10a + 25$
71. $9p^2 - 24p + 16$
72. $16m^2 + 40m + 25$
73. $a^3 - 216$
74. $b^3 + 125$
75. $8r^3 - 27$
76. $1000p^3 + 27$

1.4 QUADRATIC EQUATIONS AND INEQUALITIES

A polynomial equation whose highest exponent is 2 is called a **quadratic equation.** Examples of quadratic equations include

$$x^2 + 5x + 6 = 0, \qquad y^2 = 16, \qquad \text{and} \qquad 3p^2 + 11p = 4.$$

Some quadratic equations can be solved by factoring; this method depends on the following property.

Zero-factor property   If a and b are real numbers, with $ab = 0$, then $a = 0$ or $b = 0$. (If two numbers have a product of 0, then at least one of the numbers must be 0.)

Example 1   Solve the equation $(x - 4)(3x + 7) = 0$.

Here we are told that the product $(x - 4)(3x + 7)$ is equal to 0. By the zero-factor property, this can only be true if one of the factors is 0. That is, $x - 4 = 0$ or $3x + 7 = 0$. If we solve each of these equations separately, we will find the solutions of the original equation.

$$x - 4 = 0 \quad \text{or} \quad 3x + 7 = 0$$
$$x = 4 \qquad\qquad 3x = -7$$
$$x = -\frac{7}{3}$$

The solutions of $(x - 4)(3x + 7) = 0$ are 4 and $-7/3$. ∎

Work Problem 1 at the side.

---

1. Solve the following.
   (a) $(y - 6)(y + 2) = 0$
   (b) $(5k - 3)(k + 5) = 0$
   (c) $(2r - 9)(3r + 5) = 0$

Answer:
(a) $6, -2$
(b) $3/5, -5$
(c) $9/2, -5/3$

Example 2 Solve the quadratic equation $2m^2 + 5m = 12$.

To begin, rewrite the equation so that we have 0 alone on one side of the equals sign. We do this so that we may use the zero-factor property. Add -12 to both sides, giving

$$2m^2 + 5m - 12 = 0.$$

Use trial and error to factor on the left:

$$2m^2 + 5m - 12 = (2m - 3)(m + 4),$$

so that the given equation becomes

$$(2m - 3)(m + 4) = 0.$$

The product on the left can equal 0 only if at least one of the factors is 0.

$$2m - 3 = 0 \quad \text{or} \quad m + 4 = 0$$

Solve each of these equations separately.

$$2m = 3 \qquad\qquad m = -4$$
$$m = \frac{3}{2}$$

The solutions of $2m^2 + 5m = 12$ are 3/2 and -4. ∎

Work Problem 2 at the side.

2. Solve the following.
(a) $y^2 + 3y = 10$
(b) $2r^2 + 9r = 5$
(c) $3k^2 = 2k + 8$

Answer:
(a) 2, -5
(b) 1/2, -5
(c) $-4/3$, 2

Not all quadratic equations can be solved by this kind of factoring. For those equations that cannot, we need the more general method given by the **quadratic formula.**

Quadratic formula The solutions of the quadratic equation $ax^2 + bx + c = 0$, where $a \neq 0$, are given by

$$x = \frac{-b \pm \sqrt{b^2 - 4ac}}{2a}$$

Example 3 Use the quadratic formula to solve $x^2 - 4x - 5 = 0$.

First make sure that the equation has 0 alone on one side of the equals sign. Then identify the letters a, b, and c of the quadratic formula. The coefficient of the squared term gives the value of a; here $a = 1$. Also, $b = -4$ and $c = -5$. (Be careful to get the correct signs.) Substitute these values into the quadratic formula.

1.4 Quadratic Equations and Inequalities

$$x = \frac{-(-4) \pm \sqrt{(-4)^2 - 4(1)(-5)}}{2(1)} \qquad \text{Let } a = 1, b = -4, c = -5$$

$$= \frac{4 \pm \sqrt{16 + 20}}{2} \qquad (-4)^2 = (-4)(-4) = 16$$

$$x = \frac{4 \pm 6}{2} \qquad \sqrt{16 + 20} = \sqrt{36} = 6$$

The $\pm$ sign represents the two solutions of the equation. To find each of the solutions, first use $+$ and then use $-$.

$$x = \frac{4 + 6}{2} = \frac{10}{2} = 5 \quad \text{or} \quad x = \frac{4 - 6}{2} = \frac{-2}{2} = -1.$$

The two solutions are 5 and -1. ∎

Work Problem 3 at the side.

Example 4 Solve $x^2 + 1 = 4x$.

First add $-4x$ to both sides, to get 0 alone on the right side.

$$x^2 - 4x + 1 = 0$$

Now identify the letters a, b, and c. We have $a = 1$, $b = -4$, and $c = 1$. Substitute these numbers into the quadratic formula.

$$x = \frac{-(-4) \pm \sqrt{(-4)^2 - 4(1)(1)}}{2(1)}$$

$$= \frac{4 \pm \sqrt{16 - 4}}{2}$$

$$= \frac{4 \pm \sqrt{12}}{2}$$

To simplify the solutions, write $\sqrt{12}$ as $\sqrt{4 \cdot 3} = \sqrt{4} \cdot \sqrt{3} = 2\sqrt{3}$. Substituting $2\sqrt{3}$ for $\sqrt{12}$ gives

$$= \frac{4 \pm 2\sqrt{3}}{2}$$

$$= \frac{2(2 \pm \sqrt{3})}{2} \qquad \text{Factor } 4 \pm 2\sqrt{3}$$

$$x = 2 \pm \sqrt{3}.$$

The two solutions are $2 + \sqrt{3}$ and $2 - \sqrt{3}$.

The exact values of the solutions are $2 + \sqrt{3}$ and $2 - \sqrt{3}$. In many cases we need a decimal approximation of these solutions.

---

3. Use the quadratic formula to solve the following.

(a) $3x^2 + 11x - 4 = 0$
(b) $2z^2 - 7z - 4 = 0$

Answer:
(a) $1/3, -4$
(b) $-1/2, 4$

Review of Algebra

Use a calculator, or Table 2 in the back of the book, to find that $\sqrt{3} \approx 1.732$, so that (to the nearest thousandth) the solutions are

$$2 + \sqrt{3} \approx 2 + 1.732 = 3.732$$

or

$$2 - \sqrt{3} \approx 2 - 1.732 = .268. \blacksquare$$

4. Find exact and approximate solutions for the following.
(a) $y^2 - 2y = 2$
(b) $x^2 - 6x + 4 = 0$

Answer:
(a) exact: $1 + \sqrt{3}, 1 - \sqrt{3}$;
approximate: $2.732, -.732$
(b) exact: $3 + \sqrt{5}, 3 - \sqrt{5}$;
approximate: $5.236, .764$

Work Problem 4 at the side.

The above quadratic equations all had two different solutions. This is not always the case, as the following example shows.

Example 5 (a) Solve $9x^2 - 30x + 25 = 0$.
We have $a = 9$, $b = -30$, and $c = 25$. By the quadratic formula,

$$x = \frac{-(-30) \pm \sqrt{(-30)^2 - 4(9)(25)}}{2(9)}$$

$$= \frac{30 \pm \sqrt{900 - 900}}{18}$$

$$= \frac{30 \pm 0}{18} = \frac{30}{18} = \frac{5}{3}.$$

The given equation has only one solution.

(b) Solve $x^2 - 6x + 10 = 0$.
Since $a = 1$, $b = -6$, and $c = 10$, we have

$$x = \frac{-(-6) \pm \sqrt{(-6)^2 - 4(1)(10)}}{2(1)} = \frac{6 \pm \sqrt{36 - 40}}{2} = \frac{6 \pm \sqrt{-4}}{2}.$$

There are no real number solutions to this equation since $\sqrt{-4}$ is not a real number. $\blacksquare$

5. Solve the following equations.
(a) $9k^2 - 6k + 1 = 0$
(b) $4m^2 + 28m + 49 = 0$
(c) $2x^2 - 5x + 5 = 0$

Answer:
(a) $1/3$
(b) $-7/2$
(c) No real number solutions

Work Problem 5 at the side.

Quadratic Inequalities A **quadratic inequality** is an inequality of the form $ax^2 + bx + c > 0$ (or $<$, or $\leq$, or $\geq$). The highest exponent is always 2. Examples of quadratic inequalities include

$$x^2 - x - 12 < 0, \qquad 3y^2 + 2y \geq 0, \quad \text{and} \quad m^2 \leq 4.$$

A method of solving quadratic inequalities is shown in the next few examples.

Example 6 Solve the quadratic inequality $x^2 - x - 12 < 0$.
Since $x^2 - x - 12 = (x - 4)(x + 3)$, our given inequality is really the same as

$$(x - 4)(x + 3) < 0.$$

1.4 Quadratic Equations and Inequalities

We want the product of $x - 4$ and $x + 3$ to be negative; this product will be negative if $x - 4$ and $x + 3$ have opposite signs. The factor $x - 4$ is positive when $x - 4 > 0$, or $x > 4$. Thus, $x - 4$ is negative if $x < 4$. In the same way, $x + 3$ is positive when $x > -3$ and negative when $x < -3$. This information is shown in the **sign graph** of Figure 1.8.

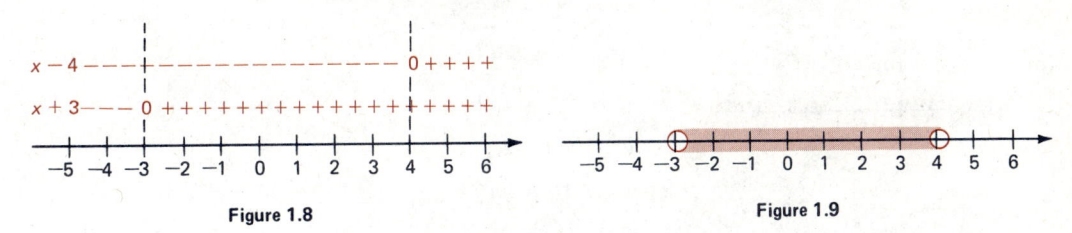

Figure 1.8

Figure 1.9

We said above that the product $(x - 4)(x + 3)$ will be negative when $x - 4$ and $x + 3$ have opposite signs. From the sign graph of Figure 1.8, these expressions have opposite signs whenever x is between -3 and 4. The solution is thus given by $-3 < x < 4$. A graph of this solution is shown in Figure 1.9. ∎

6. Solve the following and graph.
(a) $y^2 + 2y - 3 < 0$
(b) $2p^2 + 3p - 2 < 0$

Answer:
(a) $-3 < y < 1$

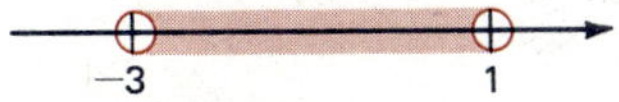

(b) $-2 < p < 1/2$

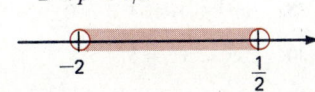

Work Problem 6 at the side.

Example 7 Solve the quadratic inequality $r^2 + 3r \geq 4$.

First rewrite the inequality so that one side is 0.

$$r^2 + 3r \geq 4$$
$$r^2 + 3r - 4 \geq 0 \quad \text{Add } -4 \text{ to both sides}$$

Factor $r^2 + 3r - 4$ as $(r + 4)(r - 1)$. The factor $r + 4$ is positive when $r > -4$ and negative when $r < -4$. Also, $r - 1$ is positive when $r > 1$ and negative when $r < 1$. We use this information to produce the sign graph in Figure 1.10. The product $(r + 4)(r - 1)$ will be positive when $r + 4$ and $r - 1$ have the same signs, either positive or negative. From Figure 1.10, the solution is seen to be made up of those numbers less than or equal to -4, together with those greater than or equal to 1. This can be written

$$r \leq -4 \quad \text{or} \quad r \geq 1.$$

A graph of the solution is given in Figure 1.11. ∎

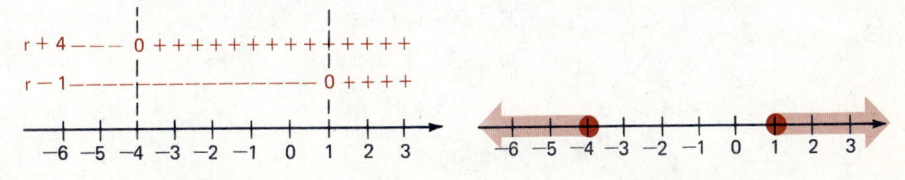

Figure 1.10

Figure 1.11

7. Solve the following and graph.
(a) $k^2 + 2k - 15 \geq 0$
(b) $3m^2 + 7m \geq 6$

Answer:
(a) $k \leq -5$ or $k \geq 3$

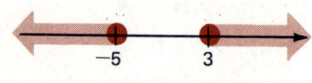

(b) $m \leq -3$ or $m \geq 2/3$

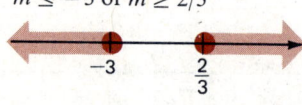

Work Problem 7 at the side.

1.4 EXERCISES

Solve each of the following equations. See Examples 1 and 2.

1. $(y - 5)(y + 4) = 0$
2. $(m + 3)(m - 1) = 0$
3. $x^2 + 5x + 6 = 0$
4. $y^2 - 3y + 2 = 0$
5. $r^2 - 5r - 6 = 0$
6. $y^2 - y - 12 = 0$
7. $a^2 + 5a = 24$
8. $y^2 = 2y + 15$
9. $m^2 + 16 = 8m$
10. $y^2 + 49 = 14y$
11. $2k^2 - k = 10$
12. $6x^2 = 7x + 5$
13. $6x^2 - 5x = 4$
14. $9s^2 + 12s = -4$
15. $m(m - 7) = -10$
16. $z(2z + 7) = 4$
17. $16x^2 - 16x = 0$
18. $12y^2 - 48y = 0$

Use the quadratic formula to solve the following equations. If the solutions involve square roots, give both the exact and approximate solutions. See Examples 3 and 4.

19. $3x^2 - 5x + 1 = 0$
20. $2x^2 - 7x + 1 = 0$
21. $2m^2 = m + 4$
22. $p^2 + p - 1 = 0$
23. $k^2 - 10k = -20$
24. $r^2 = 13 - 12r$
25. $2x^2 + 12x + 5 = 0$
26. $5x^2 + 4x = 1$
27. $6k^2 - 11k + 4 = 0$
28. $8m^2 - 10m + 3 = 0$
29. $x^2 + 3x = 10$
30. $2x^2 = 3x + 5$
31. $2x^2 - 7x + 30 = 0$
32. $3k^2 + k = 6$
33. $5m^2 + 5m = 0$
34. $8r^2 + 16r = 0$

Solve the following quadratic inequalities. Graph each solution. See Examples 5 and 6.

35. $(m + 2)(m - 4) < 0$
36. $(k - 1)(k + 2) > 0$
37. $(t + 6)(t - 1) \geq 0$
38. $(y - 2)(y + 3) \leq 0$
39. $y^2 - 3y + 2 < 0$
40. $z^2 - 4z - 5 \leq 0$
41. $2k^2 + 7k - 4 > 0$
42. $6r^2 - 5r - 4 > 0$
43. $2y^2 + 5y \leq 3$
44. $3a^2 + a > 10$
45. $x^2 < 25$
46. $y^2 \geq 4$

APPLIED PROBLEMS

Profit and loss

The commodity market is very unstable; money can be made or lost quickly when invested in soybeans, wheat, pork bellies, and so on. Suppose that an investor kept track of her total profit, P, at time t, measured in months, after she began investing and found that

$$P = 4t^2 - 29t + 30.$$

For example, 4 months after she began investing, her total profits were

$$P = 4 \cdot 4^2 - 29 \cdot 4 + 30 \qquad \text{Let } t = 4$$
$$= 4 \cdot 16 - 116 + 30$$
$$= 64 - 116 + 30$$
$$P = -22,$$

so that she was $22 *"in the hole"* after 4 months.

47. Find her total profit after 8 months.
48. Find her total profit after 10 months.
49. Find when she has just broken even. (Let $P = 0$, and solve the resulting equation.)
50. Find the time intervals when she has been ahead. (Solve the inequality $4t^2 - 29t + 30 > 0$.)

1.5 RATIONAL EXPRESSIONS

When we get into later chapters of this book, we will work with quotients of polynomials whose denominators are not 0. We call these quotients of polynomials **rational expressions.** Examples of rational expressions include

$$\frac{8}{x-1}, \quad \frac{3x^2 + 4x}{5x - 6} \quad \text{and} \quad \frac{2 + \frac{1}{y}}{y}.$$

Methods for working with rational expressions are summarized in the following list.

Properties of rational expressions For all mathematical expressions P, $Q \neq 0$, R, and $S \neq 0$, we have

(a) $\dfrac{P}{Q} = \dfrac{PS}{QS}$ (Fundamental property)

(b) $\dfrac{P}{Q} \cdot \dfrac{R}{S} = \dfrac{PR}{QS}$ (Multiplication)

Review of Algebra

(c) $\dfrac{P}{Q} + \dfrac{R}{Q} = \dfrac{P+R}{Q}$ (Addition)

(d) $\dfrac{P}{Q} - \dfrac{R}{Q} = \dfrac{P-R}{Q}$ (Subtraction)

(e) $\dfrac{P}{Q} \div \dfrac{R}{S} = \dfrac{P}{Q} \cdot \dfrac{S}{R}, \quad R \neq 0$ (Division)

Let us now look at some examples of these properties.

Example 1 Reduce each of the following rational expressions to lowest terms.

(a) $\dfrac{12m}{18}$

Both $12m$ and 18 are divisible by 6. By property (a) above, we have

$$\dfrac{12m}{18} = \dfrac{2m \cdot 6}{3 \cdot 6} = \dfrac{2m}{3}.$$

(b) $\dfrac{8x+16}{4} = \dfrac{8(x+2)}{4} = \dfrac{4 \cdot 2(x+2)}{4} = 2(x+2)$

Here we first factored $8x + 16$. The answer could also be written as $2x + 4$, if desired.

(c) $\dfrac{k^2 + 7k + 12}{k^2 + 2k - 3} = \dfrac{(k+4)(k+3)}{(k-1)(k+3)} = \dfrac{k+4}{k-1}$

The answer cannot be further reduced. ∎

Work Problem 1 at the side.

Example 2 Multiply the following.

(a) $\dfrac{2}{3} \cdot \dfrac{y}{5}$

Use property (b) above; multiply on top and multiply on the bottom.

$$\dfrac{2}{3} \cdot \dfrac{y}{5} = \dfrac{2 \cdot y}{3 \cdot 5} = \dfrac{2y}{15}$$

The result, $2y/15$, cannot be further reduced.

(b) $\dfrac{3y+9}{6} \cdot \dfrac{18}{5y+15}$

---

1. Reduce the following to lowest terms.

(a) $\dfrac{12k + 36}{18}$

(b) $\dfrac{15m + 30m^2}{5m}$

(c) $\dfrac{2p^2 + 3p + 1}{p^2 + 3p + 2}$

Answer:

(a) $\dfrac{2(k+3)}{3}$ or $\dfrac{2k+6}{3}$

(b) $3(1 + 2m)$ or $3 + 6m$

(c) $\dfrac{2p+1}{p+2}$

1.5 Rational Expressions

Factor where possible.

$$\frac{3y+9}{6} \cdot \frac{18}{5y+15} = \frac{3(y+3)}{6} \cdot \frac{18}{5(y+3)}$$

$$= \frac{3 \cdot 18(y+3)}{6 \cdot 5(y+3)} \qquad \text{Multiply on top and bottom}$$

$$= \frac{3 \cdot 6 \cdot 3(y+3)}{6 \cdot 5(y+3)} \qquad 18 = 6 \cdot 3$$

$$= \frac{3 \cdot 3}{5} \qquad \text{Reduce to lowest terms}$$

$$= \frac{9}{5}$$

(c) $\dfrac{m^2+5m+6}{m+3} \cdot \dfrac{m}{m^2+3m+2}$

$$= \frac{(m+2)(m+3)}{m+3} \cdot \frac{m}{(m+2)(m+1)} \qquad \text{Factor}$$

$$= \frac{m(m+2)(m+3)}{(m+3)(m+2)(m+1)} \qquad \text{Multiply}$$

$$= \frac{m}{m+1} \qquad \text{Reduce} \blacksquare$$

Work Problem 2 at the side.

Example 3 Divide the following.

(a) $\dfrac{8x}{5} \div \dfrac{11x^2}{20}$

As shown in property (e) above, invert the second expression and multiply.

$$\frac{8x}{5} \div \frac{11x^2}{20} = \frac{8x}{5} \cdot \frac{20}{11x^2} \qquad \text{Invert and multiply}$$

$$= \frac{8x \cdot 20}{5 \cdot 11x^2} \qquad \text{Multiply}$$

$$= \frac{32}{11x} \qquad \text{Reduce}$$

(b) $\dfrac{9p-36}{12} \div \dfrac{5(p-4)}{18}$

$$= \frac{9p-36}{12} \cdot \frac{18}{5(p-4)} \qquad \text{Invert and multiply}$$

$$= \frac{9(p-4)}{12} \cdot \frac{18}{5(p-4)} \qquad \text{Factor}$$

$$= \frac{27}{10} \qquad \text{Multiply and reduce} \blacksquare$$

2. Multiply the following.

(a) $\dfrac{3r^2}{5} \cdot \dfrac{20}{9r}$

(b) $\dfrac{y-4}{y^2-2y-8} \cdot \dfrac{y^2-4}{3y}$

Answer:

(a) $\dfrac{4r}{3}$

(b) $\dfrac{y-2}{3y}$

30 Review of Algebra

3. Divide the following.

(a) $\dfrac{5m}{16} \div \dfrac{m^2}{10}$

(b) $\dfrac{2y-8}{6} \div \dfrac{5y-20}{3}$

(c) $\dfrac{m^2-2m-3}{m(m+1)} \div \dfrac{m+4}{5m}$

Answer:

(a) $\dfrac{25}{8m}$

(b) $\dfrac{1}{5}$

(c) $\dfrac{5(m-3)}{m+4}$

Work Problem 3 at the side.

Example 4 Add or subtract as indicated.

(a) $\dfrac{4}{5k} - \dfrac{11}{5k}$

As property (d) above shows, when two rational expressions have the same denominators, we subtract by subtracting the numerators.

$$\dfrac{4}{5k} - \dfrac{11}{5k} = \dfrac{4-11}{5k} = -\dfrac{7}{5k}$$

(b) $\dfrac{7}{p} + \dfrac{9}{2p} + \dfrac{1}{3p}$

These three denominators are different; we cannot add until we make them the same. We do this by finding a common denominator into which p, $2p$, and $3p$ all divide. A common denominator here is $6p$.

Rewrite each rational expression, using property (a), so that it has a denominator of $6p$. Then use property (c) to add. We have

$$\dfrac{7}{p} + \dfrac{9}{2p} + \dfrac{1}{3p} = \dfrac{6 \cdot 7}{6 \cdot p} + \dfrac{3 \cdot 9}{3 \cdot 2p} + \dfrac{2 \cdot 1}{2 \cdot 3p}$$

$$= \dfrac{42}{6p} + \dfrac{27}{6p} + \dfrac{2}{6p}$$

$$= \dfrac{42 + 27 + 2}{6p}$$

$$= \dfrac{71}{6p}.$$

(c) $\dfrac{3}{k-1} - \dfrac{1}{k}$

The common denominator is $k(k-1)$.

$$\dfrac{3}{k-1} - \dfrac{1}{k} = \dfrac{k \cdot 3}{k(k-1)} - \dfrac{1(k-1)}{k(k-1)} \quad \text{Property (a)}$$

$$= \dfrac{3k}{k(k-1)} - \dfrac{k-1}{k(k-1)}$$

$$= \dfrac{3k - (k-1)}{k(k-1)} \quad \text{Property (d)}$$

$$= \dfrac{3k - k + 1}{k(k-1)} \qquad -(k-1) = -1(k-1)$$
$$\qquad\qquad\qquad\qquad\qquad = -k+1$$

$$= \dfrac{2k+1}{k(k-1)} \quad\blacksquare$$

1.5 Rational Expressions

4. Add or subtract.

(a) $\dfrac{1}{m} - \dfrac{7}{m}$

(b) $\dfrac{3}{4r} + \dfrac{8}{3r}$

(c) $\dfrac{1}{m-2} - \dfrac{3}{2(m-2)}$

Answer:

(a) $-\dfrac{6}{m}$

(b) $\dfrac{41}{12r}$

(c) $\dfrac{-1}{2(m-2)}$

Work Problem 4 at the side.

1.5 EXERCISES

Reduce the following to lowest terms. Factor as necessary. See Example 1.

1. $\dfrac{6m}{24}$

2. $\dfrac{8k}{56}$

3. $\dfrac{7z^2}{14z}$

4. $\dfrac{32y^2}{16y}$

5. $\dfrac{8k+16}{9k+18}$

6. $\dfrac{20r+10}{30r+15}$

7. $\dfrac{8x^2+16x}{4x^2}$

8. $\dfrac{36y^2+72y}{9y}$

9. $\dfrac{m^2-4m+4}{m^2+m-6}$

10. $\dfrac{r^2-r-6}{r^2+r-12}$

11. $\dfrac{x^2+3x-4}{x^2-1}$

12. $\dfrac{z^2-5z+6}{z^2-4}$

Multiply or divide as indicated. Reduce all answers to lowest terms. See Examples 2 and 3.

13. $\dfrac{9k^2}{25} \cdot \dfrac{5}{3k}$

14. $\dfrac{21m^3}{9m} \cdot \dfrac{12m^2}{7m}$

15. $\dfrac{15p^3}{9p^2} \div \dfrac{6p}{10p^2}$

16. $\dfrac{3r^2}{9r^3} \div \dfrac{8r^3}{6r}$

17. $\dfrac{a+b}{2p} \cdot \dfrac{12}{5(a+b)}$

18. $\dfrac{3(x-1)}{y} \cdot \dfrac{2y}{7(x-1)}$

19. $\dfrac{a-3}{16} \div \dfrac{a-3}{32}$

20. $\dfrac{9}{2(4-y)} \div \dfrac{3}{4-y}$

21. $\dfrac{2k+8}{6} \div \dfrac{3k+12}{2}$

22. $\dfrac{5m+25}{10} \cdot \dfrac{12}{6m+30}$

23. $\dfrac{4a+12}{2a-10} \div \dfrac{a^2-9}{a^2-a-20}$

24. $\dfrac{6r-18}{9r^2+6r-24} \cdot \dfrac{12r-16}{4r-12}$

25. $\dfrac{k^2-k-6}{k^2+k-12} \cdot \dfrac{k^2+3k-4}{k^2+2k-3}$

26. $\dfrac{n^2-n-6}{n^2-2n-8} \div \dfrac{n^2-9}{n^2+7n+12}$

27. $\dfrac{m^2+3m+2}{m^2+5m+4} \div \dfrac{m^2+5m+6}{m^2+10m+24}$

28. $\dfrac{y^2+y-2}{y^2+3y-4} \div \dfrac{y^2+3y+2}{y^2+4y+3}$

29. $\dfrac{2m^2-5m-12}{m^2-10m+24} \div \dfrac{4m^2-9}{m^2-9m+18}$

30. $\dfrac{6n^2-5n-6}{6n^2+5n-6} \cdot \dfrac{12n^2-17n+6}{12n^2-n-6}$

Add or subtract as indicated. Reduce all answers to lowest terms. See Example 4.

31. $\dfrac{8}{r} + \dfrac{6}{r}$

32. $\dfrac{7}{m} - \dfrac{4}{m}$

33. $\dfrac{3}{2k} + \dfrac{5}{3k}$

34. $\dfrac{8}{5p} + \dfrac{3}{4p}$

35. $\dfrac{2}{3y} - \dfrac{1}{4y}$

36. $\dfrac{6}{11z} - \dfrac{5}{2z}$

37. $\dfrac{a+1}{2} - \dfrac{a-1}{2}$

38. $\dfrac{y+6}{5} - \dfrac{y-6}{5}$

39. $\dfrac{3}{p} + \dfrac{1}{2}$

40. $\dfrac{9}{r} - \dfrac{2}{3}$

41. $\dfrac{2}{y} - \dfrac{1}{4}$

42. $\dfrac{6}{11} + \dfrac{3}{a}$

43. $\dfrac{1}{6m} + \dfrac{2}{5m} + \dfrac{4}{m}$

44. $\dfrac{8}{3p} + \dfrac{5}{4p} + \dfrac{9}{2p}$

45. $\dfrac{6}{r} - \dfrac{5}{r-2}$

46. $\dfrac{8}{a-1} - \dfrac{5}{a}$

47. $\dfrac{8}{3(a-1)} + \dfrac{2}{a-1}$

48. $\dfrac{5}{2(k+3)} + \dfrac{2}{k+3}$

49. $\dfrac{2}{5(k-2)} + \dfrac{3}{4(k-2)}$

50. $\dfrac{11}{3(p+4)} - \dfrac{5}{6(p+4)}$

Simplify the numerator and denominator separately, then simplify the final result.

51. $\dfrac{1+\dfrac{1}{x}}{1-\dfrac{1}{x}}$

52. $\dfrac{2-\dfrac{2}{y}}{2+\dfrac{2}{y}}$

53. $\dfrac{\dfrac{1}{x+1} - \dfrac{1}{x}}{\dfrac{1}{x}}$

54. $\dfrac{\dfrac{1}{y+3} - \dfrac{1}{y}}{\dfrac{1}{y}}$

1.6 EXPONENTS AND ROOTS

Earlier in this chapter we saw that $a^2 = a \cdot a$, $a^3 = a \cdot a \cdot a$, and so on. In this section we give a more general meaning to the symbol a^n. First, recall that n is the **exponent** in a^n, and a is called the **base**. If n is a natural number, then

$$a^n = a \cdot a \cdot a \cdots a,$$

where a appears as a factor n times.

1.6 Exponents and Roots

Example 1 (a) $2^3 = 2 \cdot 2 \cdot 2 = 8$
Read 2^3 as "2 cubed."
(b) $5^2 = 5 \cdot 5 = 25$
Read 5^2 as "5 squared."
(c) $4^5 = 4 \cdot 4 \cdot 4 \cdot 4 \cdot 4 = 1024$
Read 4^5 as "4 to the fifth power."
(d) $\left(\dfrac{3}{4}\right)^2 = \dfrac{3}{4} \cdot \dfrac{3}{4} = \dfrac{9}{16}$ ∎

1. Evaluate the following.
(a) 6^3
(b) 5^4
(c) 1^7
(d) $\left(\dfrac{2}{5}\right)^3$

Answer:
(a) 216
(b) 625
(c) 1
(d) 8/125

Work Problem 1 at the side.

If $n = 0$, then we define $a^n = a^0 = 1$, if $a \neq 0$. (0^0 is a meaningless symbol.) That is, if a is any nonzero real number, then

$$a^0 = 1.$$

Example 2 (a) $6^0 = 1$
(b) $(-9)^0 = 1$
(c) $-(4^0) = -(1) = -1$ ∎

2. Evaluate the following.
(a) 17^0
(b) 30^0
(c) $(-10)^0$
(d) $-(12^0)$

Answer:
(a) 1
(b) 1
(c) 1
(d) -1

Work Problem 2 at the side.

For problems having negative exponents, we need the following definition.

If n is a natural number, and if $a \neq 0$, then

$$a^{-n} = \dfrac{1}{a^n}.$$

3. Simplify the following.
(a) 6^{-2}
(b) $\left(\dfrac{5}{8}\right)^{-1}$
(c) 5^{-2}
(d) 10^{-3}
(e) $\left(\dfrac{1}{2}\right)^{-4}$

Answer:
(a) 1/36
(b) 8/5
(c) 1/25
(d) 1/1000
(e) 16

Example 3 (a) $3^{-2} = \dfrac{1}{3^2} = \dfrac{1}{9}$
(b) $5^{-4} = \dfrac{1}{5^4} = \dfrac{1}{625}$
(c) $9^{-1} = \dfrac{1}{9^1} = \dfrac{1}{9}$
(d) $\left(\dfrac{3}{4}\right)^{-1} = \dfrac{1}{\left(\dfrac{3}{4}\right)^1} = \dfrac{1}{\dfrac{3}{4}} = \dfrac{4}{3}$ ∎

Work Problem 3 at the side.

34 Review of Algebra

By using the definitions of exponents given above, we could prove the following.

Properties of exponents For any integers m and n, and any real numbers a and b for which the following exist, we have

(a) $a^m \cdot a^n = a^{m+n}$

(b) $\dfrac{a^m}{a^n} = a^{m-n}$ $(a \neq 0)$

(c) $(a^m)^n = a^{mn}$

(d) $(ab)^m = a^m \cdot b^m$

(e) $\left(\dfrac{a}{b}\right)^m = \dfrac{a^m}{b^m}$ $(b \neq 0)$.

Example 4 Use the properties of exponents to simplify each of the following. Leave answers with positive exponents.

(a) $7^4 \cdot 7^6 = 7^{4+6} = 7^{10}$ Property (a)

(b) $\dfrac{9^{14}}{9^6} = 9^{14-6} = 9^8$ Property (b)

(c) $\dfrac{3^9}{3^{17}} = 3^{9-17} = 3^{-8} = \dfrac{1}{3^8}$ Property (b)

(d) $(2^5)^3 = 2^{5 \cdot 3} = 2^{15}$ Property (c)

(e) $(3x)^4 = 3^4 \cdot x^4$ Property (d)

(f) $\left(\dfrac{9}{7}\right)^6 = \dfrac{9^6}{7^6}$ Property (e)

(g) $\dfrac{2^{-3} \cdot 2^5}{2^4 \cdot 2^{-7}} = \dfrac{2^2}{2^{-3}} = 2^{2-(-3)} = 2^5$

(h) $2^{-1} + 3^{-1} = \dfrac{1}{2} + \dfrac{1}{3} = \dfrac{5}{6}$ ■

Work Problem 4 at the side.

We have discussed and given meaning to expressions of the form a^m for all nonzero real numbers a and all *integer* values of m, both positive and negative. In the remainder of this section we define a^m for *rational* values of m. We first look at an expression of the form $a^{1/n}$, where n is a positive integer. We want any meaning that we assign to $a^{1/n}$ to be consistent with the properties given

---

4. Simplify the following. Leave answers with positive exponents.

(a) $9^6 \cdot 9^4$

(b) $\dfrac{8^7}{8^3}$

(c) $\dfrac{14^9}{14^{12}}$

(d) $(13^4)^3$

(e) $(6y)^4$

(f) $\left(\dfrac{3}{4}\right)^2$

(g) $\dfrac{3^4 \cdot 3^{-6}}{3^5 \cdot 3^{-2}}$

(h) $3^{-1} - 4^{-1}$

Answer:
(a) 9^{10}
(b) 8^4
(c) $1/14^3$
(d) 13^{12}
(e) $6^4 y^4$
(f) $3^2/4^2$ or $9/16$
(g) $1/3^5$
(h) $1/12$

above. For example, we know that for any real number a, and integers m and n, $(a^m)^n = a^{mn}$. If this property is to hold for the expression $a^{1/n}$, we must have

$$(a^{1/n})^n = a^{(1/n)n} = a^1 = a,$$
or $\quad (a^{1/n})^n = a.$

Thus, the nth power of $a^{1/n}$ must be a. For this reason, $a^{1/n}$ is called an **nth root** of a. For example, $a^{1/2}$ denotes a second root, or **square root** of a, while $a^{1/3}$ is the third root, or **cube root** of a.

There are two numbers whose square is 16, namely 4 and -4. Also, there are two possible real fourth roots of 81, namely 3 and -3. In these cases we reserve the symbol $a^{1/n}$ for the *positive* root:

$$16^{1/2} = 4 \quad \text{and} \quad 81^{1/4} = 3,$$

and so on.

Example 5 (A calculator or Tables 1 and 2 in the Appendix will be helpful here.)
(a) $121^{1/2} = 11$, since 11 is positive and $11^2 = 121$
(b) $625^{1/4} = 5$ since $5^4 = 625$
(c) $256^{1/4} = 4$
(d) $64^{1/6} = 2$ ∎

Work Problem 5 at the side.

There is no real number x such that $x^2 = -16$. Therefore, $(-16)^{1/2}$ is not a real number. In general, if $a < 0$ and n is an *even* integer then $a^{1/n}$ is not a real number.

Since $2^3 = 8$, we have $8^{1/3} = 2$. Also, $(-8)^{1/3} = -2$. In general, if a is any real number and n is an *odd* integer, then there is exactly one real number equal to $a^{1/n}$.

Example 6 (a) $27^{1/3} = 3$
(b) $(-32)^{1/5} = -2$
(c) $128^{1/7} = 2$
(d) $(-49)^{1/2}$ is not a real number ∎

Work Problem 6 at the side.

We have now defined a^m for all integer values of m and all exponents of the form $a^{1/n}$, where n is a positive integer. To extend the definition of a^m to include all rational values of m, we make the following definition.

---

5. Evaluate the following.
(a) $64^{1/2}$
(b) $256^{1/2}$
(c) $4096^{1/4}$
(d) $1296^{1/4}$

Answer:
(a) 8
(b) 16
(c) 8
(d) 6

6. Evaluate the following.
(a) $125^{1/3}$
(b) $243^{1/5}$
(c) $(-64)^{1/2}$
(d) $(-125)^{1/3}$

Answer:
(a) 5
(b) 3
(c) not a real number
(d) -5

For all real numbers a such that the following roots exist, and for any rational number m/n,

$$a^{m/n} = (a^{1/n})^m.$$

Example 7 (a) $27^{2/3} = (27^{1/3})^2 = 3^2 = 9$
(b) $32^{2/5} = (32^{1/5})^2 = 2^2 = 4$
(c) $64^{4/3} = (64^{1/3})^4 = 4^4 = 256$
(d) $25^{3/2} = (25^{1/2})^3 = 5^3 = 125$
(e) $49^{-3/2} = \dfrac{1}{49^{3/2}} = \dfrac{1}{(49^{1/2})^3} = \dfrac{1}{7^3} = \dfrac{1}{343}$ ∎

Work Problem 7 at the side.

We now summarize the following properties of rational exponents.

Properties of rational exponents For all rational numbers m and n, and all real numbers a and b for which the following exist, we have

(a) $a^m \cdot a^n = a^{m+n}$

(b) $\dfrac{a^m}{a^n} = a^{m-n} \quad (a \neq 0)$

(c) $(a^m)^n = a^{mn}$

(d) $(ab)^m = a^m b^m$

(e) $\left(\dfrac{a}{b}\right)^n = \dfrac{a^n}{b^n} \quad (b \neq 0).$

It is common to express $a^{1/2}$ as $\sqrt{a}$, where the symbol $\sqrt{}$ is called a **radical** symbol. In general, if n is an integer greater than 1, the symbol $a^{1/n}$ can be written $\sqrt[n]{a}$. Using radical symbols, we can write our definition of $a^{m/n}$ as

$$a^{m/n} = (\sqrt[n]{a})^m$$

whenever these roots exist. Using this definition, we can convert back and forth from exponential form to radical form, as the next example shows.

Example 8 Write the following in radical form.
(a) $30^{5/2}$

By the definition above,

$$30^{5/2} = (\sqrt{30})^5$$

---

7. Evaluate the following.
(a) $16^{3/4}$
(b) $25^{5/2}$
(c) $32^{7/5}$
(d) $100^{3/2}$
(e) $1000^{-2/3}$
(f) $81^{-5/4}$

Answer:
(a) 8
(b) 3125
(c) 128
(d) 1000
(e) 1/100
(f) 1/243

1.6 Exponents and Roots

(b) $7^{-3/4} = \dfrac{1}{7^{3/4}} = \dfrac{1}{(\sqrt[4]{7})^3}$

(c) $2x^{-3/2} = \dfrac{2}{x^{3/2}} = \dfrac{2}{(\sqrt{x})^3}$ ∎

Work Problem 8 at the side.

1.6 EXERCISES

Evaluate the following. Write all answers without exponents. See Examples 1–3 and 5–7.

8. Write the following in radical form.
   (a) $3^{7/2}$
   (b) $12^{-2/3}$
   (c) $9y^{-3/8}$

Answer:
(a) $(\sqrt{3})^7$
(b) $\dfrac{1}{(\sqrt[3]{12})^2}$
(c) $\dfrac{9}{(\sqrt[8]{y})^3}$

1. 7^3
2. 4^2
3. 8^{-1}
4. 9^{-2}
5. 2^{-3}
6. 3^{-4}
7. $\left(\dfrac{3}{4}\right)^3$
8. $\left(\dfrac{2}{3}\right)^4$
9. $\left(\dfrac{1}{2}\right)^{-3}$
10. $\left(\dfrac{1}{5}\right)^{-3}$
11. $81^{1/2}$
12. $16^{1/4}$
13. $27^{1/3}$
14. $81^{1/4}$
15. $8^{2/3}$
16. $9^{3/2}$
17. $1000^{2/3}$
18. $64^{3/2}$
19. $32^{2/5}$
20. $32^{6/5}$
21. $-125^{2/3}$
22. $-125^{4/3}$
23. $\left(\dfrac{4}{9}\right)^{1/2}$
24. $\left(\dfrac{64}{27}\right)^{1/3}$
25. $16^{-5/4}$
26. $625^{-1/4}$
27. $\left(\dfrac{27}{64}\right)^{-1/3}$
28. $\left(\dfrac{121}{100}\right)^{-3/2}$
29. $2^{-1} + 4^{-1}$
30. $2^{-2} + 3^{-2}$

Simplify the following. Write all answers using only positive exponents. See Example 4.

31. $\dfrac{3^8}{3^2}$
32. $\dfrac{7^5}{7^9}$
33. $\dfrac{3^{-4}}{3^2}$
34. $\dfrac{2^{-5}}{2^{-2}}$

35. $\dfrac{6^{-1}}{6}$ 36. $\dfrac{15}{15^{-1}}$

37. $4^{-3} \cdot 4^6$ 38. $5^{-9} \cdot 5^{10}$

39. $7^{-5} \cdot 7^{-2}$ 40. $9^{-1} \cdot 9^{-3}$

41. $\dfrac{8^9 \cdot 8^{-7}}{8^{-3}}$ 42. $\dfrac{5^{-4} \cdot 5^6}{5^{-1}}$

43. $\dfrac{10^8 \cdot 10^{-10}}{10^4 \cdot 10^2}$ 44. $\dfrac{2^{-4} \cdot 2^{-3}}{2^6 \cdot 2^{-5}}$

45. $\left(\dfrac{5^{-6} \cdot 5^3}{5^{-2}}\right)^{-1}$ 46. $\left(\dfrac{8^{-3} \cdot 8^4}{8^{-2}}\right)^{-2}$

47. $2^{1/2} \cdot 2^{3/2}$ 48. $5^{3/8} \cdot 5^{5/8}$

49. $27^{2/3} \cdot 27^{-1/3}$ 50. $9^{-3/4} \cdot 9^{1/4}$

51. $\dfrac{4^{2/3} \cdot 4^{5/3}}{4^{1/3}}$ 52. $\dfrac{3^{-5/2} \cdot 3^{3/2}}{3^{7/2} \cdot 3^{-9/2}}$

Write the following in radical form. See Example 8.

53. $7^{3/2}$ 54. $15^{2/3}$

55. $60^{-2/3}$ 56. $3^{-4/3}$

57. $12x^{-3/2}$ 58. $50y^{-2/3}$

59. $(3r)^{-2/3}$ 60. $(2k)^{-3/2}$

APPLIED PROBLEMS

Supply and demand

One important application of mathematics to business and management concerns *supply* and *demand*. Usually, as the price of an item increases, the supply increases and the demand decreases. By studying past records of supply and demand at different prices, economists can construct an equation which is an approximate mathematical model of supply and demand for a given item. The next two exercises show examples of this.

61. The price of a certain type of solar heater is approximated by

$$p = 2x^{1/2} + 3x^{2/3},$$

where x is the number of units supplied. Find the price when the supply is 64 units.

62. The demand for a certain commodity and the price are related by the equation

$$p = 1000 - 200x^{-2/3} \quad (x > 0),$$

where x is the number of units of the product demanded. Find the price when the demand is 27.

1.6 Exponents and Roots

Electoral college

In our system of government, the President is elected by the electoral college, and not by individual voters. Because of this, smaller states have a greater voice in the selection of a President than they would otherwise. Two political scientists have studied the problems of campaigning for President under the current system and have concluded that candidates should allot their money according to the formula

$$\text{amount for large state} = \left(\frac{E_{large}}{E_{small}}\right)^{3/2} \times \text{amount for small state.}$$

Here E_{large} represents the electoral vote of the large state, and E_{small} represents the electoral vote of the small state. Find the amount that should be spent in each of the following larger states if \$1,000,000 is spent in the small state and the following statements are true.

63. The large state has 48 electoral votes and the small state has 3.

64. The large state has 36 electoral votes and the small state has 4.

Distance to the horizon

A Delta Airlines map gives a formula for calculating the visible distance from a jet plane to the horizon. On a clear day, this distance is approximated by

$$D = 1.22x^{1/2},$$

where x is altitude in feet, and D is distance to the horizon in miles. Use a calculator or Table 2 to find D for an altitude of

65. 5000 feet; **66.** 10,000 feet;
67. 30,000 feet; **68.** 40,000 feet.

Number of plant species

(This problem requires a calculator with an x^y key.) The Galápagos Islands are a chain of islands ranging in size from .2 to 2249 square miles. A biologist has shown that the number of different land-plant species on an island in this chain is related to the size of the island by

$$S = 28.6A^{0.32},$$

where A is the area of an island in square miles and S is the number of different plant species on that island. Estimate S (rounding to the nearest whole number) for islands of area

69. 1 square mile; **70.** 25 square miles;
71. 300 square miles; **72.** 2000 square miles.

Sailboat R-rating

In competition sailboat racing, the 12-meter class is defined by the rating

$$\frac{L + 2D + \sqrt{M + .85F}}{2.37},$$

where L is the length of the boat (in meters), D is the difference between two measurements of the boat's girth (in meters), M is the area of the mainsail (in square meters), and F is the area of the fore triangle sail (in square meters).

To race in the 12-meter class, a boat must produce a number 12 or less by this formula. Find the rating for boats having

73. $L = 14.8$ meters, $D = 2.45$ meters, $M = 51.8$ square meters, and $F = 12.9$ square meters;

74. $L = 13.7$ meters, $D = 3.83$ meters, $M = 58.9$ square meters, and $F = 15.6$ square meters. Would this boat be able to compete in the 12-meter class?

CHAPTER 1 SUMMARY

Key Words

set	linear equation
set braces	common denominator
mathematical model	inequality
number line	polynomial
graph	like (unlike) terms
real number	greatest common factor
natural (counting) number	quadratic equation
whole number	quadratic inequality
integer	rational expression
rational number	exponent
irrational number	base
absolute value	nth root
equation	square root
unknown or variable	cube root
solution	radical

Things to Remember

Zero-factor property If a and b are real numbers, with $ab = 0$, then $a = 0$ or $b = 0$.

Quadratic formula The solutions of the equation $ax^2 + bx + c = 0$, where $a \neq 0$, are given by

$$x = \frac{-b \pm \sqrt{b^2 - 4ac}}{2a}.$$

Properties of rational expressions For all mathematical expressions P, $Q \neq 0$, R, and $S \neq 0$, we have

(a) $\dfrac{P}{Q} = \dfrac{PS}{QS}$ (Fundamental property)

(b) $\dfrac{P}{Q} \cdot \dfrac{R}{S} = \dfrac{PR}{QS}$ (Multiplication)

(c) $\dfrac{P}{Q} + \dfrac{R}{Q} = \dfrac{P+R}{Q}$ (Addition)

(d) $\dfrac{P}{Q} - \dfrac{R}{Q} = \dfrac{P-R}{Q}$ (Subtraction)

(e) $\dfrac{P}{Q} \div \dfrac{R}{S} = \dfrac{P}{Q} \cdot \dfrac{S}{R}$, $R \neq 0$ (Division)

Properties of rational exponents For all rational numbers m and n, and all real numbers a and b for which the following exist, we have

(a) $a^m \cdot a^n = a^{m+n}$

(b) $\dfrac{a^m}{a^n} = a^{m-n}$ $(a \neq 0)$

(c) $(a^m)^n = a^{mn}$

(d) $(ab)^m = a^m b^m$

(e) $\left(\dfrac{a}{b}\right)^n = \dfrac{a^n}{b^n}$ $(b \neq 0)$.

CHAPTER 1 TEST

[1.1] *Name all the sets of numbers that apply to the following.*

1. -9
2. $\sqrt{5}$

Graph each of the following on a number line.

3. $x \geq -3$
4. $-4 < x \leq 6$

Evaluate the following.

5. $-|-6|$
6. $|-8 - (-4)|$

[1.2] *Solve the following equations.*

7. $4k - 11 + 3k = 2k - 1$
8. $2(k+3) - 4(1-k) = 6$
9. $\dfrac{2x}{5} - \dfrac{x-3}{5} = 2$

Solve the following inequalities.

10. $-5k \leq 20$
11. $3m + 5 > 4m - 9$

12. Find the unearned interest if the original finance charge is $1200, the loan was scheduled for 24 payments, and the loan is paid off with 6 payments left.

[1.3] Factor the following as completely as possible.

13. $8y + 16$
14. $p^2 - 9p + 14$
15. $k^2 + 4k - 45$
16. $2a^2 + 7a - 15$
17. $3m^2 - 8m - 35$
18. $16p^2 + 24p + 9$

[1.4] Solve the following quadratic equations.

19. $x^2 + 3x = 18$
20. $6z^2 = 11z + 10$
21. $y^2 - 4y + 2 = 0$
22. $9k^2 - 6k = 2$

Solve the following quadratic inequalities.

23. $(x - 1)(x + 3) \leq 0$
24. $3y^2 > 5y + 2$

[1.5] Work the following problems as indicated.

25. $\dfrac{6p}{7} \cdot \dfrac{28p}{15}$
26. $\dfrac{9r^2}{16} \div \dfrac{27r}{32}$
27. $\dfrac{m^2 - 2m - 3}{m(m - 3)} \div \dfrac{m + 1}{4m}$
28. $\dfrac{5}{2r} + \dfrac{6}{r}$
29. $\dfrac{3}{8y} + \dfrac{1}{2y} + \dfrac{7}{9y}$
30. $\dfrac{8}{r - 1} - \dfrac{3}{r}$

[1.6] Simplify the following. Write all answers without exponents.

31. 4^{-2}
32. $6^{-5} \cdot 6^7$
33. $\dfrac{8^{-2}}{8^{-1}}$
34. $64^{3/2}$
35. $9^{-5/2}$
36. $\left(\dfrac{144}{49}\right)^{-1/2}$

2
Functions and Models

A common problem in many real-life situations is to describe the relationships between quantities. For example, if we assume that the number of hours a student studies each day is related to the grade the student receives, how could we express this relationship? One way is to set up pairs of symbols representing hours of study and the corresponding grades that resulted. For example, we might have the pairs

$$(3, A), \quad (2\tfrac{1}{2}, B), \quad (2, C), \quad (1, D), \quad (0, F)$$

for the relationships in a particular class.

In more complex situations, a formula of some sort can often be used to describe how one quantity changes with respect to another. For example, if a bank pays 6% interest per year, then we can write the interest, I, that a deposit of P dollars would earn in one year as

$$I = .06 \times P, \quad \text{or} \quad I = .06P,$$

to describe the relationship between interest and the amount deposited.

Since relationships between quantities often occur, the idea of a special kind of relationship called a *function* is important. We begin this chapter with a discussion of the general idea of a function. The rest of the chapter is a study of specific functions which are useful in applications.

2.1 FUNCTIONS

As we mentioned above, the relationship between interest, I, (at 6% per year) and the amount of money deposited, P, is

$$I = .06P.$$

44 Functions and Models

If $2000 is deposited, the interest earned will be

$$I = .06(\$2000) = \$120.$$

If $7000 is deposited, the interest will be $420, and so on. To keep track of the values of P and I, we can write them as **ordered pairs:** ($2000, $120) is a shortcut way of saying "a deposit of $2000 produces interest of $120." How would you interpret ($7000, $420)?

Example 1 List all ordered pairs that can be obtained from $y = x + 3$ if x can be 1, 2, 3, or 4.

For a given value of x, we find y from the equation $y = x + 3$. For example, if $x = 1$ we have

$$y = x + 3 = 1 + 3 = 4.$$

"If $x = 1$, then y equals 4" is abbreviated as the ordered pair (1, 4). In the same way, if $x = 2$, then $y = 2 + 3 = 5$; this gives the ordered pair (2, 5). When $x = 3$, we get (3, 6), and when $x = 4$ we get (4, 7). ∎

1. List all ordered pairs that can be obtained from $y = -2x + 7$ if x can be 0, 1, 2, 3, or 4.

Answer: (0, 7), (1, 5), (2, 3), (3, 1), (4, −1)

Work Problem 1 at the side.

In the equation $y = x + 3$, the value of y is determined if we know the value of x. For this reason, it is customary to call x the **independent variable** and y the **dependent variable.**

A relationship (usually a rule or equation) between two variables is a **function** if a given value of the independent variable produces exactly one value of the dependent variable.

Example 2 Decide if the following are functions.
(a) $y = x^2 + 4x$

For a given value of x, we can use the equation $y = x^2 + 4x$ to produce exactly one value of y, so that $y = x^2 + 4x$ is a function.
(b) $y^2 = x$

Suppose $x = 16$. Then

$$y^2 = x \quad \text{becomes} \quad y^2 = 16,$$

from which $y = 4$ or $y = -4$. Thus, $y^2 = x$ is *not* a function; one value of x produces more than one value of y.
(c) $y \leq x + 2$

Suppose $x = 3$, then $y \leq x + 2$ becomes $y \leq 3 + 2$, or $y \leq 5$, which is satisfied by an infinite number of values of y. For this reason, $y \leq x + 2$ is not a function. ∎

2. Decide if the following are functions.
(a) $y = -3x - 7$
(b) $y = x^2$
(c) $y > -4x + 2$
(d) $y = \dfrac{1}{x}$

Answer:
(a) yes
(b) yes
(c) no
(d) yes

Work Problem 2 at the side.

Letters such as f, g, or h are often used to name functions. For example, to show that f is used to name the function $y = 2x - 3$, write

$$f(x) = 2x - 3,$$

where $f(x)$ (read "f of x") is used instead of y. The symbol $f(x)$ has two advantages: it shows that this function is named f, and it shows that x is the independent variable.

Suppose $f(x) = 2x - 3$. If $x = 5$, then x is found by substituting 5 for x:

$$f(5) = 2(5) - 3 = 10 - 3 = 7.$$

To show that 7 is the result when $x = 5$, write

$$f(5) = 7.$$

To find $f(-4)$, replace x with -4:

$$f(-4) = 2(-4) - 3 = -8 - 3 = -11.$$

Verify that $f(0) = -3$ and $f(-8) = -19$.

Work Problem 3 at the side.

Example 3 (a) Let $g(x) = -4x + 5$. Find $g(3)$.
Replace x with 3.

$$g(3) = -4(3) + 5 = -12 + 5 = -7$$

(b) Let $f(x) = x^2 - 4x + 5$. Find $f(-1)$.

$$f(-1) = (-1)^2 - 4(-1) + 5 = 1 + 4 + 5 = 10$$

(c) Let $h(x) = |x| + 3$. Find $h(p)$.
Replace x with p.

$$h(p) = |p| + 3$$

This replacement of one variable with another has several uses later on. ■

The types of functions illustrated in Example 3 have so many applications that they are given special names. Example 3(a) is a *linear function*, a function of the form

$$f(x) = ax + b \quad \text{or} \quad y = ax + b$$

for real numbers a and b. Example 3(b) is a *quadratic function*, a function of the form

$$f(x) = ax^2 + bx + c,$$

---

3. Let $h(x) = 9x - 15$.
Find the following.
(a) $h(1)$
(b) $h(3)$
(c) $h(-5)$

Answer:
(a) -6
(b) 12
(c) -60

46 Functions and Models

for real numbers *a*, *b*, and *c*, with $a \neq 0$. Finally, Example 3(c) shows an *absolute value function*.

Work Problem 4 at the side.

The set of all possible values of the independent variable in a function is called the **domain** of the function, while the set of all possible values of the dependent variable is called the **range** of the function.

Example 4 Suppose $y = -2x + 5$ and x can be -2, -1, 0, 1, or 2. Find the domain and range for this function.

The domain is the set of all possible values of x, or the set $\{-2, -1, 0, 1, 2\}$. To find the range, replace x in turn with -2, -1, 0, 1, and 2.

If $x = -2$, we have $y = -2(-2) + 5 = 9$. If $x = -1$, we get $y = 7$. For $x = 0$, $x = 1$, and $x = 2$, we get, respectively, $y = 5$, $y = 3$, and $y = 1$. The range, the set of values of y, is the set $\{1, 3, 5, 7, 9\}$. ∎

Work Problem 5 at the side.

It is often useful to draw a graph of the ordered pairs of a function. To do this, use the perpendicular crossed number lines of a cartesian coordinate system, as shown in Figure 2.1. The horizontal number line, or ***x*-axis,** represents the first numbers of the ordered pairs, while the vertical, or ***y*-axis,** represents the second numbers. The point where the number lines cross is the zero point on both lines; this point is called the **origin.**

To plot points on the graph corresponding to ordered pairs, proceed as in the following example. Locate the point corresponding to the ordered pair $(-2, 4)$ by starting at the origin and counting two units to the left on the horizontal axis, and four units upward on a line parallel to the vertical axis. This point is labeled in Figure 2.1, as are several other sample points. The numbers -2 and 4 are called the ***x*- and *y*-coordinates** of the point $(-2, 4)$. As shown in Figure 2.1, all points above the horizontal axis and to the right of the vertical axis are in **Quadrant I.** The other three quadrants are numbered as indicated in Figure 2.1. The points on the two axes belong to no quadrant.

Work Problem 6 at the side.

---

4. Let
$f(x) = -x^2 + 7x - 2$.
Find the following.
(a) $f(1)$
(b) $f(3)$
(c) $f(-2)$
(d) $f(m)$

Answer:
(a) 4
(b) 10
(c) -20
(d) $-m^2 + 7m - 2$

5. Suppose $y = 9x - 5$, with $x = 0, 1, 2, 3,$ or 4. Find the domain and range.

Answer:
domain: $\{0, 1, 2, 3, 4\}$;
range: $\{-5, 4, 13, 22, 31\}$

6. Locate $(-3, -5)$, $(-1, 6)$, $(-5, 0)$, $(4, -3)$, and $(0, 2)$ on a coordinate system.

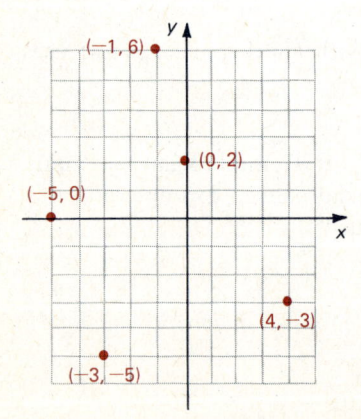

Answer:

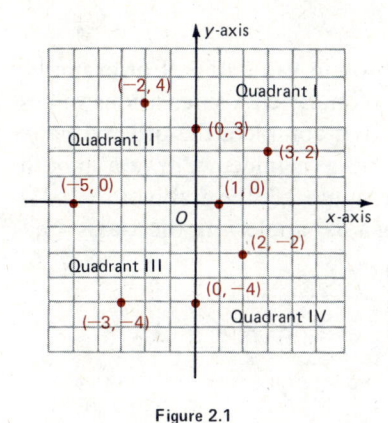

2.1 Functions   47

Figure 2.1

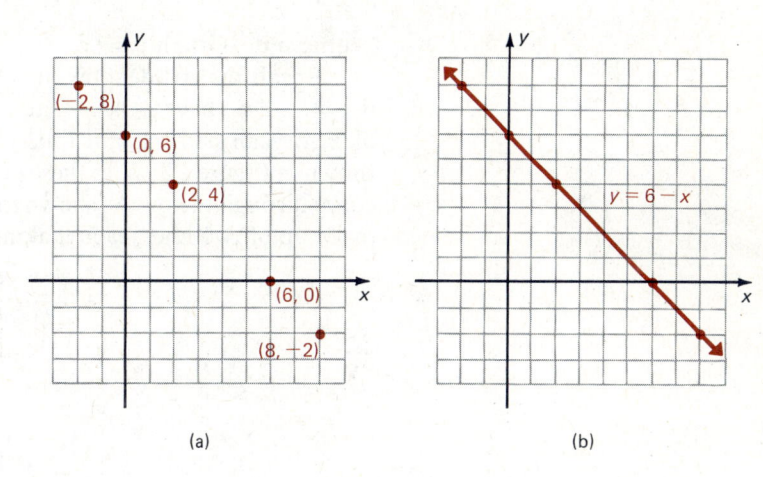

(a)   (b)

Figure 2.2

We are rarely interested in just graphing ordered pairs. In most practical problems, we are given a function and need to find the graph of the function. We shall see some specific methods of graphing certain useful functions later on, but for now we graph functions by plotting a number of ordered pairs obtained from the function. After graphing enough ordered pairs, we can decide on the general shape of the graph, and then complete it.

Example 5 Use the linear function $y = 6 - x$ and complete the ordered pairs $(-2, \;)$, $(0, \;)$, $(2, \;)$, $(6, \;)$, and $(8, \;)$. Plot these ordered pairs and then complete the graph of the function.

To complete the ordered pair $(-2, \;)$, let $x = -2$.

$$y = 6 - x = 6 - (-2) = 8$$

The ordered pair is $(-2, 8)$.

Complete the other ordered pairs in the same way, getting $(0, 6)$, $(2, 4)$, $(6, 0)$ and $(8, -2)$. These ordered pairs are graphed in Figure 2.2(a). The graph of the ordered pairs shows that a straight line would fit through them. Drawing a straight line through the various points gives the graph of Figure 2.2(b); this line is the graph of the linear function $y = 6 - x$. ∎

When we draw a straight line through the points, we are assuming that the domain of the function is the set of all real numbers that are meaningful in the function. This is an assumption that we shall make from now on.

Work Problem 7 at the side.

7. Use the linear function $y = 2x - 4$ to complete the ordered pairs $(-2, \;)$, $(0, \;)$, $(2, \;)$, and $(4, \;)$. Graph these points and draw a straight line through them.

Answer:

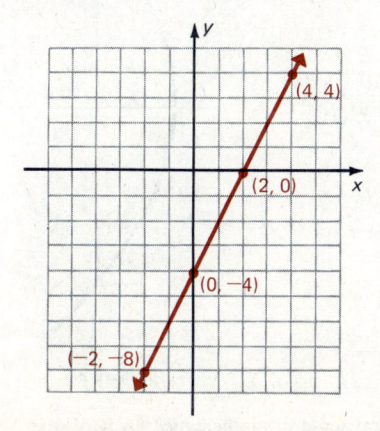

Example 6 Graph $y = |x|$.

Recall that $|x|$ denotes the absolute value of x. For example, if $x = -4$, we have $|-4| = 4$, and so on. By choosing various values of x, we can find $|x|$, and thus y, and complete the ordered pairs shown in Figure 2.3. With these points as a guide, we draw a smooth curve through them.* As shown in Figure 2.3, the graph of $y = |x|$ is made up of two lines, each making a 45° angle with the x-axis. ∎

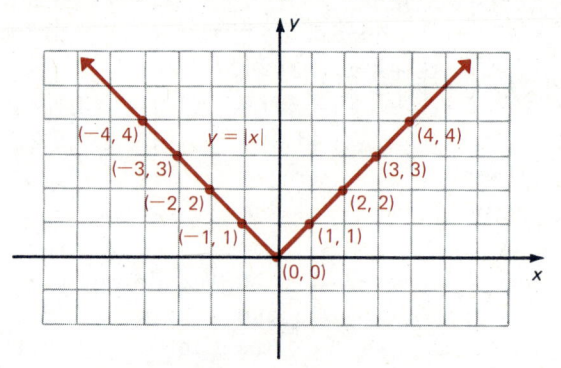

Figure 2.3

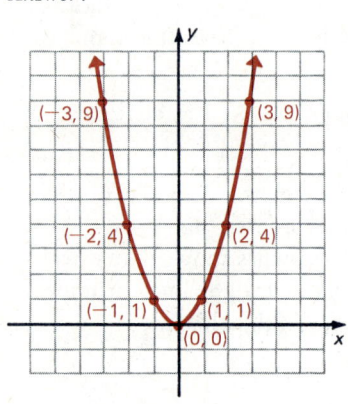

Work Problem 8 at the side.

To have a function, we must be able to assign exactly one value of y to each value of x in the domain of the function. Figure 2.4 shows a graph. Note that for the value $x = x_1$, the graph gives the two y-values $y = y_1$ and $y = y_2$. Since the x-value, x_1, corresponds to two y-values, the graph is not the graph of a function. Generalizing, we have the **vertical line test** for a function:

If a vertical line cuts a graph in more than one point, then the graph is not the graph of a function.

8. For the function $y = x^2$, complete the ordered pairs obtained when x is -3, -2, -1, 0, 1, 2, or 3. Graph them and the function.

Answer:

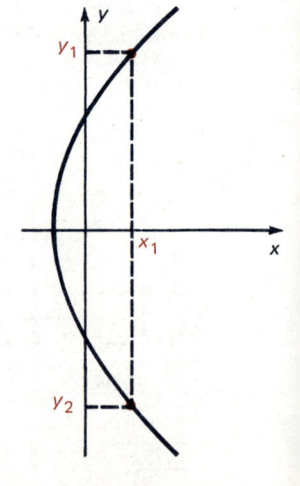

Figure 2.4

---

* The word *curve* is used in a very general sense in mathematics. In particular, a straight line is a curve.

9. Does this graph represent a function?

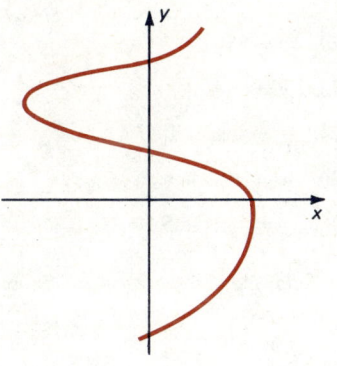

Answer: no

Work Problem 9 at the side.

Example 7 Suppose the sales of a small company are approximated by

$$S(x) = 100 + 80x,$$

where $S(x)$ represents the total sales in thousands of dollars in year x, with $x = 0$ representing 1980. Find the sales in each of the following years.

(a) 1980
In the function for $S(x)$, x represents years, where $x = 0$ represents 1980. We can find the sales in 1980 if we find $S(0)$; that is, if we substitute 0 for x.

$$S(0) = 100 + 80(0) = 100 + 0 = 100$$

Since $S(x)$ represents sales in thousands of dollars, sales totaled $100,000 in 1980.

(b) 1984
To estimate sales in 1984, let $x = 4$.

$$S(4) = 100 + 80(4) = 100 + 320 = 420,$$

so that sales are estimated at $420,000 in 1984. ∎

Work Problem 10 at the side.

10. A developer estimates that the total cost to build x large apartment complexes in a year is approximated by

$$A(x) = x^2 + 80x + 60,$$

where $A(x)$ represents costs in hundreds of thousands of dollars.
Find the total cost to build
(a) 4 complexes;
(b) 10 complexes.

Answer:
(a) $39,600,000
(b) $96,000,000

2.1 EXERCISES

For each of the following functions, find (a) $f(4)$ (b) $f(-3)$ (c) $f(0)$ *and* (d) $f(a)$. *See Example 3.*

1. $f(x) = 3x + 9$ **2.** $f(x) = 5x - 6$

3. $f(x) = -2x - 4$ **4.** $f(x) = -3x + 7$

5. $f(x) = 2x^2 + 4x$ **6.** $f(x) = x^2 - 2x$

7. $f(x) = -x^2 + 5x + 1$ **8.** $f(x) = -x^2 - x + 5$

9. $f(x) = (x-1)(x+2)$ **10.** $f(x) = (x+3)(x-4)$

Let $f(x) = 2x - 3$. Find each of the following. In Exercise 17, first find $f(2)$. See Example 3.

11. $f(0)$ **12.** $f(-1)$ **13.** $f(a)$ **14.** $f(-r)$

15. $f(m+3)$ **16.** $f(p-2)$ **17.** $f[f(2)]$ **18.** $f[f(-3)]$

List the ordered pairs obtained from the following when x is −2, −1, 0, 1, 2, or 3. Graph the ordered pairs and draw a smooth curve through them. See Examples 5 and 6.

19. $y = -x + 6$
20. $y = 2x - 4$
21. $y = |x - 2|$
22. $y = |x + 1|$
23. $y = -|x|$
24. $y = -|x - 1|$
25. $y = x^2 + 2$
26. $y = x^2 - 4$
27. $y = -x^2$
28. $y = -x^2 + 5$

Use the vertical line test to identify any of the following that represent functions.

29.

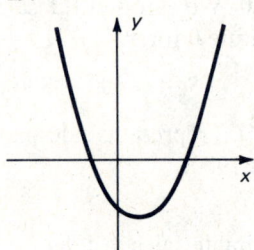

30.

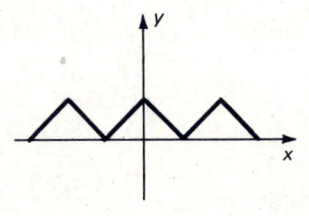

31.

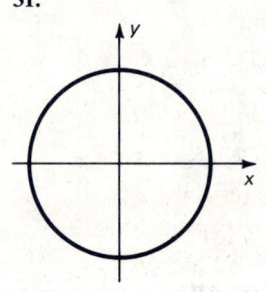

32.

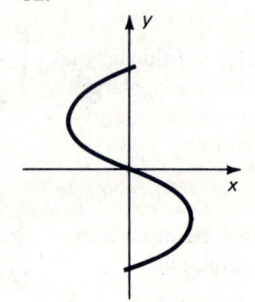

APPLIED PROBLEMS

Sales

Suppose the sales of a small company that sells by mail are approximated by

$$S(t) = 1000 + 50(t + 1),$$

where S(t) gives the sales in thousands of dollars. Here, t is time in years, with t = 0 representing the year 1977. Find the sales in each of the following years. See Example 5.

33. 1977
34. 1978
35. 1980
36. 1982

Rentals A chain-saw rental firm charges $7 per day or fraction of a day, plus a fixed fee of $4 for sharpening the blade. Let $S(x)$ represent the cost of renting a saw for x days. Find each of the following.

37. $S(\frac{1}{2})$
38. $S(1)$
39. $S(1\frac{1}{4})$
40. $S(3\frac{1}{2})$
41. $S(4)$
42. $S(4\frac{1}{10})$
43. $S(4\frac{9}{10})$

44. A portion of the graph of $y = S(x)$ is shown here. Explain how the graph could be continued.

To rent a midsize car, Avis charges $18 per day or fraction of a day. If you pick up the car in Boston and drop it off in Utica, there is a fixed $40 drop-off charge. Let $C(x)$ represent the cost of renting the car for x days. Find each of the following.

45. $C(\frac{3}{4})$
46. $C(\frac{9}{10})$
47. $C(1)$
48. $C(1\frac{5}{8})$
49. $C(2\frac{1}{9})$
50. Graph the function $y = C(x)$.

2.2 LINEAR FUNCTIONS

As we mentioned in the last section, any function of the form

$$f(x) = ax + b, \quad \text{or} \quad y = ax + b,$$

where a and b are real numbers, is a **linear function.** For example,

$$y = 6x - 11, \quad y = 4x, \quad \text{and} \quad y = 10$$

are linear functions. Also, $4x + 5y = 9$ is a linear function. To see this, solve for y.

$$4x + 5y = 9$$
$$5y = -4x + 9$$
$$y = -\frac{4}{5}x + \frac{9}{5}$$

It is shown in more advanced courses that the graph of any linear function is a straight line. Since a straight line is completely determined if we know any two distinct points through which the line passes, we really need to locate only two distinct points to draw the graph. Two points that are often useful for this purpose are the x-intercept and the y-intercept. The **x-intercept** is the x value (if any) at which the graph of the function crosses the x-axis, while the **y-intercept** is the y value (if any) at which the graph crosses the y-axis. When the graph crosses the y-axis, we have $x = 0$, and when the graph crosses the x-axis, we have $y = 0$ (see Figure 2.5).

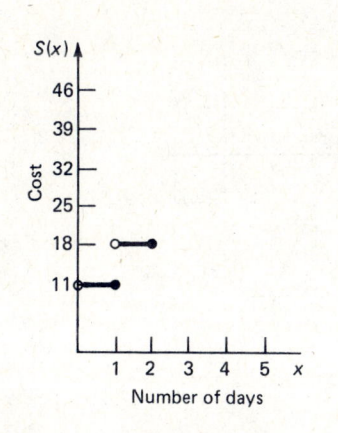

52 Functions and Models

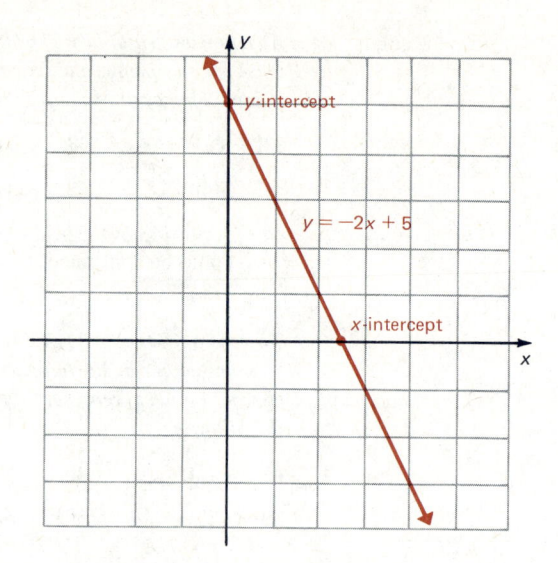

Figure 2.5

1. Find the intercepts of each of the following. Graph each line.
(a) $3x + 4y = 12$
(b) $5x - 2y = 5$

Answer:
(a) x-intercept 4,
    y-intercept 3

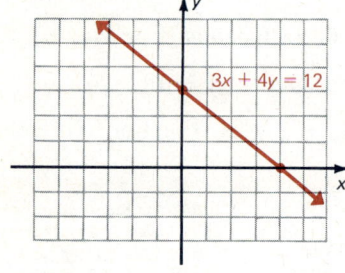

(b) x-intercept 1,
    y-intercept $-5/2$

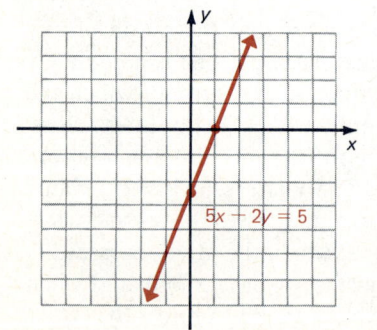

Example 1 Use intercepts to draw the graph of $y = -2x + 5$.
To find the y-intercept, the point where the line crosses the y-axis, we let $x = 0$.

$$y = -2x + 5$$
$$y = -2(0) + 5 \quad \text{Let } x = 0$$
$$y = 5 \quad\quad\quad\quad -2(0) = 0$$

leading to the ordered pair (0, 5). In the same way, we find the x-intercept by letting $y = 0$.

$$0 = -2x + 5 \quad \text{Let } y = 0$$
$$2x = 5 \quad\quad \text{Add } 2x \text{ on both sides}$$
$$x = \frac{5}{2} \quad\quad \text{Multiply both sides by } \frac{1}{2}$$

The graph goes through (5/2, 0).
Using these two intercepts, we get the graph of Figure 2.5. To check, we can find a third point on the line. We do this by choosing another value of x (or y) and finding the corresponding value of the other variable. Check that (1, 3), (2, 1), (3, -1), and (4, -3) satisfy the function $y = -2x + 5$ and also lie on the line of Figure 2.5. ∎

Work Problem 1 at the side.

2.2 Linear Functions

In the discussion of intercepts given above, we added the phrase "if any" when talking about the place where a graph crosses the x-axis. The next example shows that not every linear function has an x-intercept.

Example 2 Graph $y = -3$.

Using $y = -3$, or equivalently, $y = 0x - 3$, we always get the same y-value, -3, for any value of x. Therefore, there is no value of x that will make $y = 0$, so that the graph has no x-intercept. Since $y = -3$ is a linear function with a straight-line graph, and since the graph cannot cross the x-axis, the line must be parallel to the x-axis. For any value of x, we have $y = -3$. Therefore, the graph is the horizontal line which is parallel to the x-axis with y-intercept -3, as shown in Figure 2.6.

In general, the graph of $y = k$, where k is a real number, is the horizontal line having its y-intercept equal to k. ■

2. Graph $y = -5$. Does this graph represent a function?

Answer: yes

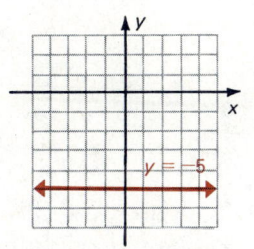

Work Problem 2 at the side.

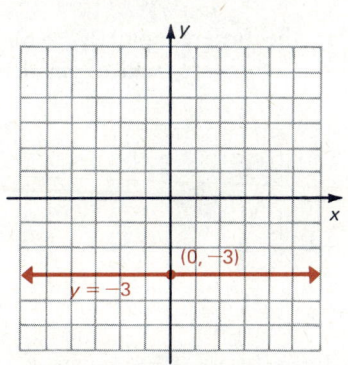

Figure 2.6

Figure 2.7

3. Graph $x = 4$. Is this a function?

Answer: no

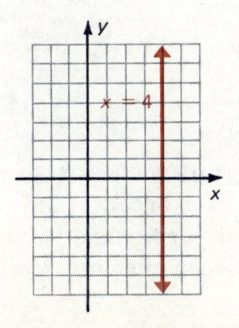

Example 3 Graph $x = -1$.

To graph $x = -1$, we complete some ordered pairs using the equivalent form, $x = 0y - 1$. For example, $(-1, 0), (-1, 1), (-1, 2)$, and $(-1, 4)$ are some ordered pairs obtained from $x = -1$. (The first number in these ordered pairs is always -1, which is what $x = -1$ says.) Here, more than one value of y corresponds to the same value of x, -1. As the graph of Figure 2.7 shows, a vertical line can cut this graph in more than one point. (In fact, a vertical line cuts the graph in an infinite number of points.) For this reason, $x = -1$ is not a function, even though the graph is a straight line. ■

Work Problem 3 at the side.

54 Functions and Models

As shown in Example 3 above, $x = -1$ is not a linear function. In fact, equations of the form $x = k$, where k is a real number, are the only equations producing straight line graphs which are not functions.

Example 4 Graph $y = -3x$.

Begin by looking for the x-intercept. If $y = 0$, then

$$y = -3x$$
$$0 = -3x$$
$$0 = x.$$

This result produces the ordered pair (0, 0). If we let $x = 0$, we end up with exactly the same ordered pair, (0, 0). Two points are needed to determine a straight line, and the intercepts have given only one point. To get a second point, we choose some other value of x (or y). For example, if $x = 2$, we have

$$y = -3x = -3(2) = -6,$$

giving the ordered pair (2, −6). These two ordered pairs, (0, 0) and (2, −6), were used to get the line of Figure 2.8. ∎

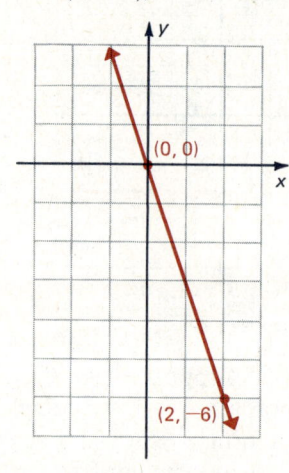

4. Graph
(a) $x = 5y$;
(b) $5x = y$.

Answer:
(a)

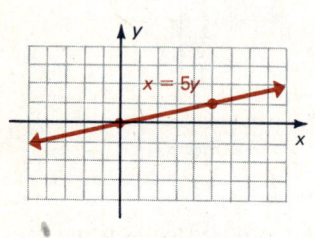

(b)

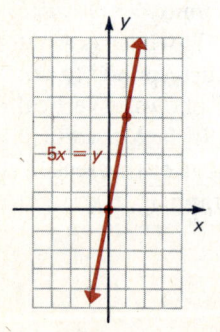

Figure 2.8

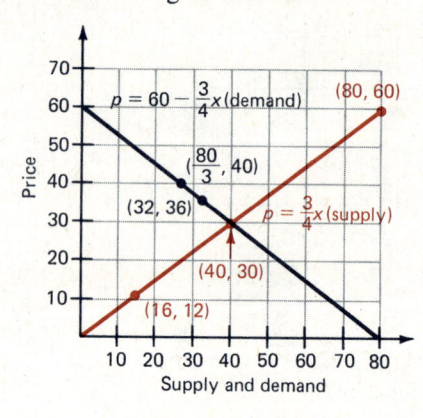

Figure 2.9

Work Problem 4 at the side.

Linear functions can be very useful in setting up a mathematical model for a real-life situation. In almost every case, linear (or any other reasonably simple) functions provide only approximations to real-world situations. However, such functions can often provide remarkably useful approximations.

2.2 Linear Functions

In particular, linear functions are often good choices for **supply and demand curves.** Typically, as the price of an item increases, the demand for the item decreases, while the supply increases. The following example shows this.

Example 5 Suppose that an economist has studied the supply and demand for aluminum siding and has come to the conclusion that unit price, p, and demand, x, in appropriate units, are related by the linear function

$$p = 60 - \frac{3}{4}x.$$

(a) Find the demand at a price of 40.
When $p = 40$, we have

$$p = 60 - \frac{3}{4}x$$
$$40 = 60 - \frac{3}{4}x \quad \text{Let } p = 40$$
$$-20 = -\frac{3}{4}x \quad \text{Add } -60 \text{ on both sides}$$
$$\frac{80}{3} = x. \quad \text{Multiply both sides by } -\frac{4}{3}$$

At a price of 40, there will be a demand for $\frac{80}{3}$ units; this gives us the ordered pair $(\frac{80}{3}, 40)$. (Since price is the dependent variable, it is customary to write the ordered pairs so that price comes second.)

(b) Find the price if the demand is 32 units.
Let $x = 32$.

$$p = 60 - \frac{3}{4}x$$
$$p = 60 - \frac{3}{4}(32)$$
$$p = 60 - 24$$
$$p = 36$$

Here we get the ordered pair (32, 36).

(c) Graph $p = 60 - \frac{3}{4}x$.

Use the ordered pairs $(\frac{80}{3}, 40)$ and (32, 36) to get the graph shown in Figure 2.9. Only the portion of the graph in Quadrant I is shown, since the model is meaningful only for positive values of p and x. ∎

Work Problem 5 at the side.

5. Suppose price and demand are related by $p = 100 - 4x$.
   (a) Find the price if the demand is 10.
   (b) Find the demand if the price is 80.

Answer:
(a) 60
(b) 5

Example 6 Suppose now that the economist of Example 5 concludes that the price, p, and supply, x, of siding are related by

$$p = \frac{3}{4}x.$$

(a) Find the supply if the price is 60.
  Let $p = 60$.

$$60 = \frac{3}{4}x$$
$$80 = x$$

Ordered pair: (80, 60).

(b) Find the price if the supply is 16.
  Let $x = 16$.

$$p = \frac{3}{4}(16) = 12$$

Ordered pair: (16, 12).

(c) Graph $p = \frac{3}{4}x$.

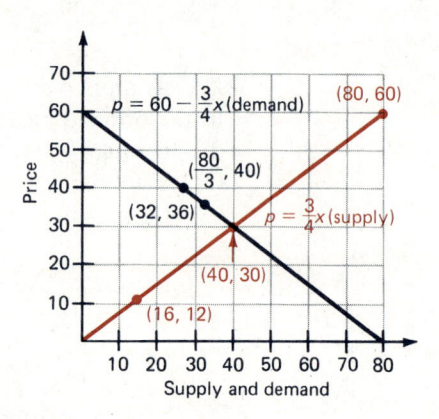

Figure 2.9

See Figure 2.9, above. ∎

As shown in the graph of Figure 2.9, both the supply and the demand functions pass through the point (40, 30). If the price of the siding is more than $p = 30$, the supply will exceed the demand. At a price less than 30, the demand will exceed the supply. Only at a price of $p = 30$ will demand and supply be equal. For this reason, $p = 30$ is called the **equilibrium price**. When the price is 30, demand and supply both equal 40 units. This number is called the **equilibrium supply**, or **equilibrium demand**.

2.2 Linear Functions   57

Example 7 Use algebra to find the equilibrium supply for the aluminum siding. (See Examples 5 and 6.)

The equilibrium supply is found when the prices from both supply and demand are equal. From Example 5, we have $p = 60 - \frac{3}{4}x$; in Example 6, we have $p = \frac{3}{4}x$. Thus,

$$60 - \frac{3}{4}x = \frac{3}{4}x$$
$$240 - 3x = 3x \quad \text{Multiply both sides by 4}$$
$$240 = 6x \quad \text{Add } 3x \text{ to both sides}$$
$$40 = x.$$

The equilibrium supply is 40, the same answer that we found above. ■

Work Problem 6 at the side.

6. The demand for a certain commodity is related to the price by $p = 80 - \frac{2}{3}x$. The supply is related to the price by $p = \frac{4}{3}x$. Find
(a) the equilibrium demand;
(b) the equilibrium price.

Answer:
(a) 40
(b) $\frac{160}{3} = 53\frac{1}{3}$

2.2 EXERCISES

Graph each of the following. Identify those lines which are horizontal or vertical See Examples 1–4.

1. $y = 2x + 1$
2. $y = 3x - 1$
3. $y = 4x$
4. $y = x + 5$
5. $3y + 4x = 12$
6. $4y + 5x = 10$
7. $y = -2$
8. $x = 4$
9. $6x + y = 12$
10. $x + 3y = 9$
11. $x - 5y = 4$
12. $2y + 5x = 20$
13. $x + 5 = 0$
14. $y - 4 = 0$
15. $5y - 3x = 12$
16. $2x + 7y = 14$
17. $8x + 3y = 10$
18. $9y - 4x = 12$
19. $y = 2x$
20. $y = -5x$
21. $y = -4x$
22. $y = x$
23. $x + 4y = 0$
24. $x - 3y = 0$

APPLIED PROBLEMS

Demand

Suppose that the demand and price for a certain model kitchen gadget are related by

$$p = 16 - \frac{5}{4}x,$$

58 Functions and Models

where p is price in dollars and x is demand in thousands. Find the price for each of the following demands. See Example 5.

25. 0   26. 4   27. 8

Find the demand for each of the following prices.

28. 6   29. 10   30. 16

31. Graph $p = 16 - \frac{5}{4}x$.

Supply

Suppose the price and supply of the item above are related by
$$p = \frac{5}{4}x.$$

See Example 6. Find the supply when the price is

32. 0;   33. 10;   34. 20.

35. Graph $p = \frac{5}{4}x$ on the same axes used for Exercise 31.

36. Find the equilibrium supply. See Example 7.
37. Find the equilibrium price.

Supply and demand

Let the supply and demand functions for a certain kind of licorice be
$$p = \frac{3}{2}x \quad \text{and} \quad p = 75 - \frac{3}{4}x,$$
where x is in thousands of pounds and p is in dollars.

38. Graph these on the same axes.
39. Find the equilibrium demand.
40. Find the equilibrium price.

Let the supply and demand functions for a certain flavor ice cream be given by
$$p = \frac{2}{5}x \quad \text{and} \quad p = 100 - \frac{2}{5}x,$$
where x is in thousands of gallons and p is in dollars.

41. Graph these on the same axes.
42. Find the equilibrium demand.
43. Find the equilibrium price.

A classic study on the supply and demand for sugar produced the following results (modified somewhat):
$$p = 1.4x - .6 \quad \text{and} \quad p = -2x + 3.2.$$

44. Graph these on the same axes.

45. Find the equilibrium demand.

46. Find the equilibrium price.

Expenditure on eating out

In a recent issue of The Wall Street Journal, *we are told that the relationship between x, the amount that an average family spends for food, and y, the amount it spends on eating out, is approximated by the model*

$$y = .36x.$$

Find y if x is

47. $40;   **48.** $80;   **49.** $120.

50. Graph the function.

Franchise fees

In an issue of Business Week, *the president of Insta-Tune, a chain of franchised automobile tune-up shops, says that people who buy a franchise and open a shop pay a weekly fee of*

$$y = .07x + \$135$$

to company headquarters. Here, y is the fee and x is the total amount of money taken in during the week by the tune-up center. Find the weekly fee if x is

51. $0;   **52.** $1000;   **53.** $2000;   **54.** $3000.

55. Graph the function.

2.3 SLOPE AND THE EQUATION OF A LINE

As we said in the previous section, the graph of a straight line is completely determined if we know two different points that the line goes through. We can also draw the graph of a straight line if we know only *one* point that the line goes through and, in addition, know the "steepness" of the line. To get a number which represents the "steepness" of a line, we define the slope of a line as in the next paragraph.

Figure 2.10 shows a line passing through the two different points $(x_1, y_1) = (-3, 5)$ and $(x_2, y_2) = (2, -4)$. The difference in the two x values,

$$x_2 - x_1 = 2 - (-3) = 5$$

in this example, is called the **change in x.** The symbol Δx (read "delta x") is used to represent the change in x. In the same way, Δy represents the **change in y.** In our example,

$$\Delta y = y_2 - y_1 = -4 - 5 = -9.$$

60 Functions and Models

The slope of a line through the two different points (x_1, y_1) and (x_2, y_2) is defined as the change in y divided by the change in x or

$$\text{slope} = \frac{\text{change in } y}{\text{change in } x} = \frac{\Delta y}{\Delta x} = \frac{y_2 - y_1}{x_2 - x_1} \quad (\Delta x \neq 0).$$

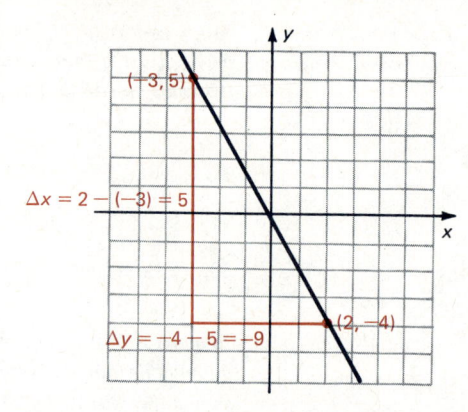

Figure 2.10

The slope of the line in Figure 2.10 is

$$\text{slope} = \frac{\Delta y}{\Delta x} = \frac{-4 - 5}{2 - (-3)} = \frac{-9}{5}.$$

Example 1 Find the slope of the line through the points $(-7, 6)$ and $(4, 5)$.

Let $(x_1, y_1) = (-7, 6)$. Then $(x_2, y_2) = (4, 5)$. Use the definition of slope.

$$\text{slope} = \frac{\Delta y}{\Delta x} = \frac{5 - 6}{4 - (-7)} = \frac{-1}{11}$$

We could have let $(x_1, y_1) = (4, 5)$. In that case,

$$\text{slope} = \frac{6 - 5}{-7 - 4} = \frac{1}{-11} = \frac{-1}{11},$$

the same answer. (The order in which coordinates are subtracted does not matter, as long as it is done consistently.) ∎

Work Problem 1 at the side.

Example 2 Find the slope of the line $3x - 4y = 12$.
To find the slope, we need two different points that the line goes through. We can find two suitable points here by first letting $x = 0$ and then letting $y = 0$.

---

1. Find the slope of the line through
(a) (6, 11) and (−4, −3)
(b) (−3, 5) and (−2, 8).

Answer:
(a) 14/10 or 7/5
(b) 3

2.3 Slope and the Equation of a Line

If $x = 0$:

$$3x - 4y = 12$$
$$3(0) - 4y = 12$$
$$-4y = 12$$
$$y = -3$$

If $y = 0$:

$$3x - 4y = 12$$
$$3x - 4(0) = 12$$
$$3x = 12$$
$$x = 4$$

Ordered pairs: $(0, -3)$ and $(4, 0)$.

Use the two ordered pairs to find the slope:

$$\text{slope} = \frac{0 - (-3)}{4 - 0} = \frac{3}{4}. \blacksquare$$

2. Find the slope of $8x + 5y = 9$.

Answer: $-8/5$

Work Problem 2 at the side.

In Example 2, we found that the slope of $3x - 4y = 12$ is $\frac{3}{4}$. We found this slope by getting two different points that the line goes through. We can get the slope more quickly if we use a shortcut. Solve the equation for y.

$$3x - 4y = 12$$
$$-4y = -3x + 12 \quad \text{Add } -3x \text{ on both sides}$$
$$y = \frac{3}{4}x - 3 \quad \text{Multiply both sides by } -\frac{1}{4}$$

In this final equation, the slope is the number in front of x (the *coefficient* of x). Also, from this last equation, if we let $x = 0$, we find the y-intercept, -3. A linear equation that is solved for y always gives the slope and y-intercept, as stated in the next theorem.

Theorem 2.1 *Slope-intercept form* If a linear function is expressed in the form

$$y = mx + b,$$

then m is the slope and b is the y-intercept.

The equation $y = mx + b$ is called the **slope-intercept form** of the equation of a line since the slope, m, and the y-intercept, b, can be read directly from the equation.

Example 3 Find the slope and y-intercept for $5x - 3y = 1$.

Solve for y.

$$5x - 3y = 1$$
$$-3y = -5x + 1$$
$$y = \frac{5}{3}x - \frac{1}{3}$$

The slope is $\frac{5}{3}$ and the y-intercept is $-\frac{1}{3}$. ∎

3. Find the slope and y-intercept for
(a) $x + 4y = 6$;
(b) $3x - 2y = 1$.

Answer:
(a) $-\frac{1}{4}, \frac{3}{2}$
(b) $\frac{3}{2}, -\frac{1}{2}$

Work Problem 3 at the side.

The slope-intercept form of the equation of a line involves the slope and the y-intercept. Sometimes, however, we know the slope of a line, together with one point that the line passes through. To find the equation of a line in this case, we use the **point-slope form** of the equation of a line, as given in the next theorem.

Theorem 2.2 *Point-slope form* If a line has slope m and passes through the point (x_1, y_1), then the equation of the line is given by

$$y - y_1 = m(x - x_1).$$

Example 4 Find the equation of the line going through the point $(3, -7)$ and having a slope of $5/4$.

Here $x_1 = 3$, $y_1 = -7$, and $m = 5/4$. Substitute these values into the point-slope form of the equation of a line.

$$y - y_1 = m(x - x_1)$$
$$y - (-7) = \frac{5}{4}(x - 3) \qquad \text{Let } y_1 = -7, m = \frac{5}{4}, x_1 = 3$$
$$y + 7 = \frac{5}{4}(x - 3)$$
$$4y + 28 = 5(x - 3) \qquad \text{Multiply both sides by 4}$$
$$4y + 28 = 5x - 15$$
$$4y = 5x - 43 \quad ∎$$

4. Find the equation of the line having the given slope and going through the given point.
(a) $-\frac{3}{5}, (5, -2)$
(b) $\frac{1}{3}, (6, 8)$

Answer:
(a) $5y = -3x + 5$
(b) $3y = x + 18$

Work Problem 4 at the side.

The point-slope form of the equation of a line can also be used to find the equation of a line if we know two different points that the line goes through. The procedure for doing this is shown in the next example.

Example 5 Find the equation of the line through $(5, 4)$ and $(-10, -2)$.

Begin by using the definition of slope to find the slope of the line through the two points.

$$\text{slope} = m = \frac{-2 - 4}{-10 - 5} = \frac{-6}{-15} = \frac{2}{5}$$

2.3 Slope and the Equation of a Line

Use $m = \frac{2}{5}$ and either of the given points in the point-slope form. If we let $(x_1, y_1) = (5, 4)$, we have

$$y - y_1 = m(x - x_1)$$
$$y - 4 = \frac{2}{5}(x - 5) \qquad \text{Let } y_1 = 4, m = \frac{2}{5}, x_1 = 5$$
$$5y - 20 = 2(x - 5) \qquad \text{Multiply both sides by 5}$$
$$5y - 20 = 2x - 10$$
$$5y = 2x + 10.$$

Check that we get the same result if we let $(x_1, y_1) = (-10, -2)$. ∎

Work Problem 5 at the side.

Following is a summary of the types of equations of lines that we have studied.

5. Find the equation of the line through
(a) $(-8, 2)$ and $(3, -6)$;
(b) $(6, -2)$ and $(4, -3)$.

Answer:
(a) $-11y = 8x + 42$
(b) $2y = x - 10$

Equation	Description
$ax + by = c$	if $a \neq 0$ and $b \neq 0$, line has x-intercept c/a and y-intercept c/b
$x = k$	**vertical line**, x-intercept k, no y-intercept, no slope
$y = k$	**horizontal line**, y-intercept k, no x-intercept, slope 0
$y = mx + b$	**slope-intercept form**, slope m, y-intercept b
$y - y_1 = m(x - x_1)$	**point-slope form**, slope m, line passes through (x_1, y_1)

2.3 EXERCISES

Find the slope, if it exists, of the line going through the given pair of points. See Example 1.

1. $(-8, 6)$, $(2, 4)$
2. $(3, 2)$, $(5, 9)$
3. $(-1, 4)$, $(2, 6)$
4. $(3, -8)$, $(4, 1)$
5. the origin and $(-4, 6)$
6. the origin and $(8, -2)$
7. $(-2, 9)$, $(-2, 11)$
8. $(7, 4)$, $(7, 12)$
9. $(3, -6)$, $(-5, -6)$
10. $(5, -11)$, $(-9, -11)$

Find the slope, if it exists, of each of the following lines. See Examples 2 and 3.

11. $y = 3x + 4$
12. $y = -3x + 2$
13. $y + 4x = 8$
14. $y - x = 3$
15. $3x + 4y = 5$
16. $2x - 5y = 8$
17. $3x + y = 0$
18. $y - 4x = 0$
19. $2x + 5y = 0$
20. $3x - 4y = 0$
21. $y = 8$
22. $y = -4$
23. $y + 2 = 0$
24. $y - 3 = 0$
25. $x = -8$
26. $x = 3$

Find the equation for each line having the given y-intercept and slope. See Example 4.

27. $4, \ m = -\frac{3}{4}$
28. $-3, \ m = \frac{2}{3}$
29. $-2, \ m = -\frac{1}{2}$
30. $\frac{3}{2}, \ m = \frac{1}{4}$
31. $\frac{5}{4}, \ m = \frac{3}{2}$
32. $-\frac{3}{8}, \ m = \frac{3}{4}$

Find equations for each of the following lines. See Example 4.

33. through $(-4, 1)$, $m = 2$
34. through $(5, 1)$, $m = -1$
35. through $(0, 3)$, $m = -3$
36. through $(-2, 3)$, $m = \frac{3}{2}$
37. through $(3, 2)$, $m = \frac{1}{4}$
38. through $(0, 1)$, $m = -\frac{2}{3}$

Find equations for each of the following lines. See Example 5. In Exercises 45–46 recall that vertical lines have no slope, and that they have equations of the form $x = k$.

39. through $(-1, 1)$ and $(2, 5)$
40. through $(4, -2)$ and $(6, 8)$
41. through $(9, -6)$ and $(12, -8)$
42. through $(-5, 2)$ and $(7, 5)$
43. through $(-8, 4)$ and $(-8, 6)$
44. through $(2, -5)$ and $(4, -5)$
45. through $(-1, 3)$ and $(0, 3)$
46. through $(2, 9)$ and $(2, -9)$

APPLIED PROBLEMS

Many real-world situations can be approximately described by a straight-line graph. One way to find the equation of this straight line is to use two typical data points from the graph and the point-slope form of the equation of a line. In each of the following exercises, assume that the data can be approximated fairly closely by a straight line. Find the slope and equation of each line. Then find the required amounts.

Cost

47. A company finds that it can make a total of 20 hot tubs for $13,900, while 10 tubs cost $7500. Let y be the total cost to produce x hot tubs. Find the cost of 18 tubs; of 27 tubs.

Sales

48. The sales of a small company were $27,000 in its second year of operation ($x = 2$), and $63,000 in its fifth year. Let y represent sales in year x. Find the sales in year 4; in year 7.

Pollution

49. When a certain industrial pollutant is introduced into a river, the reproduction of catfish declines. In a given period of time, three tons of the pollutant results in a fish population of 37,000, and 12 tons of pollutant produces a fish population of 28,000. Let y be the fish population when x tons of pollutant is introduced into the river. Find the fish population if 7 tons of pollutant is used; if 14 tons is used.

Zoology

50. In the snake *Lampropelbis polyzona*, total length, y, is related to tail length, x, in the domain 30 mm $\leq x \leq$ 200 mm, by a linear function. Find such a linear function, given the points (60, 455) and (140, 1050). Find the total length of a snake having a tail length of 80 mm; of 100 mm.

Political Science

51. According to research done by the political scientist James March, if the Democrats win 45% of the two-party vote for the House of Representatives, they win 42.5% of the seats. If the Democrats win 55% of the vote, they win 67.5% of the seats. Let y be the percent of seats won and x the percent of the two-party vote. Find the fraction of seats won by the Democrats if they win 50% of the vote; if they win 57%.

52. If the Republicans win 45% of the two-party vote, they win 32.5% of the seats (see Exercise 51). If they win 60% of the vote, they get 70% of the seats. Let y represent the percent of the seats and x the percent of the vote. Find the fraction of seats won by the Republicans if they win 40% of the vote; 50% of the vote.

2.4 LINEAR FUNCTION MODELS

In this section we briefly discuss mathematical models in general and then look at specific models which use linear equations to describe real-world situations.

A mathematical model describing an event can often be very accurate. As a generalization, the mathematical models constructed in physical science are excellent at predicting events. For example, if a body falls in a vacuum, then d, the distance the body will fall in t seconds, is given by

$$d = \frac{1}{2}gt^2,$$

where g is a constant (approximately equal to 32 feet per second per second). Using this equation, d can be predicted exactly for a known value of t. In the same way, astronomers have formulated very precise mathematical models of the movements of heavenly bodies in our solar system. When astronomers say that Halley's comet will next appear in 1986, few dispute the statement.

The situation is different in the fields of management and social science. Mathematical models in these fields tend to be only approximations, and often gross approximations at that. While an astronomer can say with certainty that Halley's comet will reappear in 1986, no corporation vice-president would dare predict that company sales will be $185,000 higher in 1986 than in 1985. So many variables enter into determining sales of a company that no mathematical model can ever hope to produce results comparable to those produced in physical science.

In spite of these limitations of mathematical models, such models have found a large and increasing acceptance in management and economic decision making. There is one main reason for this: mathematical models produce useful results.

Let us now look at some linear mathematical models of real-world situations.

Example 1 *Sales analysis* It is common to compare the change in sales of two companies by comparing the rates at which sales change. If the sales of two companies can be approximated by linear functions, we can use the work of the last section to find rates of change. For example, the chart below shows sales in two different years for companies A and B.

	Sales in 1978	Sales in 1981
Company A	$10,000	$16,000
Company B	5000	14,000

The sales of Company A increased from $10,000 to $16,000 over this three-year period, for a total increase of $6000. Thus, the average rate of change of sales is

$$\frac{\$6000}{3} = \$2000 \text{ per year.}$$

If we assume that the sales of Company A have increased linearly (that is, that the sales can be closely approximated by a linear function), then we can find the equation describing the sales by finding the equation of the line through $(-3, 10{,}000)$, and $(0, 16{,}000)$, where $x = 0$ represents 1981 and $x = -3$ represents 1978. The slope of the line is

$$\frac{16{,}000 - 10{,}000}{0 - (-3)} = 2000,$$

2.4 Linear Function Models

which is the same as the average rate of change we found above. Using the point-slope form of the equation of a line, we find that

$$y - 16{,}000 = 2000(x - 0),$$
$$y = 2000x + 16{,}000$$

gives the equation describing sales of Company A. ∎

1. Assume that the sales of Company B (see the chart above) have also increased linearly. Find the average rate of change of sales. Find the equation giving its sales.

Answer: $3000;\ y = 3000x + 14{,}000$

Work Problem 1 at the side.

As both the example and the problem show, the average rate of change is the same as the slope of the line. This is always true for data that can be modeled with a linear function.

Example 2 *Drug dosage* Suppose that a researcher has concluded that, over the short run, a dosage of x grams of a certain stimulant causes a rat to gain y grams of weight, where

$$y = 2x + 50.$$

If the researcher administers 30 grams of the stimulant, how much weight will the rat gain?

Let $x = 30$ grams; the rat will gain

$$y = 2(30) + 50 = 110$$

grams of weight. ∎

2. Find the weight gain if the researcher administers
 (a) 50 grams of the stimulant;
 (b) 80 grams.

Answer:
 (a) 150 grams
 (b) 210 grams

Work Problem 2 at the side.

As we saw from Example 1, the average rate of change is the same as the slope of the line. Then in Example 2, the average rate of change of weight gain with respect to the amount of stimulant is given by the slope of the line $y = 2x + 50$. This slope is 2. Therefore, if the researcher increases the dose of stimulant by 1 gram, the rat will gain 2 more grams of weight.

Work Problem 3 at the side.

3. When a certain new anticholesterol drug is administered, the blood cholesterol level in a certain patient is given by the linear model

$$y = 280 - 3x,$$

where x is the dosage of the drug (in grams) and y is the blood cholesterol level.
 (a) Find the blood cholesterol level if 12 grams of the drug are administered.
 (b) In general, an increase of 1 gram in the dose causes what change in the blood cholesterol level?

Answer:
 (a) 244
 (b) a decrease of 3 units

Example 3 *Cost analysis* Suppose that the cost of producing clock-radios can be approximated by the linear model

$$C(x) = 12x + 100,$$

where $C(x)$ is the **cost function** which gives the cost to produce x radios. To find the cost to produce 0 radios, we let $x = 0$.

$$C(0) = 12(0) + 100 = 100$$

68 Functions and Models

This result, $100, represents the costs involved in designing the radio, setting up a plant, training the workers, and so on. The value of $C(0)$ is called the **fixed cost.**

Once the company has invested the fixed cost into the clock-radio project, what will be the additional cost per radio? To find out, let's first find the cost of a total of 5 radios:

$$C(5) = 12(5) + 100 = 160,$$

or $160. The cost of 6 radios is

$$C(6) = 12(6) + 100 = 172,$$

or $172.

The sixth radio itself thus costs $C(6) - C(5) = \$172 - \$160 = \$12$ to produce. In the same way, the 81st radio costs $C(81) - C(80) = \$1072 - \$1060 = \$12$ to produce.

In economics, the cost to produce an additional item is called the **marginal cost** of that item. Thus, the marginal cost of the 6th radio, and the 81st radio, is $12. ∎

Work Problem 4 at the side.

4. The cost in dollars to produce x kilograms of chocolate candy is given by

$$C(x) = 3.5x + 800.$$

Find the following.
(a) total cost of 15 kilograms
(b) of 16 kilograms
(c) marginal cost of the 16th kilogram
(d) of the 30th
(e) fixed cost

Answer:
(a) $852.50
(b) $856
(c) $3.50
(d) $3.50
(e) $800

Example 4 *Break-even analysis* A firm producing chicken feed finds that the total cost, $C(x)$, to produce x units is given by

$$C(x) = 20x + 100.$$

The feed sells for $24 per unit, so that the revenue, $R(x)$, from selling x units is given by the product of the price per unit and the number of units sold,

$$R(x) = 24x.$$

The firm will break even (no profit and no loss) as long as revenue exactly equals cost. This happens whenever

$$R(x) = C(x)$$
$$24x = 20x + 100 \quad \text{Substitute for } R(x) \text{ and } C(x)$$
$$4x = 100 \quad \text{Add } -20x \text{ to both sides}$$
$$x = 25.$$

The point at which revenue exactly equals costs is called the **break-even point.** Here the break-even point is $x = 25$.

The graphs of $C(x) = 20x + 100$ and $R(x) = 24x$ are shown in Figure 2.11. The break-even point is shown on the graph. If the company produces more than 25 units (if $x > 25$), it makes a profit; if $x < 25$, it loses money. ∎

5. For a certain magazine, $C(x) = .20x + 1200$, while $R(x) = .50 x$, where x is the number of magazines sold. Find the break-even point.

Answer: 4000 magazines

Work Problem 5 at the side.

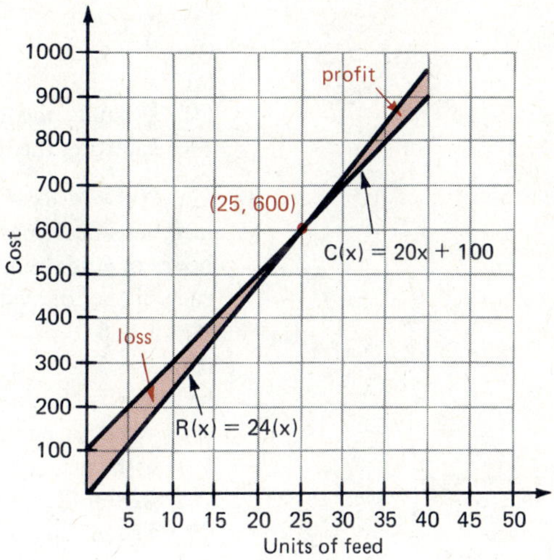

2.4 Linear Function Models 69

Figure 2.11

Graph showing $C(x) = 20x + 100$, $R(x) = 24(x)$, with point $(25, 600)$, regions marked "profit" and "loss", Cost vs. Units of feed.

2.4 APPLIED PROBLEMS

Sales

1. Suppose the sales of a particular brand of electric guitar satisfy the relationship

$$S(x) = 300x + 2000,$$

where $S(x)$ represents the number of guitars sold in year x, with $x = 0$ corresponding to 1979. Find the sales in each of the following years.
   (a) 1981
   (b) 1982
   (c) 1983
   (d) 1979
   (e) Find the annual rate of change of the sales.

Population

2. If the population of ants in an anthill satisfies the relationship

$$A(x) = 1000x + 6000,$$

where $A(x)$ represents the number of ants present at the end of month x, find the number of ants present at the end of each of the following months. Let $x = 0$ represent June.
   (a) June
   (b) July
   (c) August
   (d) December
   (e) What is the monthly rate of change of the number of ants?

3. Let $N(x) = -5x + 100$ represent the number of bacteria (in thousands) present in a certain culture at time x, measured in hours, after an antibacterial spray is introduced into the environment. Find the number of bacteria present at the following times.

(a) $x = 0$
(b) $x = 6$
(c) $x = 20$
(d) What is the hourly rate of change in the number of bacteria? Interpret the negative sign in the answer.

Education

4. Let $R(x) = -8x + 240$ represent the number of students present in a large business mathematics class, where x represents the number of hours of study required weekly. Find the number of students present at each of the following levels of required study.
   (a) $x = 0$
   (b) $x = 5$
   (c) $x = 10$
   (d) What is the rate of change of the number of students in the class with respect to the number of hours of study? Interpret the negative sign in the answer.
   (e) The professor in charge of the class likes to have exactly 16 students. How many hours of study must he require in order to have exactly 16 students? (Hint: Find a value of x such that $R(x) = 16$.)

Sales

5. Assume that the sales of a certain company are given by a linear function. Suppose the sales were $85,000 in 1968 and $150,000 in 1978. Let $x = 0$ represent 1968, with $x = 10$ representing 1978.
   (a) Find the equation giving the company's yearly sales.
   (b) What were the company's sales in 1976?
   (c) Estimate sales in 1981.

6. Assume that the sales of a certain company are given by a linear function. Suppose the sales were $200,000 in 1972 and $100,000 in 1979. Let $x = 0$ represent 1972 and $x = 7$ represent 1979.
   (a) Find the equation giving the company's yearly sales.
   (b) Find the sales in 1974.
   (c) Estimate the sales in 1981.

Supply

7. Suppose the stock of rubber dog bones, $B(x)$, on hand at the beginning of the day in a certain store is given by

$$B(x) = 600 - 20x,$$

where $x = 1$ corresponds to June 1, and x is measured in days. If the store is open every day of the month, find the supply of dog bones on hand at the beginning of each of the following days.
   (a) June 6
   (b) June 12
   (c) June 24
   (d) When will the last bone from this stock be sold?
   (e) What is the daily rate of change of this stock?

Stimulus

8. In psychology, the just-noticeable-difference (JND) for some stimulus is defined as the amount by which the stimulus must be increased so that a person will perceive it as having just barely been increased. For example, suppose a research study indicates that a line 40 centimeters in length must be increased to 42 cm before a subject thinks that it is longer.

In this case, the JND would be $42 - 40 = 2$ cm. In a particular experiment, the JND (y) is given by

$$y = .03x,$$

where x represents the original length of the line. Find the JND for lines of the following lengths.
(a) 10 cm
(b) 20 cm
(c) 50 cm
(d) 100 cm
(e) Find the rate of change in the JND with respect to the original length of the line.

Cost function

Write a cost function for each of the following. Identify all variables used.

9. A chain-saw rental firm charges $12 plus $1 per hour.

10. A trailer-hauling service charges $45 plus $2 per mile.

11. A parking garage charges 35¢ plus 30¢ per half hour.

12. For a one-day rental, a car rental firm charges $14 plus 6¢ per mile.

Assume that each of the following can be expressed as a linear cost function. Find the appropriate cost function in each case. (Hint: The cost function will be of the form $C(x) = mx + b$, where b is the fixed cost. In Exercise 13, we have $C(x) = mx + 100$. To find m, substitute 50 for x and 1600 for $C(x)$.)

13. Fixed cost, $100; 50 items cost $1600 to produce.

14. Fixed cost, $400; 10 items cost $650 to produce.

15. Fixed cost, $1000; 40 items cost $2000 to produce.

16. Fixed cost, $8500; 75 items cost $11,875 to produce.

The manager of a local McDonald's is an ex-student of one of the authors. He says that his cost function for producing coffee is $C(x) = .048x$, where $C(x)$ is the total cost to produce x cups. (He is ignoring the cost of the coffee pot and the cost of labor.) Find the total cost to produce the following numbers of cups.

17. 500   18. 1000   19. 1001

20. Find the marginal cost of the 1001st cup.

Break-even point

21. The cost to produce x units of wire is $C(x) = 50x + 5000$, while the revenue is $R(x) = 60x$. Find the break-even point. Find the revenue at the break-even point.

22. The cost to produce x units of squash is $C(x) = 100x + 6000$, while the revenue is $R(x) = 500x$. Find the break-even point.

Suppose you are the manager of a firm. You are considering the manufacture of a new product, so you ask the accounting department to produce cost estimates and the sales department to produce sales estimates. After you receive the data, you must decide whether or not to go ahead with production of the new product. Analyze the following data (find a break-even point) and then decide what you would do.

23. $C(x) = 85x + 900$; $R(x) = 105x$; not more than 38 units can be sold.

24. $C(x) = 105x + 6000$; $R(x) = 250x$; not more than 400 units can be sold.

25. $C(x) = 70x + 500$; $R(x) = 60x$. (Hint: What does a negative break-even point mean?)

26. $C(x) = 1000x + 5000$; $R(x) = 900x$.

Estimating time

Most people are not very good at estimating the passage of time. Some overestimate the rate, others underestimate. One psychologist has constructed the following mathematical model for the actual time as a function of estimated time. If y represents the actual time and x the estimated time, then

$$y = mx + b,$$

where m and b are constants that must be determined experimentally for each person. Suppose that for a particular person, $m = 1.25$ and $b = -5$. Find the actual time, y, for each of the following estimated times, x.

27. 30 minutes
28. 60 minutes
29. 120 minutes
30. 180 minutes

Suppose that for another person, $m = .85$ and $b = 1.2$. Find the actual time, y, for each of the following estimated times, x.

31. 15 minutes
32. 30 minutes
33. 60 minutes
34. 120 minutes

35. For the same person, find the estimated time, x, if the actual time, y, is 60 minutes.

36. For the same person, find the estimated time if the actual time is 90 minutes.

2.5 POLYNOMIAL FUNCTION MODELS

While linear functions provide mathematical models for many different types of problems, not all real-world situations can be adequately approximated by these functions. In many cases, we need a curve that is not a straight line.

A function of the form

$$y = a_2 x^2 + a_1 x + a_0$$

is called a **quadratic function.** In the next few examples we shall see how to graph quadratic functions.

2.5 Polynomial Function Models

Example 1 Graph the quadratic function $y = x^2$.

To graph $y = x^2$, we can choose some values for x and then find the corresponding values for y, as in the table of Figure 2.12. These points can then be plotted and a smooth curve drawn through them, as in Figure 2.12. The domain of this function is the set of real numbers, while the range is the set of nonnegative real numbers. The graph of Figure 2.12 is an example of a curve called a **parabola**. The point (0, 0), the lowest point on this parabola, is called the **vertex**. It can be proved that every quadratic function has a graph which is a parabola. ∎

Parabolas have many useful properties. Cross sections of radar dishes or spotlights form parabolas, as do the cables of suspension bridges. Circular discs often visible on the sidelines of televised football games are microphones having reflectors with parabolic cross sections. These microphones are used by the television networks to pick up the shouted signals of the quarterbacks.

1. Graph the following.
(a) $y = x^2 - 4$
(b) $y = x^2 + 5$

Answer:
(a)

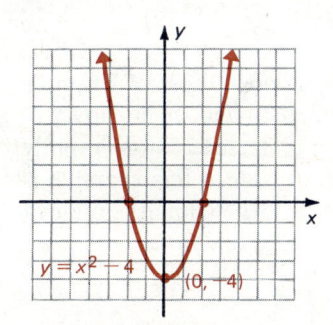

(b)

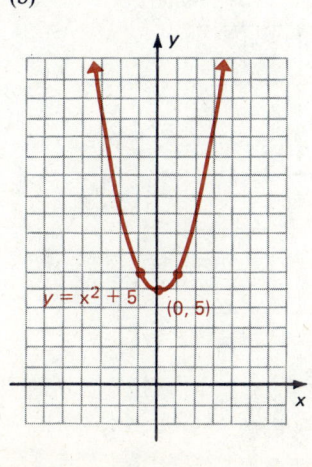

x	y
2	4
1	1
0	0
−1	1
−2	4

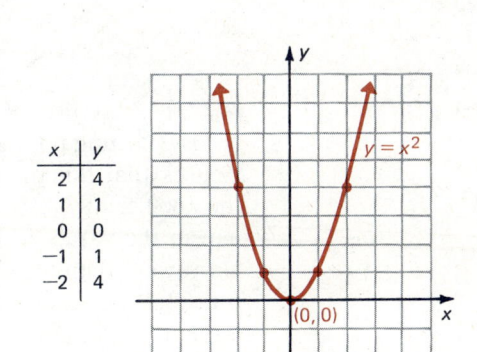

Figure 2.12

Work Problem 1 at the side.

Example 2 Graph $y = -(x + 2)^2$.

Check that the ordered pairs shown in the table of Figure 2.13 belong to the function. These ordered pairs show that the graph of $y = -(x + 2)^2$ is "upside down" in comparison to $y = x^2$, and also shifted two units to the left. Here the vertex, (−2, 0), is the *highest* point on the graph.

To verify that $y = -(x + 2)^2$ is really a quadratic function, multiply out $-(x + 2)^2$, getting $-x^2 - 4x - 4$, a function of the form defined as a quadratic function. ∎

2. Graph the following.
(a) $y = (x + 4)^2$
(b) $y = -(x - 3)^2$

Answer:
(a)

[Graph showing parabola opening upward with vertex at $(-4, 0)$, labeled $y = (x + 4)^2$]

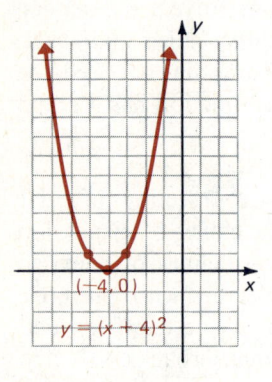

(b)

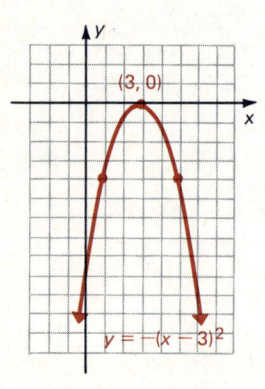

[Graph showing parabola opening downward with vertex at $(3, 0)$, labeled $y = -(x - 3)^2$]

Work Problem 2 at the side.

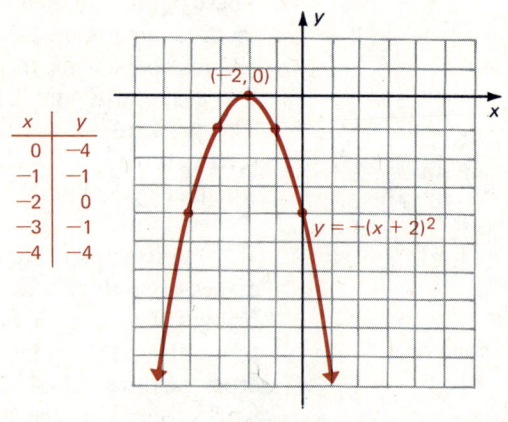

x	y
0	-4
-1	-1
-2	0
-3	-1
-4	-4

[Graph showing parabola opening downward with vertex at $(-2, 0)$, labeled $y = -(x + 2)^2$]

Figure 2.13

Example 3 Graph $y = 2(x - 3)^2 - 5$.

By plotting points and using ideas similar to those of Example 2, check that this parabola has vertex $(3, -5)$ and opens upward. The coefficient 2 causes the values of y to increase more rapidly than in the parabola $y = x^2$, so that the graph in this example is "narrower" than the graph of $y = x^2$. Some ordered pairs obtained from $y = 2(x - 3)^2 - 5$, and the graph of the function, are shown in Figure 2.14. ∎

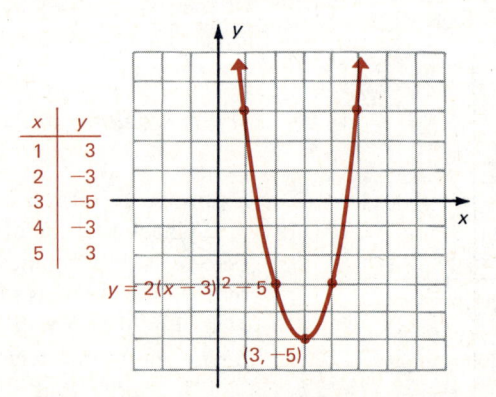

x	y
1	3
2	-3
3	-5
4	-3
5	3

[Graph showing parabola opening upward with vertex at $(3, -5)$, labeled $y = 2(x - 3)^2 - 5$]

Figure 2.14

2.5 Polynomial Function Models

In summary,

if a quadratic function is written in the form $y = a(x - h)^2 + k$, then the graph of the function is a **parabola** having vertex at (h, k). If $a > 0$, the parabola opens **"upward,"** if $a < 0$, it opens **"downward."** If $0 < |a| < 1$, the parabola is **"broader"** than $y = x^2$, while if $|a| > 1$, the parabola is **"narrower"** than $y = x^2$.

3. Graph the following.
(a) $y = (x + 4)^2 - 3$
(b) $y = -2(x - 3)^2 + 1$

Answer:
(a)

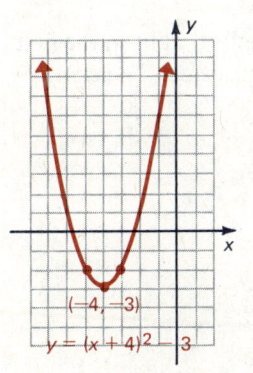

[Graph showing parabola with vertex at $(-4, -3)$, labeled $y = (x + 4)^2 - 3$]

(b)

[Graph showing downward parabola with vertex at $(3, 1)$, labeled $y = -2(x - 3)^2 + 1$]

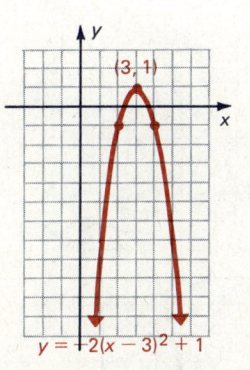

Work Problem 3 at the side.

If a quadratic function is not given in the form $y = a(x - h)^2 + k$, its graph is most easily found by using methods from calculus, as we shall see later.

Example 4 Elmyra Liskar owns and operates Aunt Elmyra's Blueberry Pie Shoppe. In an attempt to maximize her profits, she hires a consultant to analyze her business operations. The consultant tells Liskar that her profits in dollars per month are given by

$$P(x) = -(x - 60)^2 + 3600,$$

where x is the number, in hundreds, of pies that she makes. How many hundreds of pies should she make in order to maximize profit? What is the maximum profit?

Figure 2.15 shows a graph of that portion of the profit function in Quadrant I. The maximum profit can be found from the vertex of the parabola, (60, 3600). The maximum profit of $3600 is obtained when 60 hundred pies, or 6000 pies, are made. Here the profit increases as more and more pies are made, up to the point $x = 60$, and then decreases as more pies are made past this point. ∎

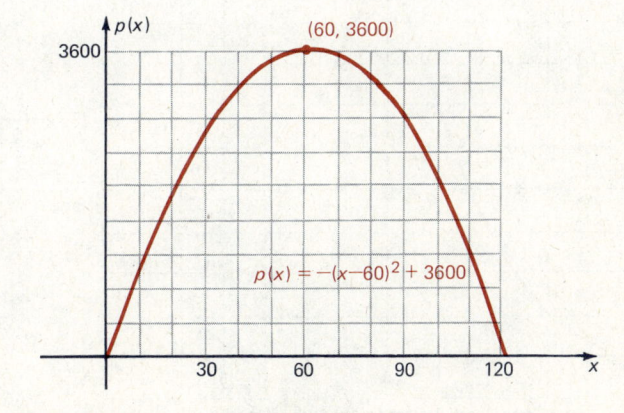

[Graph showing downward parabola with vertex at (60, 3600), labeled $p(x) = -(x-60)^2 + 3600$, x-axis marked at 30, 60, 90, 120]

Figure 2.15

76 Functions and Models

4. When a company sells x units of a product, its profit in dollars is $P(x) = -2(x - 10)^2 + 480$. Find

(a) the number of units that should be sold so that the maximum profit is received;
(b) the maximum profit.

Answer:
(a) 10
(b) $480

Work Problem 4 at the side.

Example 5 Suppose that the price and demand for an item are related by

$$p = 150 - 6x^2,$$

where p is the price (in dollars), and x is the number of items demanded (in hundreds). The supply and price are related by

$$p = 10x^2 + 2x,$$

where x is the supply of the item (in hundreds).

Both of these functions have graphs which are parabolas; these parabolas can be graphed by plotting points, giving the results shown in Figure 2.16.

We include only those portions of the graphs that lie in the first quadrant, since neither supply, demand, nor price can be negative.

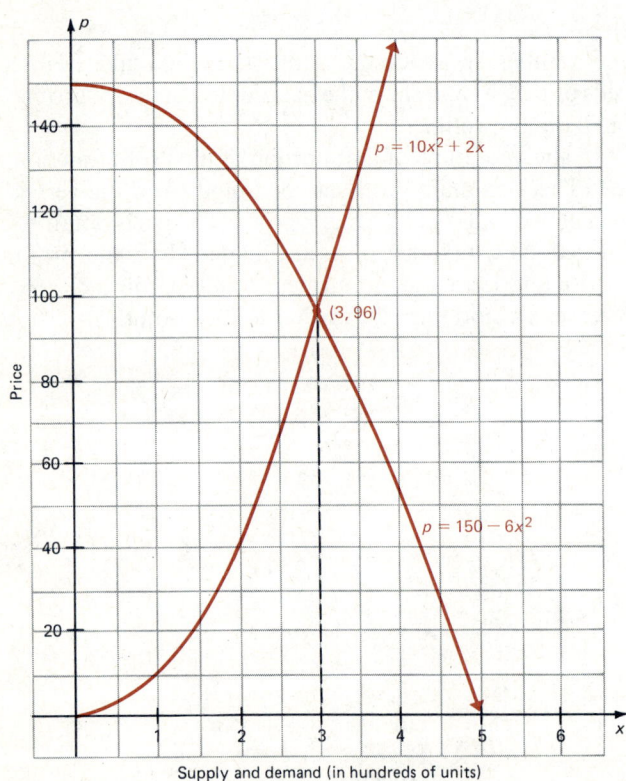

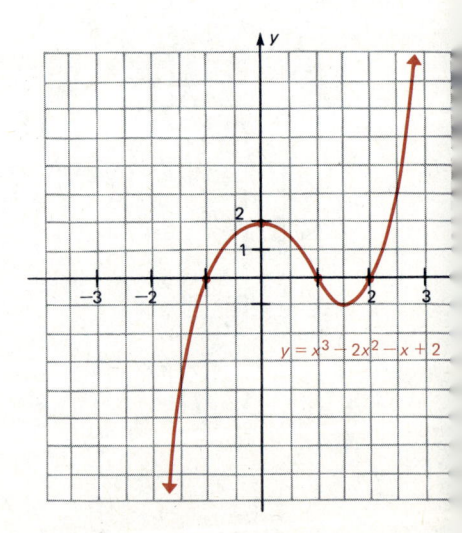

Figure 2.16

Figure 2.17

To find the equilibrium demand, we solve the equation

$$150 - 6x^2 = 10x^2 + 2x$$
$$0 = 16x^2 + 2x - 150 \quad \text{Add } -150 \text{ and } 6x^2 \text{ to both sides.}$$
$$0 = 8x^2 + x - 75. \quad \text{Multiply both sides by 1/2}$$

This equation can be solved by the quadratic formula of Chapter 1. Here $a = 8$, $b = 1$, and $c = -75$; verify that $x = 3$ or $x = -25/8$. We cannot make $-\frac{25}{8}$ units, so we discard that answer and use only $x = 3$. Since x represents supply and demand (in hundreds), we see that equilibrium demand (and supply) is 300 units. To find the equilibrium price, substitute 3 for x in either the supply or the demand function. If we use the supply function we have

$$p = 10x^2 + 2x$$
$$p = 10 \cdot 3^2 + 2 \cdot 3 \quad \text{Let } x = 3$$
$$= 10 \cdot 9 + 6$$
$$p = 96.$$

The equilibrium price is $96. ∎

Work Problem 5 at the side.

A **polynomial function** is of the form

$$y = a_n x^n + a_{n-1} x^{n-1} + \cdots + a_2 x^2 + a_1 x + a_0,$$

where $a_n, a_{n-1}, \ldots, a_2, a_1, a_0$ are real numbers and n is a non-negative integer. The function is of degree n if n is the highest exponent appearing in the function and $a_n \neq 0$.

A linear function is a polynomial function of degree 1; a quadratic function is a polynomial function of degree 2. Now we look at polynomial functions of degree three or more.

Example 6 Graph $y = x^3 - 2x^2 - x + 2$.

Once again, we start by completing several different ordered pairs, as in the table of Figure 2.17. A calculator is helpful in finding these points. Plot the points and draw a smooth curve through them. The result is shown in Figure 2.17.

This graph is typical of the graphs of polynomial functions of degree $n = 3$. As shown by the graph, both the domain and the range of this function are the set of all real numbers. ∎

---

5. The price and demand for an item are related by $p = 32 - x^2$, while price and supply are related by $p = x^2$. Find the
(a) equilibrium supply;
(b) equilibrium price.

Answer:
(a) 4
(b) 16

78 Functions and Models

6. Graph the following.
   (a) $y = x^3$
   (b) $y = x^3 - 2x^2 - 5x + 6$

Answer:
(a)

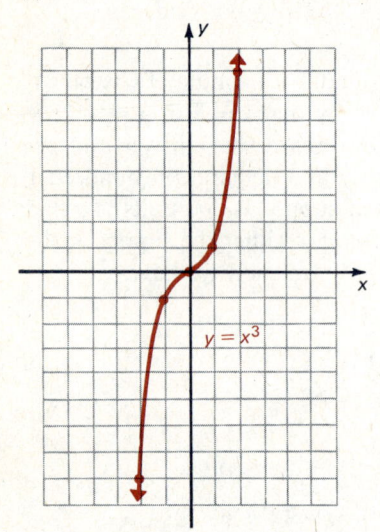

(b)

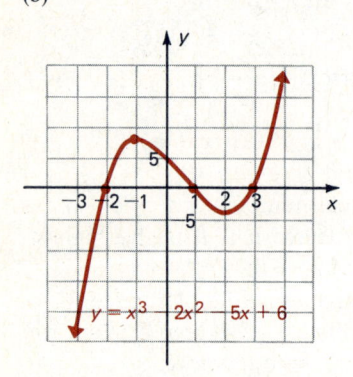

Nerve impulses

Work Problem 6 at the side.

In general, it is difficult to graph polynomial functions of degree 3 or more. In a later chapter, we discuss methods which make it much easier. However, for the graphs we need at this point, it is sufficient to plot enough points to show the basic shape of the graph.

2.5 EXERCISES

Graph the following parabolas. See Examples 1–4.

1. $y = 2x^2$
2. $y = -\frac{1}{2}x^2$
3. $y = -x^2 + 1$
4. $y = x^2 - 3$
5. $y = 3x^2 - 2$
6. $y = -2x^2 + 4$
7. $y = (x + 2)^2$
8. $y = -(x - 4)^2$
9. $y = -2(x - 3)^2$
10. $y = -3(x + 4)^2$
11. $y = (x - 1)^2 - 3$
12. $y = -(x + 4)^2 + 2$

13. José runs a sandwich shop. By studying data for his past costs, he has found that a mathematical model describing the cost of operating his shop is given by

$$C(x) = (x - 5)^2 + 15,$$

where $C(x)$ is the daily cost in dollars to make x dozen sandwiches. See Example 5.
   (a) Graph $C(x)$.
   (b) From the vertex of the parabola, find the number of dozen sandwiches José must sell to produce minimum cost.
   (c) Find the minimum cost.

14. Donna runs a taco stand. She has found that her profits are approximated by

$$P(x) = -(x - 30)^2 + 1300,$$

where $P(x)$ is the profit in hundreds of dollars per month from selling x hundred tacos in a month.
   (a) Graph $P(x)$.
   (b) Find the number of tacos that Donna should make to produce maximum profit.
   (c) What is the maximum profit?

15. A researcher in physiology has decided that a good mathematical model for the number of impulses fired after a nerve has been stimulated is given by the quadratic function

$$y = -(x - 10)^2 + 140,$$

where y is the number of responses per millisecond, and x is the number of milliseconds since the nerve was stimulated.

2.5 Polynomial Function Models

(a) When will the maximum firing rate be reached?
(b) What is the maximum firing rate?

Profit

16. French fries produce a tremendous profit (150% to 300%) for many fast-food restaurants. Management, therefore, desires to maximize the number of bags sold. Suppose that a mathematical model connecting p, the profit per week from French fries (in hundreds of dollars), and x, the price per bag (in dimes), is

$$p = -(x-3)^2 + 8.$$

(a) Find the price per bag that leads to maximum profit.
(b) What is the maximum profit?

Demand

17. Suppose the price p, in dollars, is related to the demand x, where x is measured in hundreds of items, by

$$p = 640 - 5x^2.$$

Find p when the demand is
(a) 0;
(b) 5;
(c) 10.
(d) Graph the function.

Supply

18. Suppose that the supply (in hundreds of items) and price (in dollars) of the item in Exercise 17 are related by

$$p = 5x^2.$$

(a) Graph $p = 5x^2$ on the axes used in Exercise 17.
(b) Find the equilibrium supply.
(c) Find the equilibrium price.

Supply and demand

Suppose that the supply and demand of a certain textbook are related to price by

$$\text{supply: } p = \frac{1}{5}x^2 \qquad \text{demand: } p = -\frac{1}{5}x^2 + 40,$$

where x represents the number of books sold at a price of p dollars.

19. Use the square root table in the back of the book, or a calculator, to estimate the demand at a price of
(a) 10;
(b) 20;
(c) 30;
(d) 40.
(e) Graph $p = -\frac{1}{5}x^2 + 40$.

20. Approximate the supply at a price of
(a) 5;
(b) 10;
(c) 20;
(d) 30.
(e) Graph $p = \frac{1}{5}x^2$ on the axes used in Exercise 19.

21. Find the equilibrium demand in Exercises 19 and 20.

22. Find the equilibrium price in Exercises 19 and 20.

Graph each of the following functions. See Example 6.

23. $y = x^3 + 2$
24. $y = x^3 - 1$
25. $y = x^4$
26. $y = x^4 - 2$
27. $y = (x - 2)^3$
28. $y = (x + 3)^4$
29. $y = x^3 - 2x^2 - 5x + 6$
30. $y = x^4 - x^3 - 7x^2 + x + 6$

Blood alcohol concentration

Based on a recent article that we read, we constructed the mathematical model

$$A(x) = \frac{-7}{480}x^3 + \frac{127}{120}x$$

as approximate alcohol concentration (in tenths of a percent) in an average person's bloodstream x hours after drinking about eight ounces of 100 proof whiskey. The model is approximately valid for $0 \le x < 9$. Use a calculator to find each of the following.

31. $A(0)$
32. $A(1)$
33. $A(2)$
34. $A(4)$
35. $A(6)$
36. $A(7)$
37. $A(8)$

38. Graph $y = A(x)$.

39. From the graph of Exercise 38, estimate the time of maximum alcohol concentration.

40. In one state, a person is legally drunk if the blood alcohol concentration exceeds .15%. Use the graph of Exercise 38 and estimate the period in which this average person is legally drunk.

2.6 RATIONAL FUNCTION MODELS

Functions which can be expressed as the quotient of two polynomials, such as

$$y = \frac{1}{x}, \quad y = \frac{x + 1}{x - 1}, \quad \text{and} \quad y = \frac{3x^2 + 4x - 1}{x - 1}$$

are called **rational functions.** Such functions are often useful models in applications.

Example 1 Graph the rational function $y = \dfrac{2}{1 + x}$.

2.6 Rational Function Models

First note that the function is undefined for $x = -1$, since the denominator would then become $1 + (-1) = 0$. Therefore, the graph of this function will not cross the vertical line $x = -1$. We can replace x with any number we wish, except -1, so that we can let x approach -1 as closely as we like.

From the following table, check that as x gets closer and closer to -1, $1 + x$ gets closer and closer to 0 and $|2/(1 + x)|$ gets larger and larger. When the resulting ordered pairs are graphed, as in Figure 2.18, we see that the graph gets closer and closer to the vertical line $x = -1$. This line is called a **vertical asymptote.**

x	$1 + x$	$\dfrac{2}{1 + x}$	Ordered pair
$-.5$	$.5$	4	$(-.5, 4)$
$-.8$	$.2$	10	$(-.8, 10)$
$-.9$	$.1$	20	$(-.9, 20)$
$-.99$	$.01$	200	$(-.99, 200)$
-1.01	$-.01$	-200	$(-1.01, -200)$
-1.1	$-.1$	-20	$(-1.1, -20)$
-1.5	$-.5$	-4	$(-1.5, -4)$

As $|x|$ gets larger and larger, $2/(1 + x)$ gets closer and closer to 0. Because of this, the graph has a **horizontal asymptote** on the x-axis. By using the two asymptotes and by plotting some points, we get the graph shown in Figure 2.18. ∎

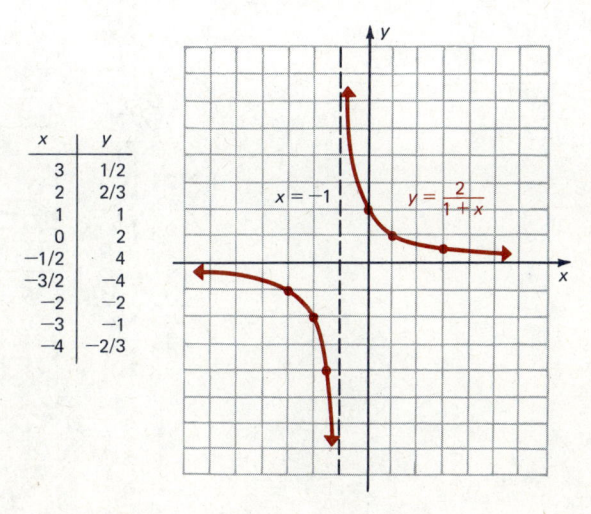

x	y
3	1/2
2	2/3
1	1
0	2
$-1/2$	4
$-3/2$	-4
-2	-2
-3	-1
-4	$-2/3$

Figure 2.18

1. Graph the following.

   (a) $y = \dfrac{3}{5-x}$

   (b) $y = \dfrac{-4}{x+4}$

Answer:

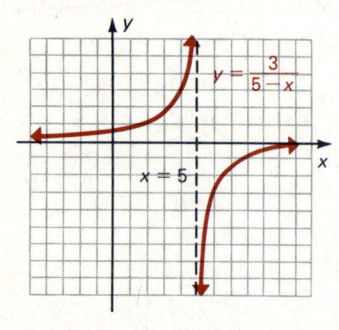

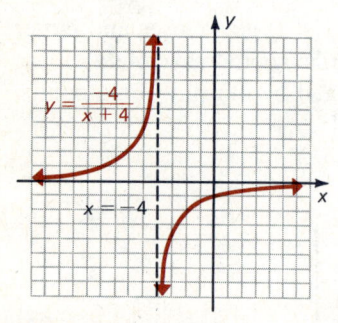

(a) [graph of $y = \dfrac{3}{5-x}$ with vertical asymptote $x = 5$]

(b) [graph of $y = \dfrac{-4}{x+4}$ with vertical asymptote $x = -4$]

Work Problem 1 at the side.

Example 2 Graph $y = \dfrac{3x+2}{2x+4}$.

The value $x = -2$ makes the denominator 0. This means that the graph has a vertical asymptote at $x = -2$.* To find a horizontal asymptote, let x get larger and larger, as shown in the chart.

x	$\dfrac{3x+2}{2x+4}$	Ordered pair
10	$\dfrac{32}{24} \approx 1.33$	(10, 1.33)
20	$\dfrac{62}{44} \approx 1.41$	(20, 1.41)
100	$\dfrac{302}{204} \approx 1.48$	(100, 1.48)
100,000	$\dfrac{300{,}002}{200{,}004} \approx 1.49998$	(100,000, 1.49998)

From this chart, we see that as x gets larger and larger, $(3x+2)/(2x+4)$ is getting closer and closer to 1.5, or 3/2. Therefore, the line $y = 3/2$ is a horizontal asymptote. Use your calculator to show that as x gets more and more negative and takes on values of -10, -100, -1000, $-100{,}000$, and so on, the graph again approaches the line $y = 3/2$. Using these asymptotes and plotting several points, we get the graph of Figure 2.19. ■

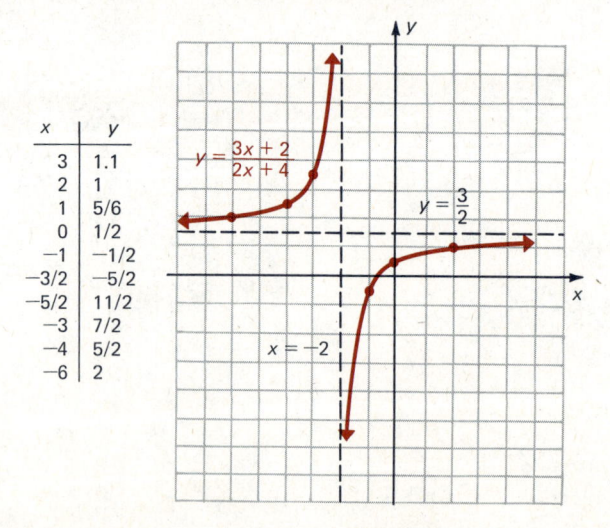

x	y
3	1.1
2	1
1	5/6
0	1/2
-1	$-1/2$
$-3/2$	$-5/2$
$-5/2$	11/2
-3	7/2
-4	5/2
-6	2

Figure 2.19

---

* Actually, we should check to see if *both* the numerator and denominator are 0, in which case there might not be a vertical asymptote.

2.6 Rational Function Models

In Example 2 above, we found that $y = 3/2$ is a horizontal asymptote for the rational function $y = (3x + 2)/(2x + 4)$. The horizontal asymptote can also be found by solving the equation $y = (3x + 2)/(2x + 4)$ for x. To do this, first multiply both sides of the equation by $2x + 4$. This gives

$$y(2x + 4) = 3x + 2,$$

or

$$2xy + 4y = 3x + 2.$$

Collect all terms containing x on one side of the equation.

$$2xy - 3x = 2 - 4y$$

Factor out x on the left, and solve for x.

$$x(2y - 3) = 2 - 4y$$

$$x = \frac{2 - 4y}{2y - 3}$$

From this form of the equation, we see that y cannot take on the value $y = 3/2$. In the same way that we identified $x = -2$ as a vertical asymptote, $y = 3/2$ is a horizontal asymptote.

Work Problem 2 at the side.

The final two examples of this section show how to construct mathematical models using rational functions.

Example 3 The U.S. Maritime Administration estimated that in a recent year the cost of building an oil tanker of 50,000 deadweight tons in the United States was $409 per ton. The cost per ton for a 100,000 ton tanker was $310, while the cost per ton for a 400,000 ton tanker was $178.

Figure 2.20 shows these values plotted on a graph, where x represents tons (in thousands) and y represents the cost per ton.

2. Graph the following.

(a) $y = \dfrac{3x + 5}{x - 3}$

(b) $y = \dfrac{2 - x}{x + 3}$

Answer:
(a)

(b)

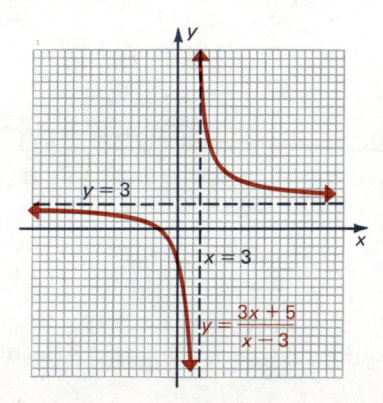

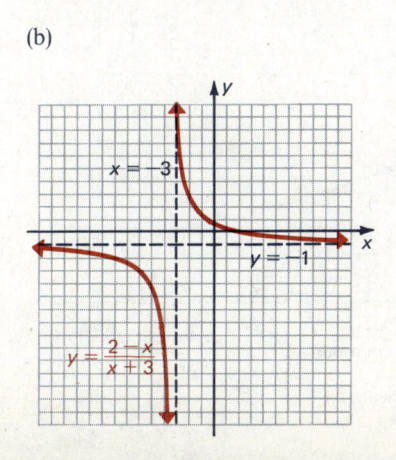

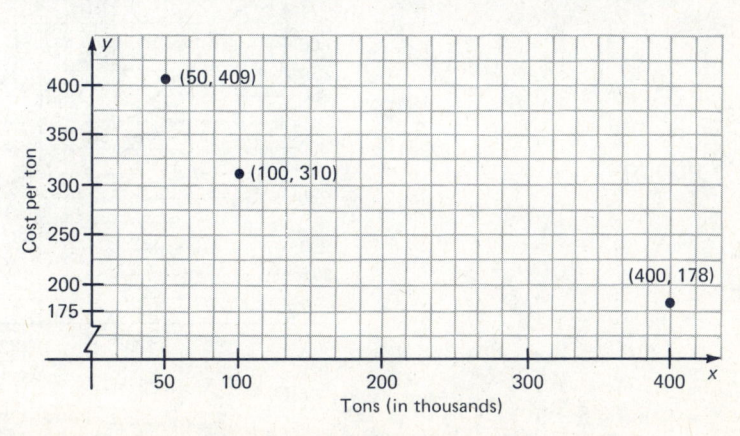

Figure 2.20

There is a gap in the information presented by the government agency: the data skips from $x = 100$ to $x = 400$, with no intermediate values. If we could fit a curve through the data points we do have, we could then approximate any desired intermediate values.

Many different functions have graphs which could give a good mathematical model for the given data. By studying the graph of Figure 2.20, we might decide that a rational function, such as the one graphed in Figure 2.18, is one such function. Using methods which we shall not discuss here, we can get,

$$y = \frac{110,000}{x + 225}$$

as a mathematical model which approximates the given data. We can substitute the known values $x = 50$, $x = 100$, and $x = 400$ to test the "goodness of fit" of the model. Doing this, we obtain the following results.

x	y (given)	y (from function)
50	409	400
100	310	338
400	178	176

As the chart shows, the rational function is a good mathematical model for the data, so that we can use it to approximate y for intermediate values of x. If we let $x = 150$, we get

$$y = \frac{110,000}{150 + 225} = 293.$$

Also, $y = \$259$ when $x = 200$, while if $x = 300$, then $y = \$210$. We can use these points to graph the function, as shown in Figure 2.21. ∎

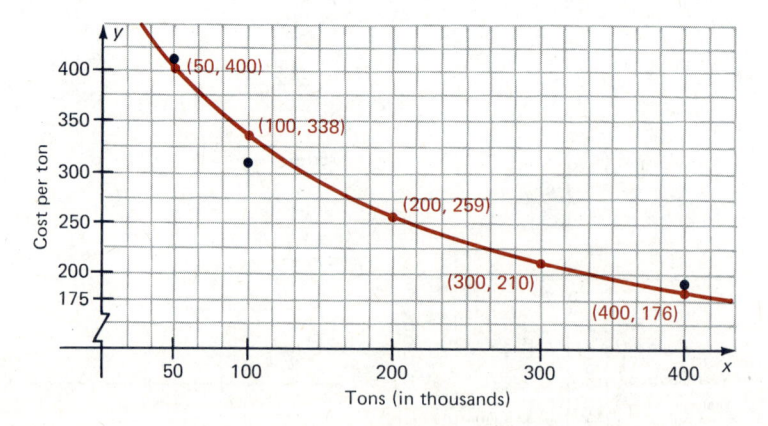

Figure 2.21

2.6 Rational Function Models

3. Using the result of Example 3, find the cost per ton to build a ship of
   (a) 75,000 tons;
   (b) 350,000 tons.

   Answer:
   (a) $367
   (b) $191

Work Problem 3 at the side.

Example 4 In many situations involving environmental pollution, it turns out that much of the pollutant can be removed from the air or water at a fairly reasonable cost, but the last, small part of the pollutant can get increasingly expensive to remove.

Cost as a function of percent of pollutant removed from the environment can be calculated for various percents of removal, with a curve fitted through the resulting data points. This curve then leads to a mathematical model of the situation. Rational functions often are a good choice for these *cost-benefit models*.

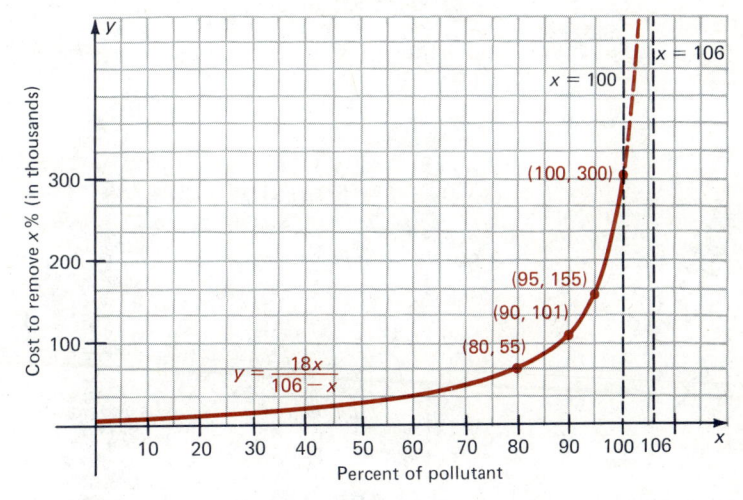

Figure 2.22

For example, suppose a cost-benefit model is given by

$$y = \frac{18x}{106 - x}$$

where y is the cost, in thousands of dollars, to remove x percent of a certain pollutant. The domain of x is the set $\{x \mid 0 \leq x \leq 100\}$; any amount of pollutant from 0% to 100% can be removed. To remove 100% of the pollutant here would cost

$$y = \frac{18(100)}{106 - 100} = 300,$$

or $300,000. Check that 95% of the pollutant can be removed for $155,000, 90% for $101,000, and 80% for $55,000. Using these points, as well as others that we could obtain from the function above, we get the graph shown in Figure 2.22. ∎

4. Using the mathematical model of Example 4, find the cost to remove the following percents of pollutants.
(a) 70%
(b) 85%
(c) 90%
(d) 98%

Answer:
(a) $35,000
(b) about $73,000
(c) about $101,000
(d) about $221,000

Work Problem 4 at the side.

2.6 EXERCISES

Graph each of the following functions. See Examples 1 and 2.

1. $y = \dfrac{1}{x+2}$
2. $y = \dfrac{-4}{x-3}$
3. $y = \dfrac{2}{x}$
4. $y = \dfrac{-3}{x}$
5. $y = \dfrac{3x}{x-1}$
6. $y = \dfrac{4x}{3-2x}$
7. $y = \dfrac{x+1}{x-4}$
8. $y = \dfrac{x-3}{x+5}$

APPLIED PROBLEMS

Cost

Suppose that the average cost in cents per pound to produce x pounds of margarine is given by

$$C(x) = \dfrac{1500}{x+30}.$$

Find the cost per pound to produce

9. 10 pounds;
10. 20 pounds;
11. 50 pounds;
12. 70 pounds.
13. Graph $C(x)$.
14. From the graph of Exercise 13, what is the trend in the average cost per unit as the number of units increases?

Estimate the cost per ton for building ships having the following weights. See Example 3.

15. 25,000 tons
16. 125,000 tons
17. 250,000 tons
18. 275,000 tons

Cost benefit

Suppose a cost-benefit model (see Example 4) is given by

$$y = \dfrac{6.5x}{102-x},$$

where y is the cost in thousands of dollars of removing x percent of a certain pollutant. Find the cost of removing the following percents of pollutants.

2.6 Rational Function Models

| 19. 0% | 20. 50% | 21. 80% | 22. 90% |
| 23. 95% | 24. 99% | 25. 100% | |

Suppose a cost-benefit model is given by

$$y = \frac{6.7x}{100 - x},$$

where y is the cost in thousands of dollars of removing x percent of a given pollutant. Find the cost of removing each of the following percents of pollutants.

| 26. 50% | 27. 70% | 28. 80% | 29. 90% |
| 30. 92% | 31. 95% | 32. 98% | 33. 99% |

34. Is it possible, according to this function, to remove *all* the pollutant?

Antique car restoration

Antique car fans often enter their cars in a concours d'élégance in which a maximum of 100 points can be awarded to a particular car. Points are awarded for the general attractiveness of the car. Based on an article in an issue of Business Week, we constructed the following mathematical model for the cost, in thousands of dollars, to restore a car so that it will win x points.

$$C(x) = \frac{10x}{49(101 - x)}$$

Find the cost to restore a car so that it will win

35. 99 points;

36. 100 points.

Product exchange functions

In management, *product exchange functions* give the relationship between quantities of two items that can be produced by the same machine or factory. For example, an oil refinery can produce gasoline, heating oil, or a combination of the two; a winery can produce red wine, white wine, or a combination. Sketch the portion of the graph of each of the following functions in Quadrant I (why this restriction?) and then use the intercepts to find the maximum quantities of each product that can be produced.

37. The product exchange function for gasoline, x, and heating oil, y, in hundreds of gallons per day is

$$y = \frac{125{,}000 - 25x}{125 + 2x}.$$

38. A drug factory has found that the product exchange function for red tranquilizers, x, and blue ones, y, is

$$y = \frac{900{,}000{,}000 - 30{,}000x}{x + 90{,}000}.$$

CHAPTER 2 SUMMARY

Key Words

ordered pair
independent variable
dependent variable
function
domain
range
vertical line test
linear function
x-intercept
y-intercept
supply and demand curves
equilibrium price

equilibrium demand
equilibrium supply
slope
fixed cost
marginal cost
break-even point
quadratic function
polynomial function
rational function
vertical asymptote
horizontal asymptote

Things to Remember

Theorem 2.1 (Slope-intercept form)   If a linear function is expressed in the form

$$y = mx + b,$$

then m is the slope and b is the y-intercept.

Theorem 2.2 (Point-slope form)   If a line has slope m and passes through the point (x_1, y_1), then the equation of the line is given by

$$y - y_1 = m(x - x_1).$$

CHAPTER 2 TEST

[2.1]   1. List the ordered pairs belonging to the function $y = 6 - 2x$ if x can be $-2, -1, 0, 1, 2,$ or 3. Give the domain and range.

Let $f(x) = -2x^2 + 3x + 1$. Find each of the following.

2. $f(-2)$ 3. $f[f(0)]$ 4. $f(p)$

[2.2]   Graph each of the following functions.

5. $x + 3y = 6$ 6. $4x - y = 8$
7. $x + 2 = 0$ 8. $y = 3$

Chapter 2 Test

[2.3] *Find the slope of each of the following lines.*

9. through $(-2, 5)$ and $(4, 7)$
10. $8x - 3y = 7$
11. $y = -5$

Find the equation of each of the following lines.

12. through $(-2, 3)$, $m = \frac{1}{2}$
13. through $(8, -1)$ and $(3, -2)$
14. through $(3, -4)$, vertical
15. A company can make 8 units of paper for \$300; 12 units cost \$420. Find the cost function. What is the fixed cost?

[2.4] *The supply and demand of a certain commodity are related to price by*

$$\text{supply: } p = 6x + 3 \qquad \text{demand: } p = 19 - 2x.$$

16. Find the equilibrium supply.
17. Find the equilibrium price.

The cost to produce x units of vodka is approximated by

$$C(x) = 20x + 100.$$

The revenue from the sale of x units is $R(x) = 40x$.

18. Find the break-even point.
19. What revenue will the company receive if it sells just that number of units?

[2.5] *Suppose that the cost to produce x units of saddles is approximated by*

$$C(x) = (x - 7)^2 + 10.$$

20. What number of units of saddles should be manufactured so as to produce the minimum total cost?
21. What is the minimum total cost?

Graph each of the following functions.

[2.5]
22. $y = -x^2$
23. $y = (x + 2)^2$
24. $y = (x - 1)^2 - 3$
25. $y = x^3 - 2$

[2.6]
26. $y = \dfrac{1}{x - 3}$
27. $y = \dfrac{-3}{2x - 4}$

CASE 1 ESTIMATING SEED DEMANDS—THE UPJOHN COMPANY*

The Upjohn Company has a subsidiary which buys seeds from farmers and then resells them. Each spring the firm contracts with farmers to grow the seeds. The firm must decide on the number of acres that it will contract for. The problem faced by the company is that the demand for seeds is not constant, but fluctuates from year to year. Also, the number of tons of seed produced per acre varies, depending on weather and other factors. In an attempt to decide the number of acres that should be planted in order to maximize profits, a company mathematician created a model of the variables involved in determining the number of acres to plant.

The analysis of this model required advanced methods that we will not go into here. (See Case 8.) We can, however, give the conclusion; the number of acres that will maximize profit in the long run is found by solving the equation

$$F(AX + Q) = \frac{(S - C_p)X - C_A}{(S - C_p + C_c)X} \tag{1}$$

for A. The function $F(z)$ represents the chances that z tons of seed will be demanded by the marketplace. The variables in the equation are:

A = number of acres of land contracted by the company
X = quantity of seed produced per acre of land
Q = quantity of seed in inventory from previous years
S = selling price per ton of seed
C_p = variable cost (production, marketing, etc.) per ton of seed
C_c = cost to carry over one ton of seed from previous year
C_A = variable cost per acre of land.

To advise management of the number of acres of seed to contract for, the mathematician studied past records to find the values of the various variables. From these records and from predictions of future trends, it was concluded that S = \$10,000 per ton, X = .1 ton per acre (on the average), Q = 200 tons, C_p = \$5000 per ton, C_A = \$100 per acre, C_c = \$3000 per ton.

By the same process, the function $F(z)$ is approximated by

$$F(z) = \frac{z}{1000} - \frac{1}{2}, \quad \text{if} \quad 500 \leq z \leq 1500 \text{ tons.} \tag{2}$$

EXERCISES

1. Evaluate $AX + Q$ using the values of X and Q given above.

2. Find $F(AX + Q)$, using equation (2) and your results from Exercise 1.

---

* Based on work by David P. Rutten, Senior Mathematician, The Upjohn Company, Kalamazoo, Michigan.

3. Solve equation (1) for A.
4. How many acres should be planted?
5. How many tons of seed will be produced?
6. Find the total revenue that will be received from the sale of the seeds.

CASE 2 MARGINAL COST—BOOZ, ALLEN AND HAMILTON*

Booz, Allen and Hamilton is a large management consulting firm. One of the services the firm provides to its client companies is profitability studies, which develop approaches with which the client can increase profit levels. The client company requesting the analysis presented in this case is a large producer of a staple food. The company buys from farmers, and then processes the food in its mills, resulting in a finished product. The company sells both at retail under its own brands, and in bulk to other companies who use the product in the manufacture of convenience foods.

The client company has been reasonably profitable in recent years, but the management retained Booz, Allen and Hamilton to see whether its consultants could suggest approaches for improving company profits. The management of the company had long operated with the philosophy of trying to process and sell as much of its product as possible, since they felt this would lower the average processing cost per unit sold. However, the consultants found that the client's fixed mill costs were quite low, and that, in fact, processing extra units made the cost per unit start to increase. (There are several reasons for this: the company must run three shifts, machines break down more often, raw material had to be transported over longer distances, and so on.)

In this case, we shall discuss the marginal cost of two of the company's products. The marginal cost (cost of producing an extra unit) of production for product A was found by the consultants to be approximated by the linear function

$$y = .133x + 10.09,$$

where x is the number of units produced (in millions) and y is the marginal cost.

For example, at a level of production of 3.1 million units, an additional unit of product A would cost about

$$y = .133(3.1) + 10.09$$
$$= \$10.50.$$

At a level of production of 5.7 million units, an extra unit costs $10.85. Figure 1 shows a graph of the marginal cost function from $x = 3.1$ to $x = 5.7$, the domain to which the function above was found to apply.

---

* "Marginal Cost," by John R. Dowdle of the Transportation Consulting Division of Booz, Allen and Hamilton. Reprinted by permission.

92 Functions and Models

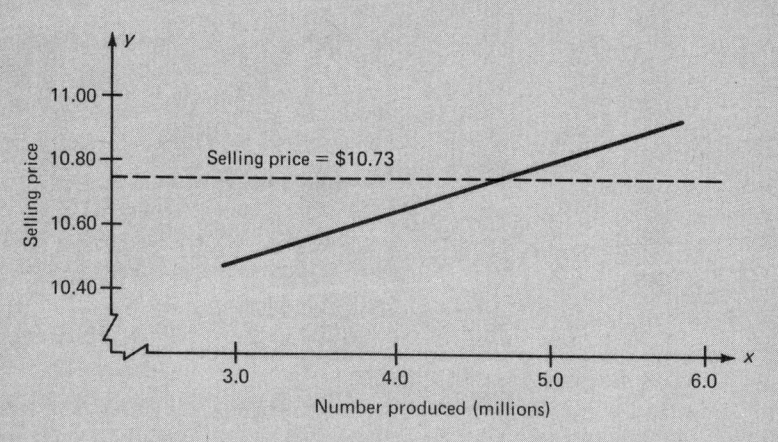

Figure 1

(Graph showing selling price vs. number produced (millions), with horizontal dashed line at Selling price = $10.73 and an upward-sloping line crossing it between 4.0 and 5.0 million units.)

The selling price for product A is $10.73 per unit, so that, as shown on the graph of Figure 1, the company was losing money on many units of the product that it sold. Since the selling price could not be raised if the company was to remain competitive, the consultants recommended that production of product A be reduced.

For product B, the Booz, Allen and Hamilton consultants found a marginal cost function given by

$$y = .0667x + 10.29,$$

with x and y as defined above. Verify that at a production level of 3.1 million units, the marginal cost is $10.50, while at a production level of 5.7 million units, the marginal cost is $10.67. Since the selling price of this product is $9.65, the consultants again recommended a cutback in production.

The consultants ran similar cost analyses of other products made by the company, and then issued their overall recommendation: The company should reduce total production by 2.1 million units. The analysts predicted that this would raise profits for the products under discussion from $8.3 million annually to $9.6 million, which is essentially what happened.

EXERCISES

1. At what level of production, x, was the marginal cost of a unit of product A equal to the selling price?
2. Graph the marginal cost function for product B from $x = 3.1$ million units to $x = 5.7$ million units.
3. Find the number of units for which marginal cost equals the selling price for product B.
4. For product C, the marginal cost of production is

$$y = .133x + 9.46.$$

   (a) Find the marginal cost at a level of production of 3.1 million units; of 5.7 million units.
   (b) Graph the marginal cost function.
   (c) For a selling price of $9.57, find the level of production for which the cost equals the selling price.

3
The Derivative

The management of a firm is willing to increase its advertising budget as long as its profit also increases. The expenditure on advertising that produces maximum profit is at the point where the rate of growth of profit slows to 0. However, if an increase in expenditure on advertising causes the rate of growth of the profit to become negative, then too much is spent on advertising.

In general, the task of finding the optimum level of expenditure is not at all easy. In this chapter we will see how one branch of calculus, called differential calculus, is used to optimize such variables. In addition, we will see how differential calculus is used to describe rates of change in business, social science, and biology.

The basic idea of differential calculus is the *derivative*, which we begin discussing in Section 4 of this chapter. The definition of a derivative requires the idea of a *limit*, which we discuss in the first section of this chapter.

3.1 LIMITS

Suppose we let the values of x in the domain of some function f get closer and closer to a fixed number, a. What will happen to the values of $f(x)$? Do they also approach a fixed number? Figure 3.1

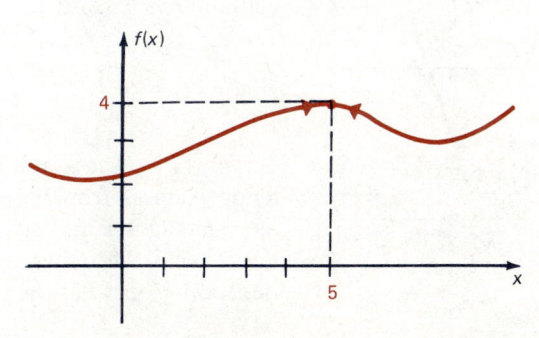

Figure 3.1

shows the graph of a function f. The arrowheads show that as x gets closer and closer to the number 5 (on the x-axis), the values of $f(x)$ get closer and closer to the number 4 (on the y-axis). In this case, we say that the **limit** of $f(x)$ as x approaches 5 is the number 4, written as

$$\lim_{x \to 5} f(x) = 4.$$

Example 1 The graph in Figure 3.2 shows a function g. The open circle at $(3, -1)$ shows that the function is not defined at $x = 3$. (That is, $g(3)$ does not exist.) What about

$$\lim_{x \to 3} g(x)?$$

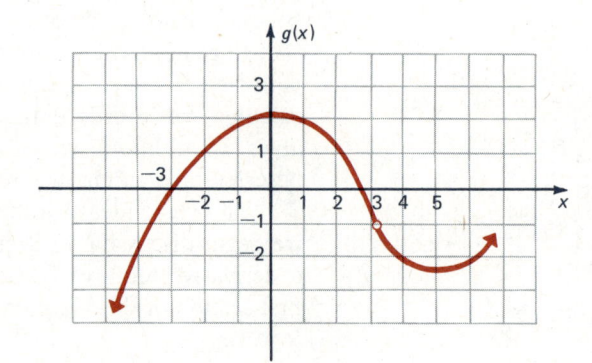

Figure 3.2

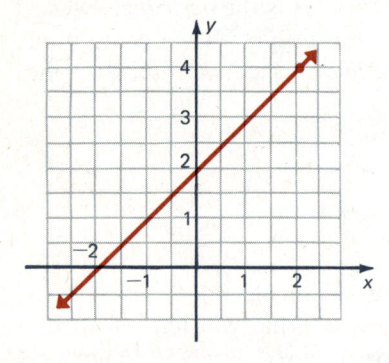

By studying the graph, we see that as x gets closer and closer to 3, $g(x)$ gets closer and closer to the number -1, so that

$$\lim_{x \to 3} g(x) = -1,$$

even though $g(3)$ does not exist. It is important to note that the phrase "x gets closer and closer to 3" does not require that x ever actually *reach* 3. ■

Work Problem 1 at the side.

Example 2 The graph of function h in Figure 3.3 shows that as x approaches 4 from the left, the values of $h(x)$ approach 3. However, as x gets closer and closer to 4 from the right, the values of $h(x)$ approach 5. Since the values of $h(x)$ approach two different numbers depending on whether x approaches 4 from the left or the right,

$$\lim_{x \to 4} h(x) \text{ does not exist.} \quad ■$$

1. Find $\lim_{x \to 2} f(x)$ for the following
(a)

(b)

Answer:
(a) 4
(b) -5

3.1 Limits

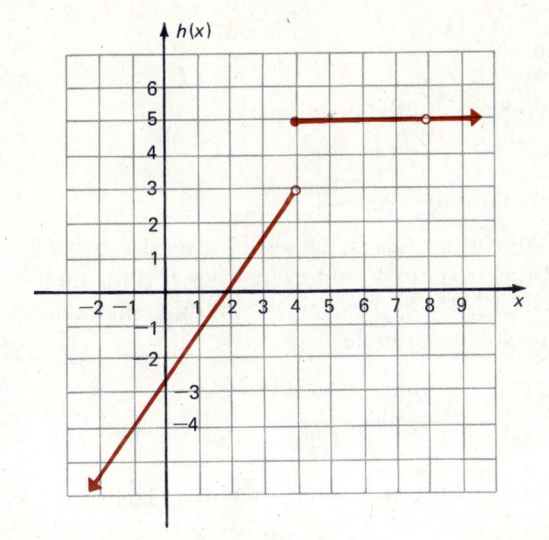

Figure 3.3

Figure 3.4

2. Use the graph of Figure 3.3 to find the following.
   (a) $\lim_{x \to 6} h(x)$
   (b) $\lim_{x \to -1} h(x)$
   (c) $\lim_{x \to 8} h(x)$

Answer:
(a) 5
(b) −4
(c) 5

Work Problem 2 at the side.

Example 3 Graph $f(x) = \dfrac{1}{x}$ and use the graph to find the following.

(a) $\lim_{x \to 4} f(x)$

The graph is shown in Figure 3.4. As x gets closer and closer to 4, the values of $f(x) = 1/x$ get closer and closer to $1/4$, so that

$$\lim_{x \to 4} f(x) = \frac{1}{4}.$$

(b) $\lim_{x \to 0} f(x)$

As x gets close to 0 from the left, the values of $f(x)$ get more and more negative. As x approaches 0 from the right, the values get larger and larger. Thus, the values of $f(x)$ approach no one fixed number, and

$$\lim_{x \to 0} f(x) \text{ does not exist.} \blacksquare$$

Work Problem 3 at the side.

3. Use Figure 3.4 to find the following.
   (a) $\lim_{x \to 1} f(x)$
   (b) $\lim_{x \to -3} f(x)$
   (c) $\lim_{x \to 2} f(x)$

Answer:
(a) 1
(b) −1/3
(c) 1/2

In the examples above, we found limits by drawing a graph. There are other methods for finding limits; for example, we can find several values of the function as x approaches a given number. The next few examples show how this works.

Example 4 Find $\lim_{x \to 1} \dfrac{x^2 - 4}{x - 2}$.

We want to find the limit of the function

$$f(x) = \frac{x^2 - 4}{x - 2}$$

as x approaches 1. We can do this by choosing several values of x on either side of 1 and close to 1, and evaluating $f(x)$ for these values. (Don't use 1 itself.) The results for several different values of x are shown in the following table.

Closer and closer to 1 ↓

x	0.8	0.9	0.99	0.9999	1.0000001	1.0001	1.001	1.01	1.05	1.1
$f(x)$	2.8	2.9	2.99	2.9999	3.0000001	3.0001	3.001	3.01	3.05	3.1

↑ Closer and closer to 3

From the table we see that the values of $f(x)$ get closer and closer to 3 as the values of x get closer and closer to 1. Thus the limit of $f(x)$ as x approaches 1 is 3, or

$$\lim_{x \to 1} \frac{x^2 - 4}{x - 2} = 3. \quad \blacksquare$$

Work Problem 4 at the side.

Example 5 Find $\lim_{x \to 2} \dfrac{x^2 - 4}{x - 2}$.

The value $x = 2$ is not in the domain of the function, since 2 makes the denominator equal 0. Even though $f(2)$ does not exist, we can still try to find the limit. (Recall Example 1 above; the limit existed, even though the function was not defined.) First, make a table of values.

Closer and closer to 2 ↓

x	1.8	1.9	1.99	1.9999	2.0000001	2.00001	2.001	2.05	2.1
$f(x)$	3.8	3.9	3.99	3.9999	4.0000001	4.00001	4.001	4.05	4.1

↑ Closer and closer to 4

From this table we see that the values of $f(x)$ get closer and closer to 4 as x gets closer and closer to 2. Therefore,

$$\lim_{x \to 2} \frac{x^2 - 4}{x - 2} = 4. \quad \blacksquare$$

---

4. Find $\lim_{x \to 4} \dfrac{x^2 - 4}{x - 2}$ by completing the following table. A calculator will be helpful.

x	$f(x)$
3.8	
3.9	
3.99	
3.999	
4.001	
4.1	
4.2	

Answer: 5.8; 5.9; 5.99; 5.999; 6.001; 6.1; 6.2; limit is 6

3.1 Limits

5. Complete the following table of values to find $\lim\limits_{x \to 3} \dfrac{x^2 - 9}{x - 3}$.

x	$f(x)$
2.9	
2.99	
2.999	
3.001	
3.01	
3.1	

Answer: 5.9, 5.99; 5.999; 6.001; 6.01; 6.1; limit is 6

6. Let $f(x) = \dfrac{2x^2 - 3x - 2}{x - 2}$.

Find the following.
(a) $\lim\limits_{x \to 3} f(x)$
(b) $\lim\limits_{x \to 0} f(x)$
(c) $\lim\limits_{x \to 2} f(x)$

Answer:
(a) 7
(b) 1
(c) 5

Work Problem 5 at the side.

In the last two examples above, you may have noticed that

$$\frac{x^2 - 4}{x - 2} = \frac{(x + 2)(x - 2)}{x - 2} = x + 2.$$

The values in the tables could have been found more easily by evaluating $x + 2$ instead of $(x^2 - 4)/(x - 2)$; the limits could also have been found in this way. For example,

$$\lim_{x \to 1} \frac{x^2 - 4}{x - 2} = \lim_{x \to 1}(x + 2).$$

As x gets closer and closer to 1, the expression $x + 2$ will get closer and closer to $1 + 2 = 3$. Thus,

$$\lim_{x \to 1} \frac{x^2 - 4}{x - 2} = \lim_{x \to 1}(x + 2) = 3.$$

Example 6 Let $f(x) = \dfrac{x^2 - x - 2}{x + 1}$. Find the following limits.

(a) $\lim\limits_{x \to 0} f(x)$

First, simplify $\dfrac{x^2 - x - 2}{x + 1}$ by factoring.

$$\frac{x^2 - x - 2}{x + 1} = \frac{(x + 1)(x - 2)}{x + 1} = x - 2$$

Now we can find the limit.

$$\lim_{x \to 0} f(x) = \lim_{x \to 0} \frac{x^2 - x - 2}{x + 1} = \lim_{x \to 0}(x - 2)$$

As x approaches 0, $x - 2$ will approach $0 - 2$, or -2.

$$\lim_{x \to 0} f(x) = \lim_{x \to 0}(x - 2) = -2.$$

(b) $\lim\limits_{x \to 5} f(x) = \lim\limits_{x \to 5}(x - 2) = 3$

As x approaches 5, $x - 2$ approaches $5 - 2 = 3$.

(c) $\lim\limits_{x \to -1} f(x) = \lim\limits_{x \to -1}(x - 2) = -3$ ∎

Work Problem 6 at the side.

7. Find the following.
(a) $\lim\limits_{x \to 4}(3x - 9)$
(b) $\lim\limits_{x \to -1}(2x^2 - 4x + 1)$
(c) $\lim\limits_{x \to 0}(9x^3 - 8x^2 + 6x + 4)$

Answer:
(a) 3
(b) 7
(c) 4

8. Find the following.
(a) $\lim\limits_{x \to 2}\sqrt{3x + 3}$
(b) $\lim\limits_{x \to -4}\sqrt{8 - 7x}$
(c) $\lim\limits_{x \to 2}\sqrt{x^2 + 3x + 4}$

Answer:
(a) 3
(b) 6
(c) $\sqrt{14}$

Example 7 Find $\lim\limits_{x \to 3}(x^2 + 2x + 5)$.

There are no denominators here that might become 0, so we can find this limit by reasoning as follows: As x approaches 3, x^2 will approach 3^2, $2x$ will approach $2 \cdot 3$, and 5 will not change. Therefore,

$$\lim_{x \to 3}(x^2 + 2x + 5) = 3^2 + 2 \cdot 3 + 5 = 9 + 6 + 5 = 20. \blacksquare$$

Work Problem 7 at the side.

Example 8 Find $\lim\limits_{x \to 9}\sqrt{2x - 2}$.

As x approaches 9, $\sqrt{2x - 2}$ approaches $\sqrt{2 \cdot 9 - 2}$. We have

$$\lim_{x \to 9}\sqrt{2x - 2} = \sqrt{2 \cdot 9 - 2} = \sqrt{18 - 2} = \sqrt{16} = 4. \blacksquare$$

Work Problem 8 at the side.

Example 9 Find $\lim\limits_{x \to 4}\dfrac{\sqrt{x} - 2}{x - 4}$.

As x approaches 4, we have

$$\lim_{x \to 4}\frac{\sqrt{x} - 2}{x - 4} = \frac{\sqrt{4} - 2}{4 - 4} = \frac{2 - 2}{4 - 4} = \frac{0}{0}.$$

The expression 0/0 is not a real number; in fact 0/0 is called an **indeterminate form.** Even though we got 0/0, the limit might still exist. We can look for the limit by making a table of values or by using a little algebra. To use algebra, multiply numerator and denominator of $(\sqrt{x} - 2)/(x - 4)$ by $\sqrt{x} + 2$. This gives

$$\frac{\sqrt{x} - 2}{x - 4} \cdot \frac{\sqrt{x} + 2}{\sqrt{x} + 2} = \frac{\sqrt{x} \cdot \sqrt{x} - 2\sqrt{x} + 2\sqrt{x} - 4}{(x - 4)(\sqrt{x} + 2)}$$

$$= \frac{x - 4}{(x - 4)(\sqrt{x} + 2)} \qquad \sqrt{x} \cdot \sqrt{x} = x$$

$$= \frac{1}{\sqrt{x} + 2}.$$

We now have

$$\lim_{x \to 4}\frac{\sqrt{x} - 2}{x - 4} = \lim_{x \to 4}\frac{1}{\sqrt{x} + 2} = \frac{1}{\sqrt{4} + 2} = \frac{1}{2 + 2} = \frac{1}{4}. \blacksquare$$

Whenever the indeterminate form 0/0 is obtained, the limit may still exist. You must do further work to find out.

9. Find the following.
(a) $\lim\limits_{x \to 1} \dfrac{\sqrt{x} - 1}{x - 1}$

(b) $\lim\limits_{x \to 25} \dfrac{\sqrt{x} - 5}{x - 25}$

Answer:
(a) 1/2
(b) 1/10

10. Use the graph of Figure 3.5 to find the following.
(a) $\lim\limits_{x \to 20} p(x)$
(b) $\lim\limits_{x \to 30} p(x)$

Answer:
(a) $300
(b) $340

Work Problem 9 at the side.

Example 10 The graph of Figure 3.5 shows the profit from producing x units of a certain item in a given day. Notice the break in the graph at $x = 25$. This break occurs because $x = 25$ items is the maximum that can be produced by one shift of workers. If more than 25 items are needed, the night shift must be called to work. This increases the fixed costs and causes a drop in profits unless many more than 25 items are needed. (By inspecting the graph, we see that 35 items must be needed before it pays to bring on the second shift.) From the graph,

$$\lim_{x \to 25} p(x) \text{ does not exist.} \blacksquare$$

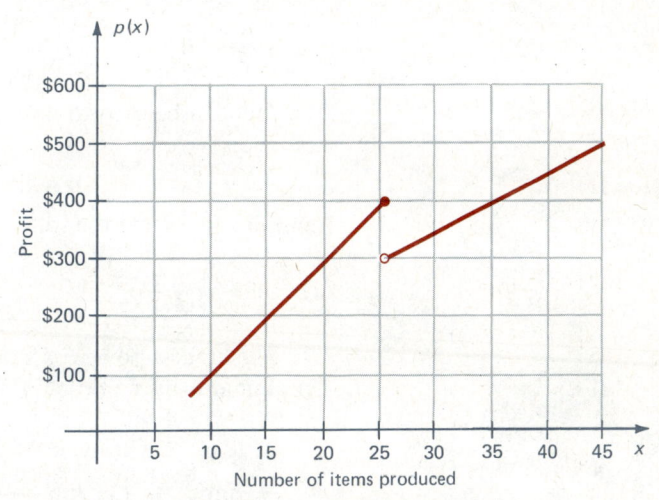

Figure 3.5

Work Problem 10 at the side.

Limits to Infinity We sometimes need to find the limit of a function as values of x get larger and larger without bound. This can be done by using **limits to infinity** as shown in the following example.

Example 11 Find the limit of $f(x) = \dfrac{6x + 1}{2x - 5}$ as x increases without bound.

This limit is written

$$\lim_{x \to \infty} \dfrac{6x + 1}{2x - 5}.$$

3.1 Limits 99

To find the limit, we can complete a table of values.

x	1000	100,000	1,000,000	1,000,000,000
$f(x)$	3.00802	3.00008	3.000008	3.000000008

From this table, $\lim\limits_{x \to \infty} \dfrac{6x + 1}{2x - 5} = 3$. ∎

Work Problem 11 at the side.

The results of parts (b) and (c) of Problem 11 can be generalized as follows:

$$\lim_{x \to \infty} \frac{k}{x^n} = 0 \quad \text{for any positive integer } n \text{ and any real number } k.$$

This fact can be used to find limits to infinity, as shown in the next example.

Example 12 Find the following limits.

(a) $\lim\limits_{x \to \infty} \dfrac{8x + 6}{3x - 1}$

The highest power of x here is x^1, so we divide numerator and denominator by x^1, or x.

$$\lim_{x \to \infty} \frac{8x + 6}{3x - 1} = \lim_{x \to \infty} \frac{\frac{8x}{x} + \frac{6}{x}}{\frac{3x}{x} - \frac{1}{x}} = \lim_{x \to \infty} \frac{8 + \frac{6}{x}}{3 - \frac{1}{x}} = \frac{8 + 0}{3 - 0} = \frac{8}{3}$$

(b) $\lim\limits_{x \to \infty} \dfrac{4x^2 - 6x + 3}{2x^2 - x + 4}$

Divide numerator and denominator by x^2, the highest power of x.

$$\lim_{x \to \infty} \frac{4x^2 - 6x + 3}{2x^2 - x + 4} = \lim_{x \to \infty} \frac{4 - \frac{6}{x} + \frac{3}{x^2}}{2 - \frac{1}{x} + \frac{4}{x^2}} = \frac{4 - 0 + 0}{2 - 0 + 0} = \frac{4}{2} = 2$$

(c) $\lim\limits_{x \to \infty} \dfrac{3x + 2}{4x^2 - 1} = \lim\limits_{x \to \infty} \dfrac{\frac{3}{x} + \frac{2}{x^2}}{4 - \frac{1}{x^2}} = \dfrac{0 + 0}{4 - 0} = \dfrac{0}{4} = 0$

---

11. Make a table of values and find the following limits.

(a) $\lim\limits_{x \to \infty} \dfrac{15x + 9}{3x - 6}$

(b) $\lim\limits_{x \to \infty} \dfrac{1}{x}$

(c) $\lim\limits_{x \to \infty} \dfrac{1}{x^2}$

Answer:
(a) 5
(b) 0
(c) 0

12. Find the following limits.

(a) $\lim\limits_{x \to \infty} \dfrac{4x^2 + 6x}{9x^2 - 3}$

(b) $\lim\limits_{x \to \infty} \dfrac{3x^2 + 4x + 5}{8x^3 + 2x + 1}$

(c) $\lim\limits_{x \to \infty} \dfrac{4x^2 + 9x + 1}{2x + 3}$

Answer:
(a) 4/9
(b) 0
(c) does not exist

(d) $\lim\limits_{x \to \infty} \dfrac{9x^2 - 1}{3x + 5} = \lim\limits_{x \to \infty} \dfrac{9 - \dfrac{1}{x^2}}{\dfrac{3}{x} + \dfrac{5}{x^2}} = \dfrac{9 - 0}{0 + 0} = \dfrac{9}{0}$

Since 9/0 is not a real number, this limit does not exist. ■

Work Problem 12 at the side.

3.1 EXERCISES

Use the following graphs to find the indicated limits if they exist. See Examples 1–3.

1. $\lim\limits_{x \to 3} f(x)$

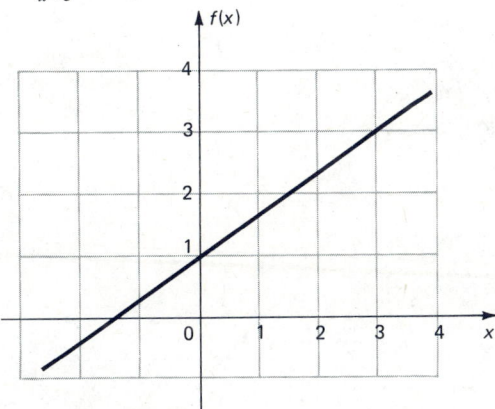

2. $\lim\limits_{x \to 2} F(x)$

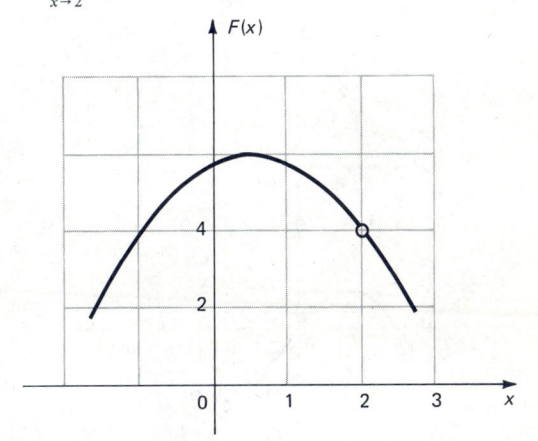

3. $\lim\limits_{x \to -2} f(x)$

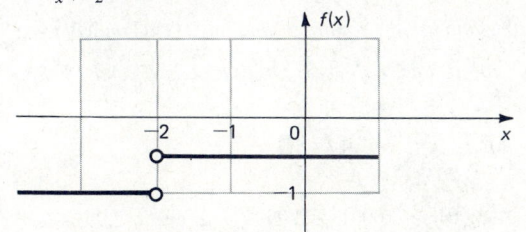

4. $\lim\limits_{x \to 3} g(x)$

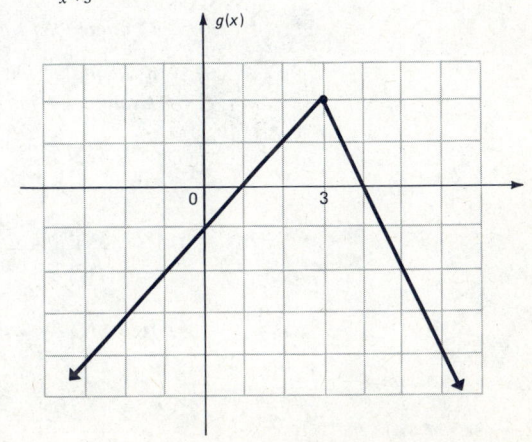

102 The Derivative

5. $\lim\limits_{x \to 0} f(x)$

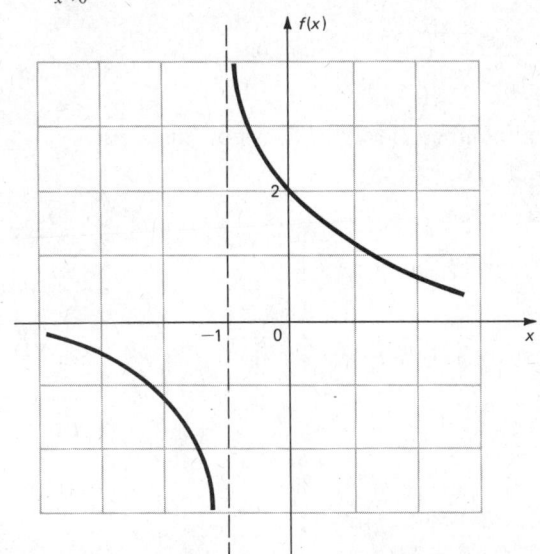

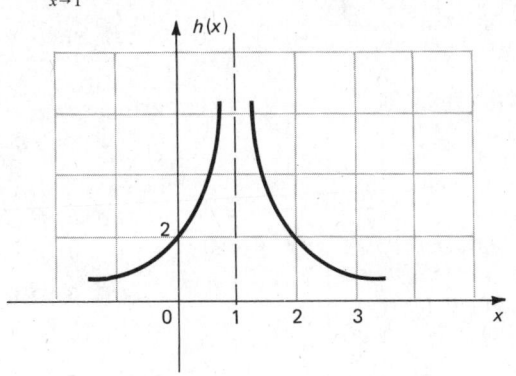

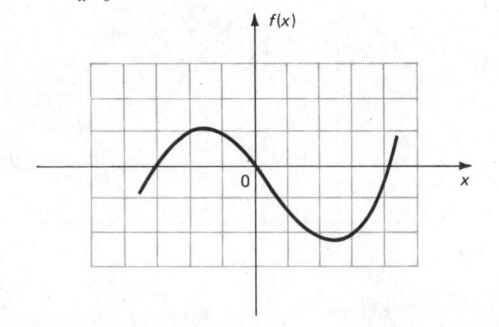

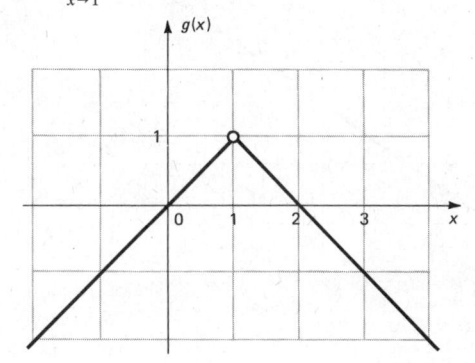

6. $\lim\limits_{x \to 1} h(x)$

7. $\lim\limits_{x \to 0} f(x)$

8. $\lim\limits_{x \to 1} g(x)$

Complete the following table and use the results to find $\lim\limits_{x \to a} f(x)$. You will need a calculator for Exercises 15–18 and a $\sqrt{x}$ key for Exercises 17 and 18. See Examples 4 and 5.

9. $f(x) = 4 + 5x$, find $\lim\limits_{x \to -3} f(x)$.

x	−3.1	−3.01	−3.001	−3.0001	−2.9999	−2.999	−2.99	−2.9
$f(x)$								

10. $g(x) = 2x - 5$, find $\lim\limits_{x \to 3} g(x)$.

x	2.9	2.99	2.999	2.9999	3.0001	3.001	3.01	3.1
$g(x)$								

3.1 Limits

11. $h(x) = x^2 + 2x + 1$, find $\lim_{x \to 2} h(x)$.

x	1.9	1.99	1.999	2.001	2.01	2.1
$h(x)$			8.994001	9.006001		

12. $f(x) = 2x^2 - 4x + 3$, find $\lim_{x \to 1} f(x)$.

x	.9	.99	.999	1.001	1.01	1.1
$f(x)$			1.000002	1.000002		

13. $g(x) = 2/x$, find $\lim_{x \to 0} g(x)$.

x	$-.1$	$-.01$	$-.001$	.001	.01	.1
$g(x)$						

14. $h(x) = -5/x$, find $\lim_{x \to 0} h(x)$.

x	$-.1$	$-.01$	$-.001$	.001	.01	.1
$h(x)$						

15. $k(x) = \dfrac{x^3 - 2x - 4}{x - 2}$, find $\lim_{x \to 2} k(x)$.

x	1.9	1.99	1.999	2.001	2.01	2.1
$k(x)$						

16. $f(x) = \dfrac{2x^3 + 3x^2 - 4x - 5}{x + 1}$, find $\lim_{x \to -1} f(x)$.

x	-1.1	-1.01	-1.001	$-.999$	$-.99$	$-.9$
$f(x)$						

17. $h(x) = \dfrac{\sqrt{x} + 2}{x - 1}$, find $\lim_{x \to 1} h(x)$.

x	.9	.99	.999	1.001	1.01	1.1
$h(x)$						

18. $f(x) = \dfrac{\sqrt{x} - 3}{x - 3}$, find $\lim_{x \to 3} f(x)$.

x	2.999	2.99	2.9	3.001	3.01	3.1
$f(x)$						

The Derivative

Use any method to find the following limits that exist. See Examples 6–9.

19. $\lim_{x \to 4} 3x$
20. $\lim_{x \to -3} -4x$
21. $\lim_{x \to -3} (4x^2 + 2x - 1)$
22. $\lim_{x \to 2} (2x^2 - 3x + 5)$
23. $\lim_{x \to 2} (2x^3 + 5x^2 + 2x + 1)$
24. $\lim_{x \to -1} (4x^3 - x^2 + 3x - 1)$
25. $\lim_{x \to 3} \dfrac{5x - 6}{2x + 1}$
26. $\lim_{x \to -2} \dfrac{2x + 1}{3x - 4}$
27. $\lim_{x \to 1} \dfrac{2x^2 - 6x + 3}{3x^2 - 4x + 2}$
28. $\lim_{x \to 2} \dfrac{-4x^2 + 6x - 8}{3x^2 + 7x - 2}$
29. $\lim_{x \to 3} \dfrac{x^2 - 9}{x - 3}$
30. $\lim_{x \to -2} \dfrac{x^2 - 4}{x + 2}$
31. $\lim_{x \to -2} \dfrac{x^2 - x - 6}{x + 2}$
32. $\lim_{x \to 5} \dfrac{x^2 - 3x - 10}{x - 5}$
33. $\lim_{x \to 0} \dfrac{x^2 - x}{x}$
34. $\lim_{x \to 0} \dfrac{3x^2 - 4x}{x}$
35. $\lim_{x \to 3} \sqrt{x^2 - 4}$
36. $\lim_{x \to 3} \sqrt{x^2 - 5}$
37. $\lim_{x \to 25} \dfrac{\sqrt{x} - 5}{x - 25}$
38. $\lim_{x \to 36} \dfrac{\sqrt{x} - 6}{x - 36}$

The graph below shows the profit from the daily production of x thousand kilograms of an industrial chemical. Use the graph to find the following limits. See Example 10.

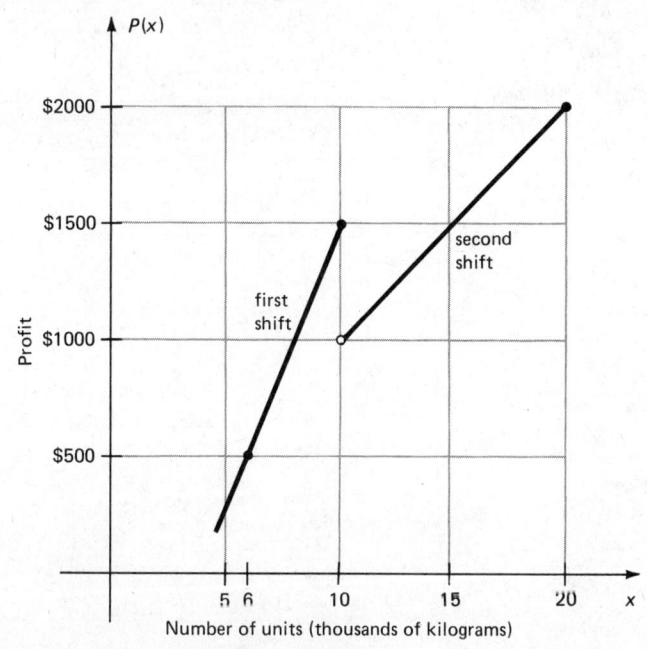

39. $\lim\limits_{x \to 6} P(x)$ 40. $\lim\limits_{x \to 10} P(x)$ 41. $\lim\limits_{x \to 15} P(x)$

42. Use the graph to estimate the number of units of the chemical that must be produced before the second shift is profitable.

Find each of the following limits that exist. See Example 12.

43. $\lim\limits_{x \to \infty} \dfrac{3x}{5x - 1}$ 44. $\lim\limits_{x \to \infty} \dfrac{5x}{3x - 1}$

45. $\lim\limits_{x \to \infty} \dfrac{2x + 3}{4x - 7}$ 46. $\lim\limits_{x \to \infty} \dfrac{8x + 2}{2x - 5}$

47. $\lim\limits_{x \to \infty} \dfrac{x^2 + 2x}{2x^2 - 2x + 1}$ 48. $\lim\limits_{x \to \infty} \dfrac{x^2 + 2x - 5}{3x^2 + 2}$

49. $\lim\limits_{x \to \infty} \dfrac{3x^3 + 2x - 1}{2x^4 - 3x^3 - 2}$ 50. $\lim\limits_{x \to \infty} \dfrac{2x^2 + 11x - 10}{5x^3 + 3x^2 + 2x}$

51. $\lim\limits_{x \to \infty} \dfrac{2x^4 + 3x + 4}{x^2 - 4x + 1}$ 52. $\lim\limits_{x \to \infty} \dfrac{2x^2 - 1}{3x^4 + 2}$

Political Science   Members of a legislature must often vote repeatedly on the same issue. As time goes on, the members may change their vote. A formula* for the chance that a legislator will vote yes on the nth roll call vote on the same issue is

$$p_n = \frac{1}{2} + \left(p_0 - \frac{1}{2}\right)(1 - 2p)^n,$$

where n is the number of roll calls taken, p_n is the chance that the member will vote yes on the nth roll call vote, p is the chance that the legislator will change his or her position on successive roll calls, and p_0 is the chance that the member favors the issue at the beginning (p_n, p, and p_0 are always between 0 and 1, inclusive). Suppose that $p_0 = .7$ and $p = .2$. Use a calculator to find p_n for

53. $n = 2$;   54. $n = 4$;   55. $n = 8$.

56. Find $\lim\limits_{n \to \infty} p_n$.

3.2 CONTINUITY

In the previous section, we saw a graph showing the profit from producing x units of a certain item. This graph is repeated in Figure 3.6. The graph is smooth and unbroken except for a sudden drop at the point $x = 25$. A break in a graph such as this is called a **discontinuity**. The function is **discontinuous** at the given point.

---

* See John W. Bishir and Donald W. Drewes, *Mathematics in the Behavioral and Social Sciences* (New York: Harcourt Brace Jovanovich, 1970), p. 538.

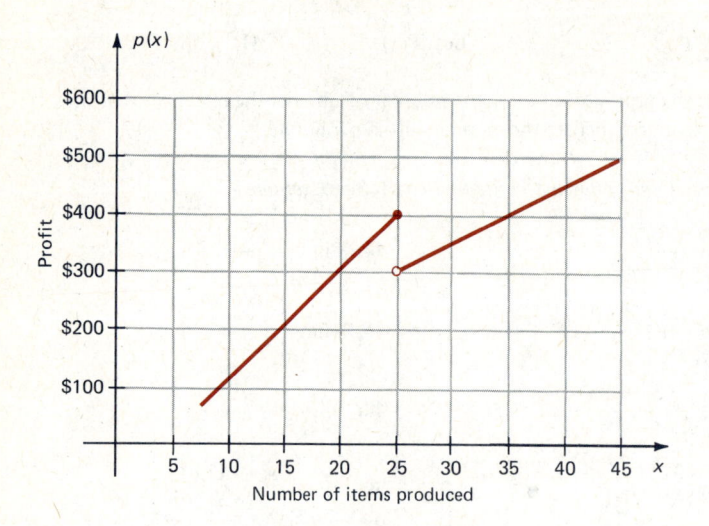

106 The Derivative

Figure 3.6

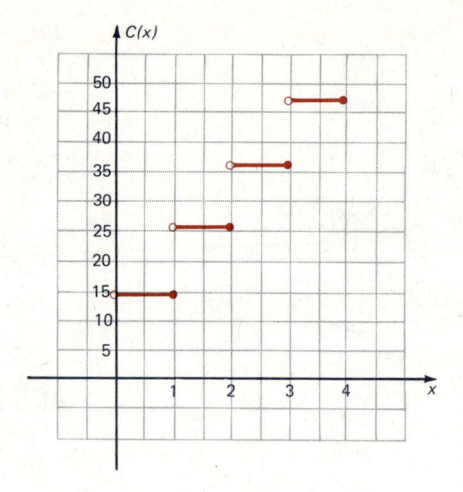

Figure 3.7

1. Find any discontinuities for the following functions.
(a)

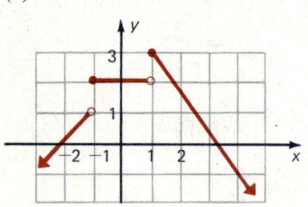

(b)

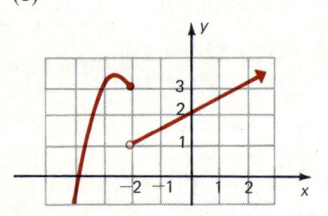

Answer:
(a) −1, 1
(b) −2

Work Problem 1 at the side.

Example 1 A trailer rental firm charges a flat $4 to rent a hitch. The trailer itself is rented for $11 per day or fraction of a day. Let $C(x)$ represent the cost of renting a hitch and trailer for x days.
(a) Graph $C(x)$.

To rent a trailer for one day, we pay $4 for the hitch and $11 for the trailer. The total rental is thus $4 + $11 = $15. In fact, if $0 < x \le 1$, then $C(x) = 15$. If we rent the trailer for more than one day, but not more than two, we pay $4 + 2 \cdot 11 = 26$ dollars. Thus, if $1 < x \le 2$, then $C(x) = 26$. Also, if $2 < x \le 3$, then $C(x) = 37$. These results lead to the graph of Figure 3.7.
(b) Find any places where the function $C(x)$ is discontinuous.

The graph of Figure 3.7 shows that the function is discontinuous at $x = 1, 2, 3, 4, 5, 6, \ldots$. ∎

Work Problem 2 at the side, on page 107 (top).

If a function has a discontinuity, then it is not possible to draw a graph of the function without lifting the pencil from the paper. On the other hand, if a function *can* be drawn without lifting the pencil from the paper, the function is *continuous*. Many useful functions are continuous. In particular, the linear, quadratic, and polynomial functions of Chapter 2 are continuous; the rational functions are usually not.

Work Problem 3 at the side.

The graph of Figure 3.8 is that of a continuous function—the graph can be drawn without lifting the pencil from the paper. Let us see why this function is continuous by studying a sample point such as the point $x = a$. First of all, $f(a)$ is defined. That is, there is a point on the graph corresponding to $x = a$.

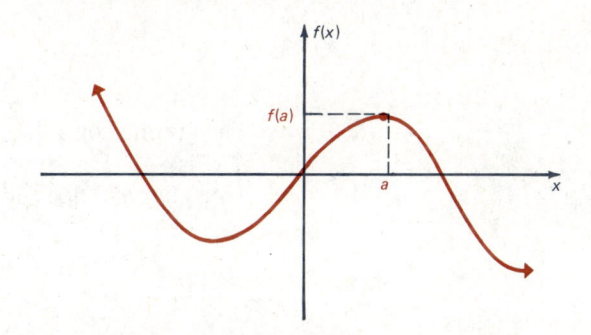

Figure 3.8

Second, notice that $\lim_{x \to a} f(x)$ exists. As x approaches a from either side of a, the values of $f(x)$ approach one particular number. The number that the values of $f(x)$ approach is in fact $f(a)$.

In general, the function f is **continuous at $x = a$** if all of the following conditions hold:

(a) $f(a)$ exists
(b) $\lim_{x \to a} f(x)$ exists
(c) $\lim_{x \to a} f(x) = f(a)$

Example 2 Tell why the following functions are discontinuous at the indicated points.
(a) $x = 3$

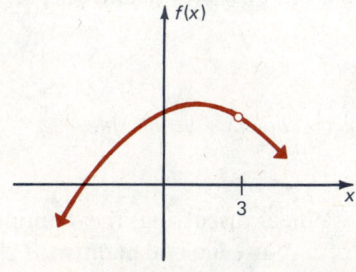

The open circle on the graph at $x = 3$ shows that $f(3)$ does not exist, and thus part (a) of the definition fails.

---

2. A chain-saw rental firm charges a flat $6 sharpening fee, plus $5 per day or fraction of a day. Let $C(x)$ represent the cost of renting a saw for x days.
   (a) Graph $C(x)$.
   (b) Find any places where the function is discontinuous.

Answer:
(a)

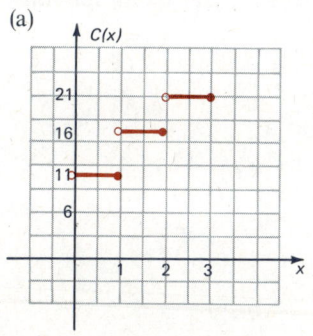

(b) at $x = 1, 2, 3, 4, 5, 6, \ldots$

3. Which of the following functions are continuous?
(a)

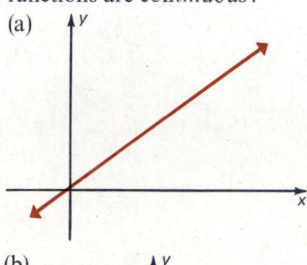

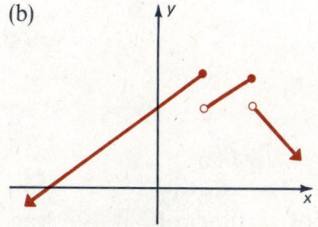

(b)

Answer:
(a) continuous
(b) not continuous

(b) $x = 0$

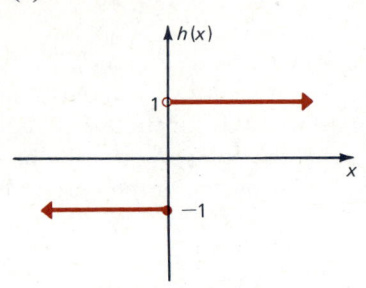

The graph shows that $f(0)$ exists. As x approaches 0 from the left, $h(x)$ is -1. However, as x approaches 0 from the right, $h(x)$ is 1. Thus, $\lim_{x \to 0} h(x)$ does not exist and part (b) of the definition fails.

(c) $x = 4$

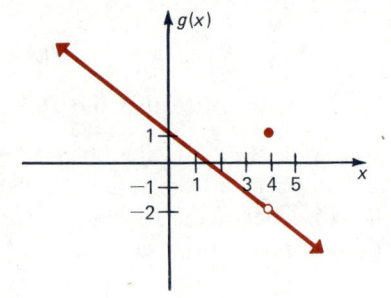

Here, the heavy dot above 4 shows that $g(4)$ exists. In fact, $g(4) = 1$. However, check that

$$\lim_{x \to 4} g(x) = -2.$$

Thus, $\lim_{x \to 4} g(x) \neq g(4)$, and part (c) of the definition fails. ∎

Work Problem 4 at the side.

When discussing the continuity of a function, it is often helpful to use **interval notation**. This notation is defined on the next page.

---

4. Tell why the following functions are discontinuous at the indicated points.
(a)

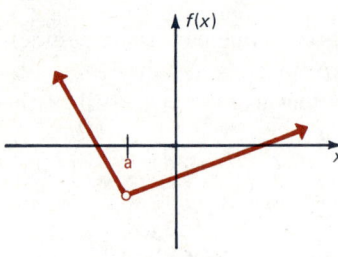

(b)

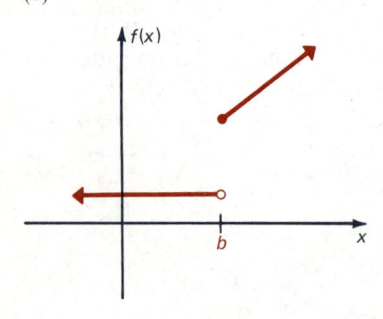

Answer:
(a) $f(a)$ does not exist
(b) $\lim_{x \to b} f(x)$ does not exist

3.2 Continuity 109

Interval	Name	Description	Interval notation
○——● between −2 and 3	Open interval	$-2 < x < 3$	$(-2, 3)$
●——● between −2 and 3	Closed interval	$-2 \leq x \leq 3$	$[-2, 3]$
←——○ at 3	Open interval	$x < 3$	$(-\infty, 3)$
○——→ at −5	Open interval	$x > -5$	$(-5, \infty)$

The symbol ∞ (read "infinity") does not represent a number; ∞ is used for convenience in interval notation.

Work Problem 5 at the side.

If a function is continuous at each point of an open interval, it is said to be **continuous on the open interval.**

Example 3 Find all open intervals where the function of Figure 3.9 is continuous.

As the graph shows, the function is discontinuous at $x = -2, 0,$ and 2. Also from the graph, the function is continuous on the open intervals $(-\infty, -2), (-2, 0), (0, 2),$ and $(2, \infty)$. ∎

5. Write each of the following using interval notation.
(a) ○——○ from −5 to 3
(b) ●——● from 4 to 7
(c) ←——● at −1

Answer:
(a) $(-5, 3)$
(b) $[4, 7]$
(c) $(-\infty, -1]$

6. Find all open intervals where the following functions are continuous.
(a) [graph of f(x)]
(b) [graph of f(x)]

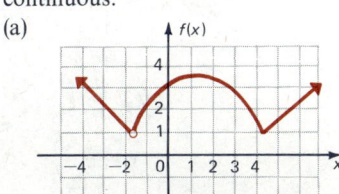

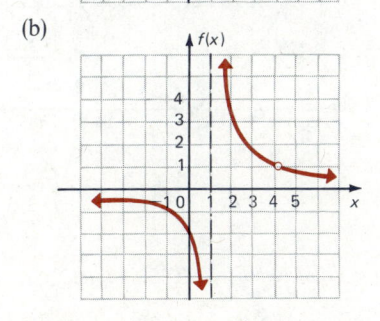

Answer:
(a) $(-\infty, -2), (-2, \infty)$
(b) $(-\infty, 1), (1, 4), (4, \infty)$

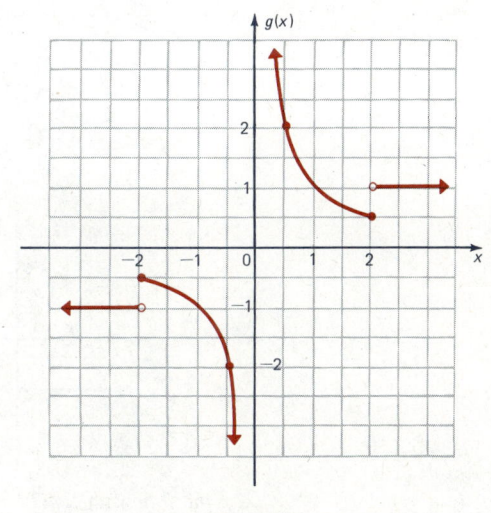

Figure 3.9

Work Problem 6 at the side.

3.2 EXERCISES

Find all points of discontinuity for the following functions. See Example 1.

1. 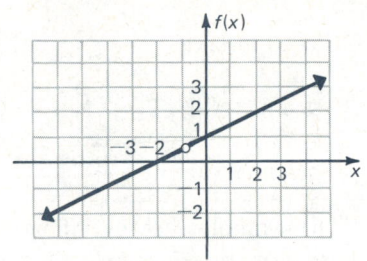
2. 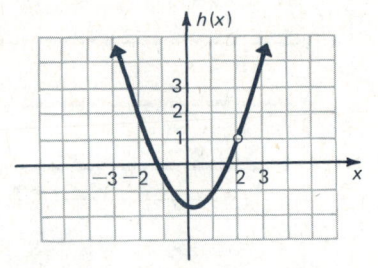
3. 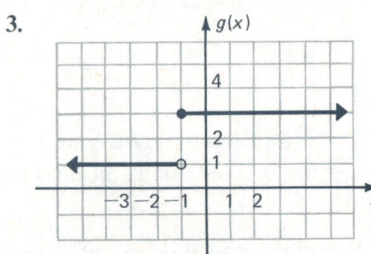
4. 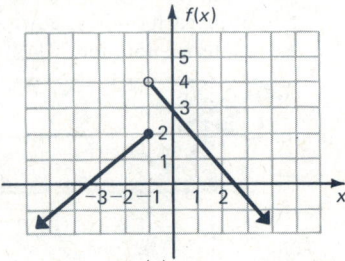
5. 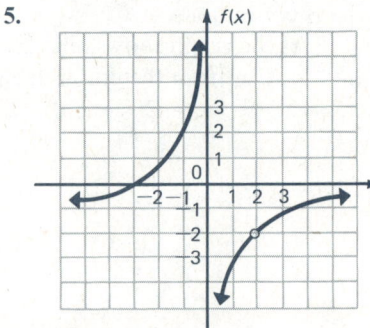
6. 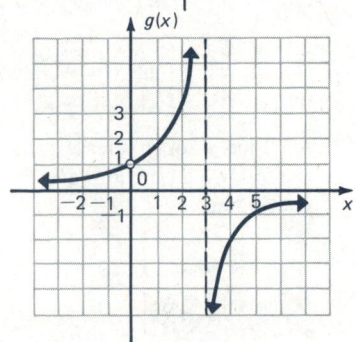
7. 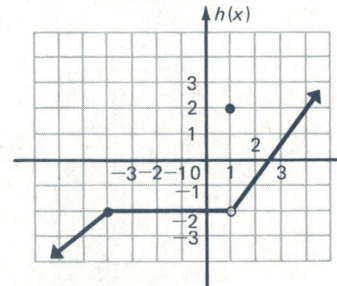
8.

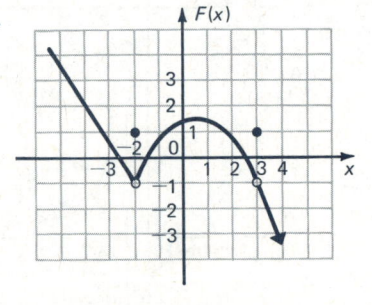

Are the following functions continuous at the given points? See Example 2.

9. $f(x) = \dfrac{2}{x-3}$; $x = 0, x = 3$

10. $f(x) = \dfrac{6}{x}$; $x = 0, x = -1$

11. $g(x) = \dfrac{1}{x(x-2)}$; $x = 0, x = 2, x = 4$

12. $h(x) = \dfrac{-2}{3x(x+5)}$; $x = 0, x = 3, x = -5$

13. $h(x) = \dfrac{1+x}{(x-3)(x+1)}$; $x = 0, x = 3, x = -1$

14. $g(x) = \dfrac{-2x}{(2x+1)(3x+6)}$; $x = 0, x = -1/2, x = -2$

15. $k(x) = \dfrac{5+x}{2+x}$; $x = 0, x = -2, x = -5$

16. $f(x) = \dfrac{4-x}{x-9}$; $x = 0, x = 4, x = 9$

17. $g(x) = \dfrac{x^2 - 4}{x - 2}$; $x = 0, x = 2, x = -2$

18. $h(x) = \dfrac{x^2 - 25}{x + 5}$; $x = 0, x = 5, x = -5$

19. $p(x) = x^2 - 4x + 11$; $x = 0, x = 2, x = -1$

20. $q(x) = -3x^3 + 2x^2 - 4x + 1$; $x = -2, x = 3, x = 1$

Cost A company charges $1.20 per pound for a certain fertilizer on all orders not over 100 pounds, and $1 per pound for orders over 100 pounds. Let $F(x)$ represent the cost for buying x pounds of the fertilizer.

21. Find $F(80)$.
22. Find $F(150)$.
23. Graph $y = F(x)$.
24. Where is F discontinuous?

The cost to transport a mobile home depends on the distance, x, in miles that the home is moved. Let $C(x)$ represent the cost to move a mobile home x miles. One firm charges as follows.

Cost per mile	Distance in miles
$2	if $0 < x \le 150$
$1.50	if $150 < x \le 400$
$1.25	if $400 < x$

25. Find $C(130)$.
26. Find $C(210)$.
27. Find $C(350)$.
28. Find $C(500)$.
29. Graph $y = C(x)$.
30. Where is C discontinuous?

112 The Derivative

31. There are certain skills (such as music) where learning is rapid at first and then levels off. All of a sudden, an insight may be had, causing learning to speed up. A typical graph of such learning is shown in the figure. Where is the function discontinuous?

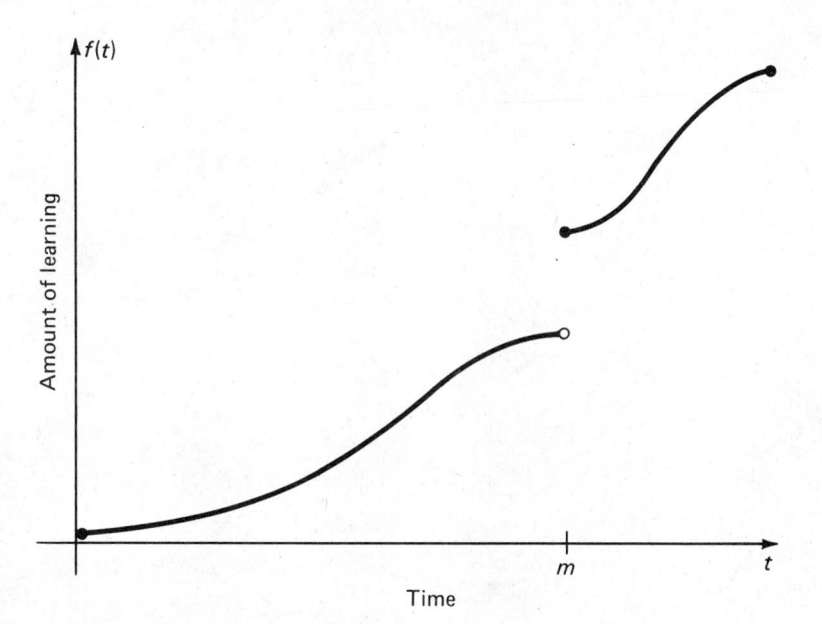

Write each of the following in interval notation.

32.  [graph: open at −4, open at 6]
33.  [graph: closed at −9, closed at 15]
34. 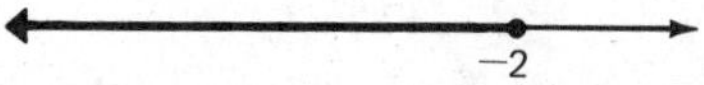 [graph: arrow to closed at −2]
35. [graph: open at 10, arrow]

36. $\{x \mid -11 \leq x \leq 9\}$
37. $\{x \mid -8 \leq x \leq 0\}$
38. $\{x \mid -6 < x < -2\}$
39. $\{x \mid 12 < x < 20\}$
40. $\{x \mid x > -4\}$
41. $\{x \mid x < 3\}$
42. $\{x \mid x < 0\}$
43. $\{x \mid x > -10\}$

Identify all open intervals where the following functions are continuous.

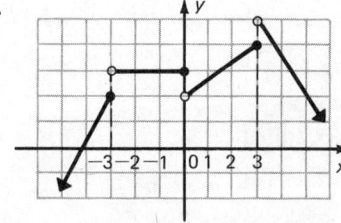

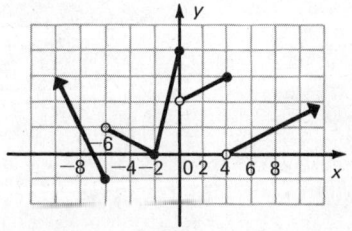

44.
45.

46.

47.

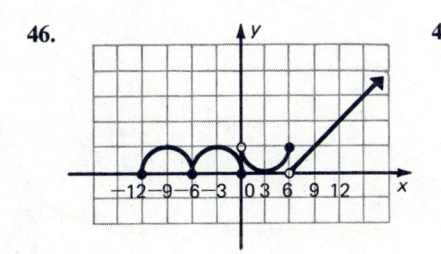

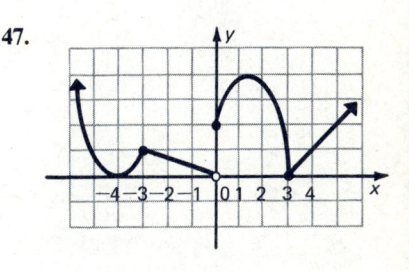

3.3 RATES OF CHANGE

One of the main applications of calculus is telling how one variable changes in relation to another. A person in business wants to know how profit changes with respect to advertising, while a person in medicine wants to know how a patient's reaction to a drug changes with respect to the dose.

For example, the graph of Figure 3.10 shows the profit $P(x)$ for a company when x thousand dollars is spent on advertising. Notice that as advertising expenditures increase, profit also increases, but at a slower rate. When $x = 1$, profit is $2000, when $x = 3$, profit is $3000, and so on. The **average rate of change** of profit with respect to advertising is defined as the change in profit divided by the change in advertising costs.

$$\text{average rate of change} = \frac{\text{change in profit}}{\text{change in advertising costs}}$$

$$= \frac{3000 - 2000}{3 - 1} = \frac{1000}{2} = 500$$

In the interval from $x = 1$ to $x = 3$, an increase of $1000 in advertising produces an average increase of $500 in profit.

Work Problem 1 at the side.

1. Use Figure 3.10 and find the average rate of change in profit if advertising increases from
(a) $x = 1$ to $x = 4$;
(b) $x = 2$ to $x = 4$;
(c) $x = 3$ to $x = 4$.

Answer:
(a)   about $430
(b)   about $350
(c)   about $300

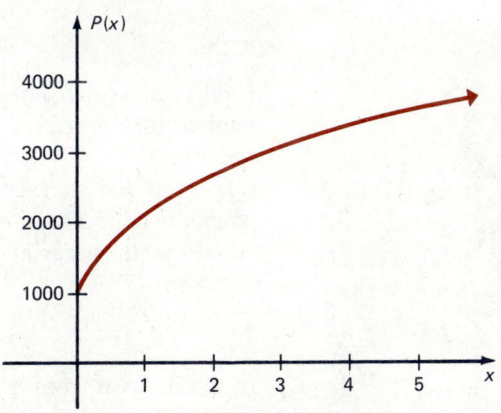

Figure 3.10

114 The Derivative

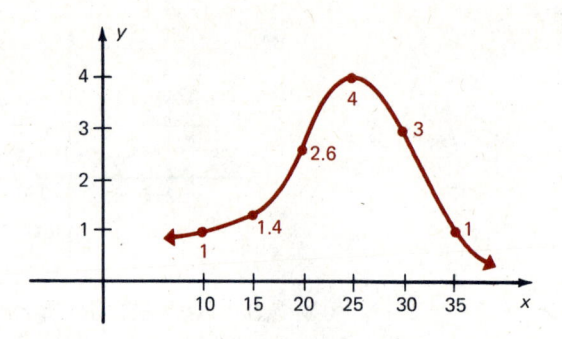

Figure 3.11

Example 1 Figure 3.11 shows a graph of the population of rodents, y, in thousands, in a given location when average midday temperatures are x degrees Celsius.

(a) Find the average rate of change in the rodent population when the temperature changes from 15° to 25°.

From the graph, there are 1.4 thousand rodents at a temperature of 15° and 4 thousand at a temperature of 25°. The average rate of change of the rodent population with respect to the temperature is given by the quotient of the change in population and the change in temperature, or

$$\frac{4 - 1.4}{25 - 15} = \frac{2.6}{10} = .26.$$

In the interval from $x = 15$ to $x = 25$, an increase of 1° in temperature produces an average of .26 thousands (or 260) additional rodents.

(b) Find the average rate of change in the rodent population when the temperature changes from 25° to 30°.

We have

$$\frac{3 - 4}{30 - 25} = \frac{-1}{5} = -.20.$$

In the interval from $x = 25$ to $x = 30$, an increase of 1° in the temperature produces an average *decrease* of .2 thousand (or 200) rodents.

(c) Find the average rate of change in the rodent population when the temperature changes from 10° to 35°.

Read the necessary values from the graph of Figure 3.11.

$$\frac{1 - 1}{35 - 10} = \frac{0}{25} = 0$$

In the interval from $x = 10$ to $x = 35$, an increase in the temperature of 1° produces an average change in the rodent population of 0. ∎

2. Use Figure 3.11 and find the average rate of change of the population if temperature changes from
(a) $x = 15$ to $x = 20$;
(b) $x = 20$ to $x = 25$;
(c) $x = 25$ to $x = 35$;
(d) $x = 30$ to $x = 35$.

Answer:
(a) .24
(b) .28
(c) −.3
(d) −.4

Work Problem 2 at the side.

As Example 1(c) showed, we can get answers that aren't very helpful if we find the average rate of change of a function over a large interval. Our results are more accurate and usable if we find the average rate of change over a fairly narrow interval. In fact, the most useful result of all comes if we take the limit as the width of the interval approaches 0.

To see how this works, let us suppose that the profit, $P(x)$, for a certain firm is related to the volume of production according to the mathematical model

$$P(x) = 16x - x^2,$$

where x represents the number of units produced. We will assume that the company can produce any nonnegative number of units. A graph of the profit function is shown in Figure 3.12. An inspection of the graph shows that an increase in production from 1 unit to 2 units will increase profit more than an increase in production from 6 units to 7 units. An increase from 7 units to 8 units produces very little increase in profit, while an increase in production from 8 units to 9 units actually produces a decline in total profit.

Using the profit function $P(x) = 16x - x^2$, we find that the profit from producing 1 unit is

$$P(1) = 16(1) - 1^2 = 15.$$

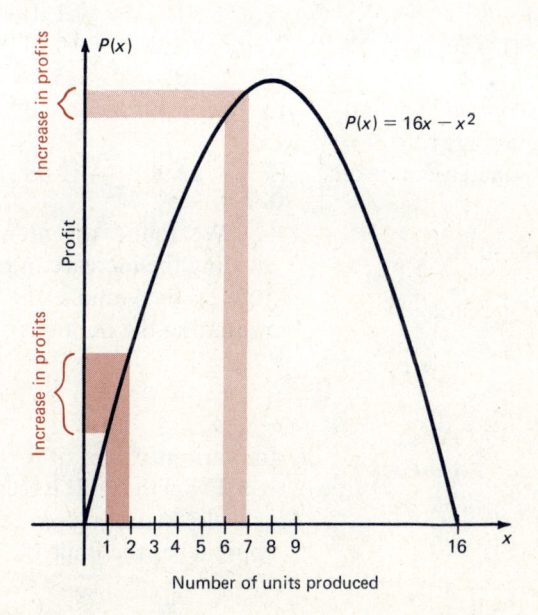

Figure 3.12

3. (a) Find the profit from producing 4 units.
(b) Find the increase in profit if production increases from 1 unit to 4 units.

Answer:
(a) 48
(b) 33

The profit from producing 2 units is

$$P(2) = 16(2) - 2^2 = 28.$$

This 1-unit increase of production (from 1 unit to 2 units) increased profit by $28 - 15 = 13$ dollars.

Work Problem 3 at the side.

As Problem 3 shows, if production increases from 1 unit to 4 units, profit increases by $33. The average rate of increase in profit for this 3-unit increase in production is found by dividing the change in profits by the change in production.

$$\text{average rate of increase} = \frac{P(4) - P(1)}{4 - 1} = \frac{48 - 15}{3} = \frac{33}{3} = 11$$

We have seen that an increase in production from 1 unit to 4 units leads to an average rate of increase in profit of 11, while an increase from 1 unit to 2 units leads to an average rate of increase in profit of 13. In general, if we let production increase from 1 unit to $1 + h$ units, where h is a small positive number, what happens to the average rate of increase in profit? To find out, we make a table which shows the average rate of increase of profit for some selected values of h.

h	1	0.1	0.01	0.001	0.0001
$1 + h$	2	1.1	1.01	1.001	1.0001
$P(1 + h)$	28	16.39	15.1399	15.013999	15.00139999
$P(1)$	15	15	15	15	15
Increase in profit	13	1.39	.1399	.013999	.00139999
Average rate of increase in profit	13	13.9	13.99	13.999	13.9999

We found the numbers in the bottom row of the table by dividing the increase in profit (row 5) by the increase in production (row 1). In symbols, the numbers in the bottom row are found by evaluating the quotient

$$\frac{P(1 + h) - P(1)}{h}$$

for various values of h.

From the table it seems that as h approaches 0 (or, as production gets closer and closer to 1), the average rate of increase in profit approaches the limit 14. This limit,

$$\lim_{h \to 0} \frac{P(1 + h) - P(1)}{h},$$

is called the **instantaneous rate of change,** or just the **rate of change,** of the profit at a production level of $x = 1$.

Think of the rate of change in profit varying all the time as the level of production changes. However, at the exact instant when production is 1 unit, the rate of change in profit is 14 dollars per unit.

Work Problem 4 at the side.

4. Make a table similar to the one in the text and find the instantaneous rate of change of profit at $x = 4$.

Answer: 8

Velocity One additional application of the idea of an instantaneous rate of change is given by velocity. Suppose a particle is moving along a straight line, with its distance from some fixed point given by a function $s(t)$, where t represents time. The quotient

$$\frac{s(t + h) - s(t)}{(t + h) - t} = \frac{s(t + h) - s(t)}{h}$$

represents the average rate of change of the distance; this is the **average velocity** of the particle. We can get the **instantaneous velocity at time t** (often called just the **velocity** at time t) by taking the limit as h approaches 0. That is, if $v(t)$ represents the velocity at time t,

$$v(t) = \lim_{h \to 0} \frac{s(t + h) - s(t)}{h}.$$

Example 2 The position of a red blood cell in the capillaries is given by

$$s(t) = 1.2t + 5,$$

where $s(t)$ gives the position of a cell in millimeters from some initial point and t is time in seconds. Find the velocity of this red blood cell.

We must evaluate the limit given above. To find $s(t + h)$, replace t in $s(t) = 1.2t + 5$ with $t + h$.

$$s(t + h) = 1.2(t + h) + 5$$

Now use the definition of velocity.

$$v(t) = \lim_{h \to 0} \frac{s(t + h) - s(t)}{h}$$

$$= \lim_{h \to 0} \frac{1.2(t + h) + 5 - (1.2t + 5)}{h}$$

$$= \lim_{h \to 0} \frac{1.2t + 1.2h + 5 - 1.2t - 5}{h}$$

$$= \lim_{h \to 0} \frac{1.2h}{h}$$

$$= 1.2$$

The velocity of the red blood cell is a constant 1.2 millimeters per second. ∎

5. Find the velocity for each of the following particles.
(a) $s(t) = 6.4t - 4$
(b) $s(t) = -2t + 12$
(c) $s(t) = 12t$

Answer:
(a) 6.4
(b) −2
(c) 12

Work Problem 5 at the side.

3.3 APPLIED PROBLEMS

Sales

The graph shows the total sales in thousands of dollars from the distribution of x thousand catalogs. Find the average rate of change of sales with respect to the number of catalogs distributed for the following changes in x. See Example 1.

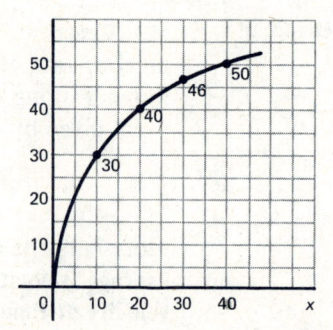

1. 10 to 20   **2.** 10 to 40

3. 20 to 30   **4.** 30 to 40

Population

The graph shows the population in millions of bacteria t minutes after a bactericide is introduced into the culture. Find the average rate of change of population with respect to time for the following time intervals.

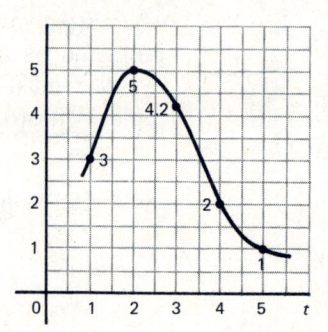

5. 1 to 2   **6.** 1 to 3

7. 2 to 3   **8.** 2 to 5

9. 3 to 4   **10.** 4 to 5

Sales

The graph shows annual sales (in units) of a typical product. Sales increase slowly at first to some peak, hold steady for a while, and then decline as the product goes out of style. (See also the exercises of Section 4.3.) Find the average annual rate of change in sales for the following changes in years.

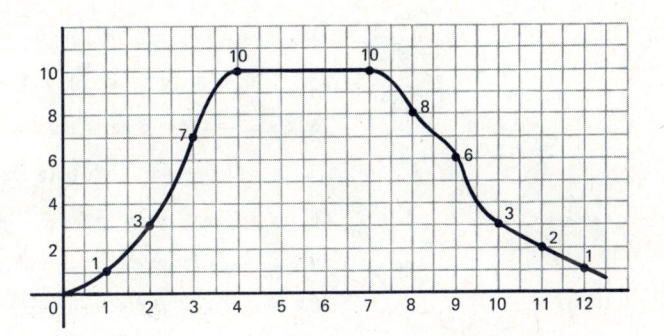

3.3 Rates of Change   119

11. 1 to 3   **12.** 2 to 4
13. 3 to 6   **14.** 5 to 7
15. 7 to 9   **16.** 8 to 11
17. 9 to 10   **18.** 10 to 12

Profit   The graph* below is from *Business Week* magazine. It shows the annual profits for Pan American World Airways for several past years. Find the average rate of change of profits for the following intervals.

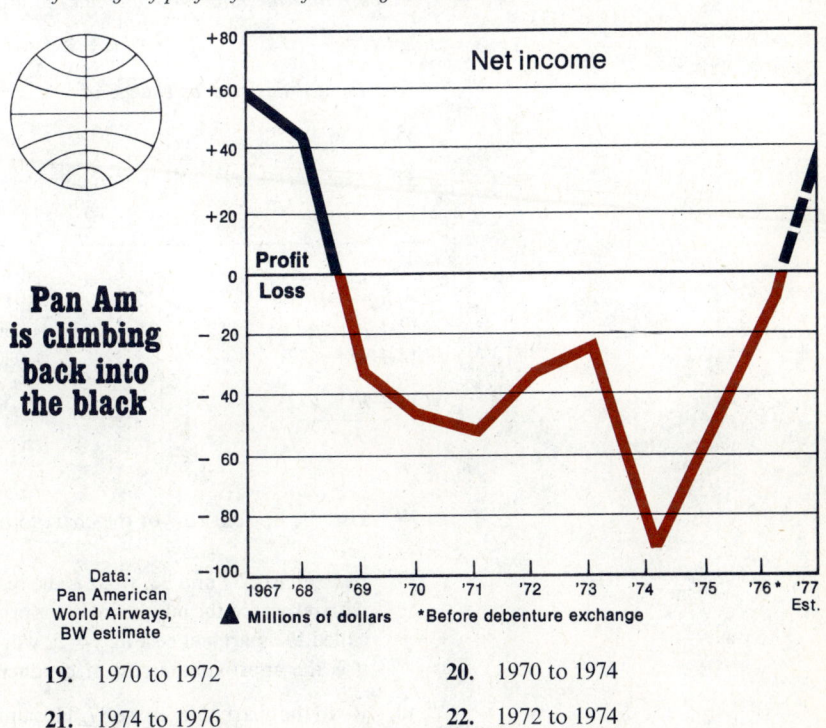

Pan Am is climbing back into the black

Data: Pan American World Airways, BW estimate

19. 1970 to 1972   **20.** 1970 to 1974
21. 1974 to 1976   **22.** 1972 to 1974
23. 1967 to 1974   **24.** 1967 to 1976

* Reprinted from the Sept. 5, 1977, issue of *Business Week* by special permission. © 1977 by McGraw-Hill, Inc. All rights reserved.

120 The Derivative

Find the average rate of change for the following functions.

25. $y = x^2 + 2x$ between $x = 0$ and $x = 3$
26. $y = -4x^2 - 6$ between $x = 2$ and $x = 5$
27. $y = 2x^3 - 4x^2 + 6x$ between $x = -1$ and $x = 1$
28. $y = -3x^3 + 2x^2 - 4x + 1$ between $x = 0$ and $x = 1$
29. $y = \sqrt{x}$ between $x = 1$ and $x = 4$
30. $y = \sqrt{3x - 2}$ between $x = 1$ and $x = 2$
31. $y = 6x - 2$ between $x = -2$ and $x = 1$
32. $y = -4x + 5$ between $x = -1$ and $x = 3$
33. $y = \dfrac{1}{x - 1}$ between $x = -2$ and $x = 0$
34. $y = \dfrac{-5}{2x - 3}$ between $x = 2$ and $x = 4$

Cost

Suppose that the total cost in dollars to produce x items is given by

$$C(x) = 2x^2 - 5x + 6.$$

Find the average rate of change of cost as x increases from

35. 2 to 4;
36. 2 to 3.

37. Use $C(x) = 2x^2 - 5x + 6$ to complete the following table.

h	1	0.1	0.01	0.001	0.0001
$2 + h$	3	2.1	2.01	2.001	2.0001
$C(2 + h)$	9	4.32	4.0302	4.003002	4.00030002
$C(2 + h) - C(2)$	5	.32	.0302	.003002	.00030002
$\dfrac{C(2 + h) - C(2)}{h}$	5	3.2	___	___	___

38. Use the bottom row of the chart to find $\lim\limits_{h \to 0} \dfrac{C(2 + h) - C(2)}{h}$.

Marginal cost

39. In Exercises 37 and 38, what is the instantaneous rate of change of cost with respect to the number of items produced when $x = 2$? (This number, called the **marginal cost** at $x = 2$, will be discussed in more detail later; it is the approximate cost of producing the third item.)

40. Redo the chart of Exercise 37. This time, change the second line to $4 + h$, the third line to $C(4 + h)$, and so on. Find

$$\lim_{h \to 0} \dfrac{C(4 + h) - C(4)}{h}.$$

Velocity

The distance of a particle from some fixed point is given by

$$s(t) = t^2 + 5t + 2,$$

where t is time measured in seconds.

Find the average velocity of the particle from

41. 4 seconds to 6 seconds;
42. 4 seconds to 5 seconds.
43. Complete the following table.

h	1	0.1	0.01	0.001	0.0001
$4 + h$	5	4.1	4.01	4.001	4.0001
$s(4 + h)$	52	39.31	38.1301	38.013001	38.00130001
$s(4)$	38	38	38	38	38
$s(4 + h) - s(4)$	14	1.31	.1301	.013001	.00130001
$\dfrac{s(4 + h) - s(4)}{h}$	14	13.1	___	___	___

44. Find $\lim\limits_{h \to 0} \dfrac{s(4 + h) - s(4)}{h}$, and give the instantaneous velocity of the particle when $x = 4$. See Example 2.

Find each of the following for $s(t) = t^2 + 5t + 2$. See Example 2.

45. $\lim\limits_{h \to 0} \dfrac{s(6 + h) - s(6)}{h}$

46. $\lim\limits_{h \to 0} \dfrac{s(1 + h) - s(1)}{h}$

3.4 DEFINITION OF THE DERIVATIVE

The idea of an instantaneous rate of change, discussed in the previous section, is very useful in applications. The methods that we used in that section to find this rate were cumbersome—if we wanted to change a value of x, we had to start all over. To get around these problems, we use the idea of a *derivative*. The derivative is one of the two main ideas of calculus (the other, the integral, is discussed in Chapter 6).

We define the derivative using the idea of a *tangent line* to a curve. Perhaps the easiest tangent line to define is the one for a circle: a tangent line to a circle at a point on the circle is a line which touches the circle in only one point. See Figure 3.13. Tangent

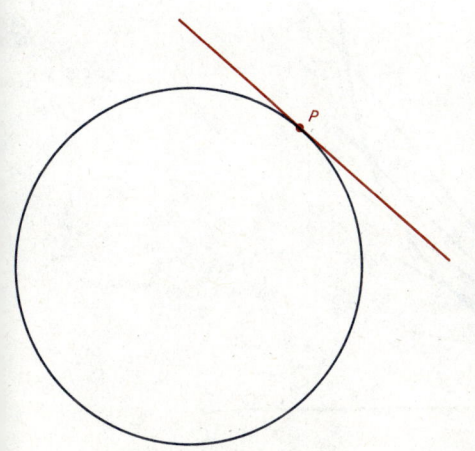

Figure 3.13

122 The Derivative

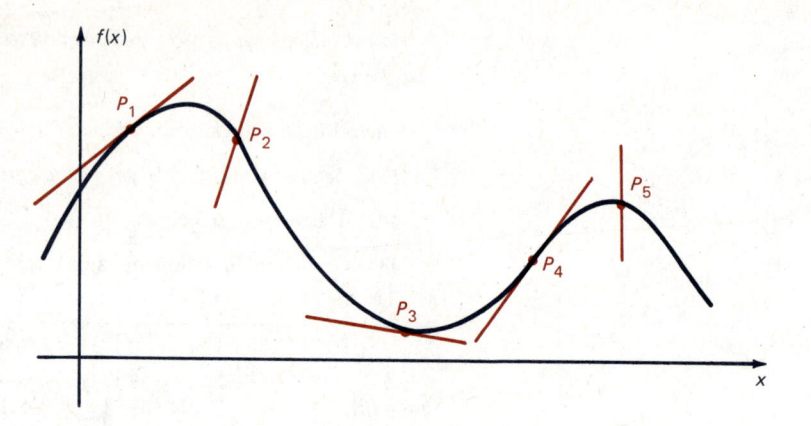

Figure 3.14

lines for more complicated curves are not as easy to define. In Figure 3.14 we would probably agree that the lines at P_1, P_3, and P_4 are tangent lines to the curve, while the ones at P_2 and P_5 are not.

We define the **tangent line** for a point on a curve by finding the slope of the tangent line. The slope and the point then determine the tangent line. To find the slope of a tangent line, let us begin with the curve $y = f(x)$ shown in Figure 3.15. The tangent line corresponding to the point x is shown in color. The tangent line goes through point R on the curve, with the coordinates of R given by $(x, f(x))$.

Figure 3.15 also shows a number of **secant lines** (lines cutting the graph in at least two points). Since the secant lines cut the

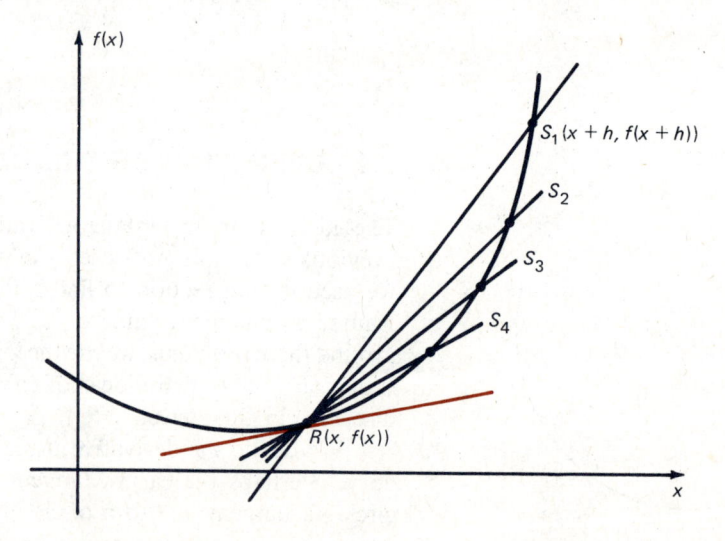

Figure 3.15

graph in two points, we can find their slope using the definition of slope from the previous chapter.

For example, let point S_1 have x-coordinate given by $x + h$ (h represents "a little change in x"). If $x + h$ represents the x-coordinate of point S_1, then its y-coordinate is $f(x + h)$; its coordinates are thus $(x + h, f(x + h))$. The slope of the secant line through R and S_1 is

$$\text{slope} = \frac{y_2 - y_1}{x_2 - x_1} = \frac{f(x + h) - f(x)}{(x + h) - x} = \frac{f(x + h) - f(x)}{h}.$$

Suppose we now let h approach 0. As h approaches 0, $x + h$ will approach $x + 0$, or x. This forces point S_1 to slide down the graph in Figure 3.15, passing, in turn, through S_2, S_3, S_4, and so on. As h gets extremely small, the secant line will get extremely close to the desired tangent line. In fact,

the **slope of the tangent line** for the curve $y = f(x)$ at the point $(x, f(x))$ is defined as

$$\lim_{h \to 0} \frac{f(x + h) - f(x)}{h}$$

provided that this limit exists.

The *slope of the tangent line* at a given point on its graph is called the *derivative* of the function at that point.

The **derivative** for the function $y = f(x)$ is written $f'(x)$ and is defined as

$$f'(x) = \lim_{h \to 0} \frac{f(x + h) - f(x)}{h}$$

provided that this limit exists.

Example 1 Let $f(x) = x^2$.
(a) Find the derivative.
The derivative is given by $f'(x)$, where

$$f'(x) = \lim_{h \to 0} \frac{f(x + h) - f(x)}{h}.$$

We can find this limit through a sequence of steps.

1. Find $f(x + h)$.
 Replace x with $x + h$ in the equation for $f(x)$.

$$f(x) = x^2$$
$$f(x + h) = (x + h)^2$$
$$= x^2 + 2xh + h^2$$

124   The Derivative

2. Find $f(x+h) - f(x)$.
   Since $f(x) = x^2$, we have
   $$f(x+h) - f(x) = x^2 + 2xh + h^2 - x^2$$
   $$= 2xh + h^2.$$

3. Form and simplify the *difference quotient* $\dfrac{f(x+h) - f(x)}{h}$.
   $$\frac{f(x+h) - f(x)}{h} = \frac{2xh + h^2}{h} = \frac{h(2x+h)}{h} = 2x + h$$

4. Take the limit as h approaches 0.
   $$f'(x) = \lim_{h \to 0} \frac{f(x+h) - f(x)}{h} = \lim_{h \to 0}(2x + h) = 2x + 0 = 2x$$

Thus, for $f(x) = x^2$, we have $f'(x) = 2x$.

(b) Calculate and interpret $f'(3)$.
$$f'(3) = 2(3) = 6$$

The number 6 is the slope of the tangent line to the graph of $y = x^2$ at the point where $x = 3$. See Figure 3.16. ∎

1. Let $f(x) = -2x^2 + 7$. Find the following.
(a) $f(x+h)$
(b) $f(x+h) - f(x)$
(c) $\dfrac{f(x+h) - f(x)}{h}$
(d) $f'(x)$
(e) $f'(4)$
(f) $f'(0)$

Answer:
(a) $-2x^2 - 4xh - 2h^2 + 7$
(b) $-4xh - 2h^2$
(c) $-4x - 2h$
(d) $-4x$
(e) -16
(f) 0

Slope of tangent line at 3 = 6

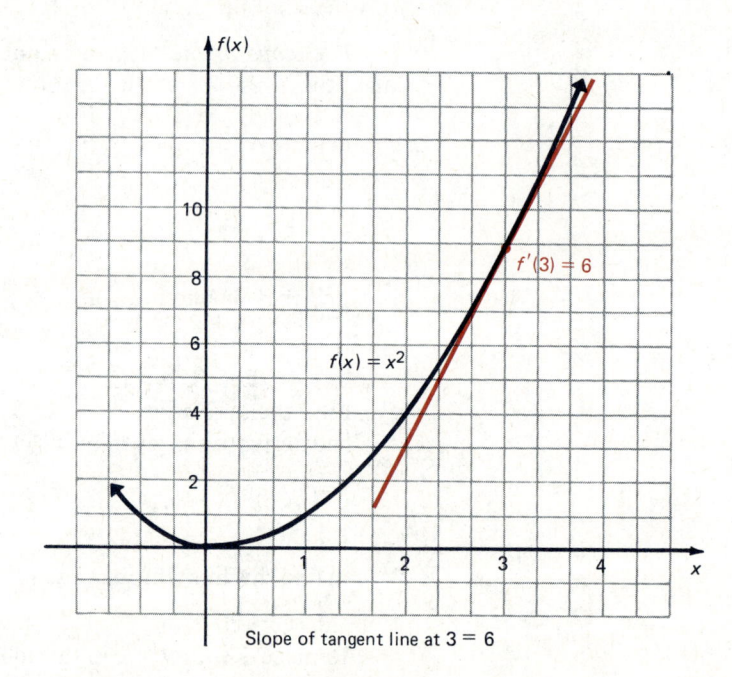

Figure 3.16

Work Problem 1 at the side.

3.4 Definition of the Derivative

Example 2 Let $f(x) = 2x^3 + 4x$. Find $f'(x)$.
Go through the four steps used above.

1. Find $f(x + h)$.

$$f(x + h) = 2(x + h)^3 + 4(x + h) \quad \text{Replace } x \text{ with } x + h$$

Now we need some algebra. Multiply $x + h$, $x + h$, and $x + h$ to get

$$(x + h)^3 = x^3 + 3x^2h + 3xh^2 + h^3.$$

Thus,

$$f(x + h) = 2(x^3 + 3x^2h + 3xh^2 + h^3) + 4(x + h)$$
$$= 2x^3 + 6x^2h + 6xh^2 + 2h^3 + 4x + 4h.$$

2. $f(x+h) - f(x) = 2x^3 + 6x^2h + 6xh^2 + 2h^3 + 4x + 4h - 2x^3 - 4x$
$$= 6x^2h + 6xh^2 + 2h^3 + 4h$$

3. $\dfrac{f(x + h) - f(x)}{h} = \dfrac{6x^2h + 6xh^2 + 2h^3 + 4h}{h}$

$$= \dfrac{h(6x^2 + 6xh + 2h^2 + 4)}{h}$$

$$= 6x^2 + 6xh + 2h^2 + 4$$

4. $f'(x) = \lim\limits_{h \to 0} \dfrac{f(x + h) - f(x)}{h} = \lim\limits_{h \to 0}(6x^2 + 6xh + 2h^2 + 4)$

$$= 6x^2 + 6x(0) + 2(0)^2 + 4$$
$$= 6x^2 + 4$$

If $f(x) = 2x^3 + 4x$, then $f'(x) = 6x^2 + 4$. ∎

Work Problem 2 at the side.

Example 3 Let $f(x) = \dfrac{4}{x}$. Find $f'(x)$.

1. $f(x + h) = \dfrac{4}{x + h}$

2. $f(x + h) - f(x) = \dfrac{4}{x + h} - \dfrac{4}{x}$

$$= \dfrac{4x - 4(x + h)}{x(x + h)} \quad \text{Find a common denominator}$$

$$= \dfrac{4x - 4x - 4h}{x(x + h)} \quad \text{Simplify the numerator}$$

$$= \dfrac{-4h}{x(x + h)}$$

---

2. Let $f(x) = -3x^3 + 2x$. Find the following.
(a) $f(x + h)$
(b) $f(x + h) - f(x)$
(c) $\dfrac{f(x + h) - f(x)}{h}$
(d) $f'(x)$
(e) $f'(-1)$

Answer:
(a) $-3x^3 - 9x^2h - 9xh^2 - 3h^3 + 2x + 2h$
(b) $-9x^2h - 9xh^2 - 3h^3 + 2h$
(c) $-9x^2 - 9xh - 3h^2 + 2$
(d) $-9x^2 + 2$
(e) -7

The Derivative

3. $\dfrac{f(x+h) - f(x)}{h} = \dfrac{\frac{-4h}{x(x+h)}}{h}$

$= \dfrac{-4h}{x(x+h)} \cdot \dfrac{1}{h}$ **Invert and multiply**

$= \dfrac{-4}{x(x+h)}$

4. $f'(x) = \lim\limits_{h \to 0} \dfrac{f(x+h) - f(x)}{h} = \lim\limits_{h \to 0} \dfrac{-4}{x(x+h)}$

$= \dfrac{-4}{x(x+0)}$

$= \dfrac{-4}{x(x)}$

$= \dfrac{-4}{x^2}$ ∎

Work Problem 3 at the side.

---

3. Let $f(x) = -5/x$. Find the following.
(a) $f(x+h)$
(b) $f(x+h) - f(x)$
(c) $\dfrac{f(x+h) - f(x)}{h}$
(d) $f'(x)$
(e) $f'(-1)$

Answer:
(a) $\dfrac{-5}{x+h}$
(b) $\dfrac{5h}{x(x+h)}$
(c) $\dfrac{5}{x(x+h)}$
(d) $\dfrac{5}{x^2}$
(e) 5

Let us summarize the four steps used when finding $f'(x)$.

1. Find $f(x+h)$.
2. Find $f(x+h) - f(x)$.
3. Find and simplify $\dfrac{f(x+h) - f(x)}{h}$.
4. $f'(x) = \lim\limits_{h \to 0} \dfrac{f(x+h) - f(x)}{h}$ if this limit exists.

Not all functions have derivatives at all points in their domains. Figure 3.17 shows a graph of the function $f(x) = |x|$. We cannot really draw a single tangent at the point $(0, 0)$ on the graph—the graph has a sharp "corner" and more than one "tangent" could be drawn there. Therefore, the function $f(x) = |x|$ does not have a derivative at $(0, 0)$. The function of Figure 3.18 does not have a derivative at points x_1, x_2, or x_3. A function which is not continuous at a point cannot have a derivative there.

Often we are given a graph and we want to find the point on the graph where the slope of the tangent line equals a given number. The next example shows how this is done.

Example 4 Find all points on the graph of $f(x) = x^2 + 6x + 5$ where the slope of the tangent line is 0.

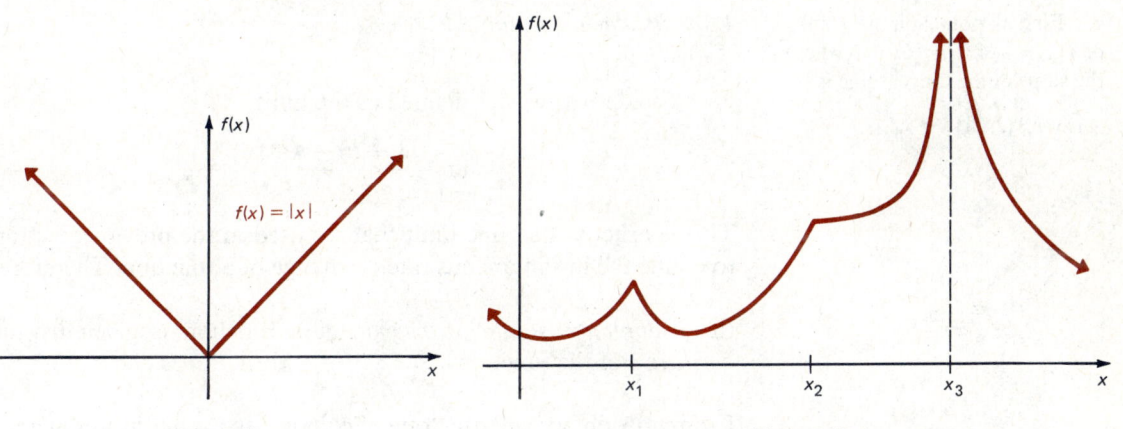

Figure 3.17

Figure 3.18

The slope of the tangent line is given by the derivative, $f'(x)$. Go through the four steps given above and verify that

$$f'(x) = 2x + 6.$$

We want to find all values of x where the slope of the tangent is 0. In other words, we want to find all values of x so that $f'(x) = 0$.

$$f'(x) = 0$$
$$2x + 6 = 0$$
$$2x = -6$$
$$x = -3$$

When $x = -3$, the slope of the tangent is 0. Recall from Chapter 2 that a horizontal line has a slope of 0. Thus, the tangent to $f(x) = x^2 + 6x + 5$ at $x = -3$ is horizontal. This is shown in Figure 3.19. When $x = -3$, $y = -4$, so the slope of the tangent line is 0 at the point $(-3, -4)$. ∎

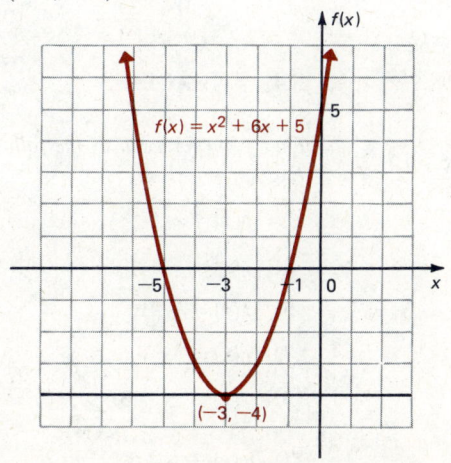

Figure 3.19

4. Find all points on the graph of $f(x) = -4x^2 + 16x + 5$ where the slope of the tangent line is 0.

Answer: (2, 21)

Work Problem 4 at the side.

The derivative was defined as the limit

$$\lim_{h \to 0} \frac{f(x+h) - f(x)}{h}.$$

This is exactly the same limit that we used in the previous section to define the instantaneous rate of change of a function. Therefore,

the instantaneous rate of change of a function is given by the derivative of the function.

From now on, we will use "rate of change" to mean "instantaneous rate of change."

Example 5 The cost in dollars to manufacture x electric pencil sharpeners is given by $C(x) = x^3 - 4x^2 + 3x$. Find the rate of change of cost when 5 pencil sharpeners are made.

The rate of change of cost is given by the derivative of the cost function. If we went through the four steps above, we would find that

$$C'(x) = 3x^2 - 8x + 3.$$

When $x = 5$,

$$C'(5) = 3(5)^2 - 8(5) + 3 = 38.$$

Thus, at the point where exactly 5 pencil sharpeners are made, the rate of change of cost is $38. ∎

5. The revenue from the sale of x can openers is given by $R(x) = 2x^3 - 4x^2 + x$. Find the rate of change of revenue when
(a) $x = 2$;
(b) $x = 8$.
(Hint: $R'(x) = 6x^2 - 8x + 1$.)

Answer:
(a) 9
(b) 321

Work Problem 5 at the side.

3.4 EXERCISES

Find $f'(x)$ for each of the following. Then find $f'(2)$, $f'(0)$, and $f'(-3)$. See Examples 1–3.

1. $f(x) = 2x^2$
2. $f(x) = -3x^2$
3. $f(x) = -4x^2 + 11x$
4. $f(x) = 6x^2 - 4x$
5. $f(x) = -9x$
6. $f(x) = 4x$
7. $f(x) = 8x + 6$
8. $f(x) = -9x - 5$
9. $f(x) = x^3 + 3x$
10. $f(x) = 2x^3 - 14x$
11. $f(x) = -\dfrac{2}{x}$
12. $f(x) = \dfrac{6}{x}$

13. $f(x) = \dfrac{4}{x-1}$
14. $f(x) = \dfrac{3}{x+2}$
15. $f(x) = \sqrt{x}$ (Hint: In Step 3, multiply numerator and denominator by $\sqrt{x+h} + \sqrt{x}$.)
16. $f(x) = -3\sqrt{x}$

Find all points where the following functions do not have derivatives.

17. 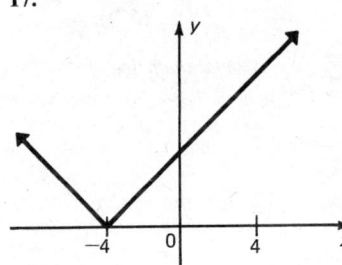
18. 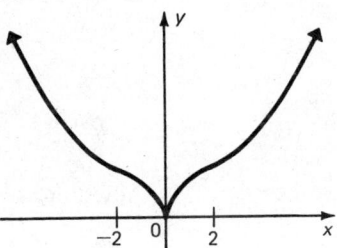
19. 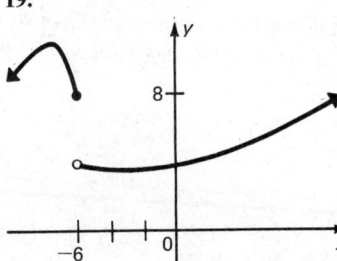
20. 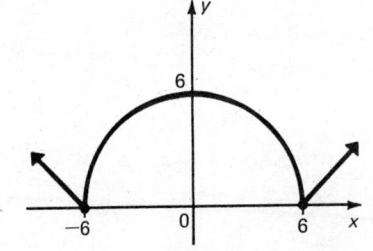
21. 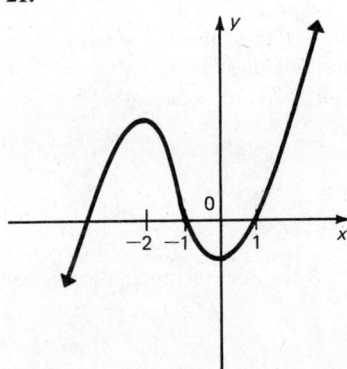
22.

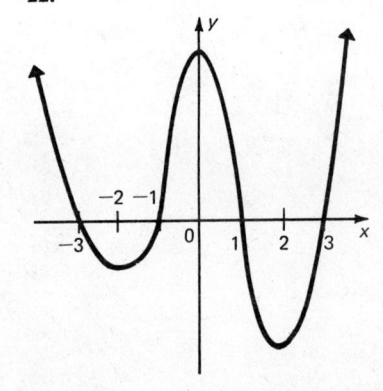

Find all points (if there are any) where the following functions have tangent lines whose slope is 0. (In Exercises 27–30, it is necessary to factor.) See Example 4.

23. $f(x) = 9x^2$
24. $f(x) = -2x^2$
25. $f(x) = 8x^2 + 32x$
26. $f(x) = 2x^2 + 6x$

27. $f(x) = 2x^3 - 3x^2 - 12x$
28. $f(x) = 2x^3 - 6x^2 + 6x$
29. $f(x) = 4x^3 + 24x^2$
30. $f(x) = 4x^3 - 18x^2$

APPLIED PROBLEMS

Demand

Suppose the demand for a certain item is given by
$$D(x) = -2x^2 + 4x + 6,$$
where x represents the price of the item. Find the rate of change of demand with respect to price for the following values of x. See Example 5.

31. $x = 3$
32. $x = 6$

Advertising

Suppose the profit from the expenditure of x thousand dollars on advertising is given by
$$P(x) = 1000 + 32x - 2x^2.$$
Find the rate of change of profit with respect to the expenditure on advertising for the following amounts. In each case, decide if the firm should increase the expenditure.

33. $x = 8$
34. $x = 6$
35. $x = 12$
36. $x = 20$

Population

A biologist estimates that when a bactericide is introduced into a culture of bacteria, the number of bacteria present is given by
$$B(t) = 1000 + 50t - 5t^2,$$
where $B(t)$ is the number of bacteria, in millions, present at time t, measured in hours. Find the rate of change of the number of bacteria with respect to time for the following values of t.

37. $t = 2$
38. $t = 3$
39. $t = 4$
40. $t = 5$
41. $t = 6$

42. When does the population of bacteria start to decline?

3.5 SOME SHORTCUTS FOR FINDING DERIVATIVES

The 4-step method for finding derivatives is hard to use—it is too easy to make an algebraic error. In the remainder of this chapter, we look at some shortcut ways of finding derivatives.

In the previous section we used $f'(x)$ to represent the derivative

3.5 Some Shortcuts for Finding Derivatives

of the function $y = f(x)$. Alternate notations for the derivative include

$$y' \quad \text{or} \quad \frac{dy}{dx} \quad \text{or} \quad \frac{d}{dx} f(x).$$

We will use these notations interchangeably.

We begin our shortcuts with the formula for the derivative of a constant; it is the simplest derivative of all.

Theorem 3.1 *Constant rule* If $y = k$, where k is any real number, then

$$y' = 0.$$

(The derivative of a constant is 0.)

Example 1 If $y = 9$, then $y' = 0$. ∎

Work Problem 1 at the side.

We can use the definition of derivative from the previous section to show that the constant rule is correct. Let $f(x) = k$, a real number. By the definition of derivative,

$$f'(x) = \lim_{h \to 0} \frac{f(x+h) - f(x)}{h}.$$

Since $f(x) = k$, we also have $f(x + h) = k$. (To see this, let $f(x) = k + 0x$.) Then

$$f'(x) = \lim_{h \to 0} \frac{k - k}{h} = \lim_{h \to 0} \frac{0}{h} = 0.$$

Functions of the form $y = x^n$, where n is a real number, occur often in applications. To find a general rule for the derivative of $y = x^n$, we work out the derivative for various special cases. Use the definition of derivative to check the entries in the table at the side. The results in the table suggest the following rule.

Theorem 3.2 *Power rule* If $y = x^n$, then

$$y' = n \cdot x^{n-1}.$$

(The derivative of $y = x^n$ is found by decreasing the exponent on x by 1 and multiplying the result by the exponent n.)

The actual proof of the power rule requires algebra which is more advanced than we have studied.

---

1. Find the derivatives of the following.
   (a) $y = -4$
   (b) $f(x) = \pi$
   (c) $y = 0$

Answer:
(a) 0
(b) 0
(c) 0

Function	Derivative
$y = x^2$	$y' = 2x^1$
$y = x^3$	$y' = 3x^2$
$y = x^4$	$y' = 4x^3$
$y = x^5$	$y' = 5x^4$
$y = x^{-1}$	$y' = -1 \cdot x^{-2}$

132 The Derivative

Example 2 Find the derivatives of the following functions.
(a) $y = x^6$
Multiply x^{6-1} by 6.
$$y' = 6 \cdot x^{6-1} = 6x^5$$
(b) $y = x^{15}$
$$y' = 15 \cdot x^{15-1} = 15x^{14}$$
(c) If $y = x^1$, then $y' = 1 \cdot x^{1-1} = 1 \cdot x^0 = 1 \cdot 1 = 1$. (Recall that $x^0 = 1$ if $x \neq 0$.)
(d) If $y = x^{-3}$, then $y' = -3 \cdot x^{-3-1} = -3 \cdot x^{-4} = -3/x^4$.
(e) If $y = x^{4/3}$, then $y' = \frac{4}{3}x^{1/3}$.
(f) $y = \sqrt{x}$
Replace $\sqrt{x}$ by the equivalent expression $x^{1/2}$. Then
$$y' = \frac{1}{2} \cdot x^{(1/2)-1} = \frac{1}{2}x^{-1/2} = \frac{1}{2x^{1/2}} = \frac{1}{2\sqrt{x}}. \blacksquare$$

2. Find the derivatives of the following.
(a) $y = x^4$
(b) $y = x^{17}$
(c) $y = x^{-2}$
(d) $y = x^{-5}$
(e) $y = x^{3/2}$

Answer:
(a) $y' = 4x^3$
(b) $y' = 17x^{16}$
(c) $y' = -2/x^3$
(d) $y' = -5/x^6$
(e) $y' = \frac{3}{2}x^{1/2}$

Work Problem 2 at the side.

The next rule tells us how to handle the derivative of the product of a constant and a function.

Theorem 3.3 *Constant times a function* Let k be a real number. Then the derivative of $y = k \cdot f(x)$ is
$$y' = k \cdot f'(x).$$

(The derivative of a constant times a function is the constant times the derivative of the function.)

Example 3 Find the derivatives of the following functions.
(a) $y = 8x^4$
Since the derivative of $f(x) = x^4$ is $f'(x) = 4x^3$, we have
$$y' = 8(4x^3) = 32x^3.$$
(b) $y = -\frac{3}{4}x^{12}$
$$y' = -\frac{3}{4}(12x^{11}) = -9x^{11}$$
(c) If $y = 15x$, then $y' = 15(1) = 15$.
(d) If $y = 10x^{3/2}$, then $y' = 10(\frac{3}{2}x^{1/2}) = 15x^{1/2}$.
(e) If $y = 1000x^{.4}$, then $y' = 1000(.4x^{-.6}) = 400x^{-.6}$.
(f) $y = \frac{6}{x}$

3.5 Some Shortcuts for Finding Derivatives

Replace $\frac{6}{x}$ with $6 \cdot \frac{1}{x}$, or $6x^{-1}$. Then

$$y' = 6(-1x^{-2}) = -6x^{-2} = \frac{-6}{x^2}.$$ ∎

Work Problem 3 at the side.

We have one final rule in this section:

Theorem 3.4   **Sum or difference rule**   If $y = f(x) + g(x)$, then

$$y' = f'(x) + g'(x);$$

if $y = f(x) - g(x)$, then

$$y' = f'(x) - g'(x).$$

(The derivative of a sum or difference of two functions is the sum or difference of the derivatives of the functions.)

Example 4   Find the derivatives of the following functions.
(a) $y = 6x^3 + 9x^2$
The derivative of $f(x) = 6x^3$ is $f'(x) = 18x^2$, and the derivative of $g(x) = 9x^2$ is $g'(x) = 18x$. Thus, if

$$y = 6x^3 + 9x^2, \quad \text{then} \quad y' = 18x^2 + 18x.$$

(b) If $y = 12x^2 + 11x - 8$, then $y' = 24x + 11 - 0 = 24x + 11$.
(c) If $y = 8x^4 - 6\sqrt{x} + \frac{5}{x}$, we find y' by first rewriting y as $y = 8x^4 - 6x^{1/2} + 5x^{-1}$. Then $y' = 32x^3 - 3x^{-1/2} - 5x^{-2}$. If desired, this result can be written as $y' = 32x^3 - 3/\sqrt{x} - 5/x^2$. ∎

Work Problem 4 at the side.

The proofs of all the rules for derivatives that we give in this chapter can be found in the books listed under "Calculus Books for Proofs" in the bibliography at the end of this book. A summary of all the formulas for derivatives is given inside the front cover of this book.

As we saw in the previous section, the derivative of a function can be used to find the rate of change of the function.

Example 5   A tumor has the approximate shape of a cone. The radius of the tumor is fixed by the bone structure at 2 centimeters, but the tumor is growing along the height of the cone. The formula

---

3. Find the derivatives of the following.
(a) $y = 12x^3$
(b) $f(x) = 30x^7$
(c) $y = -35x$
(d) $y = 10x^{.2}$
(e) $y = 5\sqrt{x}$
(f) $y = -10/x$

Answer:
(a) $36x^2$
(b) $210x^6$
(c) -35
(d) $2x^{-.8}$
(e) $\frac{5}{2}x^{-1/2}$ or $\frac{5}{2\sqrt{x}}$
(f) $10x^{-2}$ or $\frac{10}{x^2}$

4. Find the derivatives of the following.
(a) $y = -4x^5 - 8x + 6$
(b) $y = 32x^5 - 100x^2 + 12x$
(c) $y = 8x^{3/2} + 2x^{1/2}$
(d) $f(x) = -\sqrt{x} + 6/x$

Answer:
(a) $y' = -20x^4 - 8$
(b) $y' = 160x^4 - 200x + 12$
(c) $y' = 12x^{1/2} + x^{-1/2}$ or $12x^{1/2} + 1/x^{1/2}$
(d) $f'(x) = -1/[2\sqrt{x}] - 6/x^2$

for the volume of a cone is $V = \frac{1}{3}\pi r^2 h$, where r is the radius of the base and h is the height of the cone. Find the rate of change in the volume of the tumor with respect to the height.

To emphasize that we need the rate of change of the volume with respect to the height, we use the symbol dV/dh for the derivative. In our example, r is fixed at 2 cm. Thus

$$V = \frac{1}{3}\pi r^2 h \quad \text{becomes} \quad V = \frac{1}{3}\pi 2^2 \cdot h \quad \text{or} \quad V = \frac{4}{3}\pi h.$$

Since $4\pi/3$ is a constant,

$$\frac{dV}{dh} = \frac{4\pi}{3} \approx 4.2 \text{ cu cm per cm.}$$

For each additional centimeter that the tumor grows in height, its volume will increase approximately 4.2 cubic centimeters. ■

Work Problem 5 at the side.

Example 6 *Marginal cost* Suppose a function $y = C(x)$ gives the total cost to manufacture x units of an item. The derivative of $y = C(x)$ gives the rate of change of cost with respect to the number of units produced. This rate of change of cost is called the *marginal cost*; think of the marginal cost as the approximate cost of producing one more item, after x items have already been produced. Suppose that the total cost in hundreds of dollars to produce x thousand barrels of beer is given by

$$C(x) = 4x^2 + 100x + 500.$$

Find the marginal cost for the following values of x.
 (a) $x = 5$
     To find the marginal cost, we need the derivative of the total cost function.

$$C'(x) = 8x + 100$$

When $x = 5$,

$$C'(5) = 8(5) + 100 = 140.$$

After 5 thousand barrels of beer have been produced, the cost to produce 1 thousand more barrels will be approximately 140 hundred dollars, or $14,000.
 (b) $x = 30$
     After 30 thousand barrels have been produced, the cost to produce 1 thousand more barrels will be approximately

$$C'(30) = 8(30) + 100 = 340,$$

or $34,000.

---

5. A balloon is spherical. The formula for the volume of a sphere is $V = \frac{4}{3}\pi r^3$, where r is the radius of the sphere. Find the following.
(a) dV/dr
(b) the rate of change of the volume when $r = 3$ inches

Answer:
(a) $4\pi r^2$
(b) 36π cubic inches per inch

6. The cost to produce x units of wheat is given by $C(x) = 5000 + 20x + 10\sqrt{x}$. Find the marginal cost when
   (a) $x = 9$;
   (b) $x = 16$;
   (c) $x = 25$.

Answer:
(a) 65/3
(b) 85/4
(c) 21

7. The revenue from the sale of x units of ice cream is given by $R(x)$, where $R(x) = 800x + 10x^2$. Find the marginal revenue when
   (a) $x = 10$;
   (b) $x = 25$;
   (c) $x = 32$.

Answer:
(a) 1000
(b) 1300
(c) 1440

Management must be careful to keep track of marginal costs. If the marginal cost of producing an extra unit exceeds the revenue received from selling it, then the company will lose money on that unit. ∎

Work Problem 6 at the side.

The idea of marginal cost also applies to revenue or profit.

Work Problem 7 at the side.

3.5 EXERCISES

Find the derivatives of the following functions. See Examples 1–4.

1. $y = 5x^2$
2. $y = -8x^2$
3. $y = -6x^5$
4. $y = 12x^6$
5. $y = 15x^2 + 7$
6. $y = -4x^2 + 15$
7. $y = 9x^2 - 8x + 4$
8. $y = 10x^2 + 4x - 9$
9. $y = 3x^2 - 4x + 15$
10. $y = 2x^2 + 16x - 8$
11. $y = -4x^3 + 2x^2 - 6$
12. $y = 9x^3 + 8x + 5$
13. $y = 10x^3 - 9x^2 + 6x$
14. $y = 3x^3 - x^2 - 12x$
15. $y = x^4 - 5x^3 + 9x^2 + 5$
16. $y = 3x^4 + 11x^3 + 2x^2 - 4x$
17. $y = 6x^{1.5}$
18. $y = -2x^{2.5}$
19. $y = -15x^{3.2}$
20. $y = 18x^{1.6}$
21. $y = -18x^{3/2}$
22. $y = 24x^{5/2}$
23. $y = 32x^{1/2}$
24. $y = -10x^{1/2}$
25. $y = 50x^{3/2} - 6x^{1/2}$
26. $y = 32x^{3/2} + 12x^{1/2}$
27. $y = 8\sqrt{x} + 6x$
28. $y = -100\sqrt{x} - 11x$
29. $y = 6x^{-5}$
30. $y = -2x^{-4}$
31. $y = -4x^{-2} + 3x^{-1}$
32. $y = 8x^{-4} - 9x^{-2}$
33. $y = 10x^{-2} + 3x^{-4} - 6x$
34. $y = x^{-5} - x^{-2} + 5x^{-1}$
35. $y = \dfrac{6}{x} - \dfrac{8}{x^2}$
36. $y = \dfrac{4}{x} + \dfrac{2}{x^3}$
37. $y = 12x^{-1/2}$
38. $y = -30x^{-1/2}$
39. $y = -10x^{-1/2} + 8x^{-3/2}$
40. $y = x^{-1/2} - 14x^{-3/2}$

For the following functions, find all points where the tangent line is horizontal. (As we saw in the last section, we do this by finding points where the derivative is 0.)

41. $y = 6x^2 - 20x + 5$
42. $y = 9x^2 - 18x - 4$
43. $y = x^3 - \frac{5}{2}x^2 + 2x - 1$
44. $y = \frac{4}{3}x^3 - 13x^2 + 12x - 6$
45. $y = \frac{2}{3}x^3 - \frac{3}{2}x^2 - 5x + 8$
46. $y = x^3 - 2x^2 - 15x + 6$
47. $y = x^4 - 16x$
48. $y = x^4 - 256x$

APPLIED PROBLEMS

Price

49. If the price of a product is given by

$$P(x) = \frac{-1000}{x} + 1000,$$

where x represents the demand for the product, find the rate of change of price when the demand is $x = 10$. Is the price increasing or decreasing at that point?

Sales

Often sales of a new product grow rapidly at first and then level off with time. This is the case with the sales represented by the function

$$S(t) = 100 - 100t^{-1},$$

where t represents time in years. Find the rate of change of sales for the following values of t.

50. $t = 1$
51. $t = 10$

Suppose $P(x) = 100/x$ represents the percent of cheap shoes made by a company that are still wearable after x days of wearing. Find the percent of shoes wearable after the following number of days.

52. 1 day
53. 100 days
54. Find and interpret $P'(100)$.

Insulin use

Insulin affects the glucose, or blood-sugar, level of some diabetics according to the function

$$G(x) = -.2x^2 + 450,$$

where $G(x)$ is the blood-sugar level one hour after x units of insulin are injected. (This mathematical model is only approximate and valid only for values of x less than about 40.)

55. Find $G(0)$.
56. Find $G(25)$.

Find dG/dx for the following values of x. Interpret your answer. See Example 5.

57. $x = 10$ **58.** $x = 25$

Blood vessel volume   A short length of a blood vessel has the approximate shape of a cylinder. The volume of a cylinder is given by $V = \pi r^2 h$. Suppose we set up an experimental device to measure the volume of blood in a blood vessel of fixed length 80 mm.

59. Find a formula for dV/dr.

Suppose now a drug which causes a blood vessel to expand is administered. Evaluate dV/dr for the following values of r.

60. 4 mm   **61.** 6 mm   **62.** 8 mm

Profit   The profit in dollars from the sale of x expensive tape recorders is given by

$$P(x) = x^3 - 5x^2 + 7x + 10.$$

Find the marginal profit for the following values of x. See Example 6.

63. $x = 4$ **64.** $x = 8$

65. $x = 10$ **66.** $x = 12$

Cost   The total cost to produce x handcrafted weather vanes is given by

$$C(x) = 100 + 8x - x^2 + 4x^3.$$

Find the marginal cost for the following values of x.

67. $x = 0$ **68.** $x = 4$

69. $x = 6$ **70.** $x = 8$

The cost to manufacture x units of tacos is given by

$$T(x) = 500 + 2x + 4x^{1/2}.$$

Find the marginal cost for the following values of x.

71. $x = 1$ **72.** $x = 4$

73. $x = 9$ **74.** $x = 16$

Velocity   We saw in Section 3 of this chapter that the velocity of a particle is given by

$$\lim_{h \to 0} \frac{s(t + h) - s(t)}{h},$$

where $s(t)$ gives the position of the particle at time t. This limit is actually the derivative of $s(t)$, so the velocity of a particle is given by $s'(t)$. If $v(t)$ represents velocity at time t, then $v(t) = s'(t)$. For each of the following position functions, find (a) a formula for $v(t)$; (b) the velocity when $t = 0$, $t = 5$, and $t = 10$.

75. $s(t) = 6t + 5$ **76.** $s(t) = 9 - 2t$

77. $s(t) = 11t^2 + 4t + 2$ **78.** $s(t) = 25t^2 - 9t + 8$

79. $s(t) = 4t^3 + 8t^2$ **80.** $s(t) = -2t^3 + 4t^2 - 1$

3.6 DERIVATIVES OF PRODUCTS AND QUOTIENTS

To find the derivative of a sum of two functions, we find the sum of the derivatives. What about products? Is the derivative of a product equal to the product of the derivatives? Let's try an example.

$$\text{Let } g(x) = 2x + 3 \quad \text{and} \quad h(x) = 3x^2,$$
$$\text{so that } g'(x) = 2 \quad \text{and} \quad h'(x) = 6x.$$

There are two ways we might try to find the derivative of the product $g(x) \cdot h(x)$:

1. Find the product of $g(x)$ and $h(x)$ and take the derivative of this result.

    $$f(x) = g(x) \cdot h(x)$$
    $$= (2x + 3)(3x^2)$$
    $$= 6x^3 + 9x^2$$

    The derivative of this result is

    $$f'(x) = 18x^2 + 18x.$$

2. Find the derivatives of $g(x)$ and $h(x)$ and take the product of the results.

    $$g'(x) = 2$$
    $$h'(x) = 6x$$
    $$f'(x) = g'(x) \cdot h'(x)$$
    $$= 2(6x) = 12x$$

The two results are not the same. Since the method on the left uses only algebra and proven results, we can accept it as correct. The method on the right has given a wrong answer. Thus, in general, the derivative of a product *does not* equal the product of the derivatives. This fact is unfortunate, since the rule that must actually be used is a little more complicated than the other rules we have seen.

Theorem 3.5 *Product rule* If $f(x) = g(x) \cdot k(x)$, and if both g' and k' exist, then

$$f'(x) = g(x) \cdot k'(x) + k(x) \cdot g'(x).$$

(The derivative of a product of two functions is the first function times the derivative of the second, plus the second function times the derivative of the first.)

The proof of the product rule is included in Exercise 59 below.

Example 1 Let $f(x) = (2x + 3)(3x^2)$. Use the product rule to find $f'(x)$.

Here f is given as the product of $g(x) = 2x + 3$ and $k(x) = 3x^2$. Use the product rule and the fact that $g'(x) = 2$ and $k'(x) = 6x$.

3.6 Derivatives of Products and Quotients

$$f'(x) = g(x) \cdot k'(x) + k(x) \cdot g'(x)$$
$$= (2x + 3)(6x) + (3x^2)(2)$$
$$= 12x^2 + 18x + 6x^2 \qquad \text{Use algebra to simplify}$$
$$f'(x) = 18x^2 + 18x$$

This is the same result we found before. ∎

Work Problem 1 at the side.

Example 2 Let $f(x) = (x^2 - 4x)(3x + 5)$. Use the product rule to find $f'(x)$.

Let $g(x) = x^2 - 4x$, so that $g'(x) = 2x - 4$. Let $k(x) = 3x + 5$, with $k'(x) = 3$. Then

$$f'(x) = g(x) \cdot k'(x) + k(x) \cdot g'(x)$$
$$= (x^2 - 4x)(3) + (3x + 5)(2x - 4)$$
$$= 3x^2 - 12x + 6x^2 - 2x - 20$$
$$f'(x) = 9x^2 - 14x - 20. \quad ∎$$

Work Problem 2 at the side.

You may have noticed that many of the problems above could have been worked by multiplying out the original functions. The product rule would not then be needed. This is correct; however, in the next section we shall see products of functions where this cannot be done—the product rule is essential.

What about *quotients* of functions? To find the derivative of the quotient of two functions, we must use the result of the next theorem.

Theorem 3.6 *Quotient rule* If $f(x) = \dfrac{g(x)}{k(x)}$, where g and k are functions whose derivatives exist, and if $k(x) \neq 0$, then

$$f'(x) = \frac{k(x) \cdot g'(x) - g(x) \cdot k'(x)}{[k(x)]^2}.$$

(The derivative of a quotient is the denominator times the derivative of the numerator, minus the numerator times the derivative of the denominator, all over the denominator squared.)

The proof of the quotient rule is included as Exercise 60.

---

1. Use the product rule to find the derivatives of the following.
   (a) $f(x) = (5x^2 + 6)(3x)$
   (b) $g(x) = (8x)(4x^2 + 5x)$

   Answer:
   (a) $45x^2 + 18$
   (b) $96x^2 + 80x$

2. Find the derivatives of the following.
   (a) $f(x) = (x^2 - 3)(x + 5)$
   (b) $g(x) = (x + 4)(5x^2 + 6)$

   Answer:
   (a) $3x^2 + 10x - 3$
   (b) $15x^2 + 40x + 6$

140   The Derivative

Example 3   Let $f(x) = \dfrac{2x - 1}{4x + 3}$. Find $f'(x)$.

Here we can let $g(x) = 2x - 1$, with $g'(x) = 2$. Also, $k(x) = 4x + 3$, and $k'(x) = 4$. Then

$$f'(x) = \frac{k(x) \cdot g'(x) - g(x) \cdot k'(x)}{[k(x)]^2}$$

$$= \frac{(4x + 3)(2) - (2x - 1)(4)}{(4x + 3)^2}$$

$$= \frac{8x + 6 - 8x + 4}{(4x + 3)^2}$$

$$f'(x) = \frac{10}{(4x + 3)^2}. \quad \blacksquare$$

Work Problem 3 at the side.

3. Find the derivatives of the following.

(a) $f(x) = \dfrac{3x + 7}{5x + 8}$

(b) $g(x) = \dfrac{2x + 11}{5x - 1}$

Answer:

(a) $\dfrac{-11}{(5x + 8)^2}$

(b) $\dfrac{-57}{(5x - 1)^2}$

Example 4   Let $F(x) = \dfrac{6x^2 - 4x}{8x + 1}$. Find $F'(x)$.

Use the quotient rule.

$$F'(x) = \frac{(8x + 1)(12x - 4) - (6x^2 - 4x)(8)}{(8x + 1)^2}$$

$$= \frac{96x^2 - 20x - 4 - 48x^2 + 32x}{(8x + 1)^2}$$

$$F'(x) = \frac{48x^2 + 12x - 4}{(8x + 1)^2} \quad \blacksquare$$

Work Problem 4 at the side.

4. Find the derivatives of the following.

(a) $f(x) = \dfrac{9x + 2}{x^2 + 6}$

(b) $F(x) = \dfrac{4x^2 - 8x}{3x - 5}$

(c) $h(x) = \dfrac{x^3 - x^2}{x + 1}$

Answer:

(a) $\dfrac{-9x^2 - 4x + 54}{(x^2 + 6)^2}$

(b) $\dfrac{12x^2 - 40x + 40}{(3x - 5)^2}$

(c) $\dfrac{2x^3 + 2x^2 - 2x}{(x + 1)^2}$

Suppose $y = C(x)$ gives the total cost to manufacture x items. Then the **average cost per item** can be found by dividing the total cost by the number of items, or

$$\text{average cost} = \frac{C(x)}{x}.$$

A company naturally would be interested in making this average cost as small as possible. We will see in the next chapter that this can be done by taking the derivative of $C(x)/x$. This derivative can be found by using the quotient rule, as the next example shows.

Example 5   The total cost in thousands of dollars to manufacture x electrical generators is given by $C(x)$, where

$$C(x) = -x^3 + 15x^2 + 1000.$$

(a) Find the average cost per generator.

The average cost is given by the total cost divided by the number of items, or

$$\frac{C(x)}{x} = \frac{-x^3 + 15x^2 + 1000}{x}.$$

(b) Find the derivative of the average cost.

We can use the quotient rule to find the derivative of the average cost.

$$\frac{d}{dx}\left[\frac{C(x)}{x}\right] = \frac{x(-3x^2 + 30x) - (-x^3 + 15x^2 + 1000)(1)}{x^2}$$

$$= \frac{-3x^3 + 30x^2 + x^3 - 15x^2 - 1000}{x^2}$$

$$= \frac{-2x^3 + 15x^2 - 1000}{x^2} \quad \blacksquare$$

Work Problem 5 at the side.

5. The total revenue in thousands of dollars from the sale of x dozen CB radios is given by

$$R(x) = 32x^2 + 7x + 80.$$

(a) Find the average revenue.
(b) Find the derivative of the average revenue.

Answer:

(a) $\dfrac{32x^2 + 7x + 80}{x}$

(b) $\dfrac{32x^2 - 80}{x^2}$

3.6 EXERCISES

Use the product rule to find the derivative of each of the following functions. See Examples 1 and 2. (In Exercises 13–16, use the fact that $p^2 = p \cdot p$.)

1. $y = 3x(2x - 5)$
2. $y = -4x(8x + 1)$
3. $y = (2x - 5)(x + 4)$
4. $y = (3x + 7)(x - 1)$
5. $y = (8x - 2)(3x + 9)$
6. $y = (4x + 1)(7x + 12)$
7. $y = (3x^2 + 2)(2x - 1)$
8. $y = (5x^2 - 1)(4x + 3)$
9. $y = (x^2 + x)(3x - 5)$
10. $y = (2x^2 - 6x)(x + 2)$
11. $y = (9x^2 + 7x)(x^2 - 1)$
12. $y = (2x^2 - 4x)(5x^2 + 4)$
13. $y = (2x - 5)^2$
14. $y = (7x - 6)^2$
15. $y = (x^2 - 1)^2$
16. $y = (3x^2 + 2)^2$
17. $y = (x + 1)(\sqrt{x} + 2)$
18. $y = (2x - 3)(\sqrt{x} - 1)$
19. $y = (5\sqrt{x} - 1)(2\sqrt{x} + 1)$
20. $y = (-3\sqrt{x} + 6)(4\sqrt{x} - 2)$

Use the quotient rule to find the derivatives of each of the following functions. See Examples 3 and 4.

21. $y = \dfrac{x + 1}{2x - 1}$
22. $y = \dfrac{3x - 5}{x - 4}$
23. $y = \dfrac{7x + 1}{3x + 8}$
24. $y = \dfrac{6x - 11}{8x + 1}$

25. $y = \dfrac{2}{3x - 5}$

26. $y = \dfrac{-4}{2x - 11}$

27. $y = \dfrac{5 - 3x}{4 + x}$

28. $y = \dfrac{9 - 7x}{1 - x}$

29. $y = \dfrac{x^2 + x}{x - 1}$

30. $y = \dfrac{x^2 - 4x}{x + 3}$

31. $y = \dfrac{x - 2}{x^2 + 1}$

32. $y = \dfrac{4x + 11}{x^2 - 3}$

33. $y = \dfrac{3x^2 + x}{2x^2 - 1}$

34. $y = \dfrac{-x^2 + 6x}{4x^2 + 1}$

35. $y = \dfrac{x^2 - 4x + 2}{x + 3}$

36. $y = \dfrac{x^2 + 7x - 2}{x - 2}$

37. $y = \dfrac{\sqrt{x}}{x - 1}$

38. $y = \dfrac{\sqrt{x}}{2x + 3}$

39. $y = \dfrac{5x + 6}{\sqrt{x}}$

40. $y = \dfrac{9x - 8}{\sqrt{x}}$

APPLIED PROBLEMS

Cost

The total cost to produce x units of perfume is given by

$$C(x) = 9x^2 - 4x + 8.$$

Find the average cost per unit to produce (see Example 5)

41. 10 units; (Find $C(10)/10$.)

42. 20 units;

43. x units.

44. Find the derivative of the average cost function.

Profit

The total profit from selling x units of mathematics textbooks is given by

$$P(x) = 10x^2 - 5x - 18.$$

Find the average profit from selling

45. 8 units;

46. 15 units;

47. x units.

48. Find the derivative of the average profit function.

Rate of fuel use

Suppose you are the manager of a trucking firm, and one of your drivers reports that, according to her calculations, her truck burns fuel at the rate of

$$G(x) = \frac{1}{200}\left(\frac{800}{x} + x\right)$$

gallons per mile when traveling at x miles per hour on a smooth, dry road. (If a driver did report this to you, it would probably be a good idea to make the driver your mathematical modeler.)

49. If the driver tells you that she wants to travel 20 miles per hour, what should you tell her? (Hint: Take the derivative of G and evaluate it for $x = 20$. Then interpret your results.)

50. If the driver wants to go 40 miles per hour, what should you say? (Hint: Find $G'(40)$. In Example 3 of Section 4.4 we will find the speed that produces the lowest possible cost.)

Population

Assume that the total number of bacteria in millions present at a certain time t is given by

$$N(t) = (t - 10)^2(2t) + 50.$$

51. Find $N'(t)$.

At what rate is the population of bacteria changing

52. when $t = 8$;   53. when $t = 11$?

54. Answer 52 is negative, while answer 53 is positive. What does this mean in terms of the population of bacteria?

Drug concentration

When a certain drug is introduced into a muscle, the muscle responds by contracting. The amount of contraction, s, in millimeters, is related to the concentration of the drug, x, in milliliters, by

$$s(x) = \frac{x}{m + nx},$$

where m and n are constants.

55. Find $s'(x)$.

56. Evaluate $s'(x)$ when $x = 50$, $m = 10$, and $n = 3$.

Memory

According to the psychologist L. L. Thurstone, the number of facts of a certain type that are remembered after t hours is given by

$$f(t) = \frac{kt}{at - m}$$

where k, m, and a are constants. Find $f'(t)$ if $a = 99$, $k = 90$, $m = 90$, and

57. $t = 1$;   58. $t = 10$.

59. Prove the product rule. (Hint: Write $g(x + h) \cdot k(x + h) - g(x) \cdot k(x)$ as $g(x + h) \cdot k(x + h) - g(x + h) \cdot k(x) + g(x + h) \cdot k(x) - g(x) \cdot k(x)$.)

60. Prove the quotient rule.

3.7 THE CHAIN RULE

We can find the derivative of the function $y = (3 + 6x)^2$ by writing the function as $y = (3 + 6x)(3 + 6x)$ and using the product rule, as follows.

$$y' = \frac{dy}{dx} = (3 + 6x)(6) + (3 + 6x)(6)$$
$$= 18 + 36x + 18 + 36x$$
$$= 36 + 72x$$

We could find the derivative of $y = (3 + 6x)^3$ or $y = (3 + 6x)^4$ in a similar (but much longer) way. Luckily, there is a shortcut for finding derivatives of functions raised to a power.

Theorem 3.7 *Generalized power rule* Let u be some function of x. Let $y = u^n$. Then

$$y' = \frac{dy}{dx} = n \cdot u^{n-1} \cdot u'.$$

(The derivative of $y = u^n$ is found by decreasing the exponent on u by 1 and multiplying the result by the exponent n and the derivative of u.)

Example 1 Use the generalized power rule to find the derivative of $y = (3 + 6x)^2$.

Here we let $u = 3 + 6x$. Thus, $u' = 6$. By the generalized power rule, we have

$$\frac{dy}{dx} = n \cdot u^{n-1} \cdot u'$$
$$= 2(3 + 6x)^{2-1} \cdot 6 \qquad \text{Let } n = 2, u = 3 + 6x, \text{ and } u' = 6$$
$$= 12(3 + 6x)$$
$$= 36 + 72x.$$

This is the same answer we found above. ■

Example 2 Let $y = (x^3 + 5)^4$. Find dy/dx.

3.7 The Chain Rule

We have $u = x^2 + 5$ and $u' = 2x$. Thus

$$\frac{dy}{dx} = 4(x^2 + 5)^3(2x) \quad \text{Let } n = 4, u = x^2 + 5, \text{ and } u' = 2x$$

$$= 8x(x^2 + 5)^3. \quad \blacksquare$$

Work Problem 1 at the side.

Example 3 Find dy/dx if $y = 5(8x + 6)^4$.
We have

$$\frac{dy}{dx} = 5(4)(8x + 6)^3(8) = 160(8x + 6)^3. \quad \blacksquare$$

Example 4 Find dy/dx if $y = \sqrt{9x + 2}$.
Rewrite $y = \sqrt{9x + 2}$ as $y = (9x + 2)^{1/2}$. Then

$$\frac{dy}{dx} = \frac{1}{2}(9x + 2)^{-1/2}(9)$$

$$= \frac{9}{2}(9x + 2)^{-1/2}.$$

This answer can be written as

$$\frac{dy}{dx} = \frac{9}{2(9x + 2)^{1/2}} \quad (9x + 2)^{-1/2} = \frac{1}{(9x + 2)^{1/2}}$$

or

$$\frac{dy}{dx} = \frac{9}{2\sqrt{9x + 2}}. \quad \blacksquare$$

Work Problem 2 at the side.

Example 5 Suppose the revenue, $R(x)$, produced by selling x units of agar-agar to a biologist is given by

$$R(x) = 4\sqrt{3x + 1} + \frac{4}{x}.$$

Assume that $x > 0$. Find the marginal revenue when $x = 5$.
As we saw earlier, the marginal revenue is given by the derivative of the revenue function. By the generalized power rule, we have

$$R'(x) = \frac{6}{\sqrt{3x + 1}} - \frac{4}{x^2}.$$

---

1. Find dy/dx for the following functions.
   (a) $y = (2x + 5)^6$
   (b) $y = (4x^2 - 7)^3$

Answer:
(a) $12(2x + 5)^5$
(b) $24x(4x^2 - 7)^2$

2. Find dy/dx for the following functions.
   (a) $y = 12(x^2 + 6)^5$
   (b) $y = -4\sqrt{3x - 8}$
   (c) $8(4x^2 + 2)^{3/2}$

Answer:
(a) $120x(x^2 + 6)^4$
(b) $-6(3x - 8)^{-1/2}$ or $\dfrac{-6}{\sqrt{3x - 8}}$
(c) $96x(4x^2 + 2)^{1/2}$

When $x = 5$, we have

$$R'(5) = \frac{6}{\sqrt{3(5)+1}} - \frac{4}{5^2}$$

$$= \frac{6}{\sqrt{16}} - \frac{4}{25}$$

$$= \frac{6}{4} - \frac{4}{25}$$

$$= 1.34.$$

After 5 units have been sold, the sale of one more unit will increase revenue by about $1.34. ■

Work Problem 3 at the side.

3. Use the function of Example 5 and find the marginal revenue when
(a) $x = 8$;
(b) $x = 16$;
(c) $x = 20$.

Answer:
(a) 1.14
(b) .84
(c) .76

Sometimes we need to use both the generalized power rule and the product or quotient rule, as the next example shows.

Example 6 Find the derivative of $y = 4x(3x + 5)^5$.

By the product rule and the generalized power rule, we have

$$\frac{dy}{dx} = 4x[5(3x+5)^4 \cdot 3] + (3x+5)^5(4)$$

$$= 60x(3x+5)^4 + 4(3x+5)^5$$

$$= 4(3x+5)^4[15x + (3x+5)^1] \qquad \text{Factor out the greatest}$$
$$= 4(3x+5)^4(18x+5). \ ■ \qquad \text{common factor.}$$

Work Problem 4 at the side.

4. Find the derivatives of the following.
(a) $y = 6x(x+2)^2$
(b) $y = -9x(2x^2+1)^3$
(c) $y = 5x^2(3x-1)^4$

Answer:
(a) $6(x+2)(3x+2)$
(b) $-9(2x^2+1)^2(14x^2+1)$
(c) $10x(3x-1)^3(9x-1)$

Example 7 Find the derivative of $y = \dfrac{(3x+2)^7}{x-1}$.

Use the quotient rule and the generalized power rule.

$$\frac{dy}{dx} = \frac{(x-1)[7(3x+2)^6 \cdot 3] - (3x+2)^7(1)}{(x-1)^2}$$

$$= \frac{21(x-1)(3x+2)^6 - (3x+2)^7}{(x-1)^2}$$

$$= \frac{(3x+2)^6[21(x-1) - (3x+2)]}{(x-1)^2}$$

$$= \frac{(3x+2)^6[21x - 21 - 3x - 2]}{(x-1)^2}$$

$$\frac{dy}{dx} = \frac{(3x+2)^6(18x-23)}{(x-1)^2} \ ■$$

5. Find the derivatives of the following.

(a) $y = \dfrac{(2x+1)^3}{3x}$

(b) $y = \dfrac{(x-6)^5}{3x-5}$

Answer:

(a) $\dfrac{(2x+1)^2(4x-1)}{3x^2}$

(b) $\dfrac{(x-6)^4(12x-7)}{(3x-5)^2}$

6. Use the chain rule to find the derivatives of the following.
(a) $y = 10(2x^2+1)^4$
(b) $y = -8\sqrt{14x+1}$

Answer:
(a) $160x(2x^2+1)^3$

(b) $\dfrac{-56}{\sqrt{14x+1}}$

Work Problem 5 at the side.

Some functions, especially the ones we see in Chapter 5, have derivatives that cannot be found by the generalized power rule. For these functions we need the chain rule.

Theorem 3.7 *Chain rule* Let u be some function of x. Let y be a function of u. Let all indicated derivatives exist. Then

$$\frac{dy}{dx} = \frac{dy}{du} \cdot \frac{du}{dx}.$$

Note that the generalized power rule is just a special case of the chain rule.

Example 8 Use the chain rule to find the derivatives of each of the following functions.
(a) $y = (9x+2)^5$

Let $u = 9x + 2$. Then $du/dx = 9$. The expression $9x + 2$ is raised to the fifth power, so that $y = u^5$, and $dy/du = 5u^4$. By the chain rule,

$$\frac{dy}{dx} = \frac{dy}{du} \cdot \frac{du}{dx}$$
$$= 5u^4(9)$$
$$= 5(9x+2)^4(9) \quad \text{since } u = 9x + 2$$
$$\frac{dy}{dx} = 45(9x+2)^4.$$

This is the same result we would get using the generalized power rule.

(b) $y = 25\sqrt{15x^2+1}$

Here $y = 25(15x^2+1)^{1/2}$. Let $u = 15x^2 + 1$, and let $y = 25u^{1/2}$. Then $dy/du = 25/(2u^{1/2})$, and $du/dx = 30x$. Finally,

$$\frac{dy}{dx} = \frac{25}{2u^{1/2}}(30x)$$
$$= \frac{25(30x)}{2(15x^2+1)^{1/2}} \quad \text{since } u = 15x^2 + 1$$
$$= \frac{375x}{\sqrt{15x^2+1}}. \blacksquare$$

Work Problem 6 at the side.

3.7 EXERCISES

Find the derivatives of the following functions. See Examples 1–4.

1. $y = (2x + 9)^2$
2. $y = (8x - 3)^2$
3. $y = 6(5x - 1)^3$
4. $y = -8(3x + 2)^3$
5. $y = -2(12x^2 + 4)^3$
6. $y = 5(3x^2 - 5)^3$
7. $y = 9(x^2 + 5x)^4$
8. $y = -3(x^2 - 5x)^4$
9. $y = 12(2x + 5)^{3/2}$
10. $y = 45(3x - 8)^{3/2}$
11. $y = -7(4x^2 + 9x)^{3/2}$
12. $y = 11(5x^2 + 6x)^{3/2}$
13. $y = 8\sqrt{4x + 7}$
14. $y = -3\sqrt{7x - 1}$
15. $y = -2\sqrt{x^2 + 4x}$
16. $y = 4\sqrt{2x^2 + 3}$

Use the product or quotient rule to find the following derivatives. See Examples 6 and 7.

17. $y = 4x(2x + 3)^2$
18. $y = -6x(5x - 1)^2$
19. $y = (x + 2)(x - 1)^2$
20. $y = (3x + 1)^2(x + 4)$
21. $y = 5(x + 3)^2(x - 1)^2$
22. $y = -9(x + 4)^2(2x - 3)^2$
23. $y = (x + 1)^2 \sqrt{x}$
24. $y = (3x + 5)^2 \sqrt{x}$
25. $y = \dfrac{1}{(x - 4)^2}$
26. $y = \dfrac{-5}{(2x + 1)^2}$
27. $y = \dfrac{(x + 3)^2}{x - 1}$
28. $y = \dfrac{(x - 6)^2}{x + 4}$
29. $y = \dfrac{x^2 + 4x}{(x + 2)^2}$
30. $y = \dfrac{3x^2 - x}{(x - 1)^2}$

APPLIED PROBLEMS

Revenue

Assume that the total revenue from the sale of x television sets is given by

$$R(x) = 1000\left(1 - \dfrac{x}{500}\right)^2.$$

Find the marginal revenue for the following values of x. See Example 5.

31. $x = 100$
32. $x = 150$
33. $x = 200$
34. $x = 400$
35. Find the average revenue from the sale of x sets.
36. Find the derivative of the average revenue.

Population

The total number of bacteria in millions present in a culture is given by

$$N(t) = 2t(5t + 9)^{1/2} + 12,$$

where *t* represents time in hours after the beginning of an experiment. Find the rate of change of the population of bacteria with respect to time when

37. $t = 0$;
38. $t = 7/5$;
39. $t = 8$;
40. $t = 11$.

Metabolism of calcium

As a check on a person's use of calcium, a small amount of radioactive calcium is injected into the person's bloodstream with measurements of the calcium remaining in the bloodstream made each day for several days. Suppose the amount of the calcium remaining in the bloodstream in milligrams per cubic centimeter *t* days after the initial injection is approximated by

$$C(t) = \frac{1}{2}(2t + 1)^{-1/2}.$$

Find the rate of change of *C* with respect to time when

41. $t = 0$;
42. $t = 4$;
43. $t = 6$ (use a calculator);
44. $t = 7.5$.

Drug reaction

The strength of a person's reaction to a certain drug is given by

$$R(Q) = Q\left(C - \frac{Q}{3}\right)^{1/2},$$

where *Q* represents the quantity of the drug given to the patient and *C* is a constant.

45. The derivative $R'(Q)$ is called the *sensitivity* to the drug. Find $R'(Q)$.

46. Find $R'(Q)$ if $Q = 87$ and $C = 59$.

CHAPTER 3 SUMMARY

Key Words

limit
indeterminate form
limit to infinity
discontinuity
discontinuous
continuous at $x = a$
interval notation
open interval
closed interval

continuous on an open interval
average rate of change
instantaneous rate of change
tangent line
secant line
derivative
velocity
marginal cost
average cost

Things to Remember

The following theorems apply for functions *f*, *g*, *h*, and *k*, where all indicated derivatives exist.

Theorem 3.1 (*Constant function*) If $y = k$, where k is any real number, then
$$y' = 0.$$

Theorem 3.2 (*Power rule*) If $y = x^n$, then $y' = n \cdot x^{n-1}$.

Theorem 3.3 (*Constant times a function*) Let k be a real number. Then the derivative of $y = k \cdot f(x)$ is
$$y' = k \cdot f'(x).$$

Theorem 3.4 (*Sum or difference rule*) If $y = f(x) + g(x)$, then
$$y' = f'(x) + g'(x);$$
if $y = f(x) - g(x)$, then
$$y' = f'(x) - g'(x).$$

Theorem 3.5 (*Product rule*) If $f(x) = g(x) \cdot k(x)$, then
$$f'(x) = g(x) \cdot k'(x) + k(x) \cdot g'(x).$$

Theorem 3.6 (*Quotient rule*) If $f(x) = g(x)/k(x)$, and if $k(x) \neq 0$, then
$$f'(x) = \frac{k(x) \cdot g'(x) - g(x) \cdot k'(x)}{[k(x)]^2}.$$

Theorem 3.7 (*Generalized power rule*) Let u be some function of x. Let $y = u^n$. Then
$$y' = n \cdot u^{n-1} \cdot u'.$$

Theorem 3.8 (*Chain rule*) Let u be some function of x. Let y be a function of u. Then
$$\frac{dy}{dx} = \frac{dy}{du} \cdot \frac{du}{dx}.$$

CHAPTER 3 TEST

[3.1] Find each of the following limits that exist.

1. $\lim\limits_{x \to 6} \dfrac{2x + 5}{x - 3}$

2. $\lim\limits_{x \to 4} \dfrac{x^2 - 16}{x - 4}$

3. $\lim\limits_{x \to -4} \dfrac{2x^2 + 3x - 20}{x + 4}$

4. $\lim\limits_{x \to 9} \dfrac{\sqrt{x} - 3}{x - 9}$

5. $\lim\limits_{x \to \infty} \dfrac{2x + 5}{4x - 9}$

6. $\lim\limits_{x \to \infty} \dfrac{x^2 + 6x + 8}{x^3 + 2x + 1}$

[3.2] *Find all points of discontinuity for the following functions.*

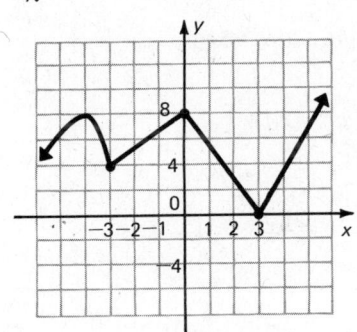

7.

8.

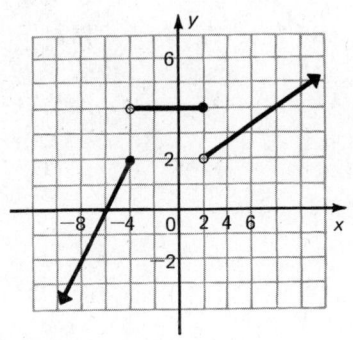

Write each of the following sets in interval notation.

9. $\{x \mid -4 \leq x \leq 2\}$ 10. $\{x \mid x > 4\}$

[3.3] *Find the average rate of change for the following functions.*

11. $y = 6x^2 + 2$, from $x = 1$ to $x = 4$

12. $y = -2x^3 - x^2 + 5$, from $x = -2$ to $x = 6$

13. $y = \dfrac{-6}{3x - 5}$, from $x = 4$ to $x = 9$

[3.4] *Use the definition of the derivative to find the derivative of each of the following functions.*

14. $y = 4x + 3$ 15. $y = 5x^2 + 6x$

[3.5] *Suppose that the profit in cents from selling x pounds of potatoes is given by*

$$P(x) = 15x + 25x^2.$$

Find the marginal profit when

16. $x = 6$; 17. $x = 20$; 18. $x = 30$.

The sales of a company are related to its expenditure on research by

$$S(x) = 1000 + 50\sqrt{x} + 10x,$$

where $S(x)$ gives sales in millions when x thousand dollars are spent on research. Find dS/dx when

19. $x = 9$; 20. $x = 16$; 21. $x = 25$.

For each of the following position functions, find (a) a formula for $v(t)$; (b) the velocity when $t = 0$, $t = 9$, and $t = 20$.

22. $s(t) = 60t + 5$ 23. $s(t) = 8t^3 - 4t^2$

Find the derivative of each of the following functions.

[3.5]   24. $y = x^3 - 4x^2$ 25. $y = 6x^{7/4}$

[3.6]   26. $y = (3x + 5)(6x - 1)$ 27. $y = 4(x + 2)\sqrt{x}$

       28. $y = \dfrac{-8}{2x + 1}$ 29. $y = \dfrac{x^2 - x + 1}{x - 1}$

[3.7]   30. $y = 6(x^2 + 2)^5$ 31. $y = \sqrt{6x - 11}$

       32. $y = 9x(x + 1)^5$ 33. $y = \dfrac{2x}{(x + 5)^2}$

4

Applications of the Derivative

While calculus has been successfully applied to mathematical models in physical science for the past three hundred years, it has been only in the last thirty years or so that calculus has been successfully applied to problems in management, social, and life sciences. One important application of calculus in these fields is finding maximum or minimum values for a function. In this chapter we discuss the basic ideas for finding maximums and minimums and look at practical examples.

4.1 RELATIVE MAXIMUMS AND MINIMUMS

Recall from Section 3.2 that an **open interval** is a set of points such as $\{x | a < x < b\}$ or $\{x | x > a\}$. The first of these open intervals would be written (a, b) and the second would be written (a, ∞). A **closed interval**, $\{x | a \leq x \leq b\}$, is written $[a, b]$. We will use this notation when we look for maximum and minimum values for a function.

A function can have two kinds of maximums: a maximum can be the largest value the function ever takes, or it can be the largest value of the function in a given interval. The largest value a function ever takes, called the *absolute maximum*, is defined in the next section. The largest value of a function in a given interval, called a *relative maximum*, is defined as follows.

A function f has a **relative maximum** at $x = a$ if $f(x) < f(a)$ for all values of x (except a) in some open interval containing a.

A **relative minimum** is defined in a very similar way. Think of a relative maximum as a peak (not necessarily the highest peak) and a relative minimum as a valley on the graph of the function. An *absolute minimum* is defined in the next section.

154 Applications of the Derivative

1. Identify the x-values of all points where the graphs here have relative maximums or relative minimums.

(a)

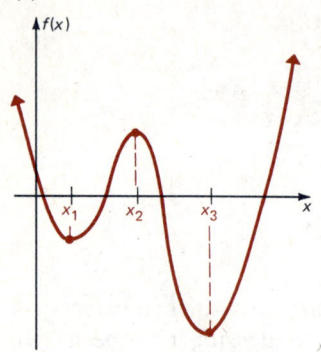

(b)

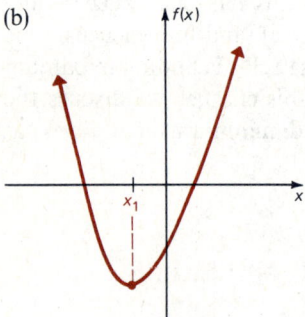

(c)

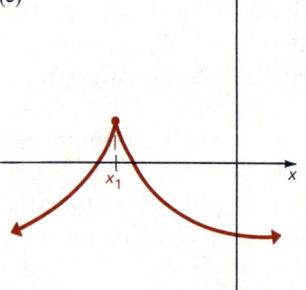

Answer:
(a) relative maximum at x_2, relative minimums at x_1 and x_3
(b) no relative maximum, relative minimum at x_1
(c) relative maximum at x_1, no relative minimum

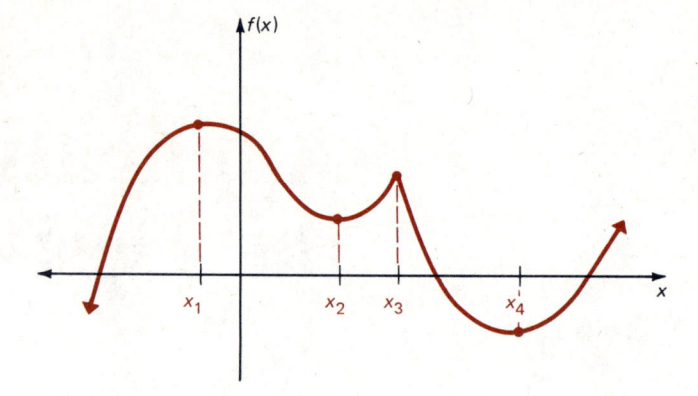

Figure 4.1

Example 1 Identify the x-values of all points where the graph of Figure 4.1 has relative maximums or relative minimums.

The graph has relative maximums at x_1 and x_3 and relative minimums at x_2 and x_4. ■

Work Problem 1 at the side.

How do we use the equation of a function to find the relative maximums or minimums for the function? We do this by using the fact that the derivative of a function can be used to find the slope of a line tangent to the graph of a function.

Figure 4.2 shows the graph of a function; this function has a relative maximum at x_2 and relative minimums at x_1 and x_3. The tangents to the curve at the points x_1, x_2, and x_3 have been drawn; all three of these tangents are horizontal lines and thus have a slope

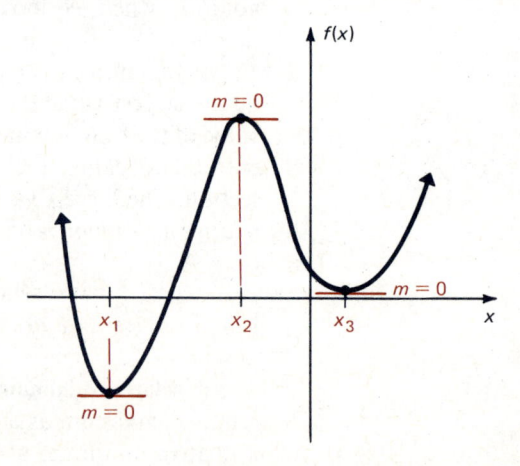

Figure 4.2

4.1 Relative Maximums and Minimums

of 0. (Recall that *m* is used to represent slope). Therefore, one way to find relative maximums and relative minimums for a function is to find all points where the derivative (the slope of the tangent) is 0. These points *might* lead to relative maximums or minimums.

Unfortunately, the process of finding relative maximums or minimums is not quite as simple as just finding values where the derivative is 0. It is possible for a function to have a relative maximum or minimum even though the derivative does not exist. Two examples of such functions are shown in Figure 4.3.

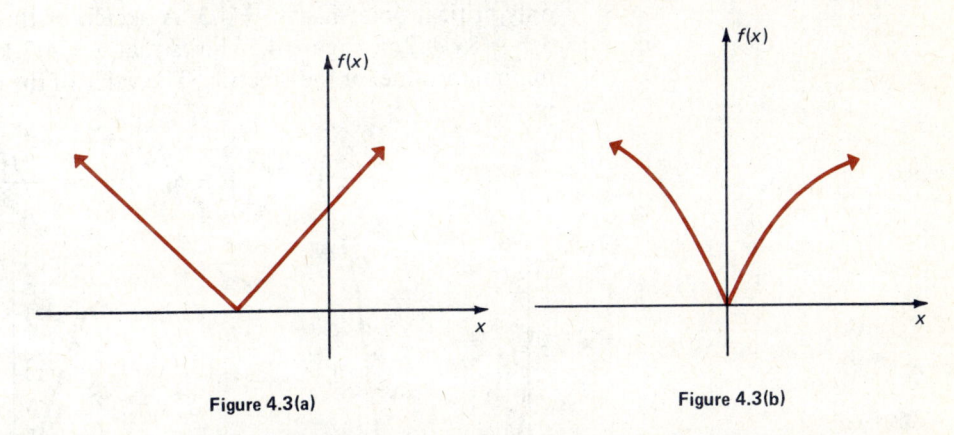

Figure 4.3(a)

Figure 4.3(b)

Because of this, when looking for relative maximums or minimums, we must check all values that make the derivative equal 0 and all values where the derivative does not exist but the function does. Any of these values might lead to relative maximums or minimums. The points where a derivative equals 0 or does not exist (but the function does exist) are called the **critical points** of the function. The following theorem summarizes these facts.

Theorem 4.1 If a function has any relative maximums or relative minimums, they occur only at critical points of the function.

Be very careful not to get the theorem backward. It does *not* say that a function has either a relative maximum or relative minimum at every critical point of the function. For example, Figure 4.4 shows the graph of $y = x^3$. This function has a critical point at $x = 0$, but neither a relative maximum nor a relative minimum at $x = 0$ (or anywhere else, for that matter).

Example 2 Find any relative maximums or minimums for the following functions.
(a) $f(x) = 3x^2 - 8x + 2$

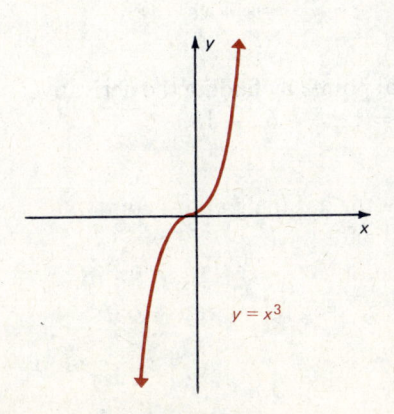

Figure 4.4

156 Applications of the Derivative

Find the x-values of the critical points. One kind of critical point is found when the derivative equals 0. Here $f'(x) = 6x - 8$. Place this derivative equal to 0.

$$6x - 8 = 0$$
$$6x = 8$$
$$x = \frac{8}{6} = \frac{4}{3}$$

There are no points where the derivative does not exist, so that the only critical point is at $x = 4/3$. A sketch of the graph of $f(x) = 3x^2 - 8x + 2$ in Figure 4.5 shows that $x = 4/3$ leads to a relative minimum value for the function. The value of the relative minimum is

$$f\left(\frac{4}{3}\right) = 3\left(\frac{4}{3}\right)^2 - 8\left(\frac{4}{3}\right) + 2 = -\frac{10}{3}.$$

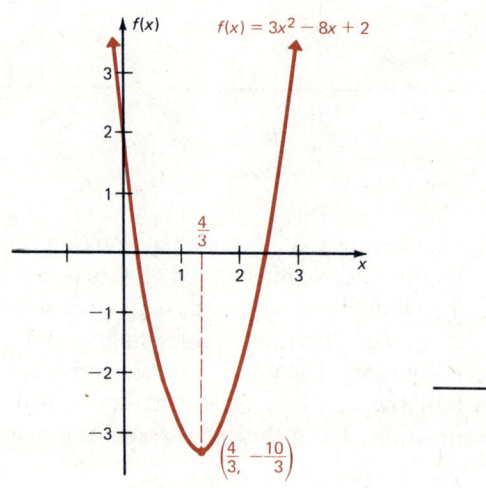

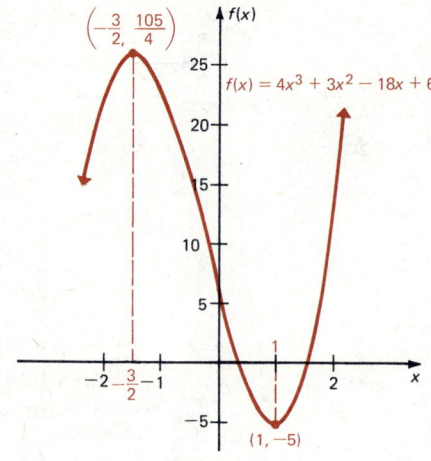

Figure 4.5

Figure 4.6

(b) $f(x) = 4x^3 + 3x^2 - 18x + 6$

Find the x-values of the critical points by finding the derivative and setting it equal to 0. Here $f'(x) = 12x^2 + 6x - 18$.

$$12x^2 + 6x - 18 = 0$$
$$6(2x^2 + x - 3) = 0 \quad \text{Factor out the greatest common factor}$$
$$6(2x + 3)(x - 1) = 0 \quad \text{Factor } 2x^2 + x - 3$$
$$2x + 3 = 0 \quad \text{or} \quad x - 1 = 0 \quad \text{Place each factor equal to 0}$$
$$2x = -3 \qquad\qquad x = 1$$
$$x = -\frac{3}{2}$$

4.1 Relative Maximums and Minimums

There are no points where the derivative does not exist, so that the only critical points are at $x = -3/2$ and $x = 1$. Figure 4.6 shows a sketch of the graph of $f(x) = 4x^3 + 3x^2 - 18x + 6$. As the graph shows, $-3/2$ leads to a relative maximum, while 1 gives a relative minimum. The relative maximum is $f(-3/2) = 105/4$, and the relative minimum is $f(1) = -5$. ∎

Work Problem 2 at the side.

In these examples we found the critical points by using algebra, but it was necessary to draw a graph of the function to see if the various critical points led to relative maximums or relative minimums. A better way to decide whether or not a given critical point leads to a relative maximum, a relative minimum, or neither will be developed in the rest of this section.

If we choose any two values x_1 and x_2 in an interval with $x_1 < x_2$ and find that $f(x_1) < f(x_2)$, we say function f is *increasing* on the interval. A function is increasing if its graph goes *up* as we move from left to right along the x-axis. Examples of increasing functions are shown in Figure 4.7.

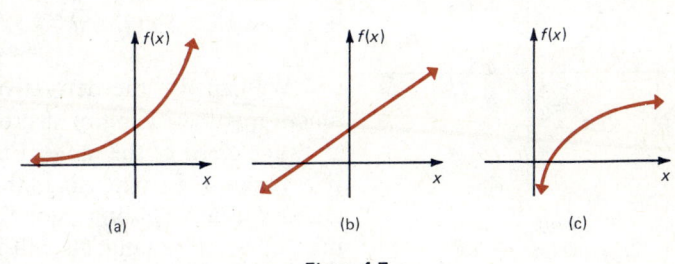

Figure 4.7

Figure 4.8

If we can choose any two values x_1 and x_2 in an interval with $x_1 < x_2$ and find that $f(x_1) > f(x_2)$, we say function f is *decreasing* on the interval. A function is decreasing if its graph goes *down* as we move from left to right along the x-axis. Examples of decreasing functions are shown in Figure 4.8.

---

2. Find all relative maximums or minimums for the following functions. Sketch the graph to decide if a critical point gives a maximum, a minimum, or neither.
(a) $f(x) = 6x^2 + 24x + 20$
(b) $f(x) = \frac{1}{3}x^3 - x^2 - 15x + 6$

Answer:
(a) relative minimum at $x = -2$, relative minimum is $f(-2) = -4$
(b) relative maximum at $x = -3$, relative maximum is $f(-3) = 33$; relative minimum at $x = 5$, minimum is $f(5) = -157/3$

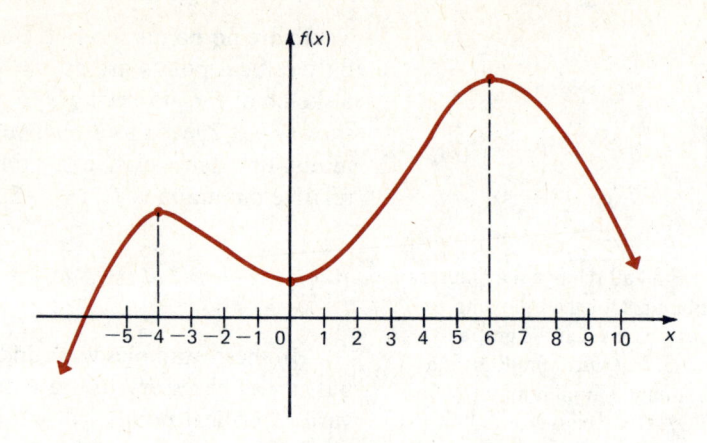

158 Applications of the Derivative

Figure 4.9

3. Find the open intervals where the function whose graph is shown here is increasing or decreasing.

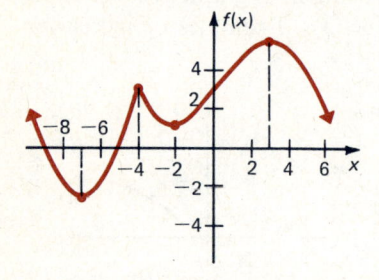

Answer: increasing on $(-7, -4)$ and $(-2, 3)$, decreasing on $(-\infty, -7)$, $(-4, -2)$, and $(3, \infty)$

Figure 4.10

Example 3 Find the open intervals where the function of Figure 4.9 is increasing or decreasing.

The function is increasing on the open intervals $(-\infty, -4)$ and $(0, 6)$. It is decreasing on $(-4, 0)$ and $(6, \infty)$. ∎

Work Problem 3 at the side.

We can use the derivative of a function to tell whether the function is increasing or decreasing. We know that the derivative gives the slope of the tangent to a curve at any point on the curve. If a function is increasing, the tangent line at any point on the graph will have *positive* slope, while if the function is decreasing, the slopes of the tangents will be *negative*. That is,

A function f is **increasing on an interval** if $f'(x) > 0$ for every value of x in the interval;

the function is **decreasing on an interval** if $f'(x) < 0$ for every x in the interval.

Example 4 Find the open intervals where the following functions are increasing or decreasing.
(a) $f(x) = 2x^2 + 6x - 5$

The derivative here is $f'(x) = 4x + 6$. This derivative equals 0 when $x = -3/2$. Do you see that for any value of x to the *left* of $-3/2$, the derivative is negative? Likewise, for any value of x to the *right* of $-3/2$, $f'(x) > 0$. Thus, the function is decreasing on the interval $(-\infty, -3/2)$ and increasing on $(-3/2, \infty)$. See Figure 4.10.

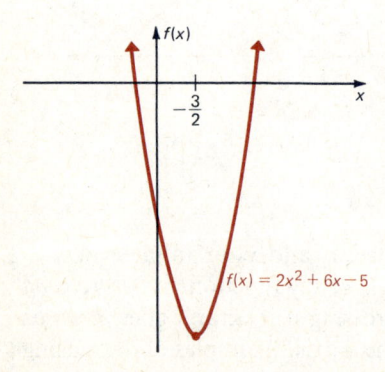

(b) $f(x) = x^3 + 3x^2 - 9x + 4$

The derivative is $f'(x) = 3x^2 + 6x - 9$. Place the derivative equal to 0 and solve the resulting equation.

4.1 Relative Maximums and Minimums 159

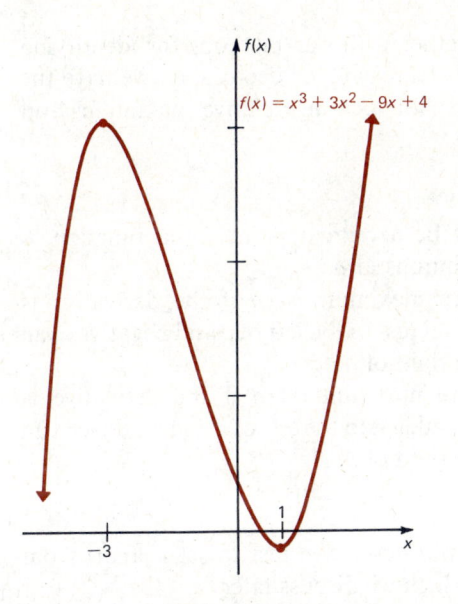

Figure 4.11

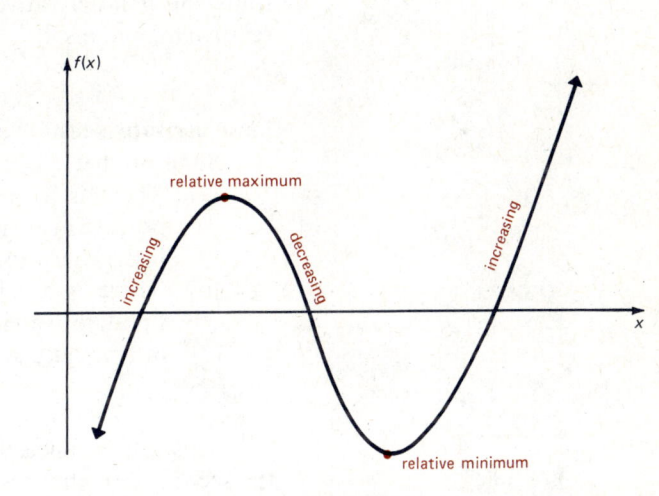

Figure 4.12

$$3x^2 + 6x - 9 = 0$$
$$3(x^2 + 2x - 3) = 0 \qquad \text{Factor}$$
$$3(x + 3)(x - 1) = 0 \qquad \text{Factor } x^2 + 2x - 3$$
$$x + 3 = 0 \quad \text{or} \quad x - 1 = 0 \qquad \text{Place each factor equal to 0}$$
$$x = -3 \quad \text{or} \quad x = 1$$

The derivative is 0 at $x = -3$ and $x = 1$. To see where the function is increasing or decreasing, make a table showing the intervals where the derivative is positive or negative.

Interval	$x + 3$	$x - 1$	$3(x + 3)(x - 1)$
$(-\infty, -3)$	−	−	+
$(-3, 1)$	+	−	−
$(1, \infty)$	+	+	+

The function $f(x) = x^3 + 3x^2 - 9x + 4$ is increasing on the intervals $(-\infty, -3)$ and $(1, \infty)$ and decreasing on $(-3, 1)$. See the graph of f in Figure 4.11. ∎

Work Problem 4 at the side.

Now we can find a method for identifying relative maximums and relative minimums. As shown in Figure 4.12, a function is increasing just to the left of a relative maximum and decreasing just to the right. A function is decreasing just to the left of a relative minimum and increasing just to the right.

---

4. Find the open intervals where the following functions are increasing or decreasing.
(a) $f(x) = -3x^2 + 18x + 4$
(b) $f(x) = \frac{1}{3}x^3 + 3x^2 + 5x - 8$

Answer:
(a) increasing on $(-\infty, 3)$, decreasing on $(3, \infty)$
(b) increasing on $(-\infty, -5)$ and $(-1, \infty)$, decreasing on $(-5, -1)$

Putting these facts together with our methods for identifying intervals where a function is increasing or decreasing, we have the following first derivative test for locating relative maximums and relative minimums.

First derivative test   Let a be a critical point for a function f. Suppose that f is continuous at a.
  (a) There is a relative maximum at a if the derivative is positive in an interval just to the left of a and negative in an interval just to the right of a.
  (b) There is a relative minimum at a if the derivative is negative in an interval just to the left of a and positive in an interval just to the right of a.

This test is called the *first* derivative test to set it apart from the second derivative test, which we discuss later.

Example 5   Find all relative maximums or relative minimums for the following functions.
(a) $f(x) = -x^2 + 8x + 3$
Here $f'(x) = -2x + 8$. This derivative is 0 when $x = 4$. To decide if 4 is a relative maximum or a relative minimum, note that $f'(x) = -2x + 8$ is positive if $x < 4$ and negative if $x > 4$. By part (a) of the first derivative test, this means that f has a relative maximum at $x = 4$. The value of this relative maximum is $f(4) = 19$.

(b) $f(x) = 2x^3 - 3x^2 - 72x + 15$
The derivative is $f'(x) = 6x^2 - 6x - 72$. Place this derivative equal to 0.

$$6x^2 - 6x - 72 = 0$$
$$6(x^2 - x - 12) = 0$$
$$6(x - 4)(x + 3) = 0$$
$$x - 4 = 0 \quad \text{or} \quad x + 3 = 0$$
$$x = 4 \quad \text{or} \quad x = -3$$

Make a table.

Interval	$x - 4$	$x + 3$	$6(x - 4)(x + 3)$
$(-\infty, -3)$	$-$	$-$	$+$
$(-3, 4)$	$-$	$+$	$-$
$(4, \infty)$	$+$	$+$	$+$

4.1 Relative Maximums and Minimums

The derivative is positive to the left of -3 and negative to the right; thus the function has a relative maximum when $x = -3$. The value of this relative maximum is $f(-3) = 150$.

The derivative is negative to the left of 4 and positive to the right. By part (b) of the first derivative test, this means that f has a relative minimum when $x = 4$. The value of this relative minimum is $f(4) = -193$. ∎

Work Problem 5 at the side.

We can now summarize the steps in finding a relative maximum or relative minimum for a function f.

1. Find the derivative, $f'(x)$.

2. Find all critical points a (those points where $f'(a) = 0$ or $f'(a)$ does not exist but $f(a)$ does exist).

3. If $f'(x) > 0$ to the left of a and if $f'(x) < 0$ to the right of a, then a leads to a relative maximum. The value of the relative maximum is $f(a)$.

4. If $f'(x) < 0$ to the left of a, and if $f'(x) > 0$ to the right of a, then a leads to a relative minimum. The value of the relative minimum is $f(a)$.

---

5. Find all relative maximums or relative minimums for the following functions.
(a) $f(x) = 5x^2 - 20x + 3$
(b) $f(x) = x^3 + 4x^2 - 3x + 5$

Answer:
(a) relative minimum at 2, minimum is $f(2) = -17$
(b) relative maximum at -3, maximum is $f(-3) = 23$; relative minimum at $1/3$, minimum is $f(1/3) = 121/27$

4.1 EXERCISES

Find the location and value of all relative maximums and relative minimums for the following functions. Identify all open intervals where the function is increasing or decreasing. See Examples 1 and 3.

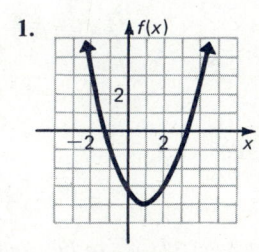

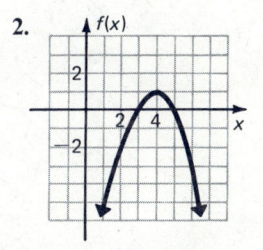

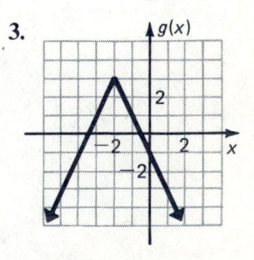

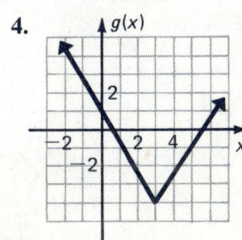

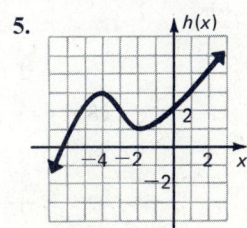

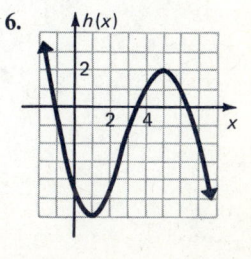

1.
2.
3.
4.
5.
6.

7. [graph of f(x)]   **8.** [graph of f(x)]   **9.** [graph of f(x)]

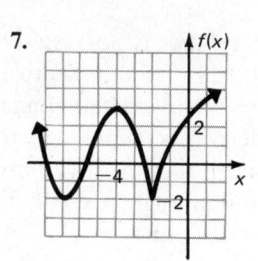

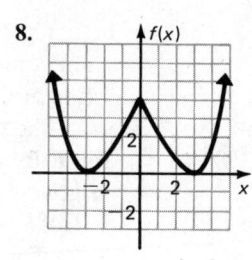

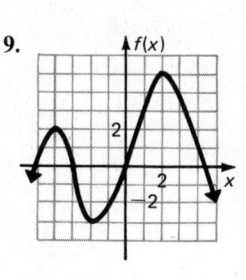

Find the x-values of all points where the following functions have any relative maximums or relative minimums. Find the value of any relative maximums or relative minimums. See Examples 2 and 5.

10. $f(x) = x^2 - 4x + 6$ **11.** $f(x) = x^2 + 12x - 8$

12. $f(x) = x^2 - 8x + 12$ **13.** $f(x) = x^2 - 14x + 3$

14. $f(x) = -x^2 + 4x - 5$ **15.** $f(x) = -x^2 - 8x + 6$

16. $f(x) = 3 - 4x - 2x^2$ **17.** $f(x) = 8 - 6x - x^2$

18. $f(x) = x^3 + 3x^2 - 24x + 2$ **19.** $f(x) = x^3 + 6x^2 + 9x - 8$

20. $f(x) = x^3 + x^2 - 5x + 1$ **21.** $f(x) = x^3 + 6x^2 + 12x - 5$

22. $f(x) = 2x^3 + 15x^2 + 36x - 4$ **23.** $f(x) = 2x^3 - 21x^2 + 60x + 5$

24. $f(x) = -\frac{2}{3}x^3 - \frac{1}{2}x^2 + 3x - 4$ **25.** $f(x) = -\frac{4}{3}x^3 - \frac{21}{2}x^2 - 5x + 8$

26. $f(x) = x^4 - 8x^2 + 9$ **27.** $f(x) = x^4 - 18x^2 - 4$

Find the vertex of each of the following parabolas. (Hint: The vertex of a parabola is the highest or lowest point on the parabola.)

28. $y = x^2 - 8x + 4$ **29.** $y = x^2 + 12x - 6$

30. $y = 2x^2 - 8x + 3$ **31.** $y = 3x^2 - 12x + 2$

32. $y = -2x^2 + 8x - 1$ **33.** $y = -x^2 - 2x + 1$

34. $y = 2x^2 - 5x + 2$ **35.** $y = 3x^2 - 8x + 4$

Find the open intervals where the following functions are increasing or decreasing. See Examples 3 and 4.

36. $f(x) = x^2 - 9x + 4$ **37.** $f(x) = x^2 + 12x - 6$

38. $f(x) = 3 + 4x - 2x^2$ **39.** $f(x) = -9 + 8x - 3x^2$

40. $f(x) = 2x^3 - 3x^2 - 12x + 2$ **41.** $f(x) = 2x^3 - 3x^2 - 72x - 4$

42. $f(x) = 4x^3 - 9x^2 - 30x + 6$ **43.** $f(x) = 4x^3 - 15x^2 - 72x + 5$

44. $f(x) = 6x - 9$ **45.** $f(x) = -3x + 6$

APPLIED PROBLEMS

Education

46. A professor has found that the number of biology students attending class is approximated by
$$S(x) = -x^2 + 20x + 80,$$
where x is the number of daily hours that the student union is open. Find the number of hours that the union should be open so that the number of students attending class is a maximum. Find the maximum number of such students.

Hotel management

47. The number of people visiting Timberline Ski Lodge on Washington's Birthday is approximated by
$$W(x) = -x^2 + 60x + 180,$$
where x is the total snowfall in inches for the previous week. Find the snowfall that will produce the maximum number of visitors. Find the maximum number of visitors.

Profit

48. The total profit in thousands of dollars from the sale of x hundred thousand automobile tires is approximated by
$$P(x) = -x^3 + 9x^2 + 120x - 400.$$
(Assume $x \geq 5$.) Find the number of hundred thousands of tires that must be sold to maximize profit. Find the maximum profit.

Salmon spawning

49. The number of salmon swimming upstream to spawn is approximated by
$$S(x) = -x^3 + 3x^2 + 360x + 5000,$$
where x represents the temperature of the water in degrees Celsius. (This function is valid only if $6 \leq x \leq 20$.) Find the water temperature that produces the maximum number of salmon swimming upstream.

Profit

50. The profit in hundreds of dollars from the sale of x hundred pounds of a certain type of chicken feed is given by
$$P(x) = \tfrac{1}{3}x^3 - 12x^2 + 23x + 500.$$
Find all open intervals (with $x > 0$) where profit is increasing.

51. The total profit in dollars from the sale of x hundred citizen's band radios is approximated by
$$P(x) = x^3 - 11x^2 + 35x + 150.$$
Find all intervals (with $x > 0$) where profit is increasing.

Advertising

52. A company has found through experience that increasing its advertising also increases its sales, up to a point. The company believes that the mathematical model connecting profit in dollars, $P(x)$, and expenditures on advertising in dollars, x, is
$$P(x) = 80 + 108x - x^3.$$
(a) Find the expenditure on advertising that leads to maximum profit.
(b) Find the maximum profit.

Profit

53. The total profit in dollars from the sale of x units of a certain prescription drug is given by
$$P(x) = -x^3 + 3x^2 + 72x + 1280.$$
   (a) Find the number of units that should be sold in order to maximize the total profit.
   (b) What is the maximum profit?

54. The total profit in dollars from the sale of x buckets of Extra Crispy fried chicken is approximated by
$$C(x) = -x^3 + 6x^2 + 288x + 500.$$
   (a) Find the number of units that should be sold in order to maximize profits.
   (b) Find the maximum profit.

Microbe concentration

55. The microbe concentration, $B(x)$, in appropriate units, of Lake Tom depends approximately on the oxygen concentration, x, again in appropriate units, according to the function
$$B(x) = x^3 - 7x^2 - 160x + 1800.$$
   (a) Find the oxygen concentration that will lead to the minimum microbe concentration.
   (b) What is the minimum concentration?

4.2 ABSOLUTE MAXIMUMS AND MINIMUMS

As we saw in the last section, a function might well have several relative maximums or relative minimums. However, a function never has more than one absolute maximum or absolute minimum. A function f has an **absolute maximum** at $x = a$ if $f(x) \le f(a)$ for all x in the domain of the function. There is an **absolute minimum** at $x = a$ whenever $f(x) \ge f(a)$ for all x in the domain of the function.

The function graphed in Figure 4.13 has an absolute maximum at x_1 and an absolute minimum at x_2. The function graphed in Figure 4.14 has neither an absolute maximum nor an absolute minimum.

A relative maximum may also be an absolute maximum, but an absolute maximum need not be a relative maximum. The function graphed in Figure 4.15 has an absolute maximum which is also a relative maximum; the function graphed in Figure 4.16 has an absolute minimum at an endpoint of the domain of the function. This absolute minimum is not a relative minimum because a relative minimum at a point where $x = a$ requires that $f(a)$ be less than all $f(x)$ in some interval containing a. Here, any open interval containing a includes points which are not in the domain of the function.

4.2 Absolute Maximums and Minimums   165

In more advanced courses it can be shown that a function defined on a closed interval and continuous on the interval will always have both an absolute maximum and an absolute minimum on that interval. These will be found either at a relative maximum or a relative minimum, or at the endpoints of the closed interval on which the function is defined.

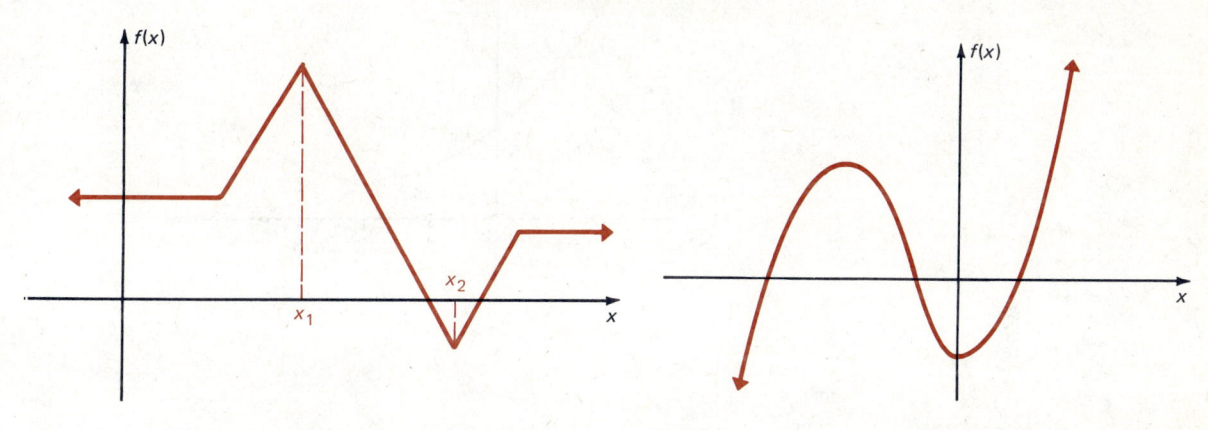

Figure 4.13

Figure 4.14

absolute maximum and relative maximum

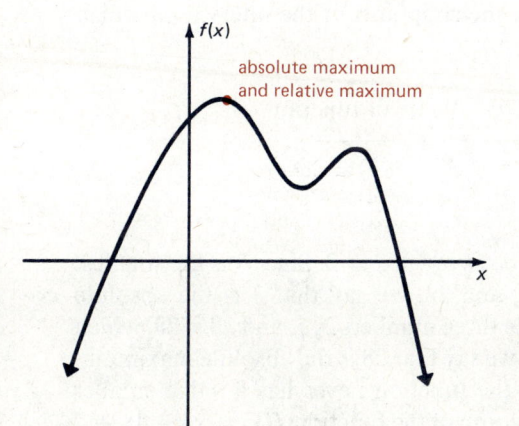

Figure 4.15

absolute minimum but not a relative minimum

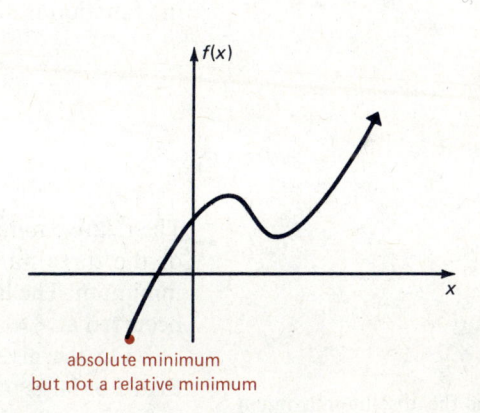

Figure 4.16

Example 1 Find the absolute maximum and absolute minimum of the function

$$f(x) = x^2 - 4x + 7$$

defined on the closed interval $[-3, 4]$.

We first look for critical points of the function in $(-3, 4)$. Here $f'(x) = 2x - 4$, and

$$2x - 4 = 0 \quad \text{if} \quad x = 2.$$

166 Applications of the Derivative

f(x)

$f(x) = x^2 - 4x + 7$

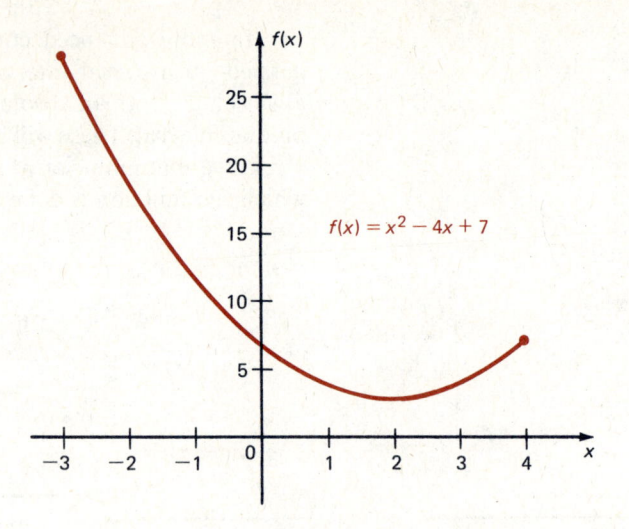

Figure 4.17

This x-value is in $(-3, 4)$. By the first derivative test, check that 2 leads to a relative minimum; the minimum is $f(2) = 3$.

Now we need to check the endpoints of the interval on which the function is defined.

Endpoint	Value of function
-3	$f(-3) = 28$
4	$f(4) = 7$

The relative minimum of 3 occurred at $x = 2$. Since 3 is the smallest of the three numbers 3, 7, and 28, we say that 3 is the absolute minimum. The largest of the three numbers 3, 7, and 28 is 28, which occurred at $x = -3$. Thus, we say that 28 is the absolute maximum.

On the given domain the function never has a value smaller than 3 or larger than 28. A graph of the function $f(x) = x^2 - 4x + 7$ defined on the closed interval $[-3, 4]$ is shown in Figure 4.17. ∎

Work Problem 1 at the side.

Example 2 Find the absolute maximum and absolute minimum of the function

$$f(x) = 2x^3 - x^2 - 20x + 40$$

defined on the interval $[-2, 5]$.

---

1. Find the absolute maximum and absolute minimum values of
   (a) $f(x) = -x^2 + 4x - 8$ on $[-4, 4]$;
   (b) $y = 2x^2 - 6x + 6$ on $[0, 5]$.

   Answer:
   (a) absolute maximum of -4 at $x = 2$, absolute minimum of -40 at $x = -4$
   (b) absolute maximum of 26 at $x = 5$, absolute minimum of $3/2$ at $x = 3/2$

4.2 Absolute Maximums and Minimums

First look for relative maximums or relative minimums in the interval $(-2, 5)$. Here $f'(x) = 6x^2 - 2x - 20$. Now place this derivative equal to 0.

$$6x^2 - 2x - 20 = 0$$
$$2(3x^2 - x - 10) = 0$$
$$2(3x + 5)(x - 2) = 0$$
$$3x + 5 = 0 \quad \text{or} \quad x - 2 = 0$$
$$3x = -5 \quad\quad\quad\quad x = 2$$
$$x = -\frac{5}{3}$$

Both x-values are in the interval $(-2, 5)$. By the first derivative test, $x = -5/3$ leads to a relative maximum and $x = 2$ leads to a relative minimum. Now evaluate the function at $-5/3$, 2, and the endpoints of its domain, -2 and 5.

x-value	Value of function	
-2	60	
$-\frac{5}{3}$	$\frac{1655}{27} \approx 61.296$	
2	12	← absolute minimum
5	165	← absolute maximum

The absolute maximum of 165 occurs when $x = 5$, and the absolute minimum of 12 occurs when $x = 2$. ∎

Work Problem 2 at the side.

2. Find the absolute maximum and absolute minimum for the function

$$f(x) = \tfrac{1}{3}x^3 + \tfrac{1}{2}x^2 - 2x + 8$$

defined on $[-3, 4]$.

Answer: absolute maximum of $88/3$ at $x = 4$, absolute minimum of $41/6$ at $x = 1$

Example 3 A company has found that its profit from the sale of x units of plastic turtles is given by

$$P(x) = x^4 - 8x^2 + 100.$$

Production bottlenecks limit the number of units that can be made to no more than 5. Find the maximum possible profit that the firm can make.

Because of the restriction, the profit function is defined only for the domain $[0, 5]$. We first look for critical points of the function in $(0, 5)$. Here $f'(x) = 4x^3 - 16x$. Now place this derivative equal to 0.

$$4x^3 - 16x = 0$$
$$4x(x^2 - 4) = 0 \quad \text{Factor out the common factor}$$
$$4x(x + 2)(x - 2) = 0$$
$$x = 0 \quad \text{or} \quad x + 2 = 0 \quad \text{or} \quad x - 2 = 0$$
$$x = 0 \quad \text{or} \quad\quad x = -2 \quad \text{or} \quad\quad x = 2$$

168 Applications of the Derivative

Since $x = 0$ and $x = -2$ are not in the interval $(0, 5)$, we disregard them. Now evaluate the function at 2 and at the endpoints of the domain, 0 and 5.

x-value	Value of function	
0	100	
2	84	
5	525	absolute maximum

Maximum profit of $525 occurs when 5 units are made. ■

Work Problem 3 at the side.

3. For a certain firm, the cost to produce x units of an item is given by

$$C(x) = x^3 - 15x^2 + 48x + 50.$$

Because of production problems, the function is defined only for the domain $[1, 6]$. Find the maximum and minimum costs that the firm will face.

Answer: maximum of 94 when $x = 2$, minimum of 14 when $x = 6$

We can now summarize the steps involved in finding absolute maximums and absolute minimums of a continuous function f defined on a closed interval $[a, b]$.

1. Find $f'(x)$ and locate all critical points in the interval (a, b).

2. Compute the values of f for all critical points from Step 1 and for the endpoints of the interval, a and b.

3. The largest value found in Step 2 is the absolute maximum and the smallest value is the absolute minimum.

4.2 EXERCISES

Find the location of all absolute maximums and absolute minimums (if any) for the following functions.

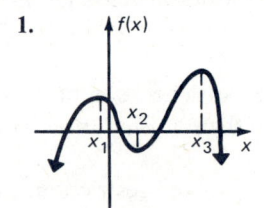

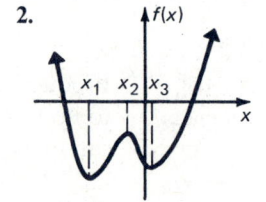

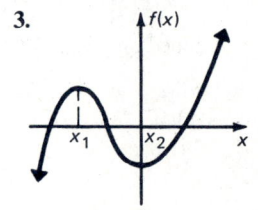

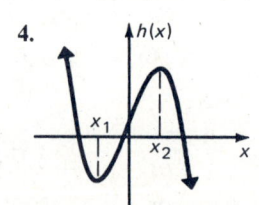

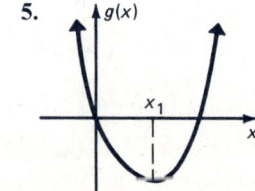

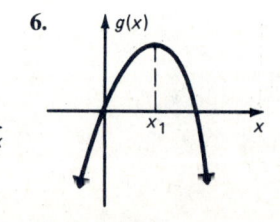

1.

2.

3.

4.

5.

6.

7. $f(x)$ graph with points x_1, x_2, x_3, x_4

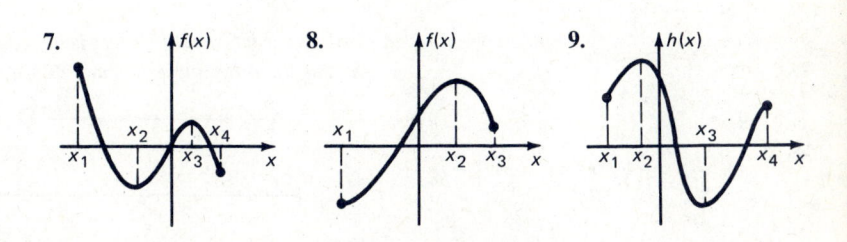

8. $f(x)$ graph with points x_1, x_2, x_3

9. $h(x)$ graph with points x_1, x_2, x_3, x_4

Find the location of all absolute maximums and absolute minimums for the following functions having domains as specified. A calculator will be helpful for many of these problems. See Examples 1 and 2.

10. $f(x) = x^2 - 4x + 1$, $[-6, 5]$
11. $f(x) = x^2 + 6x + 2$, $[-4, 0]$
12. $f(x) = 3x^2 - 8x - 6$, $[-2, 4]$
13. $f(x) = 10x^2 + 15x - 8$, $[-3, 0]$
14. $f(x) = 9 - 6x - 3x^2$, $[-4, 3]$
15. $f(x) = 5 - 8x - 4x^2$, $[-5, 1]$
16. $f(x) = x^3 - 6x^2 + 9x - 8$, $[0, 5]$
17. $f(x) = x^3 - 3x^2 - 24x + 5$, $[-3, 6]$
18. $f(x) = \frac{1}{3}x^3 + \frac{3}{2}x^2 - 4x + 1$, $[-5, 2]$
19. $f(x) = \frac{1}{3}x^3 - \frac{1}{2}x^2 - 6x + 3$, $[-4, 4]$
20. $f(x) = x^3 + 5x - 4$, $[-3, 2]$
21. $f(x) = x^3 + x + 3$, $[-1, 4]$
22. $f(x) = x^4 - 18x^2 + 1$, $[-4, 4]$
23. $f(x) = x^4 - 32x^2 - 7$, $[-5, 6]$

Auto mileage

24. From information given in a recent business publication we constructed the mathematical model

$$M(x) = -\frac{1}{45}x^2 + 2x - 20, \quad 30 \leq x \leq 65$$

to represent the miles per gallon used by a certain car at a speed of x miles per hour. Find the absolute maximum miles per gallon and the absolute minimum. See Example 3.

25. For a certain compact car,

$$M(x) = -.018x^2 + 1.24x + 6.2, \quad 30 \leq x \leq 60$$

represents the miles per gallon obtained at a speed of x miles per hour. Find the absolute maximum miles per gallon and the absolute minimum.

Geometry A piece of wire of length 12 feet is cut into two pieces. One piece is made into a circle and the other piece is made into a square. (See the figure.)

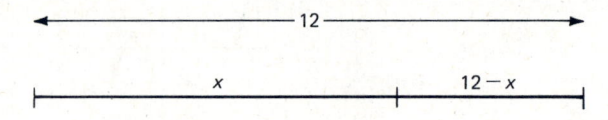

Let the piece of length x be formed into a circle;

$$\text{radius of circle} = \frac{x}{2\pi} \text{ ; area of circle} = \pi\left(\frac{x}{2\pi}\right)^2;$$

$$\text{side of square} = \frac{12-x}{4} \text{ ; area of square} = \left(\frac{12-x}{4}\right)^2.$$

26. Where should the cut be made in order to make the sum of the areas enclosed by both figures minimum? (Hint: Use 3.14 as an approximation for π. Have a calculator handy.)

27. Where should the cut be made in order to make the sum of the areas maximum? (Hint: Remember to use the endpoints of a domain when looking for absolute maximums and minimums.)

The following functions have derivatives which cannot easily be set equal to 0. To find the maximum and minimum of each function on the given domain, use a calculator to evaluate the function at intervals of .1. Then estimate the absolute maximum and absolute minimum values of the function.

28. $f(x) = x^3 - 3x^2 - 3x + 4$, $[2, 3]$
29. $f(x) = x^3 - 9x^2 + 18x - 7$, $[0, 1.5]$
30. $f(x) = x^3 - 3x^2 - 12x + 4$, $[-1.5, 0]$ and $[3, 4]$
31. $f(x) = x^3 + 6x^2 - 6x + 3$, $[-5, -4]$ and $[0, 1]$

4.3 THE SECOND DERIVATIVE TEST AND CURVE SKETCHING

The derivative of a function gives us a new function for which we can again find a derivative. We call the derivative of the derivative of a function the **second derivative** of the function. If we take the derivative of the second derivative of a function f, we get the *third derivative* of f. In the same way, we could get the *fourth derivative* and other higher derivatives. Thus, if $f(x) = x^4 + 2x^3 + 3x^2 - 5x + 7$, we have

$f'(x) = 4x^3 + 6x^2 + 6x - 5$, (the first derivative of f)
$f''(x) = 12x^2 + 12x + 6$, (the second derivative of f)
$f'''(x) = 24x + 12$, (the third derivative of f)
$f^{(4)}(x) = 24$, (the fourth derivative of f)
$f^{(5)}(x) = 0$. (the fifth derivative of f)

The second derivative of the function $y = f(x)$ can also be written as

$$y'' \quad \text{or} \quad \frac{d^2y}{dx^2}.$$

The other higher derivatives have corresponding notations.

Example 1 Find the second derivative of the following functions.
(a) $y = 8x^3 - 9x^2 + 6x + 4$

Here $y' = 24x^2 - 18x + 6$. The second derivative is the derivative of y', or

$$y'' = 48x - 18.$$

(b) $y = \dfrac{4x + 2}{3x - 1}$

Use the quotient rule to find y'.

$$y' = \frac{(3x - 1)(4) - (4x + 2)(3)}{(3x - 1)^2} = \frac{12x - 4 - 12x - 6}{(3x - 1)^2} = \frac{-10}{(3x - 1)^2}$$

We again use the quotient rule to find y''.

$$y'' = \frac{(3x - 1)^2(0) - (-10)(2)(3x - 1)(3)}{[(3x - 1)^2]^2}$$

$$= \frac{60(3x - 1)}{(3x - 1)^4}$$

$$y'' = \frac{60}{(3x - 1)^3} \quad \blacksquare$$

Work Problem 1 at the side.

Example 2 Let $f(x) = x^3 + 6x^2 - 9x + 8$. Find the following.
(a) $f''(0)$

Here $f'(x) = 3x^2 + 12x - 9$, so that $f''(x) = 6x + 12$. Then

$$f''(0) = 6(0) + 12 = 12.$$

(b) $f''(-3) = 6(-3) + 12 = -6 \quad \blacksquare$

Work Problem 2 at the side.

We defined increasing and decreasing functions in the first section of this chapter. We saw that a function is increasing on an interval if its derivative is positive on the interval. If the *derivative*

---

1. Find the second derivatives of the following.
(a) $y = -9x^3 + 8x^2 + 11x - 6$
(b) $y = -2x^4 + 6x^2$
(c) $y = \dfrac{x + 2}{5x - 1}$

Answer:
(a) $y'' = -54x + 16$
(b) $y'' = -24x^2 + 12$
(c) $y'' = \dfrac{110}{(5x - 1)^3}$

2. Let $f(x) = 4x^3 - 12x^2 + x - 1$. Find
(a) $f''(0)$;
(b) $f''(4)$;
(c) $f''(-2)$.

Answer:
(a) -24
(b) 72
(c) -72

is increasing on an interval, the function is said to be concave upward on the interval. If the derivative is decreasing, the function is concave downward. The derivative is increasing if its derivative (the second derivative) is positive and decreasing if its derivative is negative. Therefore,

a function f is **concave upward** on an interval if $f''(x) > 0$ for every value of x in the interval;
a function f is **concave downward** on an interval if $f''(x) < 0$ for every value of x in the interval.

The function itself can be either increasing or decreasing, and either concave upward or concave downward on an interval. Examples of various combinations are shown in Figure 4.18.

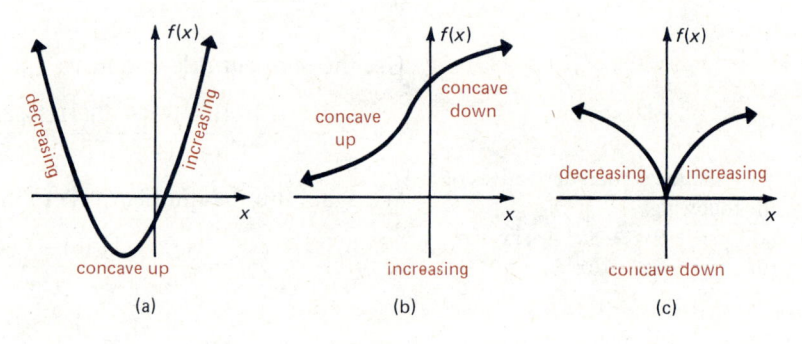

Figure 4.18

3. Find the intervals where the following functions are concave upward or concave downward. Identify any inflection points.
   (a) $f(x) = 6x^3 - 24x^2 + 9x - 3$
   (b) $f(x) = 2x^2 - 4x + 8$

Answer:
(a) concave downward on $(-\infty, 4/3)$, concave upward on $(4/3, \infty)$; point of inflection is $(4/3, -175/9)$
(b) $f''(x) = 4$, which is always positive; function is always concave upward, no inflection point

Example 3 Find the intervals where $f(x) = x^3 - 3x^2 + 5x - 4$ is concave upward or downward.

Here $f'(x) = 3x^2 - 6x + 5$ and $f''(x) = 6x - 6$. The second derivative is 0 when $x = 1$. Do you see that if we choose any value of x to the left of 1, $f''(x)$ will be negative? Thus, f is concave downward on the interval $(-\infty, 1)$. Likewise, if we choose any value of x to the right of 1, $f''(x)$ will be positive. Therefore, f is concave upward on the interval $(1, \infty)$. A graph of the function f is shown in Figure 4.19. ∎

In Example 3, the function changed from being concave downward to being concave upward at the point where $x = 1$. Such a point is called a **point of inflection.** In Example 3, the inflection point is $(1, f(1))$ or $(1, -1)$ and occurs at the value of x where $f''(x) = 0$. An inflection point may also occur where $f''(x)$ does not exist.

Work Problem 3 at the side.

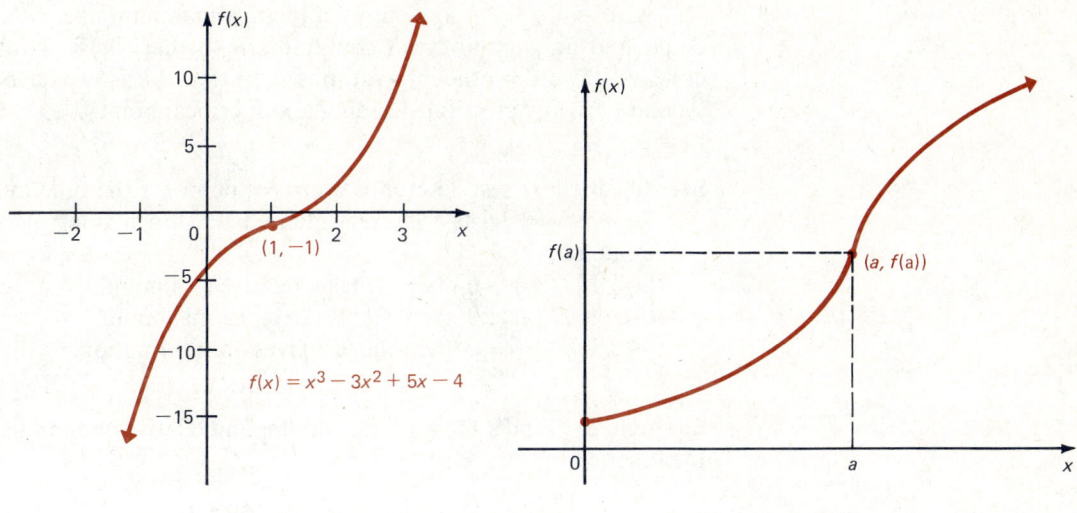

4.3 The Second Derivative Test and Curve Sketching

Figure 4.19

$f(x) = x^3 - 3x^2 + 5x - 4$

Figure 4.20

Example 4 When a new skill is being learned, it is common for learning to be slow at first, then increase, and finally level off at a fairly high degree of skill. A typical function showing such growth of learning is graphed in Figure 4.20. There is a point of inflection at the point $(a, f(a))$. ∎

Work Problem 4 at the side.

One of the main applications of second derivatives is helping us tell whether a given critical value leads to a relative maximum or a relative minimum. To see how this is done, glance at the graph of Figure 4.21. This graph shows a relative maximum at $x = a$. We could locate an open interval containing a so that the function is concave downward on this interval.

4. Find the intervals where the function of Figure 4.20 is concave upward and concave downward.

Answer: concave upward on $(0, a)$ and concave downward on (a, ∞)

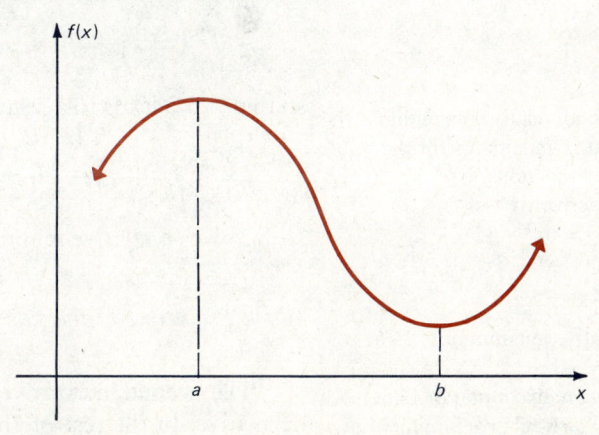

Figure 4.21

Also, since the graph shows a relative minimum at $x = b$, we could find an open interval containing b so that the function is concave upward on the interval. Based on these ideas, we have the second derivative test for deciding about critical points:

Second derivative test Let a be a critical point for the function f. Suppose f' exists for every point in an open interval containing a.
(a) If $f''(a) > 0$, then $f(a)$ is a relative minimum.
(b) If $f''(a) < 0$, then $f(a)$ is a relative maximum.
(c) If $f''(a) = 0$, then the test gives no information.

Example 5 Find all relative maximums and relative minimums for the function

$$f(x) = 4x^3 + 7x^2 - 10x + 8.$$

First, find the critical points. Here $f'(x) = 12x^2 + 14x - 10$. Place this derivative equal to 0.

$$12x^2 + 14x - 10 = 0$$
$$2(6x^2 + 7x - 5) = 0$$
$$2(3x + 5)(2x - 1) = 0$$

$$3x + 5 = 0 \quad \text{or} \quad 2x - 1 = 0$$
$$3x = -5 \quad\quad\quad 2x = 1$$
$$x = -\frac{5}{3} \quad\quad\quad x = \frac{1}{2}$$

Now we can use the second derivative test. The second derivative is $f''(x) = 24x + 14$. Since the critical points are $-5/3$ and $1/2$, we have

$$f''\left(-\frac{5}{3}\right) = 24\left(-\frac{5}{3}\right) + 14 = -40 + 14 = -26 < 0,$$

so that $-5/3$ leads to a relative maximum of $f(-\frac{5}{3}) = \frac{691}{27}$.

$$f''\left(\frac{1}{2}\right) = 24\left(\frac{1}{2}\right) + 14 = 12 + 14 = 26 > 0,$$

so $1/2$ gives a relative minimum of $f(\frac{1}{2}) = \frac{21}{4}$. ∎

Work Problem 5 at the side.

The second derivative gives the rate of change of the first derivative. In the rest of this section, we look at examples of this idea.

---

5. Find all relative maximums and relative minimums for the following functions. Use the second derivative test.
(a) $f(x) = 6x^2 + 12x + 1$
(b) $f(x) = x^3 - 3x^2 - 9x + 8$

Answer:
(a) relative minimum of -5 at $x = -1$
(b) relative maximum of 13 at $x = -1$, relative minimum of -19 at $x = 3$

4.3 The Second Derivative Test and Curve Sketching

Example 6 In the last chapter we saw that the velocity of a particle is given by the derivative of the position function for the particle. That is, if $y = s(t)$ describes the position of the particle at time t, then $v(t) = s'(t)$ gives the velocity at time t.

The rate of change of the velocity is called the **acceleration.** The acceleration is thus the derivative of the velocity; if $a(t)$ represents the acceleration at time t, then

$$a(t) = \frac{d}{dt}v(t) = s''(t).$$

For example, if a particle is moving along a line with

$$s(t) = t^3 + 2t^2 - 7t + 9,$$

then
$$v(t) = s'(t) = 3t^2 + 4t - 7,$$
and
$$a(t) = v'(t) = s''(t) = 6t + 4.$$

The acceleration is positive if $6t + 4 > 0$ (or $t > -2/3$), and negative if $t < -2/3$. ∎

Work Problem 6 at the side.

6. If $s(t) = 4t^3 + 3t^2 - 8t + 1$, find the acceleration of a particle at
(a) $t = 0$;
(b) $t = 4$;
(c) $t = 10$.

Answer:
(a) 6
(b) 102
(c) 246

Example 7 Suppose the revenue from the sale of x units of a certain product is

$$R(x) = 6x^3 + 8x^2 + 9.$$

Then $R'(x) = 18x^2 + 16x$ is called the *velocity of the revenue*, and $R''(x) = 36x + 16$ is called the *acceleration of the revenue*. ∎

Work Problem 7 at the side.

7. In Example 7, find
(a) $R'(3)$;
(b) $R'(10)$;
(c) $R''(3)$;
(d) $R''(10)$.

Answer:
(a) 210
(b) 1960
(c) 124
(d) 376

All the methods of this section, together with previous methods such as point plotting and finding asymptotes, permit us to quickly sketch the graphs of many functions. This process is called **curve sketching.**

Example 8 Sketch the graph of $f(x) = x^3 - 2x^2 - 4x + 3$.

To find all intervals for which the function is increasing or decreasing, we find $f'(x)$.

$$f'(x) = 3x^2 - 4x - 4$$
$$= (3x + 2)(x - 2)$$

Verify that $f'(x) = 0$ when $x = -2/3$ or when $x = 2$. Use the second derivative test to check for maximums or minimums. Here

$$f''(x) = 6x - 4,$$

which is positive when $x = 2$ and negative when $x = -2/3$. Thus, $f(2)$ is a relative minimum and $f(-2/3)$ is a relative maximum.

The first derivative is positive on the intervals $(-\infty, -2/3)$ and $(2, \infty)$, so the function is increasing on these intervals. The first derivative is negative on the interval $(-2/3, 2)$, which shows that the function is decreasing on this interval.

The second derivative, $f''(x) = 6x - 4$, is 0 when $x = 2/3$. Using this fact, verify that the second derivative is negative for all values of x in the interval $(-\infty, 2/3)$ and positive for all values of x in the interval $(2/3, \infty)$. By this result, the graph is concave downward on $(-\infty, 2/3)$ and concave upward on $(2/3, \infty)$. A summary of this information is given in the chart below. Using these facts, and plotting a few points, we get the graph shown in the figure.

Interval	f'	f''	f
$(-\infty, -2/3)$	+		increasing
$(-2/3, 2)$	−		decreasing
$(2, \infty)$	+		increasing
$(-\infty, 2/3)$		−	concave downward
$(2/3, \infty)$		+	concave upward

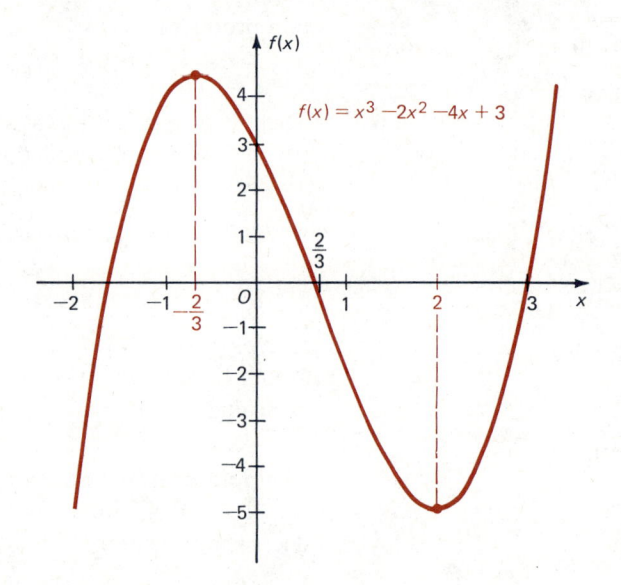

In the example above, we found that the function f was concave upward on the interval $(2/3, \infty)$ and concave downward on $(-\infty, 2/3)$. At the point where $x = 2/3$ there is a point of inflection. As shown above, if $(x_0, f(x_0))$ is a point of inflection, then $f''(x_0) = 0$. However, just because $f''(x_0) = 0$, we have no assurance

4.3 The Second Derivative Test and Curve Sketching

that a point of inflection has been located. For example, if $f(x) = (x-1)^4$, then $f''(x) = 12(x-1)^2$, which is 0 at $x = 1$. However, the graph of $f(x) = (x-1)^4$ is always concave upward and thus has no point of inflection.

The steps involved in sketching the graph of a function are as summarized below.

1. Find f' and f''.
2. Find all values of x such that $f'(x) = 0$. Check these points to identify maximums or minimums.
3. Find any intervals for which $f'(x) < 0$ or $f'(x) > 0$.
4. Find any values of x for which $f''(x) = 0$.
5. Find any intervals for which $f''(x) < 0$ or $f''(x) > 0$.
6. Locate as many points of the graph as needed and sketch the graph, using the facts summarized in the following table.

| $f'(x) > 0$, f is increasing | $f''(x) > 0$, f is concave upward |
| $f'(x) < 0$, f is decreasing | $f''(x) < 0$, f is concave downward |

8. Sketch the graphs.
(a) $f(x) = x^3 - 3x^2$
(b) $f(x) = x + 1/x$

Answer:
(a)

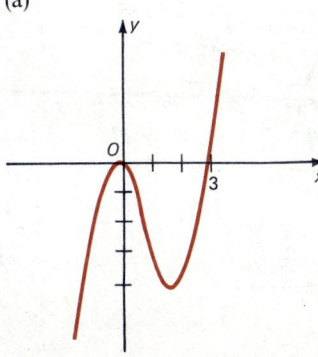

(b)

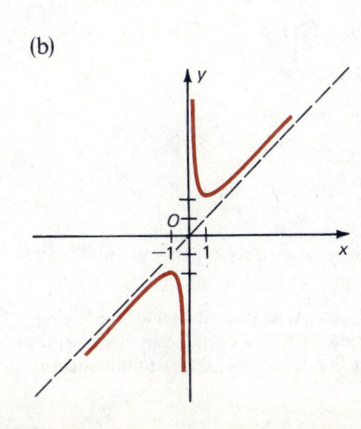

Work Problem 8 at the side.

4.3 EXERCISES

For each of the following functions, find f'''. Then find $f'''(0)$, $f'''(2)$, and $f'''(-3)$. See Examples 1 and 2.

1. $f(x) = x^3 + 4x^2 + 2$
2. $f(x) = 3x^3 - 4x + 5$
3. $f(x) = -x^4 + 2x^3 - x^2$
4. $f(x) = 3x^4 - 5x^3 + 2x^2$
5. $f(x) = 8x^2 + 6x + 5$
6. $f(x) = 3x^2 - 4x + 8$
7. $f(x) = (x-2)^3$
8. $f(x) = (x+4)^3$
9. $f(x) = \dfrac{x+1}{x-1}$
10. $f(x) = \dfrac{2x+1}{x-2}$

For each of the following functions, find $f''''(x)$ and $f^{(4)}(x)$.

11. $f(x) = 2x^4 - 3x^3 + x^2$
12. $f(x) = -x^4 + 2x^2 + 8$
13. $f(x) = 3x^5 - x^4 + 2x^3 - 7x$
14. $f(x) = 4x^5 + 6x^4 - x^2 + 2$
15. $f(x) = \dfrac{x+1}{x}$
16. $f(x) = \dfrac{x-1}{x+2}$
17. $f(x) = \dfrac{x}{2x+1}$
18. $f(x) = \dfrac{3x}{x-2}$

Find any critical points for the following functions and then use the second derivative test to find out if the critical points lead to relative maximums or relative minimums. If $f''(a) = 0$ for a critical point $x = a$, then the second derivative test gives no information. In this case, use the first derivative test instead. Sketch the graph of each function. See Examples 5 and 8.

19. $f(x) = x^2 - 12x + 36$
20. $f(x) = -x^2 - 10x - 25$
21. $f(x) = 12 - 8x + 4x^2$
22. $f(x) = 6 + 4x + x^2$
23. $f(x) = 2x^3 - 4x^2 + 2$
24. $f(x) = 3x^3 - 3x^2 + 1$
25. $f(x) = -2x^3 - 9x^2 + 60x - 8$
26. $f(x) = -2x^3 - 9x^2 + 108x - 10$
27. $f(x) = x^3 - \frac{15}{2}x^2 - 18x - 1$
28. $f(x) = 2x^3 + \frac{7}{2}x^2 - 5x + 3$
29. $f(x) = x^3$
30. $f(x) = (x - 1)^4$
31. $f(x) = x^4 - 8x^2$
32. $f(x) = x^4 - 18x^2 + 5$
33. $f(x) = 2x + \dfrac{8}{x}$
34. $f(x) = x + \dfrac{1}{x}$
35. $f(x) = \dfrac{x^2 + 4}{x}$
36. $f(x) = \dfrac{x^2 + 25}{x}$
37. $f(x) = \dfrac{x}{1 + x}$
38. $f(x) = \dfrac{x - 1}{x + 1}$

Find all intervals where the following functions are concave upward or concave downward. Find the location of any points of inflection. See Example 3.

39. $f(x) = x^2 - 4x + 3$
40. $f(x) = x^2 + 10x - 9$
41. $f(x) = 8 - 6x - x^2$
42. $f(x) = -3 + 8x - x^2$
43. $f(x) = 2x^3 - 3x^2 - 12x + 1$
44. $f(x) = x^3 + 3x^2 - 45x - 3$
45. $f(x) = -x^3 - 12x^2 - 45x + 2$
46. $f(x) = -2x^3 + 9x^2 + 168x - 3$
47. $f(x) = -2/(x + 1)$
48. $f(x) = 3/(x - 5)$

APPLIED PROBLEMS

Product life The figure on the following page shows the *product life cycle graph*, with typical products marked on it.*

49. Where would you place home-video tape recorders on this graph?

---

* John E. Smallwood, "The Product Life Cycle: A Key to Strategic Marketing Planning," pp. 29–35, *MSU Business Topics* (Winter 1973). Reprinted by permission of the publisher, Division of Research, Graduate School of Business Administration, Michigan State University.

4.3 The Second Derivative Test and Curve Sketching

50. Where would you place light bulbs?

51. Which products are closest to the point of inflection on the left of the graph? What does the point of inflection mean here?

52. Which products are closest to the point of inflection on the right of the graph? What does the point of inflection mean here?

Mosquito population

53. The number of mosquitos, $M(x)$, in millions, in a certain area depends on the April rainfall, x, measured in inches, approximately as follows:

$$M(x) = 50 - 32x + 14x^2 - x^3.$$

Find the amounts of rainfall that will produce the maximum and the minimum number of mosquitos.

Cigar production

54. Because of raw material shortages, it is increasingly expensive to produce fine cigars. In fact, the profit in thousands of dollars from producing x hundred thousand cigars is approximated by

$$P(x) = x^3 - 23x^2 + 120x + 60.$$

Find the level of production that will lead to maximum profit and to minimum profit.

Chemical reaction rate

55. An autocatalytic chemical reaction is one in which the product being formed causes the rate of formation to increase. The rate of a certain autocatalytic reaction is given by

$$V(x) = 12x(100 - x),$$

where x is the quantity of the product present and 100 represents the quantity of chemical present initially. For what value of x is the rate of the reaction a maximum?

Drug concentration in blood

56. The percent of concentration of a certain drug in the bloodstream x hours after the drug is administered is given by

$$K(x) = \frac{3x}{x^2 + 4}.$$

For example, after one hour the concentration is given by $K(1) = 3(1)/(1^2 + 4) = \frac{3}{5}\% = .6\% = .006$.
(a) Find the time at which concentration is a maximum.
(b) Find the maximum concentration.

57. The percent of concentration of another drug in the bloodstream x hours after the drug is administered is given by

$$K(x) = \frac{4x}{3x^2 + 27}.$$

(a) Find the time at which the concentration is a maximum.
(b) Find the maximum concentration.

Bacteria population

58. Assume that the number of bacteria, in millions, present in a certain culture at time t is given by

$$R(t) = t^2(t - 18) + 96t + 1000.$$

(a) At what time before 8 hours will the population be maximized?
(b) Find the maximum population.

Velocity-Acceleration

Each of the following functions gives the displacement at time t of a particle moving along a line. Find the velocity and acceleration functions. Then find the velocity and acceleration at $t = 0$ and $t = 4$. Assume that time is measured in seconds and distance is measured in centimeters. Velocity will be in centimeters per second (cm/sec) and acceleration in centimeters per second per second (cm/sec^2). See Example 6.

59. $s(t) = -3t^2 - 6t + 2$ 60. $s(t) = 8t^2 + 4t$

61. $s(t) = 3t^3 - 4t^2 + 8t - 9$ 62. $s(t) = -5t^3 - 8t^2 + 6t - 3$

63. $s(t) = \dfrac{1}{t + 3}$ 64. $s(t) = \dfrac{-2}{3t + 4}$

When an object is dropped straight down, the distance in feet that it travels in t seconds is given by

$$s(t) = -16t^2.$$

Find the velocity of an object

65. after 3 seconds;

66. after 5 seconds;

67. after 8 seconds.

68. Find the acceleration. (The answer here is a constant, the acceleration due to the influence of gravity alone.)

Revenue

The revenue of a small firm, in thousands of dollars, is given by

$$R(x) = 2x^3 - 8x^2 + 4x,$$

where x is the number of units of product that the firm sells. Find the following. See Example 7.

69. Velocity of revenue when $x = 8$

70. When $x = 12$

71. Acceleration of revenue when $x = 4$

72. When $x = 8$

4.4 APPLICATIONS OF MAXIMUMS AND MINIMUMS

To apply our methods of finding maximums and minimums to realistic problems, we first need a mathematical model of the realistic situation. For example, suppose we find that the number of units that can be produced on a production line can be approximated by the mathematical model

$$T(x) = 8x^{1/2} + 2x^{3/2} + 50,$$

where x is the number of employees on the line. Once this mathematical model has been established, we can use it to produce information about the production line. For example, we could use the derivative of $T(x)$ to estimate the marginal production resulting from the addition of an extra worker to the line.

However, in writing the mathematical model itself, we must be aware of restrictions on the values of the variables involved. For example, since x represents the number of employees on a production line, x must certainly be restricted to the positive integers, or perhaps to a few common fractional values (we can conceive of half-time employees, but probably not 1/32-time employees). Certainly, in this example we cannot have $x = \sqrt{2}$ or $x = 3 + \sqrt{7}$.

On the other hand, if we wish to apply the tools of calculus to obtain a maximum value or minimum value for some function, it is necessary that the function be defined and be meaningful at every real-number point in some interval. Because of this, the answer obtained by using a mathematical model of a practical problem might be a number that is not feasible in the setting of the problem.

Usually, the requirement that we use a continuous function instead of a function which can take on only certain selected values is of theoretical interest only. In most cases, calculus gives results which *are* acceptable in a given situation. And if the methods of calculus should be used on a function f and lead to the conclusion that $80\sqrt{2}$ units should be produced in order to get the lowest possible cost, it is usually only necessary to calculate $f(80\sqrt{2})$, and then compare this result to various values of $f(x)$, where x is an acceptable number. The lowest of these values of $f(x)$ then gives minimum cost. In most cases, the result obtained will be very close to the theoretical minimum.

In the rest of this section we give several examples showing applications of calculus to maximum and minimum problems.

Example 1 For a chicken farmer to sell x roosters in a small farming community, the price per rooster must be given by

$$p(x) = 4 - \frac{x}{25}.$$

182 Applications of the Derivative

(a) Find an expression for the total revenue from the sale of x roosters. Let $T(x)$ represent this revenue.

The total revenue is given by the product of the number of roosters sold, x, and the price per rooster, $4 - x/25$. Thus,

$$T(x) = x\left(4 - \frac{x}{25}\right) = 4x - \frac{x^2}{25}.$$

(b) Find the value of x that leads to maximum revenue. Here

$$T'(x) = 4 - \frac{2x}{25}.$$

Place this derivative equal to 0.

$$4 - \frac{2x}{25} = 0$$

$$-\frac{2x}{25} = -4$$

$$2x = 100$$

$$x = 50$$

We must decide if $x = 50$ leads to maximum revenue or minimum revenue. Use the second derivative test. Since $T'(x) = 4 - 2x/25$, we have $T''(x) = -2/25$, which is negative. Thus, $x = 50$ leads to the maximum revenue, as desired.

(c) Find the maximum revenue.

$$T(50) = 4(50) - \frac{50^2}{25} = 200 - 100 = 100 \text{ dollars.}$$

The function $T(x) = 4x - x^2/25$ is valid only for x in the interval $[0, 100]$, since $p(x)$ is meaningful only for these values. Check the endpoints of the interval and verify that both 0 and 100 lead to a *minimum* revenue of 0. ■

Work Problem 1 at the side.

Example 2 When Eastside University charges $600 for a class that includes a tour of the home where Calvin Coolidge was born, it attracts 1000 students. For each $20-decrease in the charge, an additional 100 students will attend the class.

(a) Find an expression for the total revenue if there are x $20-decreases in the price.

The price charged by the university will be

$$\text{price per student} = 600 - 20x,$$

---

1. A total of x earthmovers will be sold if the price, in thousands of dollars, is given by

$$p(x) = 32 - \frac{x}{8}.$$

Find
(a) an expression for the total revenue $R(x)$;
(b) the value of x that leads to maximum revenue;
(c) the maximum revenue.

Answer:
(a) $R(x) = x(32 - x/8)$
    $= 32x - x^2/8$
(b) $x = 128$
(c) $2048 thousand, or $2,048,000

and the number of students taking the class will be

$$\text{number of students} = 1000 + 100x.$$

The total revenue, $R(x)$, is given by the product of the price and the number of students, or

$$R(x) = (600 - 20x)(1000 + 100x)$$
$$= 600{,}000 + 40{,}000x - 2000x^2.$$

(b) Find the value of x that maximizes revenue.
Here $R'(x) = 40{,}000 - 4000x$. Place this derivative equal to 0.

$$40{,}000 - 4000x = 0$$
$$-4000x = -40{,}000$$
$$x = 10$$

Since $R''(x) = -4000$, $x = 10$ leads to maximum revenue.

(c) Find the maximum revenue.

$$R(10) = 600{,}000 + 40{,}000(10) - 2000(10)^2 = 800{,}000$$

dollars. There will be $1000 + 100(10) = 2000$ students in the class; each student will pay $600 - 20(10) = 400$ dollars. ∎

Work Problem 2 at the side.

Example 3 A truck burns fuel at the rate of

$$G(x) = \frac{1}{200}\left(\frac{800}{x} + x\right)$$

gallons per mile when traveling x miles per hour on a straight, level road. If fuel costs \$1 per gallon, find the speed that will produce the minimum total cost for a 1000-mile trip. Find the minimum total cost.

The total cost of the trip, in dollars, is the product of the number of gallons per mile, the number of miles, and the cost per gallon. If we let $C(x)$ represent this cost, then

$$C(x) = \left[\frac{1}{200}\left(\frac{800}{x} + x\right)\right](1000)(1)$$
$$= \frac{1000}{200}\left(\frac{800}{x} + x\right)$$
$$= 5\left(\frac{800}{x} + x\right)$$
$$= \frac{4000}{x} + 5x.$$

2. An investor has built a series of self-storage units near a group of apartment houses. She must now decide on the monthly rental. From past experience, she feels that 200 units will be rented for \$15 per month, with 5 additional rentals for each \$.25-reduction in the rental price. Let x be the number of \$.25-reductions in the price and find
(a) an expression for the number of units rented;
(b) an expression for the price per unit;
(c) an expression for the total revenue;
(d) the value of x leading to maximum revenue;
(e) the maximum revenue.

Answer:
(a) $200 + 5x$
(b) $15 - .25x$
(c) $(200 + 5x)(15 - .25x)$
  $= 3000 + 25x - 1.25x^2$
(d) $x = 10$
(e) \$3125

Find the value of x that will minimize cost. Here

$$C'(x) = \frac{-4000}{x^2} + 5.$$

Place this derivative equal to 0.

$$-\frac{4000}{x^2} + 5 = 0$$

$$-\frac{4000}{x^2} = -5$$

$$4000 = 5x^2$$

$$800 = x^2$$

$$x \approx \pm 28.3 \text{ mph} \quad \text{Take the square root of both sides}$$

We reject -28.3 as a speed, leaving $x = 28.3$ as the only critical value. To see if we get a minimum when $x = 28.3$, find $C''(x)$.

$$C''(x) = \frac{8000}{x^3}$$

Since $C''(28.3) > 0$, we see that 28.3 leads to a minimum. (We need not actually calculate $C''(28.3)$; we just check and see that $C''(28.3) > 0$.)

The minimum total cost is $C(28.3)$, or

$$C(28.3) = \frac{4000}{28.3} + 5(28.3) \approx 141.34 + 141.50 = 282.84$$

dollars. ■

3. A diesel generator burns fuel at the rate of

$$G(x) = \frac{1}{48}\left(\frac{300}{x} + 2x\right)$$

gallons per hour when producing x thousand kilowatt hours of electricity. Suppose that fuel costs $.75 a gallon and find the value of x that leads to minimum total cost if the generator is operated for 32 hours. Find the minimum cost.

Answer: $x = \sqrt{150} \approx 12.2$, minimum cost is $24.50

Work Problem 3 at the side.

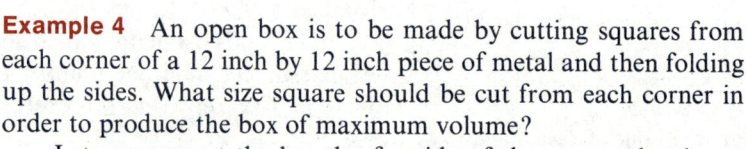

Example 4 An open box is to be made by cutting squares from each corner of a 12 inch by 12 inch piece of metal and then folding up the sides. What size square should be cut from each corner in order to produce the box of maximum volume?

Let x represent the length of a side of the square that is cut from each corner, as shown in Figure 4.22. The width of the box that is to be built will be $12 - 2x$, while the length is also $12 - 2x$. The volume of the box is given by the product of the length, width, and height. In our example, the volume, $V(x)$, depends on x:

$$V(x) = x(12 - 2x)(12 - 2x)$$
$$= 144x - 48x^2 + 4x^3.$$

Find the value of x that will maximize volume.

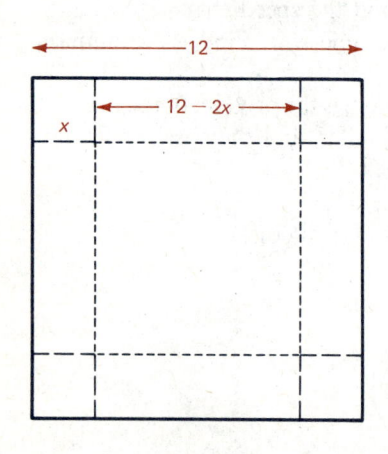

Figure 4.22

Here $V'(x) = 144 - 96x + 12x^2$. Place this derivative equal to 0.

$$12x^2 - 96x + 144 = 0$$
$$12(x^2 - 8x + 12) = 0$$
$$12(x - 2)(x - 6) = 0$$
$$x - 2 = 0 \quad \text{or} \quad x - 6 = 0$$
$$x = 2 \quad \text{or} \quad x = 6$$

Since $V''(x) = -96 + 24x$, we have $V''(2) = -96 + 24(2) = -48$, which is negative. Thus, $x = 2$ leads to maximum volume; the maximum volume is

$$V(2) = 144(2) - 48(2)^2 + 4(2)^3 = 128$$

cubic inches. The function $V(x)$ is defined only on the interval $[0, 6]$. Both the endpoints of this domain, $x = 0$ and $x = 6$, lead to *minimum* volumes of 0. ∎

Work Problem 4 at the side.

4.4 EXERCISES

Exercises 1–8 involve maximizing and minimizing products and sums of numbers. Work all these problems using the steps shown in Exercise 1.

1. We want to find two numbers x and y such that $x + y = 100$ and the product $P = xy$ is as large as possible.
   (a) We know $x + y = 100$. Solve this equation for y.
   (b) Substitute this result for y into $P = xy$.
   (c) Find P'. Solve the equation $P' = 0$.
   (d) What are the two numbers?
   (e) What is the maximum value of P?

2. Find two numbers whose sum is 250 and whose product is as large as possible. What is the maximum product?

3. Find two numbers whose sum is 200 such that the sum of the squares of the two numbers is minimized.

4. Find two numbers whose sum is 30 such that the sum of the squares of the numbers is minimized.

5. Find numbers x and y such that $x + y = 150$ and $x^2 y$ is maximized.

6. Find numbers x and y such that $x + y = 45$ and xy^2 is maximized.

7. Find numbers x and y such that $x - y = 10$ and xy is minimized.

8. Find numbers x and y such that $x - y = 3$ and xy is minimized.

---

4. An open box is to be made by cutting squares from each corner of a 20 cm by 32 cm piece of metal and folding up the sides. Let x represent the length of the side of the square to be cut out. Find
   (a) an expression for the volume of the box, $V(x)$;
   (b) $V'(x)$;
   (c) the value of x that leads to maximum volume; (Hint: the solutions of the equation $V'(x) = 0$ are 4 and 40/3.)
   (d) the maximum volume.

Answer:
(a) $V(x) = 640x - 104x^2 + 4x^3$
(b) $V'(x) = 640 - 208x + 12x^2$
(c) $x = 4$
(d) $V(4) = 1152$ cubic centimeters

Applications of the Derivative

APPLIED PROBLEMS

Revenue

9. If the price charged for a candy bar is $p(x)$ cents, then x thousand candy bars will be sold in a certain city, where

$$p(x) = 100 - \frac{x}{10}.$$

   (a) Find an expression for the total revenue from the sale of x thousand candy bars. (Hint: Find the product of $p(x)$, x, and 1000.) See Examples 1 and 2.
   (b) Find the value of x that leads to maximum revenue.
   (c) Find the maximum revenue.

10. The sale of cassette tapes of "lesser" performers is very sensitive to price. If a tape manufacturer charges $p(x)$ dollars per tape, where

$$p(x) = 6 - \frac{x}{8},$$

then x thousand tapes will be sold.
   (a) Find an expression for the total revenue from the sale of x thousand tapes. (Hint: Find the product of $p(x)$, x, and 1000.)
   (b) Find the value of x that leads to maximum revenue.
   (c) Find the maximum revenue.

Fuel cost

11. A truck burns fuel at the rate of $G(x)$ gallons per mile, where

$$G(x) = \frac{1}{32}\left(\frac{64}{x} + \frac{x}{50}\right),$$

while traveling x miles per hour.
   (a) If fuel costs \$.80 per gallon, find the speed that will produce minimum total cost for a 400-mile trip. See Example 3.
   (b) Find the minimum total cost.

12. A rock and roll band travels from engagement to engagement in a large bus. This bus burns fuel at the rate of $G(x)$ gallons per mile, where

$$G(x) = \frac{1}{50}\left(\frac{200}{x} + \frac{x}{15}\right),$$

while traveling x miles per hour.
   (a) If fuel costs \$1 per gallon, find the speed that will produce minimum total cost for a 250-mile trip.
   (b) Find the minimum total cost.

Area

13. A farmer has 1200 meters of fencing. He wants to enclose a rectangular field bordering a river, with no fencing needed along the river. (See the sketch.) Let x represent the width of the field.

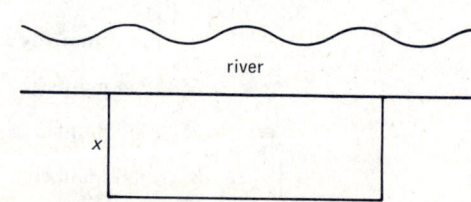

(a) Write an expression for the length of the field.
(b) Find the area of the field (area = length × width).
(c) Find the value of x leading to the maximum area.
(d) Find the maximum area.

14. Find the dimensions of the rectangular field of maximum area that can be made from 200 meters of fencing material. (This fence has four sides.)

Pheasant breeding

15. An ecologist is conducting a research project on breeding pheasants in captivity. She first must construct suitable pens. She wants a rectangular area with two additional fences down the center, as shown in the sketch. Find the maximum area she can enclose with 3600 meters of fencing.

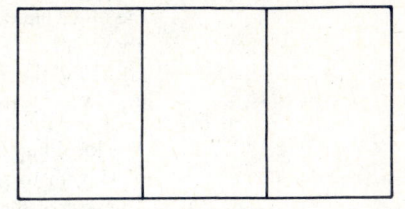

Geometry

16. A rectangular field is to be enclosed with a fence. One side of the field is against an existing fence, so that no fence is needed on that side. If material for the fence costs $2 per foot for the two ends and $4 per foot for the side parallel to the existing fence, find the dimensions of the field of largest area that can be enclosed for $1000.

17. A rectangular field is to be enclosed on all four sides with a fence. Fencing material costs $3 per foot for two opposite sides, and $6 per foot for the other two sides. Find the maximum area that can be enclosed for $2400.

Rent charges

18. The manager of an 80-unit apartment complex is trying to decide on the rent to charge. It is known from experience that at a rent of $200, all the units will be full. However, on the average, one additional unit will remain vacant for each $10 increase in rent.
    (a) Let x represent the number of $10 increases. Find the amount of rent per apartment. See Example 2.
    (b) Find the number of apartments rented.
    (c) Find the total revenue from all rented apartments.
    (d) What value of x leads to maximum revenue?
    (e) What is the maximum revenue?

Timing fruit picking

19. The manager of a peach orchard is trying to decide when to arrange for picking the peaches. If they are picked now, the average yield per tree will be 100 pounds, which can be sold for 40¢ per pound. Past experience shows that the yield per tree will increase about 5 pounds per week, while the price will decrease about 2¢ per pound per week.
    (a) Let x represent the number of weeks that the manager should wait. Find the income per pound.
    (b) Find the number of pounds per tree.
    (c) Find the total revenue from a tree.
    (d) When should the peaches be picked in order to produce maximum revenue?
    (e) What is the maximum revenue?

Applications of the Derivative

Revenue

20. A local group of scouts has been collecting old aluminum beer cans for recycling. The group has already collected 12,000 pounds of cans, for which they could currently receive $4 per hundred pounds. The group can continue to collect cans at the rate of 400 pounds per day. However, a glut in the old-beer-can market has caused the recycling company to announce that it will lower its price, starting immediately, by $.10 per hundred pounds per day. The scouts can make only one trip to the recycling center. Find the best time for that single trip. What total income will be received?

Profit

21. In planning a small restaurant, it is estimated that a profit of $5 per seat will be made if the number of seats is between 60 and 80, inclusive. On the other hand, the profit on each seat will decrease by 5¢ for each seat above 80.
    (a) Find the number of seats that will produce the maximum profit.
    (b) What is the maximum profit?

Revenue

22. The local hamburger fan club is arranging a charter flight to the island of Hawaii to see the southernmost McDonald's in the United States. The cost of the trip is $425 each for 75 passengers, with a refund of $5 per passenger for each passenger in excess of 75.
    (a) Find the number of passengers that will maximize the revenue received from the flight.
    (b) Find the maximum revenue.

Minimizing materials

23. A television manufacturing firm needs to design an open-topped box with a square base. The box must hold 32 cubic inches. Find the dimensions of the box that can be built with the minimum amount of materials.

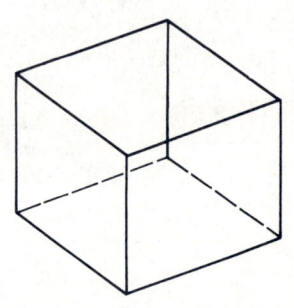

24. A closed box with a square base is to have a volume of 16,000 cubic centimeters. The material for the top and bottom of the box costs $3 per square centimeter, while the material for the sides costs $1.50 per square centimeter. Find the dimensions of the box that will lead to minimum total cost. What is the minimum total cost?

25. A company wishes to manufacture a box with a volume of 36 cubic feet which is open on top and which is twice as long as it is wide. Find the dimensions of the box produced from the minimum amount of material.

4.4 Applications of Maximums and Minimums

Paper use
26. A mathematics book is to contain 36 square inches of printed matter per page, with margins of 1 inch along the sides, and $1\frac{1}{2}$ inches along the top and bottom. Find the dimensions of the page that will lead to the minimum amount of paper being used for a page.

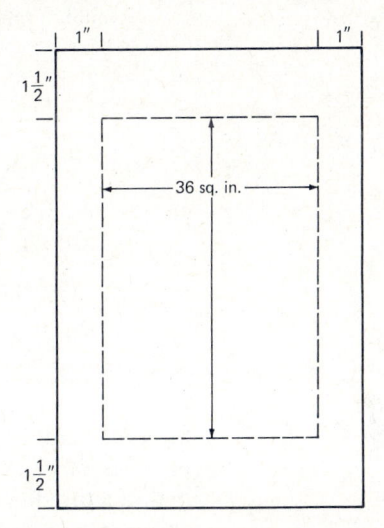

Cost
27. In Example 3, we found the speed in miles per hour that minimized cost when we considered only the cost of the fuel. Rework the problem taking into account the driver's salary of $8 per hour. (Hint: If the trip is 1000 miles at x miles per hour, the driver will be paid for $1000/x$ hours.)

Revenue
28. Decide what you would do if your assistant brought you the following contract for your signature:
> Your firm offers to deliver 300 tables to a dealer, at $90 per table, and to reduce the price per table on the entire order by 25¢ for each additional table over 300.

Find the dollar total involved in the largest possible transaction between the manufacturer and the dealer; find the smallest possible dollar amount.

Cost
29. A company wishes to run a utility cable from point A on the shore (see the figure) to an installation at point B on the island. The island is 6 miles from the shore. It costs $400 per mile to run the cable on land and $500 per mile underwater. Assume that the cable starts at A and runs along the shoreline, then angles and runs underwater to the island. Find the point at which the line should begin to angle in order to yield the minimum total cost. (Hint: The length of the line underwater is $\sqrt{x^2 + 36}$.)

Time 30. A hunter is at a point on a river bank. He wants to get to his cabin, located three miles north and eight miles west. (See the figure.) He can travel 5 miles per hour on the river but only 2 miles per hour on this very rocky land. How far up river should he go in order to reach the cabin in minimum time? (Hint: Distance = rate × time.)

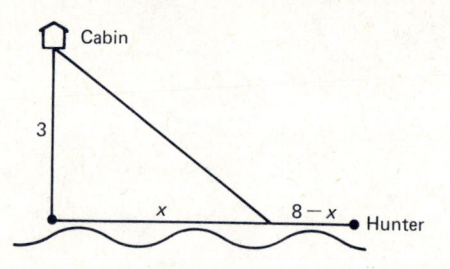

4.5 ECONOMIC LOT SIZE (OPTIONAL)

Suppose that a company manufactures a certain constant number of units of a product per year. Assume that the product can be manufactured in a number of batches of equal size during the year. The company could manufacture the item only once per year to minimize setup costs, but they would then have much higher warehouse costs. On the other hand, they might make many small batches, but this would increase setup costs. We can use calculus to find the number of batches per year that should be manufactured in order to minimize total cost.

Figure 4.23 shows several of the possibilities for a product having an annual demand of 12,000 units. The top graph shows the results if only one batch of the product is made annually: an average of 6000 items will be held in a warehouse. If four batches (of 3000 each) are made at equal time intervals during a year, the average number of units in the warehouse falls to only 1500. If twelve batches are made, an average of 500 items will be in the warehouse.

In this section, we use the following variables:

x = number of batches to be manufactured annually
k = cost of storing one unit of the product for one year
f = fixed setup cost to manufacture the product
g = variable cost of manufacturing a single unit of the product
M = total number of units produced annually.

The company has two types of costs associated with the production of its product: a cost associated with manufacturing the item, and a cost associated with storing the finished product.

Let us first consider manufacturing costs. During a year the company will produce x batches of the product. The company will

4.5 Economic Lot Size (Optional)

[Figure 4.23 — three graphs showing number of units per batch vs. months: top graph declining from 12,000 to 0 over 12 months with average 6000; middle graph sawtooth from 3000 with average 1500 over 3-month cycles; bottom graph sawtooth from 1000 with average 500 over 1-month cycles.]

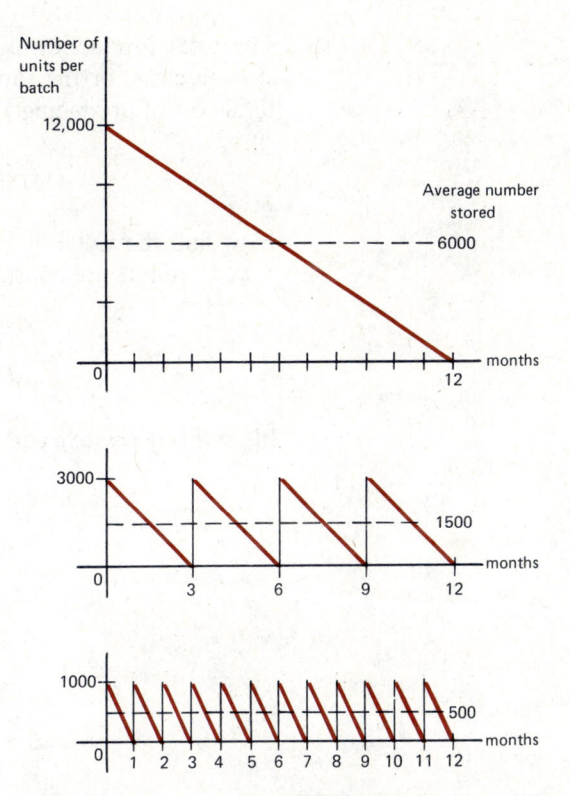

Figure 4.23

thus produce M/x units of the product per batch. Each batch has a fixed cost f and a variable cost g per unit, so that the manufacturing cost per batch is

$$f + g\left(\frac{M}{x}\right).$$

Since there are x batches per year, the total annual manufacturing cost is

$$\left[f + g\left(\frac{M}{x}\right)\right]x. \qquad (1)$$

Now we must find the storage cost. Since each batch consists of M/x units, and since we have assumed demand is constant, it is common to assume an average inventory of

$$\frac{1}{2}\left(\frac{M}{x}\right) = \frac{M}{2x}$$

units per year. It costs k to store one unit of the product for a year, making a total storage cost of

$$k\left(\frac{M}{2x}\right) = \frac{kM}{2x}. \qquad (2)$$

Applications of the Derivative

The total production cost is the sum of the manufacturing and storage costs, or the sum of expressions (1) and (2). If $T(x)$ is the total cost of producing x batches, we have

$$T(x) = \left[f + g\left(\frac{M}{x}\right) \right] x + \frac{kM}{2x}.$$

Now find the value of x that will minimize $T(x)$. Remember that f, g, k, and M are constants.
Here

$$T'(x) = f - \frac{kM}{2} x^{-2}.$$

Place this derivative equal to 0.

$$f - \frac{kM}{2} x^{-2} = 0$$

$$f = \frac{kM}{2x^2}$$

$$2fx^2 = kM$$

$$x^2 = \frac{kM}{2f}$$

$$x = \sqrt{\frac{kM}{2f}} \qquad (3)$$

Thus, $\sqrt{kM/2f}$ is the annual number of batches which will give minimum total production cost.

Example 1 A paint company has a steady annual demand for 24,500 cans of automobile primer. The cost accountant for the company says that it costs $2 to hold one can of paint for one year and $500 to set up the plant for the production of the primer. Find the number of batches of primer that should be produced for the minimum total production cost.

Use equation (3) above.

$$x = \sqrt{\frac{kM}{2f}}$$

$$x = \sqrt{\frac{2(24{,}500)}{2(500)}} \qquad \text{Let } k = 2, M = 24{,}500, f = 500$$

$$x = \sqrt{49} = 7$$

Seven batches of primer per year will lead to minimum production costs. ∎

Work Problem 1 at the side.

---

1. A manufacturer of business forms has an annual demand for 30,720 units of form letters to people delinquent in their payments of installment debt. It costs $5 per year to store one unit of the letters and $1200 to set up the machines to produce them. Find the number of batches that should be made annually to minimize total cost.

Answer: 8 batches

4.5 Economic Lot Size (Optional)

The analysis above also applies to reordering an item that is used at a constant rate throughout the year, as the next example shows.

Example 2 A large pharmacy has an annual need for 200 units of a certain antibiotic. It costs $10 to store one unit for one year. The fixed cost of placing an order (clerical time, mailing, and so on) amounts to $40. Find the number of orders that should be placed annually.

Here $k = 10$, $M = 200$, and $f = 40$. We have

$$x = \sqrt{\frac{10(200)}{2(40)}} = \sqrt{25} = 5.$$

Order the drug 5 times a year. ∎

Work Problem 2 at the side.

2. An office uses 576 cases of copy-machine paper during the year. It costs $3 per year to store one case. Each reorder costs $24. Find the number of orders that should be placed annually.

Answer: 6

4.5 APPLIED PROBLEMS

Production

1. Find the approximate number of batches that should be produced annually if 100,000 units are to be manufactured. It costs $1 to store a unit for one year and it costs $500 to set up the factory to produce each batch.

2. How many units per batch will be manufactured in Exercise 1?

Ordering

3. A market has a steady annual demand for 16,800 cases of sugar. It costs $3 to store one case for one year. The market pays $7 for each order that is placed. Find the number of orders for sugar that should be placed each year.

4. Find the number of cases per order in Exercise 3.

Printing

5. The publisher of a best-selling book has an annual demand for 100,000 copies. It costs 50¢ to store one copy for one year. Setup of the press for a new printing costs $1000. Find the number of batches that should be printed annually.

Ordering

6. A restaurant has an annual demand for 810 bottles of a California wine. It costs $1 to store one bottle for one year, and it costs $5 to place a reorder. Find the number of orders that should be placed annually.

7. Use the second derivative to show that the value of x obtained in the text [equation (3)] really leads to the minimum cost.

8. Why do you think that the variable cost g does not appear in the answer for x [equation (3)]?

194 Applications of the Derivative

4.6 IMPLICIT DIFFERENTIATION (OPTIONAL)*

In almost every example so far, the functions that we have used have been written in the form $y = f(x)$, in which y is expressed as a function of x. Where $y = f(x)$, the variable y is said to be expressed as an **explicit function** of x. For example,

$$y = 3x - 2,$$
$$y = x^2 + x + 6,$$

and

$$y = -x^3 + 2$$

are all explicit functions of x. We can express $4xy - 3x = 6$ as an explicit function of x by solving for y. This gives

$$4xy - 3x = 6$$
$$4xy = 3x + 6$$
$$y = \frac{3x + 6}{4x}.$$

Work Problem 1 at the side.

On the other hand, some functions are expressed by equations which cannot be solved readily for y, and some by equations which cannot be solved for y at all. For example, it would be possible (but tedious) to use the quadratic formula to solve for y in the equation $y^2 + 2yx + 4x^2 = 0$. On the other hand, it would not be possible to solve for y in the equation $y^5 + 8y^3x + 6y^2x^2 + 2yx^3 + 6 = 0$. Functions such as these last two are said to be expressed as **implicit functions**.

Even though a function is written implicitly, it is still often possible to calculate dy/dx, the derivative of the function. In doing so, we assume that there exists some function f (which we may or may not be able to find) such that $y = f(x)$ and dy/dx exists. (We will use dy/dx here rather than $f'(x)$ for convenience.)

For example, to find dy/dx for the function $3xy + 4y^2 = 10$, first take the derivative of both sides of the equation. The first term $3xy$ can be written as the product $(3x)(y)$. To use the product rule, we need

$$\frac{d(3x)}{dx} = 3 \quad \text{and} \quad \frac{d(y)}{dx} = \frac{dy}{dx}.$$

---

1. Write as explicit functions of x.
   (a) $2y + 4x = 7$
   (b) $xy = 10$
   (c) $5x^2y - 3y = 5$

Answer:
   (a) $y = \dfrac{-4x + 7}{2}.$
   (b) $y = \dfrac{10}{x}$
   (c) $y = \dfrac{5}{5x^2 - 3}$

---

* The material in this section is used only in Section 5.4 to obtain the derivative of e^x.

4.6 Implicit Differentiation (Optional)

By the product rule, the derivative of the first term is

$$3x \cdot \frac{dy}{dx} + y \cdot 3 = 3x\frac{dy}{dx} + 3y.$$

To differentiate the second term, $4y^2$, we use the chain rule, since y is some function of x.

$$\frac{d(4y^2)}{dx} = 4 \cdot 2y^1 \cdot \frac{dy}{dx} = 8y\frac{dy}{dx}$$

On the other side of the equation, the derivative of 10 is 0.
Taking the indicated derivatives term by term, we have

$$3x\frac{dy}{dx} + 3y + 8y\frac{dy}{dx} = 0.$$

To complete the process, solve this result for dy/dx.

$$(3x + 8y)\frac{dy}{dx} = -3y$$

$$\frac{dy}{dx} = \frac{-3y}{3x + 8y}$$

Finding the derivative by this procedure is called **implicit differentiation**.

Example 1 Find dy/dx for $x^2 + 2xy^2 + 3x^2y = 0$.
We have

$$\frac{d(x^2 + 2xy^2 + 3x^2y)}{dx} = \frac{d(0)}{dx}.$$

Now we treat $2xy^2$ and $3x^2y$ as products and differentiate term by term to get

$$2x + 2x\left(2y\frac{dy}{dx}\right) + 2y^2 + 3x^2\left(\frac{dy}{dx}\right) + 6xy = 0$$

or

$$2x + 4xy\left(\frac{dy}{dx}\right) + 2y^2 + 3x^2\left(\frac{dy}{dx}\right) + 6xy = 0,$$

from which

$$(4xy + 3x^2)\frac{dy}{dx} = -2x - 2y^2 - 6xy,$$

or, finally,

$$\frac{dy}{dx} = \frac{-2x - 2y^2 - 6xy}{4xy + 3x^2}. \blacksquare$$

Work Problem 2 at the side.

2. Find $\frac{dy}{dx}$ for the following.

(a) $5xy + 6y = 12$
(b) $9x^2y^2 + 2xy = 1$

Answer:

(a) $\frac{dy}{dx} = \frac{-5y}{5x + 6}$

(b) $\frac{dy}{dx} = \frac{-9xy^2 - y}{9x^2y + x}$

Applications of the Derivative

Example 2 Find dy/dx for $x + \sqrt{xy} = y^2$.

We have

$$\frac{d(x + \sqrt{xy})}{dx} = \frac{d(y^2)}{dx}.$$

Since $\sqrt{xy} = \sqrt{x} \cdot \sqrt{y} = x^{1/2} \cdot y^{1/2}$, we have, upon taking derivatives,

$$1 + x^{1/2}\left(\frac{1}{2}y^{-1/2} \cdot \frac{dy}{dx}\right) + y^{1/2}\left(\frac{1}{2}x^{-1/2}\right) = 2y\frac{dy}{dx}$$

$$1 + \frac{x^{1/2}}{2y^{1/2}} \cdot \frac{dy}{dx} + \frac{y^{1/2}}{2x^{1/2}} = 2y\frac{dy}{dx}.$$

Multiply both sides by $2x^{1/2} \cdot y^{1/2}$.

$$2x^{1/2} \cdot y^{1/2} + x\frac{dy}{dx} + y = 4x^{1/2} \cdot y^{3/2} \cdot \frac{dy}{dx}.$$

Upon combining terms we have

$$2x^{1/2} \cdot y^{1/2} + y = (4x^{1/2} \cdot y^{3/2} - x)\frac{dy}{dx},$$

or

$$\frac{dy}{dx} = \frac{2x^{1/2} \cdot y^{1/2} + y}{4x^{1/2} \cdot y^{3/2} - x}. \quad \blacksquare$$

Work Problem 3 at the side.

Example 3 Find an equation of the tangent line to the curve $x^2 + 4y^2 = 17$ at the point (1, 2).

To find the equation of the tangent line, we must first find the slope of the tangent to the curve at the point (1, 2). This can be done by finding dy/dx. Using implicit differentiation, we have

$$2x + 8y\frac{dy}{dx} = 0$$

$$\frac{dy}{dx} = \frac{-2x}{8y}$$

$$= \frac{-x}{4y}.$$

The slope of the tangent line to the curve at the point (1, 2) is thus

$$m = \frac{-x}{4y} = \frac{-1}{4(2)} = \frac{-1}{8}.$$

---

3. Find $\dfrac{dy}{dx}$ for the following.

(a) $\sqrt{x^2y} + y^2 = 1$
(b) $y/\sqrt{x} = x + y^4$
(c) $\sqrt{y} + \sqrt{x} = 4$

Answer:

(a) $\dfrac{dy}{dx} = \dfrac{-2y}{x + 4y^{3/2}}$

(b) $\dfrac{dy}{dx} = \dfrac{2x^{3/2} + y}{2x - 8x^{3/2}y^3}$

(c) $\dfrac{dy}{dx} = \dfrac{-y^{1/2}}{x^{1/2}}$

4.6 Implicit Differentiation (Optional)

The equation of the tangent line is then found by using the point-slope form of the equation of a line.

$$y - y_1 = m(x - x_1)$$
$$y - 2 = \frac{-1}{8}(x - 1)$$
$$8y - 16 = -x + 1$$
$$8y + x = 17 \quad \blacksquare$$

Work Problem 4 at the side.

4. Find the equation of the tangent line to the given curve at the given point.
(a) $y^2 + x^2 = 25$ at $(-3, 4)$
(b) $x^2 y^3 = 27$ at $(1, 3)$
(c) $y^3 - x = 121$ at $(4, 5)$

Answer:
(a) $4y = 3x + 25$
(b) $2x + y = 5$
(c) $75y = x + 371$

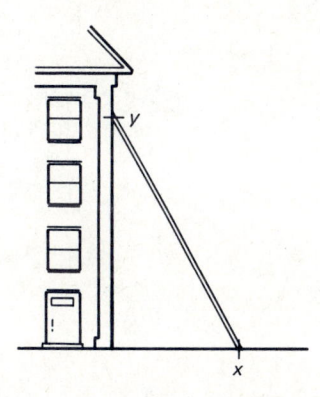

Figure 4.24

Example 4 A biologist has placed a 50-foot ladder against a large building. The base of the ladder is resting on an oil spill, and slips (to the right in Figure 4.24) at the rate of 3 feet per minute. Find the rate of change of the height of the top of the ladder above the ground at the instant when the base of the ladder is 30 feet from the base of the building.

Let y be the height of the top of the ladder above the ground, and let x be the distance of the base of the ladder from the base of the building. By the Pythagorean theorem we have

$$50^2 = x^2 + y^2. \tag{1}$$

Both x and y are functions of time, t, measured from the moment that the ladder starts slipping. If we now take the derivative of both sides of equation (1) with respect to time, we get

$$\frac{d(50^2)}{dt} = \frac{d(x^2 + y^2)}{dt},$$

or

$$0 = 2x\frac{dx}{dt} + 2y\frac{dy}{dt}. \tag{2}$$

In the statement of the problem, we are told that the base is sliding at the rate of 3 feet per minute, so that

$$\frac{dx}{dt} = 3.$$

We are told that the base of the ladder is 30 feet from the base of the building. This fact can be used to find y. We know that $50^2 = x^2 + y^2$. Thus,

$$50^2 = 30^2 + y^2$$
$$2500 = 900 + y^2$$
$$1600 = y^2$$
$$y = 40 \text{ feet.}$$

We now know that $y = 40$, $x = 30$, and $dx/dt = 3$. Substituting these values into equation (2), we have

$$0 = 2x\frac{dx}{dt} + 2y\frac{dy}{dt}$$

$$0 = 2(30)(3) + 2(40)\frac{dy}{dt}$$

$$0 = 180 + 80\frac{dy}{dt}$$

$$80\frac{dy}{dt} = -180$$

$$\frac{dy}{dt} = \frac{-180}{80} = \frac{-9}{4} = -2.25 \text{ feet per minute.}$$

Thus, at the instant when the base of the ladder is 30 feet from the base of the building, the top of the ladder is sliding down the building at the rate of 2.25 feet per minute. ■

Work Problem 5 at the side.

5. In Example 4, find the rate at which the ladder is moving down at the instant that the top of the ladder is
(a) 30 feet above the ground;
(b) 20 feet above the ground.

Answer:
(a) 4 feet per minute
(b) 6.9 feet per minute

4.6 EXERCISES

Find dy/dx by implicit differentiation for the following. See Examples 1 and 2.

1. $4x^2 + 3y^2 = 6$
2. $2x^2 - 5y^2 = 4$
3. $2xy + y^2 = 8$
4. $-3xy - 4y^2 = 2$
5. $y^2 = 4x + 1$
6. $y^2 - 2x = 6$
7. $6xy^2 - 8y + 1 = 0$
8. $-4y^2x^2 - 3x + 2 = 0$
9. $x^2 + 2xy = 6$
10. $?x^2 - 3xy = 10$
11. $6x^2 + 8xy + y^2 = 6$
12. $8x^2 = 6y^2 + 2xy$

13. $x^3 = y^2 + 4$
14. $x^3 - 6y^2 = 10$
15. $x^2y = 4$
16. $-2x^2y = 3$
17. $x^2y^2 = 6$
18. $5 - x^2y^2 = 0$
19. $x^2y + y^3 = 4$
20. $2xy^2 + 2y^3 + 5x = 0$
21. $\sqrt{x} + \sqrt{y} = 4$
22. $2\sqrt{x} - \sqrt{y} = 1$
23. $\sqrt{xy} + y = 1$
24. $\sqrt{2xy} - 1 = 3y^2$

Find the equation of the tangent line at the given point on each of the following curves. See Example 3.

25. $x^2 + y^2 = 25;\ (-3, 4)$
26. $x^2 + y^2 = 100;\ (8, -6)$
27. $x^2y^2 = 1;\ (-1, 1)$
28. $x^2y^3 = 8;\ (-1, 2)$
29. $x^2 + \sqrt{y} = 7;\ (2, 9)$
30. $2y^2 - \sqrt{x} = 4;\ (16, 2)$
31. $y + \dfrac{\sqrt{x}}{y} = 3;\ (4, 2)$
32. $x + \dfrac{\sqrt{y}}{3x} = 2;\ (1, 9)$

APPLIED PROBLEMS

For Exercises 33–40, see Example 4.

Rate of change of distance

33. A 25-foot ladder is placed against a building. The base of the ladder is slipping away from the building at the rate of 4 feet per minute. Find the rate at which the top of the ladder is sliding down the building at the instant when the bottom of the ladder is 7 feet from the base of the building.

34. One car leaves a given point and travels north at 30 miles per hour. Another car leaves the same point at the same time and travels west at 40 miles per hour. At what rate is the distance between the two cars changing at the instant when the cars have traveled 2 hours?

Rate of change of area

35. A rock is thrown into a still pond. The circular ripples move outward from the point of impact of the rock so that the radius of the area of ripples increases at the rate of 2 feet per minute. Find the rate at which the area is changing at the instant the radius is 3/2 feet.

Rate of change of volume

36. A spherical snowball is placed in the sun. The sun melts the snowball so that its radius decreases 1/4 inch per hour. Find the rate of change of the volume with respect to time at the instant the radius is 4 inches.

Use the ideas of similar triangles from geometry to work the following problems.

37. A sand storage tank used by the highway department for winter storms is leaking. As the sand leaks out, it forms a conical pile. The radius of the base of the pile increases at the rate of 1 inch per minute. The height of the pile is always twice the radius of the base. Find the rate at which the volume of the pile is increasing at the instant the radius of the base is 5 inches. (Hint: The volume of a cone is given by $V = \tfrac{1}{3}\pi r^2 h$.)

Rate of change of length

38. A man 6 feet tall is walking away from a lamp post at the rate of 50 feet per minute. When the man is 8 feet from the lamp post, his shadow is 10 feet long. Find the rate at which the length of the shadow is increasing when he is 25 feet from the lamp post.

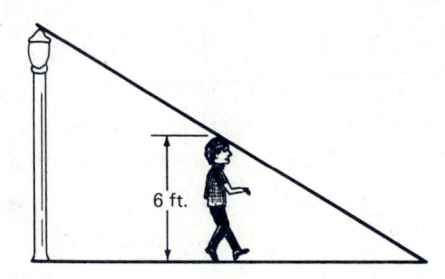

39. A trough has a triangular cross section. The trough is 6 feet across the top, 6 feet deep, and 16 feet long. Water is being pumped into the trough at the rate of 4 cubic feet per minute. Find the rate at which the height of water is increasing at the instant that the height is 4 feet.

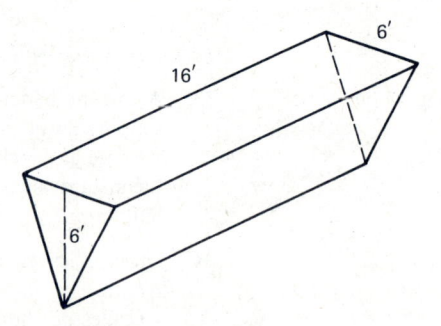

40. A pulley is on the edge of a dock, 8 feet above the water level. A rope is being used to pull in a boat. The rope is attached to the boat at water level. The rope is being pulled in at the rate of 1 foot per second. Find the rate at which the boat is approaching the dock at the instant the boat is 8 feet from the dock.

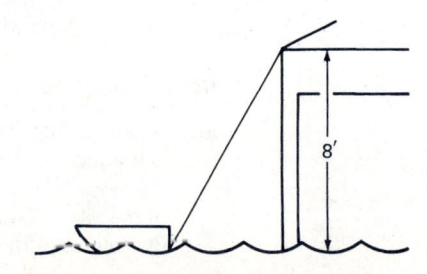

4.7 NEWTON'S METHOD (OPTIONAL)

Most of the equations we have seen so far in this book have been linear (such as $3x + 8y = 15$) or quadratic (such as $x^2 - 4x - 6 = 0$). We know how to solve these equations. More complicated equations cannot be solved at all or can only be solved in more complicated ways.

Derivatives can be used to find approximate solutions of many different equations. Suppose we want to solve the equation $f(x) = 0$, where f is a continuous function defined on some closed interval $[a, b]$. Assume that $f(a)$ and $f(b)$ are of opposite signs (one positive and one negative). This forces the existence of a value c in the interval (a, b) such that $f(c) = 0$. The number c is a solution of the equation $f(x) = 0$.

We would like to find an approximate value for c. To do so, we first make a guess for c. Let c_1 be our initial guess. (See Figure 4.25.) Then locate the point $(c_1, f(c_1))$ on the graph of $y = f(x)$ and draw the tangent line at this point. This tangent line will cut the x-axis at a point c_2. The number c_2 is often a better approximation to c than was c_1. It can be shown that

$$c_2 = c_1 - \frac{f(c_1)}{f'(c_1)},$$

provided $f'(c_1) \neq 0$.

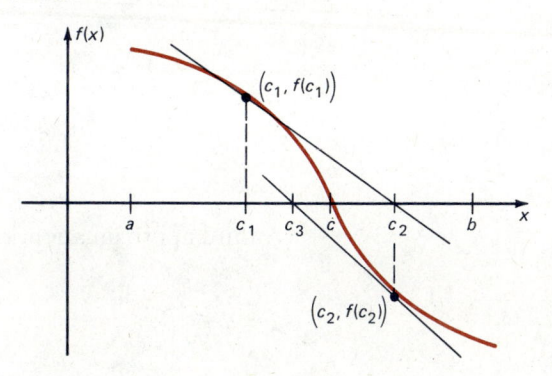

Figure 4.25

To try to improve the approximation to c, locate the point $(c_2, f(c_2))$ on the graph and draw the tangent line at this point. This tangent line will cut the x-axis at the point c_3, which is usually a better approximation to c than was c_2. (See Figure 4.25.) Also, c_3 is given by

$$c_3 = c_2 - \frac{f(c_2)}{f'(c_2)},$$

provided $f'(c_2) \neq 0$.

202   Applications of the Derivative

We can often improve the approximation to c by repeating this process as many times as desired. In general, if we have a value c_n which is an approximation to c, we can often find a better approximation c_{n+1} by the formula

$$c_{n+1} = c_n - \frac{f(c_n)}{f'(c_n)},$$

provided $f'(c_n) \neq 0$.

This process of obtaining a rough approximation for c, and replacing it successively by approximations that are often better, is called **Newton's method**, named after Sir Isaac Newton, the co-discoverer of calculus.

Example 1   Find a solution for the equation

$$3x^3 - x^2 + 5x - 12 = 0$$

in the interval $[1, 2]$. Approximate the solution to the nearest hundredth.

Let $f(x) = 3x^3 - x^2 + 5x - 12$, so that $f'(x) = 9x^2 - 2x + 5$. Check that $f(1) < 0$ with $f(2) > 0$. Since $f(1)$ and $f(2)$ have opposite signs, there is a solution for the equation in the interval $(1, 2)$. As an initial guess, let $c_1 = 1$. A better guess, c_2, can be found as follows.

$$c_2 = c_1 - \frac{f(c_1)}{f'(c_1)}$$

$$= 1 - \frac{-5}{12}$$

$$= \frac{17}{12}$$

$$c_2 = 1.42$$

A third approximation, c_3, can now be found.

$$c_3 = c_2 - \frac{f(c_2)}{f'(c_2)}$$

$$= 1.42 - \frac{1.67}{20.3}$$

$$c_3 = 1.34$$

Work Problem 1 at the side.

In Problem 1 we see that $c_4 = c_5 = 1.33$. Then we can say that to two decimal places of accuracy, $x = 1.33$ is a solution of $3x^3 - x^2 + 5x - 12 = 0$. (The exact solution is $4/3$.) ∎

Work Problem 2 at the side.

1. Find c_4 and c_5.

Answer:

$c_4 = 1.34 - \dfrac{.122}{18.5} = 1.33$

$c_5 = 1.33 - \dfrac{-.06}{18.3} = 1.33$

2. Find a solution to the nearest hundredth for the equation

$$3x^3 - 11x^2 + 5x + 8 = 0$$

in the interval $[2, 3]$.

Answer: 2.67

In Example 1 we had to go through five steps in order to get the degree of accuracy that we needed. The solutions of similar polynomial equations can usually be found in about that many steps, while other types of equations might require more steps.

In any case, if a solution can be found by Newton's method, it can usually be found by a computer in a few seconds of running time.* Some functions, such as the one of Figure 4.26, have solutions which cannot be found by Newton's method. Because of the symmetry of the graph of Figure 4.26, c_1 will always equal c_3 and c_5, while c_2 will equal c_4 and c_6. Newton's method would not give a solution here. However, these cases are rare in practice.

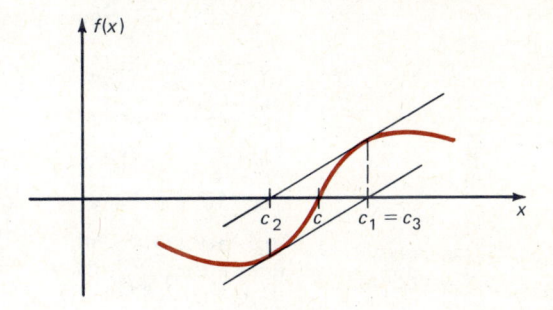

Figure 4.26

Newton's method can also be used to approximate roots, as shown by the next example.

Example 2 Approximate $\sqrt{12}$ to the nearest thousandth.

First, note that $\sqrt{12}$ is a solution of the equation $x^2 - 12 = 0$. Therefore, we let $f(x) = x^2 - 12$, so that $f'(x) = 2x$. We know that $3 < \sqrt{12} < 4$, so let us use $c_1 = 3$ as the first approximation to $\sqrt{12}$. A better approximation is given by c_2:

$$c_2 = 3 - \frac{-3}{6} = 3.5.$$

Now we can find c_3. $\quad c_3 = 3.5 - \dfrac{.25}{7} = 3.464$

To find c_4, we have $\quad c_4 = 3.464 - \dfrac{-.0007}{6.928} = 3.464.$

Since $c_3 = c_4 = 3.464$, we can say that, to the nearest thousandth, $\sqrt{12} = 3.464.$ ∎

Work Problem 3 at the side.

3. Approximate the following to the nearest thousandth.
(a) $\sqrt{20}$
(b) $\sqrt[3]{7}$

Answer:
(a) 4.472
(b) 1.913

* See Margaret L. Lial, *Study Guide with Computer Problems* (Glenview, Ill.: Scott, Foresman, 1979) pp. 68–73.

4.7 EXERCISES

Use Newton's method to find a solution for the given equations in the given interval. Find all solutions to the nearest hundredth. Use a calculator if available. See Example 1.

1. $x^2 - 2x - 2 = 0$, $[2, 3]$
2. $x^2 - 6x + 4 = 0$, $[5, 6]$
3. $3x^2 + 2x - 1 = 0$, $[0, 1]$
4. $6x^2 - 13x + 2 = 0$, $[0, 1]$
5. $3x^3 - 5x^2 + 8x - 4 = 0$, $[0, 1]$
6. $5x^3 - 2x^2 - 2x - 7 = 0$, $[1, 2]$
7. $3x^3 - 4x^2 - 4x - 7 = 0$, $[2, 3]$
8. $3x^3 - 14x^2 + 17x - 22 = 0$, $[3, 4]$
9. $5x^3 + 6x^2 + 4x + 33 = 0$, $[-3, -2]$
10. $7x^3 + 2x^2 - 30x - 27 = 0$, $[2, 3]$
11. $x^3 - 3x^2 - 1 = 0$, $[3, 4]$
12. $x^3 + 4x^2 + 6 = 0$, $[-4, -5]$
13. $6x^3 - 13x^2 - 36x + 28 = 0$, $[-2, -1]$, $[0, 1]$, $[3, 4]$
14. $6x^3 + x^2 - 46x + 15 = 0$, $[0, 1]$, $[2, 3]$
15. $3x^4 - 5x^3 - 16x^2 - 19x - 11 = 0$, $[-2, -1]$, $[3, 4]$
16. $2x^4 + 7x^3 + 6x^2 + 7x - 6 = 0$, $[0, 1]$, $[-3, -2]$

Use Newton's method to find the following roots to the nearest thousandth. See Example 2.

17. $\sqrt{2}$
18. $\sqrt{3}$
19. $\sqrt{11}$
20. $\sqrt{15}$
21. $\sqrt{250}$
22. $\sqrt{300}$
23. $\sqrt[3]{9}$
24. $\sqrt[3]{15}$
25. $\sqrt[3]{100}$
26. $\sqrt[3]{121}$

Use Newton's method to find the critical points for the following functions; approximate them only to the nearest tenth. Decide whether each critical point leads to a relative maximum or a relative minimum.

27. $f(x) = x^3 - 3x^2 - 18x + 4$
28. $f(x) = x^3 + 9x^2 - 6x + 4$
29. $f(x) = x^4 - 3x^3 + 6x - 1$
30. $f(x) = x^4 + 2x^3 - 5x + 2$

APPLIED PROBLEMS

Interest Federal government regulations require that people loaning money to consumers disclose the true annual interest of the loan. The formulas for calculating this interest rate are very complex. For example, suppose P dollars is loaned, with the money to be repaid in n monthly payments of M dollars each. Then the true annual interest rate is found by solving the equation

$$\frac{1-(1+i)^{-n}}{i} - \frac{P}{M} = 0$$

for i, the monthly interest rate, and then multiplying i by 12 to get the true annual rate. This equation can best be solved by Newton's method, as explained in the next few exercises. (You will need a calculator with an x^y key.)

31. Let $f(i) = \dfrac{1-(1+i)^{-n}}{i} - \dfrac{P}{M}$. Find $f'(i)$.

32. Form the quotient $f(i)/f'(i)$.

33. Suppose that $P = \$4000$, $n = 24$, and $M = \$197$. Let the initial guess for i be $i_1 = .01$. Use Newton's method and find i_2.

34. Find i_3. (Note: For the accuracy required by the federal law, it is usually sufficient to keep going until two successive values of i differ by no more than 10^{-7}.)

Find i_2 and i_3 for each of the following.

35. $P = \$600$, $M = \$57$, $n = 12$, $i_1 = .02$

36. $P = \$15{,}000$, $M = \$337$, $n = 60$, $i_1 = .01$

4.8 APPROXIMATION BY DIFFERENTIALS

In this section we look at differentials, which are useful in finding approximate values of functions. To define a differential, look first at Figure 4.27.

Recall that Δx represents a change in x. In particular, Δx is often used to represent a small change in x. In Figure 4.27, PM is the line tangent to the graph of the function $y = f(x)$ at the point P. Line PM then has slope $f'(x)$. Let $PR = \Delta x = dx$ and $MR = dy$. Then by the definition of slope, the slope of line PM is

$$\frac{\text{the change in } y}{\text{the change in } x} = \frac{dy}{dx}.$$

Since both $f'(x)$ and dy/dx represent the slope of line PM, we have

$$f'(x) = \frac{dy}{dx}.$$

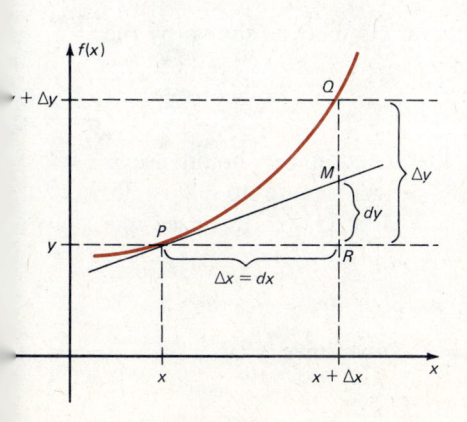

Figure 4.27

206 Applications of the Derivative

Until now we have treated the expression dy/dx as a single quantity which represented the derivative of y with respect to x. Now, by defining dy and dx as we have, we are also able to treat dy/dx as a fraction which represents the quotient of two numbers. This allows us to rewrite $f'(x) = dy/dx$ as

$$dy = f'(x)\,dx.$$

We call dy the **differential of y** and dx the **differential of x**.

In Figure 4.27 we see that if we let Δx get smaller and smaller, the value of dy gets closer and closer to Δy, so that for small values of Δx (but $\Delta x \neq 0$),

$$dy \approx \Delta y.$$

Example 1 Find dy for the following functions.
(a) $y = 6x^2$
The derivative is $y' = 12x$, so that

$$dy = 12x \cdot dx.$$

(b) If $y = 8x^{-3/4}$, then $dy = -6x^{-7/4}\,dx.$ ∎

Work Problem 1 at the side.

1. Find dy for the following.

(a) $y = 9x^{2/3}$

(b) $y = \dfrac{2 + x}{2 - x}$

(c) $y = \sqrt{8x - 7}$

Answer:

(a) $dy = 6x^{-1/3}\,dx$

(b) $dy = \dfrac{4}{(2 - x)^2}\,dx$

(c) $dy = 4(8x - 7)^{-1/2}\,dx$

As shown in Figure 4.27, if a change of Δx is made in the value of x, then the corresponding change in y is Δy, where

$$\Delta y = f(x + \Delta x) - f(x).$$

For small values of Δx, we know that $\Delta y \approx dy$, so that

$$dy \approx f(x + \Delta x) - f(x),$$

or

$$f(x) + dy \approx f(x + \Delta x). \tag{1}$$

This result can be used to approximate numbers as shown by the next examples.

Example 2 Approximate $\sqrt{38}$.

Since we want to find $\sqrt{38}$, the function we should use is $y = f(x) = \sqrt{x} = x^{1/2}$. The nearest perfect square to 38 is 36. Therefore, in equation (1), we let $x = 36$ and $\Delta x = 2$. Reversed, the equation becomes

$$\sqrt{36 + 2} = \sqrt{38} \approx \sqrt{36} + dy,$$

where $dy = f'(x) \cdot dx$. Since $f(x) = x^{1/2}$, we have $f'(x) = \tfrac{1}{2}x^{-1/2}$, and

$$dy = \frac{1}{2}x^{-1/2}\,dx.$$

In this example, $x = 36$ and $dx = \Delta x = 2$;

$$dy = \frac{1}{2}(36)^{-1/2}(2) = \frac{1}{6}.$$

Finally,

$$\sqrt{38} \approx \sqrt{36} + dy = 6 + \tfrac{1}{6} = 6\tfrac{1}{6} \approx 6.167.$$

From a square root table or calculator, $\sqrt{38} \approx 6.164$, so that our approximation is fairly close. ∎

2. Approximate the following.
(a) $\sqrt{10}$
(b) $\sqrt{8}$
(c) $\sqrt[3]{9}$

Answer:
(a) $3\tfrac{1}{6}$
(b) $2\tfrac{5}{6}$
(c) $2\tfrac{1}{12}$

Work Problem 2 at the side.

Example 3 A tumor is in the shape of a cone of height 3 centimeters. Find the approximate increase in the volume of the tumor if the radius of the base increases from 1 cm to 1.1 cm.

The volume of a cone is given by $V = \tfrac{1}{3}\pi r^2 h$. In our example, $h = 3$, so that

$$V = \frac{1}{3}\pi r^2 \cdot 3 = \pi r^2.$$

The increase in volume is approximated by dV, where

$$dV = 2\pi r \cdot dr.$$

In our example, $r = 1$, and $dr = 1.1 - 1 = .1$.

$$dV = 2\pi(1)(.1) \approx 2(3.14)(1)(.1) = .628$$

The approximate increase in volume is .628 cubic centimeters. ∎

3. The volume of a sphere is given by $V = \tfrac{4}{3}\pi r^3$. Find the approximate change in the volume of a snowball if the radius decreases from 8 cm to 7 cm.

Answer: volume decreases approximately 804 cubic centimeters

Work Problem 3 at the side.

4.8 EXERCISES

Find dy for the following functions. See Example 1.

1. $y = 6x^2$
2. $y = -8x^4$
3. $y = 2\sqrt{x}$
4. $y = 8\sqrt{2x - 1}$
5. $y = 7x^2 - 9x + 6$
6. $y = -3x^3 + 2x^2$
7. $y = \dfrac{8x - 2}{x - 3}$
8. $y = \dfrac{-4x + 7}{3x - 1}$

For the following functions, find $f(x)$, $f(x + \Delta x)$, and $f'(x)$ for the given values of x and Δx. (A calculator will be helpful for finding $f(x + \Delta x)$.) Then
(a) *find Δy by using the result $\Delta y = f(x + \Delta x) - f(x)$;*
(b) *find dy by the result $dy = f'(x) \cdot \Delta x$.*

9. $f(x) = 6x - 8; \quad x = 4, \Delta x = .1$

10. $f(x) = -5x + 7; \quad x = 3, \Delta x = .2$

11. $f(x) = 2x^2; \quad x = -2, \Delta x = .2$

12. $f(x) = x^2 - 1; \quad x = 3, \Delta x = .1$

13. $f(x) = x^3 - 2x^2 + 3; \quad x = 1, \Delta x = -.1$

14. $f(x) = 2x^3 + x^2 - 4x; \quad x = 2, \Delta x = -.2$

15. $f(x) = \sqrt{3x}; \quad x = 1, \Delta x = .15$

16. $f(x) = \sqrt{4x - 1}; \quad x = 5, \Delta x = .08$

17. $f(x) = \dfrac{2x - 5}{x + 1}; \quad x = 2, \Delta x = -.03$

18. $f(x) = \dfrac{6x - 3}{2x + 1}; \quad x = 3, \Delta x = -.04$

Find approximations for the following numbers. See Example 2.

19. $\sqrt{10}$ 20. $\sqrt{26}$ 21. $\sqrt{15}$ 22. $\sqrt{63}$

23. $\sqrt{123}$ 24. $\sqrt{146}$ 25. $\sqrt[3]{7}$ 26. $\sqrt[3]{25}$

27. $\sqrt[3]{65}$ 28. $\sqrt[3]{127}$ 29. $\sqrt[4]{17}$ 30. $\sqrt[4]{83}$

31. $\sqrt{4.02}$ 32. $\sqrt{16.08}$ 33. $\sqrt[3]{7.9}$ 34. $\sqrt[3]{124.75}$

APPLIED PROBLEMS

For Exercises 35–38, see Example 3.

Volume

35. A spherical beachball is being inflated. Find the approximate change in volume if the radius increases from 4 cm to 4.2 cm.

36. A spherical snowball is melting; find the approximate change in volume if the radius decreases from 3 cm to 2.8 cm.

Area of an oil slick

37. An oil slick is in the shape of a circle. Find the approximate increase in the area of the slick if its radius increases from 1.2 miles to 1.4 miles.

Area of a bacterial colony

38. The shape of a colony of bacteria on a Petri dish is circular. Find the approximate increase in its area if the radius increases from 20 mm to 22 mm.

CHAPTER 4 SUMMARY

Key words

open interval
closed interval
relative maximum
relative minimum
critical point
increasing function
decreasing function
first derivative test
absolute maximum
absolute minimum
second derivative

concave upward
concave downward
point of inflection
second derivative test
acceleration
explicit function
implicit function
implicit differentiation
Newton's method
differentials

Things to Remember

Theorem 4.1 If a function has any relative maximums or relative minimums, they occur only at critical points of the function.

First derivative test Let a be a critical point for a function f. Suppose that f is continuous at a.
  (a) There is a relative maximum at a if the derivative is positive in an interval just to the left of a and negative in an interval just to the right of a.
  (b) There is a relative minimum at a if the derivative is negative in an interval just to the left of a and positive in an interval just to the right of a.

Second derivative test Let a be a critical point for the function f. Suppose f' exists for every point in an open interval containing a.
  (a) If $f''(a) > 0$, then $f(a)$ is a relative minimum.
  (b) If $f''(a) < 0$, then $f(a)$ is a relative maximum.
  (c) If $f''(a) = 0$, then the test gives no information.

Properties
(a) A function f is increasing on an interval if $f'(x) > 0$ for every value of x in the interval; the function is decreasing on an interval if $f'(x) < 0$ for every x in the interval.

(b) A function f is concave upward on an interval if $f''(x) > 0$ for every value of x in the interval; a function f is concave downward on an interval if $f''(x) < 0$ for every value of x in the interval.

CHAPTER 4 TEST

[4.1] Find the location of all relative maximums and relative minimums for each of the following functions. Find the value of any of these maximums or minimums.

1. $f(x) = -x^2 + 4x - 8$
2. $f(x) = -2x^3 - \frac{1}{2}x^2 + x - 3$
3. $f(x) = \dfrac{x - 1}{2x + 1}$

Find the open intervals where the following functions are increasing or decreasing.

4. $y = x^2 - 5x + 3$
5. $y = -x^3 - 5x^2 + 8x - 6$

[4.2] Find the location of all absolute maximums and absolute minimums for the following functions defined on the given intervals.

6. $f(x) = -x^2 + 5x + 1$, $[1, 4]$
7. $f(x) = x^3 + 2x^2 - 15x + 3$, $[-4, 2]$

[4.3] Each of the following functions gives the displacement at time t of a particle moving along a line. Find the velocity and acceleration functions. Then find the velocity and acceleration at time $t = 0$ and at time $t = 3$.

8. $s(t) = 9t^2 - 7t + 8$
9. $s(t) = -2t^3 + 4t^2 - 6t - 1$

[4.3] Find $f''(x)$ for each of the following. Then find $f''(0)$ and $f''(-5)$.

10. $f(x) = 6x^3 - 9x^2 + 2x - 3$
11. $f(x) = \dfrac{5x - 1}{2x + 3}$

Find $f'''(x)$ and $f^{(4)}(x)$ for each of the following functions.

12. $y = 9x^3 - 8x^2 + 6x - 5$
13. $y = \dfrac{-2}{x - 4}$

[4.4] The total profit in dollars from the sale of x hundred boxes of candy is given by

$$P(x) = -x^3 + 10x^2 - 12x + 106.$$

14. Find the number of boxes of candy that should be sold in order to produce maximum profit.
15. Find the maximum profit.
16. The city park department is planning an enclosed play area in a new park. One side of the area will be against an existing building, with no fence needed there. Find the dimensions of the rectangular field of maximum area that can be made with 900 meters of fence.

[4.5]   17. A very large camera store sells 320,000 rolls of film annually. It costs 10¢ to store one roll for one year and $10 to place a reorder. Find the number of orders that should be placed annually.

Find dy/dx for the following functions.

[4.7]   18. $x^2 y^3 + 4xy = 2$ 19. $\dfrac{x}{y} - 4y = 3x$

20. Find the equation of the tangent to the graph of $\sqrt{2x} - 4yx = -22$ at the point (2, 3).

[4.6]   21. Use Newton's method to find a solution for the equation $x^3 - 8x^2 + 18x - 12 = 0$ in the interval $[4, 5]$. Find the solution to the nearest hundredth.

22. Use Newton's method to find $\sqrt{27}$ to the nearest thousandth.

[4.8]   *Find dy for the following functions.*

23. $y = 8x^3 - 2x^2$ 24. $y = 4(x^2 - 1)^3$

25. Use differentials to approximate $\sqrt{166}$.

CASE 3 COMPRESSION OF THE TRACHEA AND BRONCHI DUE TO COUGHING*

It has been shown clinically that when a person coughs, the diameters of the trachea and bronchi shrink. It is assumed that the pressure which serves to empty the lungs during exhalation has an unwanted side effect of compressing the bronchi and windpipe. The simplest assumption states that constriction of the tubes increases in direct proportion to the pressure. Let r_P represent the radius under a pressure of P (measured as the excess above atmospheric pressure). If r_0 is the radius in the absence of pressure on the lungs, $r_0 - r_P = aP$ where a is a constant. This formula provides a satisfactory approximation only when P lies between 0 and $r_0/2a$. For P greater than this, the resistance of the tube to compression becomes greater. If it were not for this we would suffocate whenever we cough. **Poiseuille's law** states that resistance to the flow of air through a tube increases in proportion to the reciprocal of r_P^4. Thus, $R = k/r_P^4$, where R is the resistance and k is a constant. The flow equals the pressure divided by the resistance:

$$\text{flow} = \frac{P}{R} = \frac{Pr_P^4}{k} = \frac{(r_0 - r_P)r_P^4}{ak}.$$

The velocity is given by the flow divided by the area of a cross section of the tube, or

$$V(r_P) = \frac{\text{flow}}{\pi r_P^2}.$$

EXERCISES

1. Find the radius at which the flow will be a maximum.
2. Find the radius at which the velocity will be a maximum.

CASE 4 PRICING AN AIRLINER— THE BOEING COMPANY[†]

The Boeing Company is one of the United States' largest producers of civilian airliners. It has developed a series of successful jet aircraft, starting with the 707 in 1955. In this case we discuss the mathematics involved in determining the optimum price for a new model jet airliner. It is helpful to summarize here the variables to be used in this case.

---

* From *Mathematics in Biology* by Duane J. Clow and N. Scott Urquhart. Copyright © 1974 by W. W. Norton & Company, Inc. Used by permission.

† Adaptation from "Pricing, Investment, and Games of Strategy" by George Brigham from *Management Sciences, Models and Techniques* by C. W. Churchman and M. Verhulst. Copyright © 1960 by Pergamon Press Ltd. Reprinted by their permission.

Case 4 Pricing an Airliner—The Boeing Company

p = price per airliner, in millions of dollars
$N(p)$ = total number of airliners that will be sold by the industry at a price of p
x = total number of airliners to be produced by Boeing
$C(x)$ = total cost to manufacture x airliners
h = share of the total market to be won by Boeing
P^* = total profit for Boeing

For the airliner in question, Boeing had only one competitor. The price charged by both Boeing and its competitor would have to be the same—any attempt by one firm to lower the price would of necessity be met by the other firm. Thus, the price charged by Boeing would have no effect on the share of the total market to be won by Boeing. However, the price charged by Boeing (and the competitor) would have considerable effect on the *size* of the total market. In fact, Boeing sales analysts made predictions of the total market at various price levels and found that the function

$$N(p) = -78p^2 + 655p - 1125$$

gave a reasonable estimate of the total market; that is, $N(p)$ is the total number of planes that will be sold, by both Boeing and its competitor, at a price p, in millions of dollars per plane. A graph of $N(p)$ is shown in Figure 1.

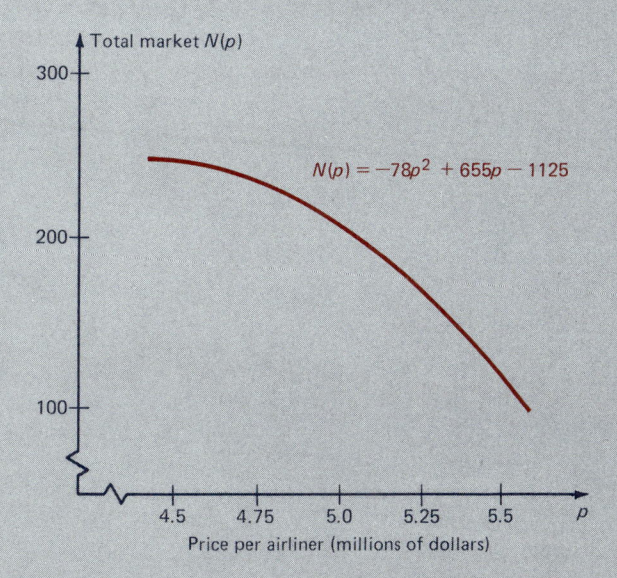

Figure 1

Production analysts at Boeing estimated that if $C(x)$ is the total cost to manufacture x airplanes, then

$$C(x) = 50 + 1.5x + 8x^{3/4},$$

where $C(x)$ is measured in millions of dollars.

The company desires to know the price, p, it should charge per plane so that the total profit, P^*, will be a maximum. Profit is given as the numerical product of the price per plane, p, and the total number of planes sold by Boeing, x, minus the cost to manufacture x planes, $C(x)$. Thus, the profit function is

$$P^* = p \cdot x - C(x). \tag{1}$$

If h is the fractional share of the total market for this plane that will be won by Boeing (note: $0 \leq h \leq 1$), then

$$x = h \cdot N(p)$$

(The number of planes sold by Boeing equals its share of the market times the total market.) Substituting $h \cdot N(p)$ for x in the profit function, equation (1), gives

$$P^* = p \cdot h \cdot N(p) - C[h \cdot N(p)].$$

To find the maximum profit, we must take the derivative of this function with respect to p. Assume h is a constant; treat $p \cdot h \cdot N(p)$ as the product $(ph) \cdot N(p)$, and use the chain rule on $C[h \cdot N(p)]$. This gives

$$(P^*)' = (ph) \cdot N'(p) + N(p) \cdot h - C'[h \cdot N(p)] \cdot h \cdot N'(p).$$

For convenience, replace $h \cdot N(p)$ by x in the expression $C'[h \cdot N(p)]$. Then put the derivative equal to 0 and simplify:

$$(ph) \cdot N'(p) + N(p) \cdot h - C'(x) \cdot h \cdot N'(p) = 0,$$
$$p \cdot N'(p) + N(p) = C'(x) \cdot N'(p),$$

from which

$$p + \frac{N(p)}{N'(p)} = C'(x). \tag{2}$$

Thus, the optimum price, p, must satisfy equation (2). (It is necessary to verify that this value of p leads to a *maximum* profit and not a minimum. This calculation, using the second derivative test, is left for the energetic reader.)

Returning to the functions $N(p)$ and $C(x)$ given above, verify that

$$N'(p) = -156p + 655 \quad \text{and} \quad C'(x) = 1.5 + 6x^{-1/4}.$$

Using p, $N(p)$, and $N'(p)$, we can sketch a graph of the left-hand side of equation (2), as shown by the left-hand graph of Figure 2. On the right in Figure 2 is the graph of the right-hand side of equation (2), $C'(x) = 1.5 + 6x^{-1/4}$, for various values of x considered feasible by the company. We know that the maximum profit is produced when the left-hand and right-hand sides of equation (2) are equal. Using this fact, we can read the price that leads to the maximum profit from Figure 2. If $x = 60$ (the company sells a total of 60 airplanes), the price per plane will be about $5.1 million, while if $x = 120$, the price should be a little less than $5 million.

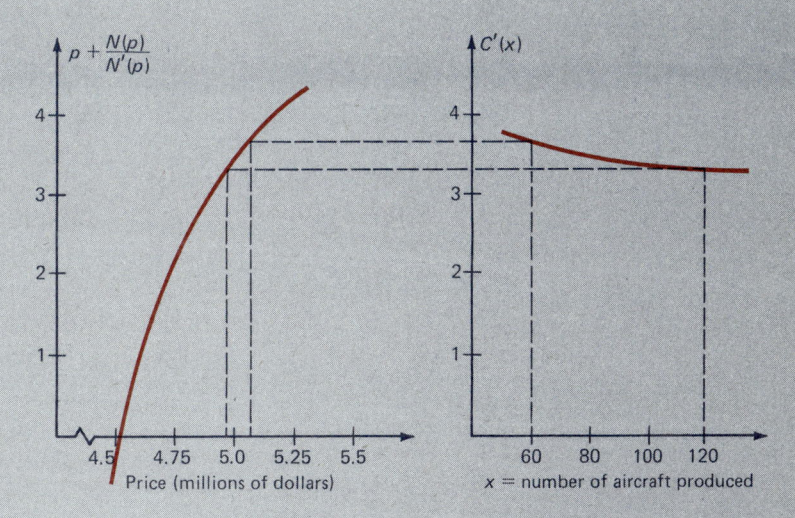

Figure 2

EXERCISES

1. (a) Find the total cost of manufacturing 120 planes. (Hint: $8 \cdot 120^{3/4} \approx 290$)
   (b) Assume each plane sells for $5 million, and find the total profit from the sale of 120 planes, using equation (1).

2. (a) Find the total market at a price of $5 million per plane.
   (b) If Boeing sells 120 planes at a price of $5 million each, how many will be sold by the competitor?

3. (a) Find $C(60)$. (Hint: $8 \cdot 60^{3/4} \approx 170$)
   (b) Assume each plane is sold for $5.1 million, and find the total profit from the sale of 60 planes, using equation (1).

CASE 5 A TOTAL COST MODEL FOR A TRAINING PROGRAM*

In this case, we set up a mathematical model for determining the total costs in setting up a training program. Then we use calculus to find the time between training programs that produces the minimum total cost. The model assumes that the demand for trainees is constant and that the fixed cost of training a batch of trainees is known. Also, it is assumed that people who are trained, but for whom no job is readily available, will be paid a fixed amount per month while waiting for a job to open up.

---

*Based on "A Total Cost Model for a Training Program" by P. L. Goyal and S. K. Goyal, Department of Mathematics and Computer Science, The Polytechnic of Wales, Treforest, Pontypridd. Used with permission.

216 Applications of the Derivative

The model uses the following variables:

D = demand for trainees per month
N = number of trainees per batch
C_1 = fixed cost of training a batch of trainees
C_2 = variable cost of training per trainee per month
C_3 = salary paid monthly to a trainee who has not yet been given a job after training
m = time interval in months between successive batches of trainees
t = length of training program in months
$Z(m)$ = total monthly cost of program

The total cost of training a batch of trainees is given by $C_1 + NtC_2$. However, $N = mD$, so that the total cost per batch is $C_1 + mDtC_2$.

After training, personnel are given jobs at the rate of D per month. Thus, $N - D$ of the trainees will not get a job the first month, $N - 2D$ will not get a job the second month, and so on. The $N - D$ trainees who do not get a job the first month produce total costs of $(N - D)C_3$, those not getting jobs during the second month produce costs of $(N - 2D)C_3$, and so on. Since $N = mD$, the costs during the first month can be written as

$$(N - D)C_3 = (mD - D)C_3 = (m - 1)DC_3,$$

while the costs during the second month are $(m - 2)DC_3$, and so on. The total cost for keeping the trainees without a job is thus

$$(m - 1)DC_3 + (m - 2)DC_3 + (m - 3)DC_3 + \cdots + 2DC_3 + DC_3,$$

which can be factored to give

$$DC_3[(m - 1) + (m - 2) + (m - 3) + \cdots + 2 + 1].$$

The expression in brackets is the sum of the terms of an arithmetic sequence, discussed in most algebra texts. Using formulas for arithmetic sequences, the expression in brackets can be shown to equal $m(m - 1)/2$, so that we have

$$DC_3\left[\frac{m(m - 1)}{2}\right] \tag{1}$$

as the total cost for keeping jobless trainees.

The total cost per batch is the sum of the training cost per batch, $C_1 + mDtC_2$, and the cost of keeping trainees without a proper job, given by (1). Since we assume that a batch of trainees is trained every m months, the total cost per month, $Z(m)$, is given by

$$Z(m) = \frac{C_1 + mDtC_2}{m} + \frac{DC_3\left[\frac{m(m - 1)}{2}\right]}{m} = \frac{C_1}{m} + DtC_2 + DC_3\left(\frac{m - 1}{2}\right).$$

EXERCISES

1. Find $Z'(m)$.

2. Solve the equation $Z'(m) = 0$.

   As a practical matter, it is usually required that m be a whole number. If m does not come out to be a whole number in Exercise 2, then m^+ and m^-, the two whole numbers closest to m, must be chosen. Calculate both $Z(m^+)$ and $Z(m^-)$; the smaller of the two provides the optimum value of Z.

3. Suppose a company finds that its demand for trainees is 3 per month, that a training program requires 12 months, that the fixed cost of training a batch of trainees is $15,000, that the variable cost per trainee per month is $100, and that trainees are paid $900 per month after training but before going to work. Use your result from Exercise 2 and find m.

4. Since m is not a whole number, find m^+ and m^-.

5. Calculate $Z(m^+)$ and $Z(m^-)$.

6. What is the optimum time interval between successive batches of trainees? How many trainees should be in a batch?

5
Exponential and Logarithmic Functions

In Chapter 2 we studied several types of functions that are useful in setting up mathematical descriptions or models of practical situations. While the functions that we have already studied are certainly important, it is probably a safe generalization to say that exponential functions are the single most important type of function in practical applications. Exponential functions and the closely related logarithmic functions are often used in management, social science, and biology to describe growth of populations, increase in sales, growth of money with time, as well as decay of radioactive samples and decay of sales in the absence of advertising.

5.1 EXPONENTIAL FUNCTIONS

In Chapter 1, we defined the symbol a^x where a is a real number and x is a *rational number*. In more advanced courses, a^x is defined for *any* real number value of x, rational or irrational. Irrational exponents, such as $2^{\sqrt{2}}$, 3^{π}, or $5^{-\sqrt{7}}$, can be evaluated as decimals with logarithms, calculators with an x^y key, or computers. After a^x is defined for all real-number values of x, we can then define an exponential function. If a is a positive real number and $a \neq 1$, then

$$y = a^x$$

is an **exponential function**.

5.1 Exponential Functions   219

1. Graph the following.
(a) $y = 4^x$
(b) $y = (\frac{7}{2})^x$

Answer:
(a)

x	y
−2	1/16
−1	1/4
0	1
1	4
2	16

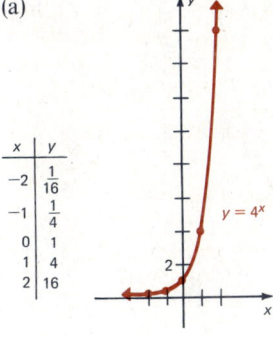

$y = 4^x$

(b)

x	y
−1	.29
0	1
1	3.5
2	12.25

$y = (\frac{7}{2})^x$

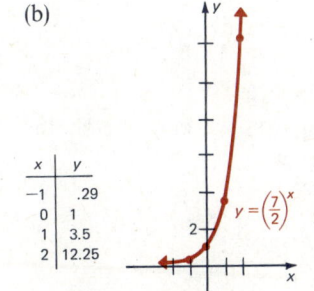

2. Graph the following.
(a) $y = (\frac{1}{3})^x$
(b) $y = (\frac{9}{10})^x$

Answer:
(a)

x	y
−2	9
−1	3
0	1
1	1/3

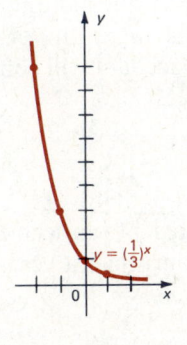

$y = (\frac{1}{3})^x$

(b)

x	y
−2	1.23
−1	1.1
0	1
1	.9
2	.81

$y = (\frac{9}{10})^x$

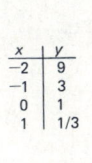

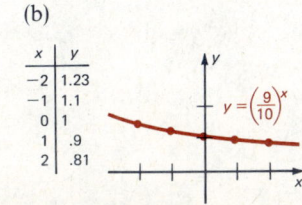

x	y
−3	1/8
−2	1/4
−1	1/2
0	1
1	2
2	4
3	8
4	16

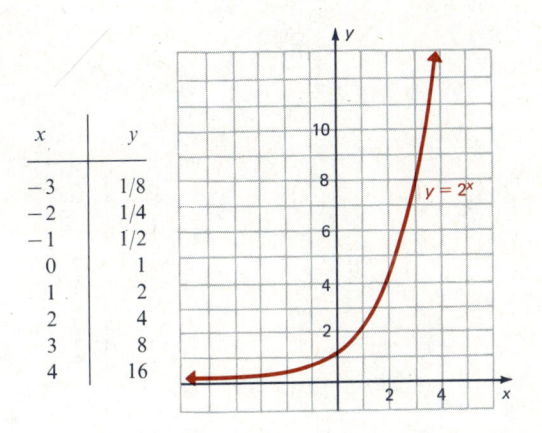

Figure 5.1

Example 1 Graph $y = 2^x$.

Make a table of values of x and y as shown in Figure 5.1. Then plot these points and draw a smooth curve through them to get the graph of Figure 5.1. This graph is typical of the graphs of exponential functions of the form $y = a^x$, where $a > 1$. As the values of x get smaller and smaller, the graph approaches the x-axis. Since the expression 2^x can never equal zero, the graph will never reach the axis. Thus, the x-axis is a horizontal asymptote. ∎

Work Problem 1 at the side.

x	y
−4	16
−3	8
−2	4
−1	2
0	1
1	1/2
2	1/4
3	1/8

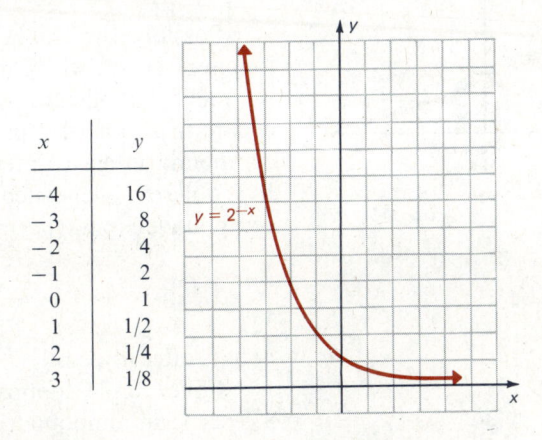

Figure 5.2

Example 2 Graph $y = 2^{-x} = 1/2^x = (1/2)^x$.

Once again we construct a table of values of x and y and draw a smooth curve through the resulting points (see Figure 5.2). This graph is typical of the graphs of exponential functions of the form $y = a^x$, where $0 < a < 1$. ∎

Work Problem 2 at the side.

220 Exponential and Logarithmic Functions

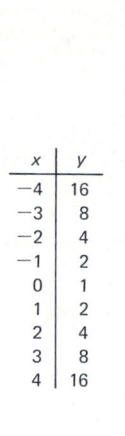

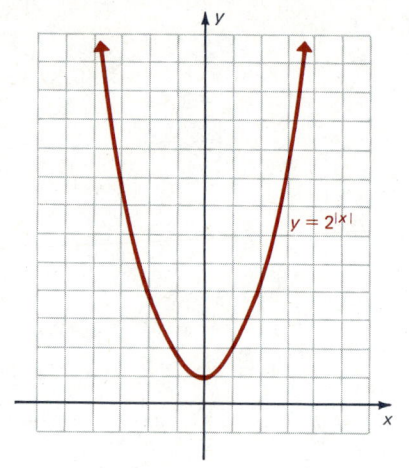

x	y
−4	16
−3	8
−2	4
−1	2
0	1
1	2
2	4
3	8
4	16

Figure 5.3

3. Graph $y = 2^{-|x|}$.

Answer:

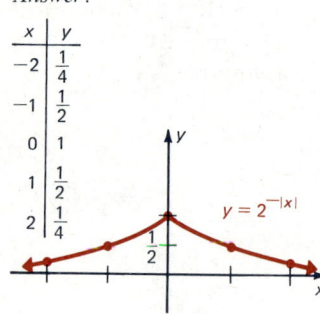

x	y
−2	$\frac{1}{4}$
−1	$\frac{1}{2}$
0	1
1	$\frac{1}{2}$
2	$\frac{1}{4}$

$y = 2^{-|x|}$

Example 3 Graph $y = 2^{|x|}$.

By completing a table of ordered pairs, we can sketch the graph of $y = 2^{|x|}$, as shown in Figure 5.3. ∎

Work Problem 3 at the side.

Possibly the single most useful exponential function is the function $y = e^x$, where e is an irrational number that occurs often in practical applications, as we shall see. To see how the number e occurs in a practical problem, let us begin with the formula for compound interest (interest paid on both principal and interest). If P dollars is deposited in an account paying a rate of interest i compounded (paid) m times per year, the account will contain

$$P\left(1 + \frac{i}{m}\right)^{mn}$$

dollars after n years.

For example, suppose $1000 is deposited into an account paying 8% per year compounded quarterly, or four times a year. After 10 years the account will contain

$$P\left(1 + \frac{i}{m}\right)^{mn} = 1000\left(1 + \frac{.08}{4}\right)^{4(10)} \qquad 8\% = .08$$
$$= 1000(1 + .02)^{40}$$
$$= 1000(1.02)^{40}$$

5.1 Exponential Functions

dollars. The number $(1.02)^{40}$ can be found in financial tables, or with a calculator. To five decimal places, $(1.02)^{40} = 2.20804$. So the amount on deposit after 10 years is

$$1000(1.02)^{40} = 1000(2.20804) = 2208.04 \text{ dollars.}$$

4. Find the final amounts on deposit.
(a) $P = 2000$, $i = 6\%$, $n = 4$, compounded semiannually (Hint: $(1.03)^8 = 1.26677$.)
(b) $P = 15,000$, $i = 12\%$, $n = 5$, compounded monthly (Hint: $(1.01)^{60} = 1.81670$.)

Answer:
(a) $2533.54
(b) $27,250.50

Work Problem 4 at the side.

Suppose now that a lucky investment produces annual interest of 100%, so that $i = 1.00$, or $i = 1$. Suppose also that you can deposit only $1 at this rate, and for only one year. Then $P = 1$ and $n = 1$. Substituting into our formula for compound interest, we then have

$$P\left(1 + \frac{i}{m}\right)^{mn} = 1\left(1 + \frac{1}{m}\right)^{m(1)} = \left(1 + \frac{1}{m}\right)^m.$$

As interest is compounded more and more often, the value of this expression will increase. If $m = 1$ (interest is compounded annually), we have

$$\left(1 + \frac{1}{m}\right)^m = \left(1 + \frac{1}{1}\right)^1 = (1 + 1)^1 = 2^1 = 2,$$

so that your $1 will become $2 in one year.

5. Use a calculator to evaluate $(1 + 1/m)^m$ for the following values of m.
(a) 2
(b) 4
(c) 12

Answer:
(a) 2.25
(b) 2.44
(c) 2.61

m	$\left(1 + \dfrac{1}{m}\right)^m$
1	2
2	2.25
5	2.48832
10	2.59374
25	2.66584
50	2.69159
100	2.70481
500	2.71557
1000	2.71692
10,000	2.71815
1,000,000	2.71828

Work Problem 5 at the side.

With a calculator or computer, we get the results, rounded to five decimal places, shown at the side. It appears that as m increases, the expression $(1 + 1/m)^m$ gets closer and closer to some fixed number. It turns out that this is indeed the case; this fixed number is called e and

$$e = \lim_{m \to \infty} \left(1 + \frac{1}{m}\right)^m.$$

To nine decimal places,

$$e = 2.718281828.$$

Table 4 in the back of the book gives various powers of e. Also, some calculators have e keys.

222 Exponential and Logarithmic Functions

Example 4 Graph $y = e^x$.

Use Table 4 to verify the ordered pairs shown in the table of Fig 5.4. These ordered pairs were plotted and a smooth curve drawn through them to get the graph of Figure 5.4. ∎

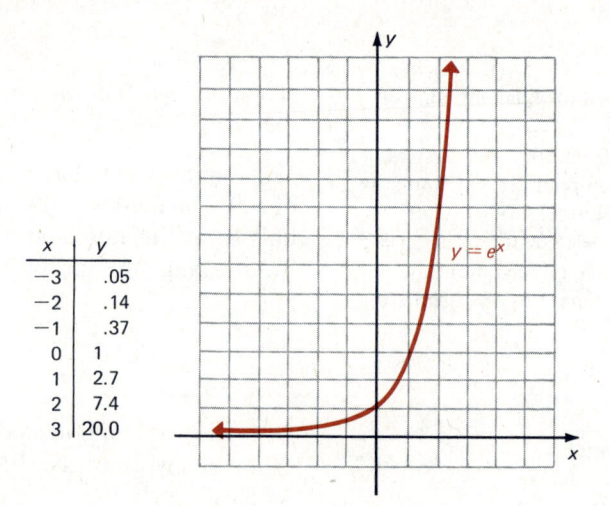

x	y
−3	.05
−2	.14
−1	.37
0	1
1	2.7
2	7.4
3	20.0

Figure 5.4

6. Graph $y = e^{-x}$.

Answer:

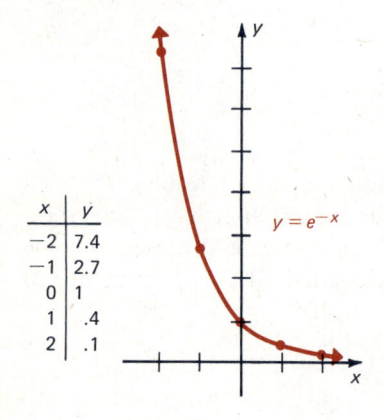

x	y
−2	7.4
−1	2.7
0	1
1	.4
2	.1

Work Problem 6 at the side.

Example 5 Graph $y = e^{-x^2}$.

If we use Table 4 to help plot several pairs of values of x and y, we get the graph shown in Figure 5.5. This graph is very important in the study of probability. In fact, this graph, with a slight modification,* is the normal curve, one of the major topics of probability and statistics. ∎

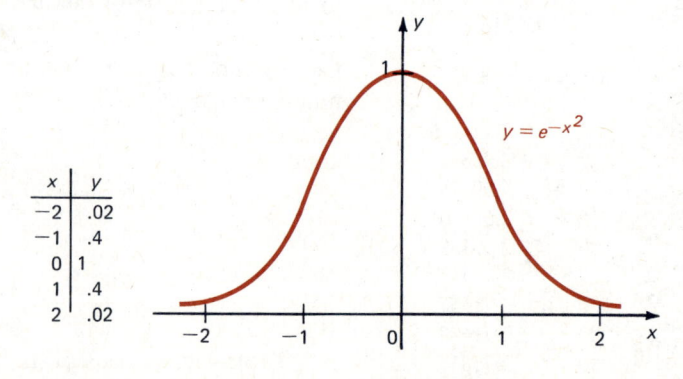

x	y
−2	.02
−1	.4
0	1
1	.4
2	.02

Figure 5.5

---

*The normal curve has equation $y = e^{-n^2}/\sqrt{2\pi}$.

To solve exponential equations, such as $2^{3x} = 2^7$, we need the following property of exponents:

If $a \neq 1$, and if $a^x = a^y$, then $x = y$.

Example 6 Solve $2^{3x} = 2^7$.

Since $2^{3x} = 2^7$ and the base is 2 on both sides, use the property above to get

$$3x = 7$$
$$x = \frac{7}{3}. \blacksquare$$

Example 7 Solve $9^{x+1} = 27$.

To use the property above, both bases must be the same. Here, we can write 9 as 3^2 and 27 as 3^3. This gives

$$9^{x+1} = 27$$
$$(3^2)^{x+1} = 3^3$$
$$3^{2x+2} = 3^3$$
$$2x + 2 = 3$$
$$2x = 1$$
$$x = \frac{1}{2}.$$

Verify that $9^{(1/2)+1} = 27$. $\blacksquare$

Work Problem 7 at the side.

7. Solve the following.
   (a) $3^5 = 3^{m+2}$
   (b) $8^{2x} = 4$
   (c) $100^{p+1} = 1000$

Answer:
(a) $m = 3$
(b) $x = \frac{1}{3}$
(c) $p = \frac{1}{2}$

5.1 EXERCISES

Graph the following exponential functions. See Examples 1–5.

1. $y = 3^x$
2. $y = 4^x$
3. $y = 3^{-x}$
4. $y = 4^{-x}$
5. $y = (1/4)^x$
6. $y = (1/3)^x$
7. $y = (1/3)^{-x}$
8. $y = (1/4)^{-x}$
9. $y = 4^{-|x|}$
10. $y = 3^{-|x|}$
11. $y = 3^{-x^2}$
12. $y = 2^{-x^2}$
13. $y = 2^{1-x}$
14. $y = 3^{1-x}$
15. $y = e^{x+1}$
16. $y = e^{2x}$
17. $y = e^{-3x}$
18. $y = 2e^{-x}$

Solve the following equations for x. See Examples 6 and 7.

19. $5^x = 25$
20. $3^x = \frac{1}{9}$
21. $2^x = \frac{1}{8}$
22. $4^x = 64$
23. $4^x = 8$
24. $25^x = 125$
25. $(\frac{1}{2})^x = 8$
26. $(\frac{3}{4})^x = \frac{16}{9}$
27. $5^{-2x} = \frac{1}{25}$
28. $3^{x-1} = 9$
29. $16^{-x+1} = 8$
30. $25^{-3x} = 3125$

APPLIED PROBLEMS

Compound interest

If $1 is deposited into an account paying 6% compounded annually, then the account will contain
$$y = (1 + .06)^t = (1.06)^t$$
dollars after t years.

31. Use a calculator to help you complete the following table.

t	0	1	2	3	4	5	6	7	8	9	10
y	1					1.34					1.79

32. Graph $y = (1.06)^t$.

Inflation

If money loses value at the rate of 8% per year, the value of $1 in t years is given by
$$y = (1 - .08)^t = (.92)^t.$$

33. Use a calculator to help you complete the following table.

t	0	1	2	3	4	5	6	7	8	9	10
y	1						.66				.43

34. Graph $y = (.92)^t$.

Use the results of Exercise 33 to answer the following questions.

35. Suppose a house costs $50,000 today. Estimate the cost of a similar house in 10 years. (Hint: Solve the equation $.43x = 50,000$.)

36. Find the cost in 8 years of a textbook costing $14 today.

Compound interest

Use the expression $P(1 + i/m)^{mn}$ and a calculator to find the final amount on deposit when the following deposits are made.

37. $1000 for 5 years at 6% compounded annually
38. $5800 for 9 years at 7% compounded annually

39. $41,500 at 8% compounded quarterly for 3 years

40. $6400 at 4% compounded quarterly for 7 years

41. $5180 at 6% compounded semiannually for 9 years

42. $38,580 at 8% compounded semiannually for 7 years

Population growth

Under certain conditions, the number of individuals of a species that is newly introduced into an area can double every year. That is, if t represents the number of years since the species was introduced into the area, and y represents the number of individuals, then

$$y = 6 \cdot 2^t$$

if 6 animals were introduced into the area originally.

43. Complete the following table.

t	0	1	2	3	4	5	6	7	8	9	10
y	6					192					6144

44. Graph $y = 6 \cdot 2^t$.

Growth of bacteria

Escherichia coli is a strain of bacteria that occurs naturally in many different situations. Under certain conditions, the number of these bacteria present in a colony is given by

$$E(t) = E_0 \cdot 2^{t/30},$$

where $E(t)$ is the number of bacteria present t minutes after the beginning of an experiment, and E_0 is the number present when $t = 0$. Let $E_0 = 1,000,000$ and use a calculator with an x^y key to find the number of bacteria at the following times.

45. $t = 5$ 46. $t = 10$ 47. $t = 15$ 48. $t = 20$

49. $t = 30$ 50. $t = 60$ 51. $t = 90$ 52. $t = 120$

Medical school

The higher a student's grade-point average, the fewer applications he or she has to send to medical schools (other things being equal). Using information given in a guidebook for prospective medical school students, we constructed the following mathematical model for the number of applications that a student should send out:

$$y = 540e^{-1.3x},$$

where y is the number of applications that should be sent out by a person whose grade-point average is x. Here $2.0 \le x \le 4.0$. Use a calculator with an x^y key to find the number of applications that should be sent out by students having the following grade-point averages.

53. 2.0   54. 2.5   55. 3.0

56. 3.5   57. 3.9   58. 4.0

Atmospheric pressure

The atmospheric pressure h meters above sea level is approximately

$$P = 10{,}000 e^{-.00001h}$$

kilograms per square meter. Use a calculator with an x^y key to find the pressure when

59. $h = 1000$;

60. $h = 5000$.

5.2 LOGARITHMIC FUNCTIONS

Exponential functions can be used as mathematical models in situations involving an increasing rate of growth, as we shall see in the next section. Logarithmic functions can be used as mathematical models involving a decreasing rate of growth.

Logarithmic functions can be obtained from exponential functions by using the following definition. For any real numbers x, y, $a > 0$ and $a \neq 1$,

$$x = a^y \quad \text{means the same as} \quad y = \log_a x.$$

(Read $y = \log_a x$ as "y is the logarithm of x to the base a.") For example, since $16 = 2^4$, we can write $4 = \log_2 16$. Also, since $10^3 = 1000$, we have $\log_{10} 1000 = 3$.

From the definition of $y = \log_a x$, we have the restriction that $a > 0$. Since $x = a^y$, this means that x must also be a positive number. Thus, we can only find logarithms of positive numbers. This is shown in the graphs of Figures 5.6(a) and 5.6(b).

Example 1 Several pairs of equivalent statements, written in both exponential and logarithmic forms, are shown below.

Exponential form	Logarithmic form
$3^2 = 9$	$\log_3 9 = 2$
$(\frac{1}{5})^{-2} = 25$	$\log_{1/5} 25 = -2$
$10^5 = 100{,}000$	$\log_{10} 100{,}000 = 5$
$4^{-3} = 1/64$	$\log_4 1/64 = -3$ ∎

Work Problem 1 at the side.

The exponential function $f(x) = 2^x$ and the logarithmic function $g(x) = \log_2 x$ are closely related. Note that $f(3) = 2^3 = 8$, while $g(8) = \log_2 8 = 3$. Hence, $f(3) = 8$ and $g(8) = 3$. Also, $f(5) = 32$ and $g(32) = 5$. In fact, for any number m, if $f(m) = n$, then $g(n) = m$.

1. (a) Write $6^2 = 36$ in logarithmic form.
(b) Write $\log_{10} .01 = -2$ in exponential form.

Answer:
(a) $\log_6 36 = 2$
(b) $10^{-2} = .01$

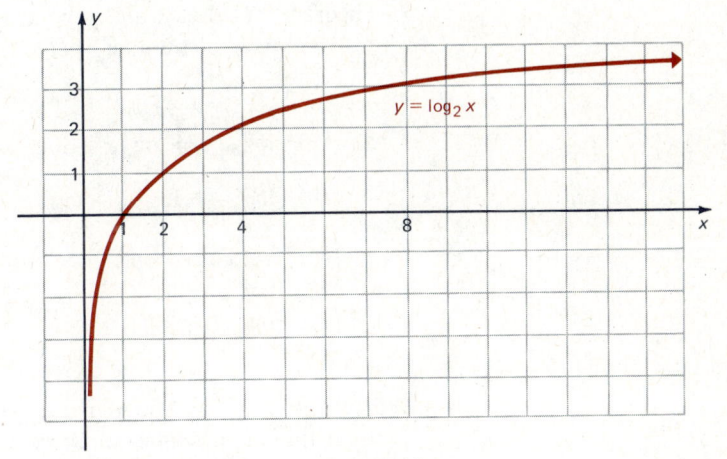

Figure 5.6(a)

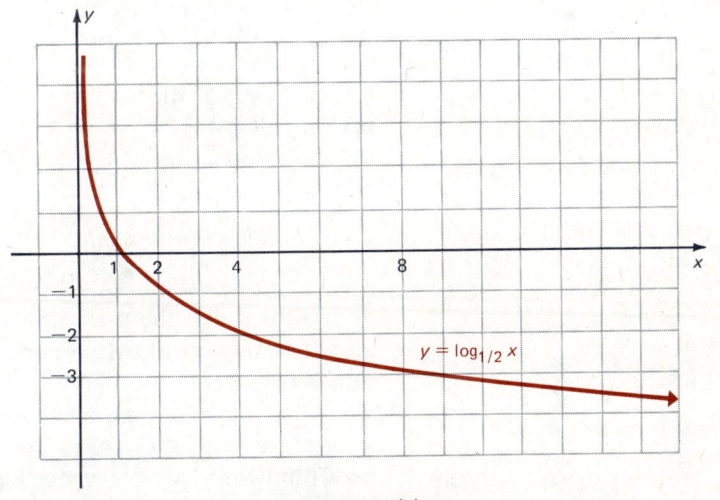

Figure 5.6(b)

Functions related in this way are called **inverses** of each other. In general, $f(x) = a^x$ and $g(x) = \log_a x$ are inverses of each other. A general discussion of inverse functions would carry us too far afield; such a discussion can be found in books listed in the bibliography.

By finding a series of points that satisfy $y = \log_2 x$ and another series satisfying $y = \log_{1/2} x$, we get the graphs shown in Figure 5.6. Note that both graphs are functions. The graph of Figure 5.6(a) is typical of the graphs of logarithmic functions of base $a > 1$, while the graph of Figure 5.6(b) is typical of the graphs of logarithmic functions of base a where $0 < a < 1$. In both graphs the y-axis is a vertical asymptote.

One of the main reasons for the usefulness of logarithmic functions comes from the properties of logarithms given in the following theorem.

228 Exponential and Logarithmic Functions

Theorem 5.1 Let x and y be any positive real numbers, and let r be any real number. Let a be a positive real number, $a \neq 1$. Then

(a) $\log_a xy = \log_a x + \log_a y$

(b) $\log_a \dfrac{x}{y} = \log_a x - \log_a y$

(c) $\log_a x^r = r \log_a x$

(d) $\log_a a = 1$

(e) $\log_a 1 = 0.$

Example 2 Use Theorem 5.1 to write the following as a single logarithm or a single number.

(a) $\log_6 7 + \log_6 9 = \log_6 7 \cdot 9 = \log_6 63$

(b) $\log_2 5 - \log_2 7 = \log_2 \dfrac{5}{7}$

(c) $3.2 \log_{10} 5 = \log_{10} 5^{3.2}$

(d) $\log_8 8 = 1$

(e) $\log_4 1 = 0.$ ∎

2. Use Theorem 5.1 to write the following in simpler form.
(a) $\log_3 35 - \log_3 5$
(b) $\log_5 2 + \log_5 7$
(c) $4 \log_4 2$
(d) $\log_9 9$
(e) $\log_{15} 1$

Answer:
(a) $\log_3 7$
(b) $\log_5 14$
(c) $\log_4 16$ or 2
(d) 1
(e) 0

Work Problem 2 at the side.

Historically, the main application of logarithms has been as an aid to numerical calculations. Using the properties above and tables of logarithms, many numerical problems can be greatly simplified. Since our number system is base 10, logarithms to base 10 are the most convenient for numerical calculations. Base 10 logarithms are called **common logarithms.** For simplicity, $\log_{10} x$ is abbreviated $\log x$. Using this notation we have, for example,

$$\log 1000 = 3$$
$$\log 100 = 2$$
$$\log 1 = 0$$
$$\log 0.01 = -2$$
$$\log 0.0001 = -4.$$

3. Find the following.
(a) $\log 10,000$
(b) $\log 1,000,000$
(c) $\log .001$
(d) $\log .0000001$

Answer:
(a) 4
(b) 6
(c) -3
(d) -7

Work Problem 3 at the side.

Other common logarithms can be found from Table 3 in the Appendix. Common logarithms have few applications other than numerical calculations, and especially few applications involving calculus. Most applications of logarithms use the number e rather

5.2 Logarithmic Functions

than 10 as the base. (Recall that, to seven decimal places, $e = 2.7182818$.) Logarithms to base e are called **natural logarithms** and written $\ln x$. A table of natural logarithms is given as Table 5 in the Appendix. From this table, for example, we find that

$$\ln 55 \approx 4.0073$$
$$\ln 1.9 \approx 0.6419$$
$$\ln 0.4 \approx -0.9163$$

and so on.

Work Problem 4 at the side.

While common logarithms may seem more "natural" than logarithms to base e, there are several good reasons for using natural logarithms instead. In fact, many advanced textbooks use "log" as an abbreviation for natural logarithm, with no abbreviation used for common logarithms. In this text, $\log x$ represents the common logarithm of x. A major reason for the importance of e and natural logarithms is given in Section 5.4.

Example 3 Use Table 5 and the properties of logarithms in Theorem 5.1 to evaluate the following.

(a) $\ln 350$

Since 350 is not listed in Table 5, we use Theorem 5.1.

$$\ln 350 = \ln(35)(10)$$
$$= \ln 35 + \ln 10$$
$$\approx 3.5553 + 2.3026$$
$$= 5.8579$$

(b) $\ln 27$

Again, $\ln 27$ is not given in Table 5. By Theorem 5.1, we have

$$\ln 27 = \ln 3^3$$
$$= 3 \ln 3$$
$$\approx 3(1.0986)$$
$$= 3.2958. \quad \blacksquare$$

Work Problem 5 at the side.

When a table of natural logarithms is not available at all, we can still find the value of natural logarithms by using the following theorem.

---

4. Use Table 5 to find the following.
(a) $\ln 70$
(b) $\ln 8.2$
(c) $\ln .8$

Answer:
(a) 4.2485
(b) 2.1041
(c) $-.2231$

5. Evaluate the following using Table 5 and Theorem 5.1.
(a) $\ln 26$
(b) $\ln 49$
(c) $\ln 8,000,000$
(d) $\ln .076$

Answer:
(a) 3.2580
(b) 3.8918
(c) 15.8950
(d) -2.5771

Theorem 5.2 For any positive number x,

$$\ln x \approx 2.3026 \log x,$$

where $\log x$ is the common logarithm of x.*

Example 4 Use Theorem 5.2 to evaluate the following.
(a) $\ln 83$
  By Theorem 5.2 above, $\ln 83 \approx 2.3026 \log 83$. Using Table 3, Common Logarithms, we have

$$\begin{aligned}\ln 83 &\approx 2.3026 \log 83 \\ &\approx 2.3026(1.9191) \\ &\approx 4.4189.\end{aligned}$$

(b) $\ln 6000 \approx 2.3026(3.7782)$
$\approx 8.6997.$ ■

Work Problem 6 at the side.

6. Evaluate the following using Theorem 5.2.
(a) $\ln 642$
(b) $\ln 2300$

Answer:
(a) 6.4645
(b) 7.7407

Example 5 The cost in dollars of manufacturing x picture frames, where x is measured in thousands, is given by

$$C(x) = 5000 + 2000 \ln(x + 1).$$

(a) Find the fixed cost.
  The fixed cost represents the cost of setting up the plant, designing the items, and so on. To find the fixed cost, let $x = 0$ in the cost function above. This gives

$$\begin{aligned}C(0) &= 5000 + 2000 \ln(0 + 1) \\ &= 5000 + 2000 \ln 1 \\ &= 5000 + 2000 \cdot 0 \quad \text{Recall that } \ln 1 = 0 \\ &= 5000.\end{aligned}$$

The fixed cost is $5000.
(b) Find the cost of manufacturing 19,000 frames.
  To find the cost of producing 19,000 frames, let $x = 19$. This gives

$$\begin{aligned}C(19) &= 5000 + 2000 \ln(19 + 1) \\ &\approx 5000 + 2000(2.9957)\end{aligned}$$

(From Table 5, $\ln 20 = 2.9957$.)

$$\begin{aligned}&\approx 5000 + 5991 \\ &= 10{,}991.\end{aligned}$$

Thus, 19,000 frames cost a total of $10,991. ■

---
* $2.3026 \approx 1/\log e$.

7. In Example 4, find the cost of manufacturing the following.
(a) 9000 frames
(b) 49,000 frames

Answer:
(a) about $9600
(b) about $12,800

Work Problem 7 at the side.

5.2 EXERCISES

Write the following using logarithms. See Example 1.

1. $2^3 = 8$
2. $5^2 = 25$
3. $3^4 = 81$
4. $6^3 = 216$
5. $(1/3)^{-2} = 9$
6. $(3/4)^{-2} = 16/9$

Write the following using exponents. See Example 1.

7. $\log_2 8 = 3$
8. $\log_3 27 = 3$
9. $\log_{10} 100 = 2$
10. $\log_2 1/8 = -3$
11. $\log 100{,}000 = 5$
12. $\ln e^5 = 5$

Evaluate the following. (Hint: Use the definition of logarithm.)

13. $\log_{10} 10{,}000$
14. $\log_9 81$
15. $\log_4 64$
16. $\log_2 1/4$
17. $\log 0.01$
18. $\log 0.00001$

Write each of the following as a single logarithm or, if possible, as a single number. See Example 2.

19. $\log_4 8 + \log_4 2$
20. $\log_9 3 + \log_9 27$
21. $\log_7 15 - \log_7 11$
22. $\log_8 21 - \log_8 40$
23. $0.2(\log_{10} 30)$
24. $-0.8(\log_5 12)$
25. $2 \ln e + \ln 1/e$
26. $-\ln e^{-1} - \ln e$
27. $0.3(\ln 5) - 0.4(\ln 6)$
28. $-0.2(\ln 12) + 0.5(\ln 8)$

Evaluate the following natural logarithms using either Theorem 5.1 with Table 5 or Theorem 5.2. See Examples 3 and 4.

29. $\ln 20$
30. $\ln 35$
31. $\ln 60$
32. $\ln 50$
33. $\ln 800$
34. $\ln 920$
35. $\ln 532$
36. $\ln 255$
37. $\ln 768$
38. $\ln 324$
39. $\ln 58{,}500$
40. $\ln 12{,}400$

Complete the following tables of ordered pairs and then graph the given functions.

41. $y = \log_3 x$

x	27	9	3	1	1/3	1/9	1/27
y							

42. $y = \log_4 x$

x	64	16	4	1	1/4	1/16	1/64
y							

43. $y = \log_{10} x$

x	100	10	1	1/10	1/100	1/1000
y						

44. $y = \ln x$ (Hint: Use Table 5.)

x	100	10	e	1	0.1
y					

45. $y = \log_{1/4} x$

x	64	16	4	1	1/4	1/16	1/64
y							

Graph this function and the function of Exercise 42 on the same axes. Compare the two graphs. How are they related?

46. $y = \log_{1/3} x$

x	27	9	3	1	1/3	1/9	1/27
y							

Graph this function and the function of Exercise 41 on the same axes. Compare the two graphs. How are they related?

APPLIED PROBLEMS

Solve the following problems. See Example 5.

Sales   **47.** A company finds that its total sales in dollars from the distribution of x catalogs, where x is measured in thousands, is approximated by

$$T(x) = 5000 \ln(x + 1).$$

Find the total sales resulting from the distribution of
(a) 0 catalogs;
(b) 5000 catalogs;
(c) 24,000 catalogs;
(d) 49,000 catalogs.

Population   **48.** The population of an animal species that is introduced in a certain area may grow rapidly at first and then more slowly as time goes on. A logarithmic function can provide an excellent description of such growth. Suppose that the population of foxes in an area t months after the species is introduced is given by

$$F(t) = 500 \ln(2t + 2).$$

Find the population of foxes
(a) when they are first released into the area (that is, when $t = 0$);
(b) after 3 months;
(c) after 7 months;
(d) after 10 months.

5.3 APPLICATIONS OF EXPONENTIAL AND LOGARITHMIC FUNCTIONS

In this section we discuss several applications of exponential and logarithmic functions. A common application of exponential functions involves growth or decay of a population and depends on the fact that the amount or number present at time t can be closely approximated by a function of the form

$$y = y_0 e^{kt},$$

where y_0 is the amount or number present at time $t = 0$ and k is a constant. (The reason for using this particular function is given in Section 8.1.)

Example 1   Suppose that the population of a city is given by

$$P(t) = 10{,}000 e^{0.04t},$$

where t represents time measured in years.
(a) Find the population at time $t = 0$.

$$\begin{aligned} P(0) &= 10{,}000 e^{(0.04)\cdot 0} \\ &= 10{,}000 e^0 \\ &= 10{,}000 \cdot 1 \qquad \text{Recall that } e^0 = 1 \\ &= 10{,}000. \end{aligned}$$

Thus, the population of the city is 10,000 at time $t = 0$. The initial number present is often expressed with the subscript 0. For example, here we could express the fact that the population is 10,000 at time $t = 0$ by saying that $P_0 = P(0) = 10{,}000$.
(b) Find the population of the city in year $t = 5$.

$$\begin{aligned} P(5) &= 10{,}000 e^{0.04(5)} \\ &= 10{,}000 e^{0.2}. \end{aligned}$$

234 Exponential and Logarithmic Functions

From Table 4, $e^{0.2} = 1.22140$, so that

$$P(5) \approx 10{,}000(1.22140) = 12214.$$

Thus, in 5 years the population of the city will be about 12,200. ■

1. In Example 1, find the population at the following times.
(a) $t = 2$
(b) $t = 10$

Answer:
(a) about 10,800
(b) about 14,900

Work Problem 1 at the side.

Example 2 Sales of a new product often grow rapidly at first and then begin to level off with time. For example, suppose that the sales, in some appropriate units, of a new model typewriter are approximated by the exponential function

$$S(x) = 1000 - 800e^{-x},$$

where x represents the number of years that the typewriter has been on the market. Calculate S_0, $S(1)$, $S(2)$, and $S(4)$. Graph $y = S(x)$.

Verify that $S_0 = S(0) = 1000 - 800 \cdot 1 = 200$. Using Table 4, we have

$$S(1) = 1000 - 800e^{-1}$$
$$\approx 1000 - 800(0.36787)$$
$$\approx 1000 - 294$$
$$= 706.$$

In the same way, verify that $S(2) = 892$, and $S(4) = 985$. By plotting several such points we get the graph shown in Figure 5.7. Note the line $y = 1000$ is an asymptote for the graph. This shows that in this case sales will tend to level off with time and gradually approach a level of 1000 units. ■

2. Use Example 2 to find the following.
(a) $S(3)$
(b) $S(5)$
(c) $S(10)$

Answer:
(a) about 960
(b) about 995
(c) about 1000

Work Problem 2 at the side.

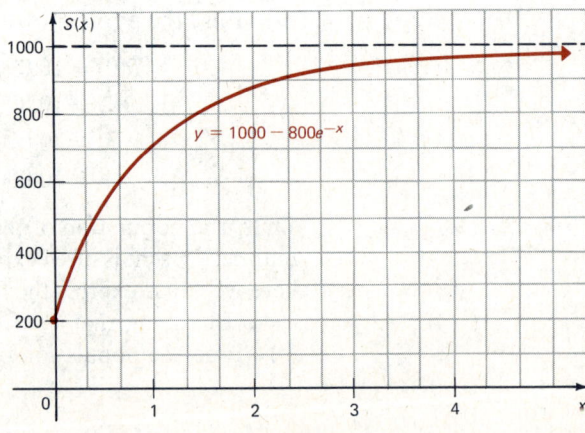

Figure 5.7

5.3 Applications of Exponential and Logarithmic Functions

Example 3 Workers new to a certain task on an assembly line will produce items according to the function

$$P(x) = 25 - 25e^{-0.3x},$$

where $P(x)$ items are produced on day x. Using this function, how many items will be produced by a new worker on the eighth day?
We must evaluate $P(8)$.

$$\begin{aligned} P(8) &= 25 - 25e^{-0.3(8)} \\ &= 25 - 25e^{-2.4} \\ &\approx 25 - 25(0.09071) \\ &\approx 25 - 2.3 \\ &= 22.7 \end{aligned}$$

Thus, on the eighth day, a new worker can be expected to produce about 23 items. By finding several other values of $P(x)$, we can graph $y = P(x)$ as shown in Figure 5.8. Such a graph is called a *learning curve*. According to the graph, a new worker tends to learn quickly at first, then learning tapers off and approaches some upper limit. This is characteristic of the learning of certain skills involving the repetitive performance of the same task. ■

3. Suppose the value of the assets of a certain company at time t is

$$A(t) = 100,000 - 75,000e^{-.2t},$$

where t is measured in years. Find the following
(a) A_0
(b) $A(5)$
(c) $A(10)$
(d) $A(25)$
(e) Graph the function.

Answer:
(a) 25,000
(b) 72,400
(c) 89,800
(d) 99,500
(e)

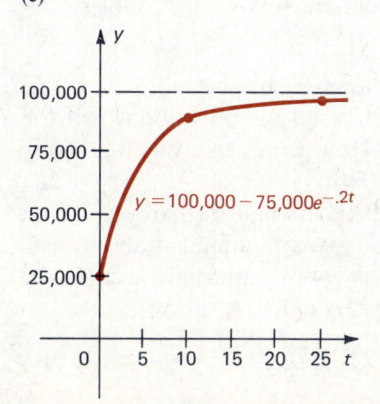

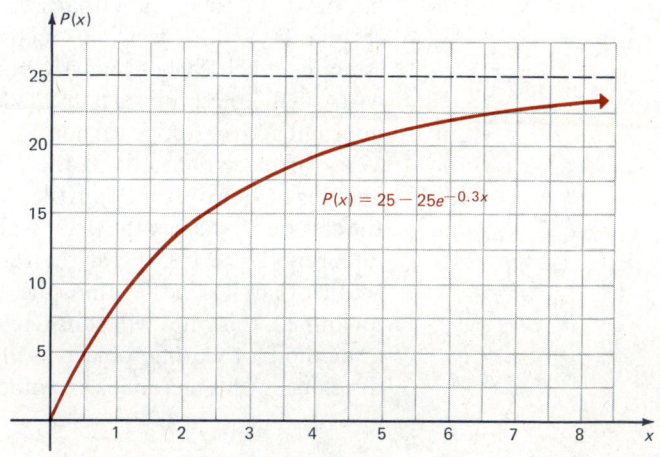

Figure 5.8

Work Problem 3 at the side.

Some applications of exponential and logarithmic functions require us to solve exponential equations such as $3^x = 5$, which have different bases. To solve such equations, we use the following property. If x and y are both positive numbers and

$$\text{if } x = y, \quad \text{then } \ln x = \ln y.$$

Exponential and Logarithmic Functions

(The property holds true for logarithms to any base, not just for natural logarithms. However, natural logarithms are most useful in applications.) The following example illustrates how we use the property to solve equations.

Example 4 Solve $3^x = 5$.

$$3^x = 5$$
$$\ln 3^x = \ln 5$$
$$x \ln 3 = \ln 5 \quad \text{By Theorem 5.1(c)}$$
$$x = \frac{\ln 5}{\ln 3}$$

From Table 5, we find $\ln 5 \approx 1.6094$ and $\ln 3 \approx 1.0986$. Thus,

$$x = \frac{1.6094}{1.0986} \approx 1.5. \quad \blacksquare$$

Work Problem 4 at the side.

4. Solve the following.
   (a) $2^x = 7$
   (b) $4^x = 12$
   (c) $e^{.04x} = 8$
   (d) $e^{-.14x} = .3$

Answer:
   (a) 2.81
   (b) 1.79
   (c) 52.0
   (d) 8.60

Carbon 14 is a radioactive isotope of carbon which has a half-life of about 5600 years (that is, about half of a given quantity of carbon 14 would decay in 5600 years). The earth's atmosphere contains much carbon, mostly in the form of carbon dioxide gas, with small traces of carbon 14. Most atmospheric carbon is the nonradioactive isotope carbon 12. The ratio of carbon 14 to carbon 12 is virtually constant in the atmosphere. However, as a plant absorbs carbon dioxide from the air in the process of photosynthesis, the carbon 12 stays in the plant while the carbon 14 is converted into nitrogen. Thus, in a plant the ratio of carbon 14 to carbon 12 is smaller than it is in the atmosphere. Even when the plant is eaten by an animal, this ratio will continue to decrease. Based on these facts, a method of dating objects, called *carbon-14 dating*, has been developed. This method is explained in the following examples.

Example 5 Suppose an Egyptian mummy has been discovered in which the ratio of carbon 14 to carbon 12 is only about half the ratio found in the atmosphere. How long ago was the body mummified?

To solve this, note that in 5600 years half the carbon 14 in a specimen will decay. Thus, the body was mummified about 5600 years ago. (If the ratio in an object is 1/4 the atmospheric ratio, the object is about $2 \cdot 5600 = 11{,}200$ years old.) $\blacksquare$

5. In Example 5, if the ratio is 1/8 the normal ratio, how old is the object?

Answer: about 16,800 years

Work Problem 5 at the side.

5.3 Applications of Exponential and Logarithmic Functions

Example 6 Let R be the (nearly constant) ratio of carbon 14 to carbon 12 found in the atmosphere, and let r be the ratio of carbon 14 to carbon 12 as found in an observed specimen. It can then be shown that the relationship between R and r is given by

$$R = r \cdot e^{(t \ln 2)/5600},$$

where t is the age of the specimen in years. Verify the formula for $t = 0$.

To do this, substitute 0 for t in the formula. This gives

$$R = r \cdot e^{(0 \cdot \ln 2)/5600}$$
$$= r \cdot e^0$$
$$= r \cdot 1$$
$$= r.$$

This result is correct—when $t = 0$ the specimen has just died, so that R and r would be the same. ∎

Example 7 Suppose a specimen has been found in which $r = \frac{2}{3}R$. Estimate the age of the specimen.

Here we use the formula given in Example 6 above.

$$R = r \cdot e^{(t \ln 2)/5600}$$
$$= \frac{2}{3} R \cdot e^{(t \ln 2)/5600}$$

Dividing through by R and multiplying through by 3/2, we have

$$\frac{3}{2} = e^{(t \ln 2)/5600}.$$

Taking natural logarithms of both sides of this last result gives

$$\ln \frac{3}{2} = \ln e^{(t \ln 2)/5600}.$$

Using properties of logarithms, we have

$$\ln \frac{3}{2} = \left[\frac{t \ln 2}{5600} \right] \ln e.$$

Since $\ln e = 1$, we have

$$\ln \frac{3}{2} = \frac{t \ln 2}{5600}.$$

To solve this equation for t, the age of the specimen, multiply both sides by $5600/\ln 2$. This gives

$$\frac{5600 \ln(3/2)}{\ln 2} = t.$$

238 Exponential and Logarithmic Functions

6. Suppose the amount of a substance present at time t (in hours) is given by
$$A(t) = 530e^{-.2t},$$
where $A(t)$ is measured in grams.
(a) Find A_0.
(b) Find the *half-life* of the substance, the time in hours when half the original amount is left.

Answer:
(a) 530 grams
(b) about 3.5 hours

Using Table 5, we have $\ln(3/2) = \ln 1.5 \approx 0.4055$ and $\ln 2 \approx 0.6931$. Thus,
$$t \approx \frac{5600(0.4055)}{0.6931} \approx 3276.$$
The specimen is about 3276 years old. ∎

Work Problem 6 at the side.

5.3 EXERCISES

Solve the following equations for x. See Example 4.

1. $5^{2x} = 6$
2. $3^{-x} = 8$
3. $e^{-.3x} = 5$
4. $e^{.1x} = 40$
5. $e^{.02x} = 10$
6. $e^{-.01x} = 12$
7. $e^{x+2} = 4$
8. $e^{1-x} = 7$
9. $e^{2x/5} = 8$
10. $e^{.1x/2} = 5$

APPLIED PROBLEMS

Population growth

Suppose that the population of a city is given by
$$P(t) = 1,000,000e^{0.02t},$$
where t represents time measured in years. Find the following values. See Example 1.

11. P_0 12. $P(2)$ 13. $P(4)$ 14. $P(10)$

Radioactive decay

Suppose the quantity, measured in grams, of a radioactive substance present at time t is given by
$$Q(t) = 500e^{-0.05t},$$
where t is measured in days. Find the quantity present at the following times.

15. $t = 0$ 16. $t = 4$ 17. $t = 8$ 18. $t = 20$

Bacteria growth

Let the number of bacteria present in a certain culture be given by
$$B(t) = 25,000e^{0.2t},$$
where t is measured in hours, and $t = 0$ corresponds to noon. Find the number of bacteria present at the following times.

19. noon 20. 1 p.m. 21. 2 p.m. 22. 5 p.m.

When a bactericide is introduced into a certain culture, the number of bacteria present is given by
$$D(t) = 50,000e^{0.01t},$$

where t is time measured in hours. Find the number of bacteria present at the following times.

23. $t = 0$ **24.** $t = 5$ **25.** $t = 20$ **26.** $t = 50$

Sales growth Sales of a new model can opener are approximated by

$$S(x) = 5000 - 4000e^{-x},$$

where x represents the number of years that the can opener has been on the market, and $S(x)$ represents the sales in thousands. Find the following. See Example 2.

27. S_0 **28.** $S(1)$ **29.** $S(2)$ **30.** $S(5)$

31. $S(10)$ **32.** Graph $y = S(x)$

Learning theory Assume that a person new to an assembly line will produce

$$P(x) = 500 - 500e^{-x}$$

items per day, where x is measured in days. Find the following. See Example 3.

33. P_0 **34.** $P(1)$ **35.** $P(2)$ **36.** $P(5)$

37. $P(10)$ **38.** Graph $y = P(x)$.

Languages The number of years since two independently evolving languages split off from a common ancestral language is approximated by

$$N(r) = -5000 \ln r,$$

where r is the proportion of the words from the ancestral language that are common to both languages now. Find the following.

39. $N(0.9)$ **40.** $N(0.5)$ **41.** $N(0.3)$

42. How many years have elapsed since the split if 80 percent of the words of the ancestral language are common to both languages today?

Specimen aging In Example 7, find the age of a specimen in which

43. $r = 0.8R$;   **44.** $r = 0.4R$;

45. $r = 0.1R$;   **46.** $r = 0.01R$.

Continuous compounding of interest In Section 5.1 we saw the formula $P\left(1 + \dfrac{i}{m}\right)^{mn}$ for compound interest. In some applications in the mathematics of finance, it is necessary to study **continuous compounding,** in which interest is compounded every instant. To find a formula for continuous compounding, take the limit as m approaches infinity (why m?).

$$\lim_{m \to \infty} P\left(1 + \frac{i}{m}\right)^{mn}$$

By certain algebraic manipulations, this limit becomes

$$\lim_{m \to \infty} P\left(1 + \frac{i}{m}\right)^{mn} = P \cdot \lim_{m \to \infty} \left[\left(1 + \frac{1}{m}\right)^m\right]^{ni}.$$

Using the definition $e = \lim\limits_{m \to \infty} \left(1 + \dfrac{1}{m}\right)^m$, we have

$$P \cdot \lim_{m \to \infty} \left[\left(1 + \frac{1}{m}\right)^m\right]^{ni} = P \cdot e^{ni}$$

as the final amount on deposit if P dollars is deposited for n years at a rate of interest i compounded continuously. Find the final amount on deposit if the following amounts are deposited at the stated interest rate, compounded continuously.

47. $1000, $i = .05$, 1 year

48. $5000, $i = .06$, 5 years

49. $18,500, $i = .04$, 10 years

50. $25,000, $i = .08$, 5 years

51. Find the number of years it will take for $1000 compounded continuously at 5% annual interest to double.

52. Find the annual interest rate at which $1000 compounded continuously for 10 years will double.

Learning theory

Under certain conditions, the total number of facts of a certain kind that are remembered is approximated by

$$N(t) = 1000\left(\frac{1 + e}{1 + e^{t+1}}\right),$$

where $N(t)$ is the number of facts remembered at time t measured in days. Find the following.

53. N_0 54. $N(1)$ 55. $N(5)$

56. Graph $y = N(t)$. This graph is called a *forgetting curve*.

It is fairly common for sales of a product, or the population of a species, to grow very slowly at first, then grow quickly, and then grow more slowly again, approaching some fixed upper limit. Such growth is often shown by **logistic curves,** having equations of the form

$$y = \frac{M}{1 + ce^{-kt}}.$$

Use a calculator to help you graph each of the following logistic curves.

57. $y = \dfrac{1000}{1 + 100e^{-t}}$ 58. $y = \dfrac{50}{1 + 5e^{-.2t}}$

A sociologist has shown that the fraction of a population who have heard a rumor after t days is approximated by

$$y(t) = \frac{y_0 e^{kt}}{1 - y_0(1 - e^{kt})},$$

where y_0 is the fraction of people who heard the rumor at time $t = 0$ and k is a constant. A graph of $y(t)$ for a particular value of k is shown in the figure.

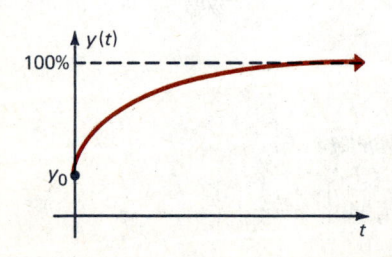

59. If $k = 0.1$ and $y_0 = 0.05$, find $y(10)$.
60. If $k = 0.2$ and $y_0 = 0.10$, find $y(5)$.

Evaporation in plants

(*A hand calculator will be helpful for the following problems.*) Helms* reports that evaporation of water from Ponderosa pines can be approximated by

$$W = 4.6 e^z,$$

where $z = \dfrac{17.3C}{C + 237}$. C is the temperature of the air surrounding the tree, in degrees Celsius. Estimate W, using Table 4, for the following temperatures.

61. $C = 15°$ **62.** $C = 20°$ **63.** $C = 22°$ **64.** $C = 25°$

5.4 DERIVATIVES OF EXPONENTIAL AND LOGARITHMIC FUNCTIONS

We complete our discussion of exponential and logarithmic functions for this chapter by discussing methods of finding derivatives of these functions.

We begin by finding the derivative of the function $y = \ln x$, where we assume that $x > 0$. This derivative can be found by going back to the definition of the derivative given in Chapter 3: if $y = f(x)$, then

$$y' = f'(x) = \lim_{h \to 0} \frac{f(x + h) - f(x)}{h}$$

provided this limit exists.

Here $y = \ln x$, so that

$$y' = f'(x) = \lim_{h \to 0} \frac{\ln(x + h) - \ln x}{h}.$$

This limit can be found by using various properties of logarithms.

$$y' = \lim_{h \to 0} \frac{\ln(x + h) - \ln x}{h} = \lim_{h \to 0} \frac{\ln\left(\dfrac{x + h}{x}\right)}{h}$$

$$= \lim_{h \to 0} \frac{1}{h} \ln\left(\frac{x + h}{x}\right)$$

$$= \lim_{h \to 0} \ln\left(\frac{x + h}{x}\right)^{1/h}$$

$$= \lim_{h \to 0} \ln\left(1 + \frac{h}{x}\right)^{1/h}$$

---

* John A. Helms, "Environmental Control of Net Photosynthesis in Naturally Growing *Pinus Ponderosa* Laws." *Ecology*, Winter, 1972, p. 92.

Now make a substitution: let $m = x/h$, so that $h/x = 1/m$. As $h \to 0$, the fact that $h/x = 1/m$ means that $m \to \infty$. So

$$\lim_{h \to 0} \ln\left(1 + \frac{h}{x}\right)^{1/h} = \lim_{m \to \infty} \ln\left(1 + \frac{1}{m}\right)^{m/x}$$

$$= \lim_{m \to \infty} \ln\left[\left(1 + \frac{1}{m}\right)^m\right]^{1/x}$$

$$= \lim_{m \to \infty} \frac{1}{x} \cdot \ln\left(1 + \frac{1}{m}\right)^m$$

In Section 1 of this chapter, we defined e as $\lim_{m \to \infty}\left(1 + \frac{1}{m}\right)^m$, so

$$\lim_{m \to \infty} \frac{1}{x} \cdot \ln\left(1 + \frac{1}{m}\right)^m = \frac{1}{x} \cdot \ln e = \frac{1}{x} \cdot 1 = \frac{1}{x}.$$

That is, if $y = \ln x$, then $y' = \frac{1}{x}$. From the chain rule, we get a more general statement. If $y = \ln g(x)$, then $y' = \frac{1}{g(x)} \cdot g'(x) = \frac{g'(x)}{g(x)}$. The next theorem summarizes the results. Absolute value bars have been used because the domain of $\ln x$ includes only positive values of x.

Theorem 5.3 (a) If $y = \ln|x|$, then $y' = \frac{1}{x}$.

(b) If $y = \ln|g(x)|$, then $y' = \frac{g'(x)}{g(x)}$.

Example 1 Find the derivative of each of the following functions.
(a) $y = \ln|5x|$
Let $g(x) = 5x$, so that $g'(x) = 5$. From Theorem 5.3(b),

$$y' = \frac{g'(x)}{g(x)} = \frac{5}{5x} = \frac{1}{x}.$$

(b) $y = \ln|3x^2 - 4x|$

$$y' = \frac{6x - 4}{3x^2 - 4x} \quad \blacksquare$$

Work Problem 1 at the side.

Now we need to find the derivative of the function $y = e^x$. To do so, first take the natural logarithm of both sides.

$$y = e^x$$
$$\ln y = \ln e^x$$

1. Find y' for the following.
(a) $y = \ln|7 + x|$
(b) $y = \ln|4x^2|$
(c) $y = \ln|8x^3 - 3x|$

Answer:
(a) $y' = \dfrac{1}{7 + x}$

(b) $y' = \dfrac{2}{x}$

(c) $y' = \dfrac{24x^2 - 3}{8x^3 - 3x}$

5.4 Derivatives of Exponential and Logarithmic Functions

By properties of logarithms,

$$\ln y = x \cdot \ln e$$
$$\ln y = x \qquad \ln e = 1$$

Use implicit differentiation to take the derivative of each side.

$$\frac{1}{y} \cdot \frac{dy}{dx} = 1$$

or

$$\frac{dy}{dx} = y.$$

Since $y = e^x$, we end up with

$$\frac{dy}{dx} = e^x, \qquad \text{or} \qquad y' = e^x.$$

This result shows one of the main reasons for the widespread use of e as a base—the function $y = e^x$ is its own derivative. Theorem 5.4 summarizes this result and the more general result using the chain rule.

Theorem 5.4   (a)   If $y = e^x$, then $y' = e^x$.

               (b)   If $y = e^{f(x)}$, then $y' = f'(x) \cdot e^{f(x)}$.

Example 2   Find derivatives of the following.
(a) $y = 4e^{5x}$
Let $f(x) = 5x$, with $f'(x) = 5$. Then

$$y' = 4 \cdot 5e^{5x} = 20e^{5x}.$$

(b) $y = 3e^{-4x}$

$$y' = 3(-4)e^{-4x} = -12e^{-4x}$$

(c) $y = 10e^{3x^2}$

$$y' = 10(6x)e^{3x^2} = 60xe^{3x^2} \quad \blacksquare$$

Work Problem 2 at the side.

Example 3   Let $y = e^x \cdot \ln|x|$. Find y'.
Use the product rule.

$$y' = e^x \cdot \frac{1}{x} + \ln|x| \cdot e^x$$

$$= e^x \left(\frac{1}{x} + \ln|x| \right) \quad \blacksquare$$

---

2. Find y' for the following.
(a) $y = 3e^{12x}$
(b) $y = -6e^{-10x+1}$
(c) $y = e^{-x^2}$

Answer:
(a) $y' = 36e^{12x}$
(b) $y' = 60e^{-10x+1}$
(c) $y' = -2xe^{-x^2}$

Exponential and Logarithmic Functions

Example 4 Let $y = \dfrac{100{,}000}{1 + 100e^{-0.3x}}$. Find y'.

Use the quotient rule.

$$y' = \frac{(1 + 100e^{-0.3x})(0) - 100{,}000(-30e^{-0.3x})}{(1 + 100e^{-0.3x})^2}$$

$$= \frac{3{,}000{,}000e^{-0.3x}}{(1 + 100e^{-0.3x})^2} \quad \blacksquare$$

Work Problem 3 at the side.

Example 5 Often a population, or the sales of a certain product, will start growing slowly, then grow more rapidly, and then gradually level off. Such growth can often be approximated by a function of the form

$$f(x) = \frac{b}{1 + ae^{kx}}$$

for appropriate constants a, b, and k. For example, suppose that population planners predict that the population of a certain city will be approximated for the next few years by the function

$$P(x) = \frac{100{,}000}{1 + 100e^{-0.3x}},$$

where x is time in years. Find the rate of change of the population at time $x = 4$.

Notice that the derivative of $P(x)$ was given in the previous example. Using this we have

$$P'(4) = \frac{3{,}000{,}000e^{-0.3(4)}}{[1 + 100e^{-0.3(4)}]^2}$$

$$= \frac{3{,}000{,}000e^{-1.2}}{(1 + 100e^{-1.2})^2}.$$

From Table 4, we see that $e^{-1.2} \approx 0.301$. Using this value, we have

$$P'(4) \approx \frac{3{,}000{,}000(0.301)}{[1 + 100(0.301)]^2}$$

$$= \frac{903{,}000}{(1 + 30.1)^2}$$

$$= \frac{903{,}000}{967}$$

$$= 934.$$

Thus the rate of change of the population at time $x = 4$ is an increase of about 934 per year. $\blacksquare$

Work Problem 4 at the side.

---

3. Find y' for the following.
(a) $y = x^2 \cdot e^x$

(b) $y = \dfrac{e^x}{1 + x}$

(c) $y = \dfrac{\ln|x|}{2 - 3x}$

Answer:
(a) $y' = x(x + 2)e^x$

(b) $y' = \dfrac{xe^x}{(1 + x)^2}$

(c) $y' = \dfrac{2 - 3x + 3x \cdot \ln|x|}{x(2 - 3x)^2}$

4. In Example 5, find the following.
(a) $P'(0)$
(b) $P'(2)$

Answer:
(a) 294
(b) 527

Example 6 Find all relative maximums or relative minimums for the function $y = (2x + 1)e^{-x}$.

First take the derivative, using the product rule.

$$y' = (2x + 1)(-e^{-x}) + e^{-x}(2)$$
$$= -2x \cdot e^{-x} - e^{-x} + 2e^{-x}$$
$$= -2x \cdot e^{-x} + e^{-x}$$
$$y' = e^{-x}(-2x + 1)$$

Place the derivative equal to 0.

$$e^{-x}(-2x + 1) = 0$$

Since e^{-x} is never 0, this derivative can equal 0 only when $-2x + 1 = 0$, or when $x = 1/2$.

Use the second derivative test to see that $x = 1/2$ leads to a relative maximum of $2/e^{1/2}$ or about 1.2. ∎

5. Find any relative maximums or relative minimums for the following.
   (a) $y = 3e^x$
   (b) $y = x^2 e^{5x}$

Answer:
(a) none
(b) relative minimum of 0 at $x = 0$; relative maximum of about .02 at $x = -2/5$

Work Problem 5 at the side.

5.4 EXERCISES

Find derivatives of the following. See Examples 1–4.

1. $y = e^{4x}$
2. $y = e^{-2x}$
3. $y = -6e^{-2x}$
4. $y = 8e^{4x}$
5. $y = -8e^{2x}$
6. $y = 0.2e^{5x}$
7. $y = -16e^{x+1}$
8. $y = -4e^{-0.1x}$
9. $y = e^{x^2}$
10. $y = e^{-x^2}$
11. $y = 3e^{2x^2}$
12. $y = -5e^{4x^3}$
13. $y = 4e^{2x^2-4}$
14. $y = -3e^{3x^2+5}$
15. $y = xe^x$
16. $y = x^2 e^{-2x}$
17. $y = (x - 3)^2 e^{2x}$
18. $y = (3x^2 - 4x)e^{-3x}$
19. $y = \ln|3 - x|$
20. $y = \ln|1 + x^2|$
21. $y = \ln|2x^2 - 7x|$
22. $y = \ln|-8x^2 + 6x|$
23. $y = \ln\sqrt{x + 5}$
24. $y = \ln\sqrt{2x + 1}$
25. $y = \dfrac{x^2}{e^x}$
26. $y = \dfrac{e^x}{2x + 1}$
27. $y = x^2 \cdot \ln|x|$
28. $y = \dfrac{\ln|x|}{x^3}$
29. $y = \dfrac{\ln|x|}{4x + 7}$
30. $y = \dfrac{-2\ln|x|}{3x - 1}$

31. $y = \dfrac{e^x}{\ln|x|}$

32. $y = \dfrac{e^x - 1}{\ln|x|}$

33. $y = \dfrac{e^x + e^{-x}}{x}$

34. $y = \dfrac{e^x - e^{-x}}{x}$

35. $y = \dfrac{5000}{1 + 10e^{0.4x}}$

36. $y = \dfrac{600}{1 - 50e^{0.2x}}$

37. $y = \dfrac{10{,}000}{9 + 4e^{-0.2x}}$

38. $y = \dfrac{500}{12 + 5e^{-0.5x}}$

39. $y = \ln|(\ln|x|)|$

40. $y = \ln 4 \cdot (\ln|3x|)$

APPLIED PROBLEMS

For Exercises 41–45, see Example 5.

Population

41. Suppose that the population of a certain collection of rare Brazilian ants is given by
$$P(t) = 1000e^{0.2t},$$
where t represents the time in days. Find the rate of change of the population when $t = 2$; when $t = 8$. (Hint: Use Table 4 of the Appendix.)

Sales

42. Often, sales of a new product grow rapidly and then level off with time. Such a situation can be represented by an equation of the form
$$S(t) = 100 - 90e^{-0.3t},$$
where t represents time in years and $S(t)$ represents sales. Find the rate of change of sales when $t = 1$; when $t = 10$.

Durability of goods

43. Suppose $P(x) = e^{-0.02x}$ represents the proportion of shoes manufactured by a given company that are still wearable after x days of use. Find the proportion of shoes wearable after
   (a) 1 day;
   (b) 10 days;
   (c) 100 days.
   (d) Calculate and interpret $P'(100)$.

Insect mating

44. Consider an experiment in which equal numbers of male and female insects of a certain species are permitted to intermingle. Assume that
$$M(t) = (e^{0.1t} + 1)\ln\sqrt{t}$$
represents the number of matings observed among the insects in an hour, where t is the temperature in degrees Celsius. (Note: The formula is an approximation at best and holds only for specific temperature intervals.) Find
   (a) $M(15)$;
   (b) $M(25)$.
   (c) Find the rate of change of $M(t)$ when $t = 15$.

Pollution

45. The concentration of pollutants, in grams per liter, in the east fork of the Big Weasel River is approximated by
$$P(x) = 0.04e^{-4x},$$

where x is the number of miles downstream from a paper mill that the measurement is taken. Find

(a) $P(0.5)$;
(b) $P(1)$;
(c) $P(2)$.

Find the rate of change of the concentration with respect to distance at

(d) $x = 0.5$;
(e) $x = 1$;
(f) $x = 2$.

Find all relative maximums or relative minimums for the following functions. See Example 6.

46. $y = xe^{-x}$
47. $y = -xe^x$
48. $y = (x + 4)e^{-2x}$
49. $y = x^2 e^{-x}$
50. $y = -x^2 e^x$
51. $y = e^x + e^{-x}$
52. $y = x - \ln|x|$

Drug concentration in blood

When a drug is injected into a muscle, the bloodstream concentration increases to some maximum and then declines. According to one mathematical model, the concentration of the drug in the bloodstream, in milliliters per cubic centimeter, is given by

$$C(t) = \frac{k}{b-a}(e^{-at} - e^{-bt}),$$

when a, b, and k are constants. In these exercises, we shall find a value of t (time) that leads to maximum drug concentration.

53. Find $C'(t)$, remembering that $k/(b - a)$ is a constant.

54. Put $C'(t) = 0$, getting

$$\frac{k}{b-a}(-ae^{-at} + be^{-bt}) = 0.$$

Multiply both sides by $(b - a)/k$, then put ae^{-at} on one side of the equals sign and be^{-bt} on the other side.

55. Divide both sides by ae^{-bt}.

56. On one side, you should have e^{-at}/e^{-bt}. Use properties of exponents to show that this equals e^{bt-at}, or $e^{(b-a)t}$.

57. Take the natural logarithm of both sides; you should get

$$\ln e^{(b-a)t} = \ln \frac{b}{a}.$$

By a property of logarithms, we rewrite this as $(b - a)t \cdot \ln e = \ln \frac{b}{a}$. Use the fact that $\ln e = 1$ to simplify this equation.

58. Solve for t.

59. Find t if $b = 50$ and $a = 10$.

60. Find t if $b = 8$ and $a = 4$.

CHAPTER 5 SUMMARY

Key Words

exponential function
logarithmic function
inverse of a function
common logarithms
natural logarithms

Things to Remember

Properties For any real numbers $x, y, a > 0$ and $a \neq 1$,

(a) If $a^x = a^y$, then $x = y$.

(b) If $x = y$, then $\ln x = \ln y$.

Definition For any real numbers $x > 0, y, a > 0$ and $a \neq 1$,

$$x = a^y \quad \text{means the same as} \quad y = \log_a x.$$

Theorem 5.1 Let x and y be any positive real numbers, and let r be any real number. Let a be a positive real number, $a \neq 1$. Then

(a) $\log_a xy = \log_a x + \log_a y$

(b) $\log_a \dfrac{x}{y} = \log_a x - \log_a y$

(c) $\log_a x^r = r \log_a x$

(d) $\log_a a = 1$

(e) $\log_a 1 = 0$.

Theorem 5.2 For any positive number x, $\ln x \approx 2.3026 \log x$, where $\log x$ is the common logarithm of x.

Theorem 5.3 (a) If $y = \ln|x|$, then $y' = \dfrac{1}{x}$.

(b) If $y = \ln|g(x)|$, then $y' = \dfrac{g'(x)}{g(x)}$.

Theorem 5.4 (a) If $y = e^x$, then $y' = e^x$.

(b) If $y = e^{f(x)}$, then $y' = f'(x) \cdot e^{f(x)}$.

CHAPTER 5 TEST

[5.1]&[5.3] *Solve the following equations.*

1. $3^x = 81$
2. $2^{3x} = \frac{1}{8}$
3. $2^x = 7$

Graph the following functions.

4. $y = 5^x$
5. $y = 5^{-x}$
6. $y = \log_5 x$

[5.2] *Write the following using logarithms.*

7. $2^6 = 64$
8. $3^{1/2} = \sqrt{3}$
9. $1000^{-1} = .001$

Write the following using exponents.

10. $\log_2 32 = 5$
11. $\log_{10} 100 = 2$
12. $\log_{27} 3 = \frac{1}{3}$

Evaluate the following.

13. $\log_7 49$
14. $\log_{25} 5$

Write each of the following as a single logarithm or, if possible, as a single number.

15. $\log_5 3 + \log_5 6$
16. $\log_3 8 - \log_3 2$
17. $6 \log_4 2$

Evaluate the following natural logarithms.

18. $\ln 6.2$
19. $\ln 700$
20. $\ln 483$

[5.3] *Solve each of the following problems.*

21. Sales of a new product are approximated by

$$S(x) = 100{,}000 e^{.2x}$$

where x represents the number of years since the product was introduced. Find
   (a) S_0;
   (b) $S(2)$;
   (c) $S(5)$.

22. Let $A(t)$ represent the amount of a certain radioactive substance present at time t (in days), where

$$A(t) = 5000 e^{-.2t}.$$

Find
   (a) A_0;
   (b) $A(2)$;
   (c) the time t when half of the substance is left.

23. A company finds that its new workers produce
$$P(x) = 100 - 100e^{-.8x}$$
items per day, after x days on the job. Find
   (a) P_0;
   (b) $P(1)$;
   (c) $P(5)$;
   (d) $P(30)$.
   (e) How many items per day would you expect an experienced worker to produce?

[5.4] Find derivatives of the following functions.

24. $y = e^{6x^2}$

25. $y = \ln|3x - 2|$

26. $y = \ln\sqrt{3-x}$

27. $y = \dfrac{e^{2x}}{5x}$

28. $y = e^x \cdot \ln x^2$

29. $y = \dfrac{6x + 4}{\ln|x|}$

30. Find the location of any relative maximums or relative minimums for $y = (5x + 3)e^{-2x}$.

CASE 6 THE VAN MEEGEREN ART FORGERIES*

After the liberation of Belgium at the end of World War II, officials began a search for Nazi collaborators. One person arrested as a collaborator was a minor painter, H. A. Van Meegeren; he was charged with selling a valuable painting by the Dutch artist Vermeer (1632–1675) to the Nazi Hermann Goering. He defended himself from the very serious charge of collaboration by claiming that the painting was a fake—he had forged it himself.

He also claimed that the beautiful and famous painting "Disciples at Emmaus," as well as several other supposed Vermeers, was his own work. To prove this, he did another "Vermeer" in his prison cell. An international panel of experts was assembled, which pronounced as forgeries all the "Vermeers" in question.

Many people would not accept the verdict of this panel for the painting "Disciples at Emmaus"; it was felt to be too beautiful to be the work of a minor talent such as Van Meegeren. In fact, the painting was declared genuine by a noted art scholar and sold for $170,000. The question of the authenticity of this painting continued to trouble art historians, who began to insist on conclusive proof one way or the other. This proof was given by a group of scientists at Carnegie-Mellon University, using the idea of radioactive decay.

The dating of objects is based on radioactivity; the atoms of certain radioactive elements are unstable, and within a given time period a fixed fraction of such atoms will spontaneously disintegrate, forming atoms of a new element. If t_0 represents some initial time, N_0 represents the number of atoms present at time t_0, and N represents the number present at some later time t, then it can be shown (using physics and calculus) that

$$t - t_0 = \frac{1}{\lambda} \cdot \ln \frac{N_0}{N}$$

where λ is a "decay constant" that depends on the radioactive substance under consideration.

If t_0 is the time that the substance was formed or made, then $t - t_0$ is the age of the item. Thus, the age of an item is given by

$$\frac{1}{\lambda} \cdot \ln \frac{N_0}{N}.$$

The decay constant λ can be readily found, as can N, the number of atoms present now. The problem is N_0—we can't find a value for this variable. However, it is possible to get reasonable ranges for the values of N_0. This is done by studying the white lead in the painting. This pigment has been used by artists for over 2000 years. It contains a small amount of the radioactive substance lead 210 and an even smaller amount of radium 226.

Radium 226 disintegrates through a series of intermediate steps to produce lead 210. The lead 210, in turn, decays to form polonium 210. This

---

* From "The Van Meegeren Art Forgeries" by Martin Braun from *Applied Mathematical Sciences*, Vol. 15. Copyright © 1975. Published by Springer-Verlag New York, Inc. Reprinted by permission.

last process, lead 210 to polonium 210, has a half life of 22 years. That is, in 22 years half the initial quantity of lead 210 will decay to polonium 210.

When lead ore is processed to form white lead, most of the radium is removed with other waste products. Thus, most of the supply of lead 210 is cut off, with the remainder beginning to decay very rapidly. This process continues until the lead 210 in the white lead is once more in equilibrium with the small amount of radium then present. Let $y(t)$ be the amount of lead 210 per gram of white lead present at time of manufacture of the pigment, t_0. Let r represent the number of disintegrations of radium 226 per minute per gram of white lead. (Actually, r is a function of time, but the half life of radium 226 is so long in comparison to the time interval in question that we assume it to be a constant.) If λ is the decay constant for lead 210, then it can be shown that

$$y(t) = \frac{r}{\lambda}[1 - e^{-\lambda(t-t_0)}] + y_0 e^{-\lambda(t-t_0)}. \qquad (1)$$

All variables in this result can be evaluated except y_0. To get around this problem, we use the fact that the original amount of lead 210 was in radioactive equilibrium with the larger amount of radium 226 in the ore from which the metal was extracted. We therefore take samples of different ores and compute the rate of disintegration of radium 226. The results are as shown in the table.

Location of ore	Disintegrations of radium 226 per minute per gram of white lead
Oklahoma	4.5
S.E. Missouri	.7
Idaho	.18
Idaho	2.2
Washington	140
British Columbia	.4

The numbers in the table vary from .18 to 140—quite a range. Since the number of disintegrations is proportional to the amount of lead 210 present originally, we must conclude that y_0 also varies over a tremendous range. Thus, equation (1) cannot be used to obtain even a crude estimate of the age of a painting. However, we want to distinguish only between a modern forgery and a genuine painting that would be 300 years old.

To do this, we observe that if the painting is very old compared to the 22 year half of lead 210, then the amount of radioactivity from the lead 210 will almost equal the amount of radioactivity from the radium 226. On the other hand, if the painting is modern, then the amount of radioactivity from the lead 210 will be much greater than the amount from the radium 226.

We want to know if the painting is modern or 300 years old. To find out, let $t - t_0 = 300$ in equation (1), getting

$$\lambda y_0 = \lambda \cdot y(t) \cdot e^{300\lambda} - r(e^{300\lambda} - 1) \qquad (2)$$

after some rearrangement of terms. If the painting is modern, then λy_0 should be a very large number; λy_0 represents the number of disintegrations of the lead 210 per minute per gram of white lead at the time of manufacture. By studying samples of white lead, we can conclude that λy_0 should never be anywhere near as large as 30,000. Thus, we use equation (2) to calculate λy_0; if our result is greater than 30,000 we conclude that the painting is a modern forgery. The details of this calculation are left for the exercises.

EXERCISES

1. To calculate λ, use the formula

$$\lambda = \frac{\ln 2}{\text{half life}}$$

   Find λ for lead 210, whose half life is 22 years.

2. For the painting "Disciples at Emmaus," the current rate of disintegration of the lead 210 was measured and found to be $\lambda \cdot y(t) = 8.5$. Also, r was found to be .8. Use this information and equation (2) to calculate λy_0. Based on your results, what do you conclude about the age of the painting?

The table below lists several other possible forgeries. Decide which of them must be modern forgeries.

	Title	$\lambda \cdot y(t)$	r
3.	"Washing of Feet"	12.6	.26
4.	"Lace Maker"	1.5	.4
5.	"Laughing Girl"	5.2	6
6.	"Woman Reading Music"	10.3	.3

6
Integration

Calculus is divided into two broad areas—differential calculus, which we discussed in Chapters 3 and 4, and integral calculus, which we consider in this chapter. Like the derivative of a function, the definite integral of a function is a special limit with many applications. Geometrically, the definite integral is related to the area under a curve. In Section 6.3 we shall see how differential and integral calculus are related by the fundamental theorem of calculus.

6.1 THE ANTIDERIVATIVE

In Chapter 3 we saw that the marginal cost to produce x items was given by the derivative $f'(x)$ of the cost function $f(x)$. Now suppose we know the marginal cost function and wish to find the cost function. For example, suppose the marginal cost at a level of production of x units is

$$f'(x) = 2x - 10.$$

Can we find the cost function? In other words, can we find a function whose derivative is $2x - 10$?

By trial and error, with a little thought on how the derivative of $f(x)$ is found, we have

$$f(x) = x^2 - 10x,$$

which has $2x - 10$ as its derivative. Is $x^2 - 10x$ the only function with derivative $2x - 10$? No, there are many; for example,

$$f(x) = x^2 - 10x + 2,$$
$$f(x) = x^2 - 10x + 5,$$
$$f(x) = x^2 - 10x - 8,$$

and so on. (Verify that each of these functions has $2x - 10$ as its derivative.) In fact, if we add any real number to $x^2 - 10x$ we have a function whose derivative is $2x - 10$. To express this fact, we write

$$f(x) = x^2 - 10x + C,$$

where C represents any constant. The function $f(x)$ is called an **antiderivative** of $f'(x)$. It can be shown that every antiderivative of $f'(x)$ is of the form $f(x) = x^2 - 10x + C$.

6.1 The Antiderivative

The trial and error method of finding an antiderivative used in the example above is not very satisfactory. We need some rules for finding antiderivatives. First, recall that to take the derivative of x^n we reduce the exponent on x by 1 and multiply by n:

$$\frac{d(x^n)}{dx} = nx^{n-1}.$$

Thus, to antidifferentiate, that is, to undo what was done, we should increase the exponent by 1 and divide by the new exponent $n + 1$. In general, we have the following rule.

Theorem 6.1 If $n \neq -1$, then an antiderivative of $f(x) = x^n$ is

$$F(x) = \frac{1}{n+1}x^{n+1} + C.$$

Note that we are labeling the antiderivative of a function $f(x)$ as $F(x)$. We can check that $F(x)$ is the antiderivative of $f(x)$ by showing that the derivative of $F(x)$ is $f(x)$, or

$$\frac{d}{dx}F(x) = f(x).$$

Here we would have

$$\frac{d}{dx}F(x) = \frac{d}{dx}\left(\frac{1}{n+1}x^{n+1} + C\right) = \frac{n+1}{n+1}x^n + 0 = x^n.$$

Why must we exclude $n = -1$?

Example 1 Find an antiderivative of each of the following.
(a) $f(x) = x^3$
By Theorem 6.1, an antiderivative of $f(x) = x^3$ is

$$F(x) = \frac{1}{3+1}x^{3+1} + C = \frac{1}{4}x^4 + C = \frac{x^4}{4} + C.$$

(b) $f(x) = \sqrt{x}$
First write $\sqrt{x}$ as $x^{1/2}$. Then use Theorem 6.1 to write

$$F(x) = \frac{1}{1/2 + 1}x^{3/2} + C = \frac{2}{3}x^{3/2} + C.$$

To check this result, show that the derivative of $F(x)$ is $f(x)$.
(c) $f(x) = 1$
To use Theorem 6.1 here, write 1 as x^0, so that the antiderivative of $f(x) = 1 = 1 \cdot x^0$ is given by

$$F(x) = \frac{1}{1}x^1 + C = x + C. \quad \blacksquare$$

Work Problem 1 at the side

1. Find an antiderivative of each of the following.
(a) $f(x) = x^5$
(b) $f(x) = \sqrt[3]{x}$
(c) $f(x) = 5$

Answer:

(a) $\dfrac{x^6}{6} + C$

(b) $\dfrac{3x^{4/3}}{4} + C$

(c) $5x + C$

256 Integration

In Chapter 3 we saw that the derivative of the product of a constant and a function is the product of the constant and the derivative of the function. We could expect a similar rule to apply to antidifferentiation and it does. Also, since we differentiate term by term, it seems reasonable to antidifferentiate term by term. The next two theorems state these properties of antiderivatives.

Theorem 6.2 If $F(x)$ is an antiderivative of $f(x)$ and if k is any constant, then the antiderivative of $k \cdot f(x)$ is $k \cdot F(x) + C$, where C is a constant.

Example 2 Find an antiderivative of $f(x) = 2x^3$.
By Theorems 6.1 and 6.2, we have the antiderivative

$$F(x) = 2\left(\frac{1}{4}x^4\right) + C = \frac{x^4}{2} + C. \quad \blacksquare$$

2. Find an antiderivative of each of the following.
(a) $f(x) = -6x^4$
(b) $f(x) = 9x^{2/3}$
(c) $f(x) = 8/x^3$

Answer:
(a) $\dfrac{-6x^5}{5} + C$
(b) $\dfrac{27x^{5/3}}{5} + C$
(c) $\dfrac{-4}{x^2} + C$

Work Problem 2 at the side.

Theorem 6.3 If $F(x)$ is an antiderivative of $f(x)$ and $G(x)$ is an antiderivative of $g(x)$, then an antiderivative of $f(x) \pm g(x)$ is $F(x) \pm G(x) + C$.

Example 3 Find an antiderivative of $f(x) = 3x^2 - 4x + 5$. Using all the theorems of this section, we have

$$F(x) = 3\left(\frac{1}{3}x^3\right) - 4\left(\frac{1}{2}x^2\right) + 5x + C$$
$$= x^3 - 2x^2 + 5x + C,$$

as an antiderivative of $f(x) = 3x^2 - 4x + 5$. $\quad \blacksquare$

3. Find an antiderivative of each of the following.
(a) $f(x) = 5x^4 - 3x^2 + 6$
(b) $f(x) = -2x^3 + 6x^2 - 3$
(c) $f(x) = 3\sqrt{x} + 2/x^2$

Answer:
(a) $x^5 - x^3 + 6x + C$
(b) $\dfrac{-x^4}{2} + 2x^3 - 3x + C$
(c) $2x^{3/2} - \dfrac{2}{x} + C$

Work Problem 3 at the side.

In Chapter 5, we saw that the derivative of $f(x) = e^x$ was $f'(x) = e^x$. Thus, we have the following theorem.

Theorem 6.4 An antiderivative of $f(x) = e^x$ is

$$F(x) = e^x + C.$$

Example 4 Find an antiderivative of $f(x) = 5e^x$.
By Theorems 6.2 and 6.4,

$$F(x) = 5e^x + C. \quad \blacksquare$$

The restriction $n \neq -1$ in Theorem 6.1 was necessary because $n = -1$ makes the denominator of the antiderivative 0. Recall from Chapter 5 that the derivative of $f(x) = \ln|x|$, where $x \neq 0$, is $f'(x) = 1/x = x^{-1}$. From that, we have the next theorem.

Theorem 6.5 An antiderivative of $f(x) = \dfrac{1}{x} = x^{-1}$ is

$$F(x) = \ln|x| + C.$$

The domain of the logarithmic function is the set of positive numbers. However, in the function $1/x$, x can be any nonzero real number. Thus, we must use absolute value in the antiderivative.

Example 5 Find an antiderivative of $f(x) = \dfrac{4}{x}$.

We use Theorems 6.2 and 6.5.

$$F(x) = 4\ln|x| + C. \quad \blacksquare$$

Work Problem 4 at the side.

We know that a marginal function is the derivative of some function $F(x)$. The antiderivative can be used to find $F(x)$ when the marginal function $F'(x)$ is known.

Example 6 Suppose a company has found that the marginal cost at a level of production of x thousand books is given by

$$C'(x) = \frac{50}{\sqrt{x}}$$

and the fixed cost (the cost to produce 0 books) is \$25,000. Find the cost function $C(x)$.

Writing $50/\sqrt{x}$ as $50x^{-1/2}$, we find that an antiderivative of $C'(x)$ is

$$C(x) = 50(2x^{1/2}) + k = 100x^{1/2} + k,$$

where k is a constant. We can find k by using the fact that $C(x)$ is 25,000 when $x = 0$.

$$C(x) = 100x^{1/2} + k$$
$$25{,}000 = 100 \cdot 0 + k$$
$$k = 25{,}000$$

Therefore, the cost function is $C(x) = 100x^{1/2} + 25{,}000$. $\quad\blacksquare$

Work Problem 5 at the side.

---

4. Find an antiderivative of each of the following.
(a) $f(x) = -5e^x$
(b) $f(x) = -6/x$
(c) $f(x) = 3e^x - 2x^{-1}$

Answer:
(a) $-5e^x + C$
(b) $-6\ln|x| + C$
(c) $3e^x - 2\ln|x| + C$

5. The marginal cost at a level of production of x items is

$$C'(x) = 2x^3 + 6x - 5.$$

The fixed cost is \$800. Find the cost function $C(x)$.

Answer:
$C(x) = \tfrac{1}{2}x^4 + 3x^2 - 5x + 800$

Example 7 If the function $s(t)$ gives the position of a particle at time t, then the velocity of the particle, $v(t)$, and its acceleration, $a(t)$, are given by

$$v(t) = s'(t)$$
$$a(t) = v'(t) = s''(t).$$

(a) Suppose the velocity of an object is $v(t) = 6t^2 - 8t$, with a position of -5 when time is 0. Find $s(t)$.

Since $v(t) = s'(t)$, we find $s(t)$ by identifying the antiderivative of $v(t)$:

$$v(t) = 6t^2 - 8t$$

so that
$$s(t) = 2t^3 - 4t^2 + C$$

for some constant C. We are told that $s = -5$ when $t = 0$.

$$s(t) = 2t^3 - 4t^2 + C$$
$$-5 = 2(0)^3 - 4(0)^2 + C$$
$$-5 = C$$

Finally,
$$s(t) = 2t^3 - 4t^2 - 5.$$

(b) Many experiments have shown that when an object is dropped, its acceleration (ignoring air resistance) is constant. This constant has been found to be approximately 32 feet per second every second, or

$$a(t) = -32.$$

The negative sign shows that the object is falling. Suppose an object is dropped so that its initial velocity is -20 feet per second. Find $v(t)$ by taking the antiderivative of $a(t)$:

$$v(t) = -32t + k.$$

When $t = 0$, $v(t) = -20$:

$$-20 = -32(0) + k$$
$$-20 = k$$

and
$$v(t) = -32t - 20.$$

Also, suppose the object was dropped from the top of the 1100-foot tall Sears Tower in Chicago. Then $s(0) = 1100$. Take the antiderivative of $v(t)$ to find $s(t)$.

$$s(t) = -16t^2 - 20t + C$$

We are told that $s(t) = 1100$ when $t = 0$. By substituting these values into our equation for $s(t)$, we end up with

$$s(t) = -16t^2 - 20t + 1100$$

as the distance of the object from the top of the building after t seconds. ∎

6.1 The Antiderivative

6. In Example 7, find the distance of the object from the top of the building after the following times.
   (a) 1 second
   (b) 4 seconds
   (c) 6 seconds
   (d) Use the quadratic formula to find the approximate time when the object will hit the ground.

Answer:
(a) 1064 feet
(b) 764 feet
(c) 404 feet
(d) about 7.7 seconds

Work Problem 6 at the side.

In Chapters 7 and 8 we will discuss applications of antiderivatives in more detail.

6.1 EXERCISES

Find an antiderivative of each of the following. See Examples 1–5.

1. $f(x) = 5x^2$
2. $f(x) = 6x^3$
3. $f(x) = 6$
4. $f(x) = 2$
5. $f(x) = 2x + 3$
6. $f(x) = 3x - 5$
7. $f(x) = x^2 - 4x + 5$
8. $f(x) = 5x^2 - 6x + 3$
9. $f(x) = 4x^3 + 3x^2 + 2x - 6$
10. $f(x) = 12x^3 + 6x^2 - 8x + 5$
11. $f(x) = \sqrt{x}$
12. $f(x) = x^{1/3}$
13. $f(x) = x^{1/2} + x^{3/2}$
14. $f(x) = 4\sqrt{x} - 3x^{3/2}$
15. $f(x) = 10x^{3/2} - 14x^{5/2}$
16. $f(x) = 56x^{5/2} + 18x^{7/2}$
17. $f(x) = 9x^{1/2} + 4$
18. $f(x) = -18x^{1/2} - 6$
19. $f(x) = 1/x^2$
20. $f(x) = 4/x^3$
21. $f(x) = 1/x^3 - 1/\sqrt{x}$
22. $f(x) = \sqrt{x} + 1/x^2$
23. $f(x) = -4e^x$
24. $f(x) = -2e^x$
25. $f(x) = 30e^x + 5x^2$
26. $f(x) = 25e^x - 10x^3$
27. $f(x) = 2x^{-1}$
28. $f(x) = -4x^{-1}$
29. $f(x) = \dfrac{3}{x} + e^x$
30. $f(x) = \dfrac{-2}{x} - 3e^x$

Find the cost function for each of the following marginal cost functions. See Example 6.

31. $C'(x) = 4x - 5$,   fixed cost is $8
32. $C'(x) = 2x + 3x^2$,   fixed cost is $15
33. $C'(x) = 0.2x^2$,   fixed cost is $10
34. $C'(x) = x^2 + 3x$,   fixed cost is $5
35. $C'(x) = \sqrt{x}$,   16 units cost $40
36. $C'(x) = x^{2/3} + 2$,   8 units cost $58
37. $C'(x) = x^2 - 2x + 3$,   3 units cost $15
38. $C'(x) = x + 1/x^2$,   2 units cost $5.50

APPLIED PROBLEMS

Profit

39. The marginal profit of Henrietta's Hamburgers is
$$P'(x) = -2x + 20,$$
where x is the sales volume in thousands of hamburgers. Henrietta knows that her profit is -50 dollars when she sells no hamburgers. What is her profit function?

40. Suppose the marginal profit on x hundred items is
$$P'(x) = 4 - 6x + 3x^2,$$
and the profit on 0 items is -40 dollars. Find the profit function.

Tangent lines

41. The slope of the tangent line to a curve is given by
$$f'(x) = 6x^2 - 4x + 3.$$
If the point $(0, 1)$ is on the curve, find the equation of the curve.

42. Find the equation of a curve whose tangent line has a slope of
$$f'(x) = x^{2/3},$$
if the point $(1, 3/5)$ is on the curve.

Velocity

43. For a particular object, $a(t) = t^2 + 1$ and $v(0) = 6$. Find $v(t)$. See Example 7.

44. Suppose $v(t) = 6t^2 - 2/t^2$ and $s(1) = 8$. Find $s(t)$.

45. Repeat the analysis of Example 7(b). Assume that an object is dropped from a small plane flying at 6400 feet, and assume that $v(0)$ is 0. Find $s(t)$. How long will it take the object to hit the ground?

Cell membranes

46. According to Fick's Law, the diffusion of a solute across a cell membrane is given by
$$c'(t) = \frac{kA}{V}(C - c(t)), \qquad (1)$$
where A is the area of the cell membrane, V is the volume of the cell, $c(t)$ is the concentration inside the cell at time t, C is the concentration outside the cell, and k is a constant. If c_0 represents the concentration of the solute inside the cell when $t = 0$, then it can be shown that
$$c(t) = (c_0 - C)e^{-kAt/V} + C. \qquad (2)$$
(a) Use this last result to find $c'(t)$.
(b) Substitute back into equation (1) to show that (2) is indeed the correct antiderivative of (1).

6.2 AREA AND THE DEFINITE INTEGRAL

In this section, we consider a method for finding the area of a region bounded on one side by a curve. For example, Figure 6.1 shows the region between the lines $x = 1$, $x = 3$, the x-axis, and the graph of the function defined by

$$f(x) = x^2 + 1.$$

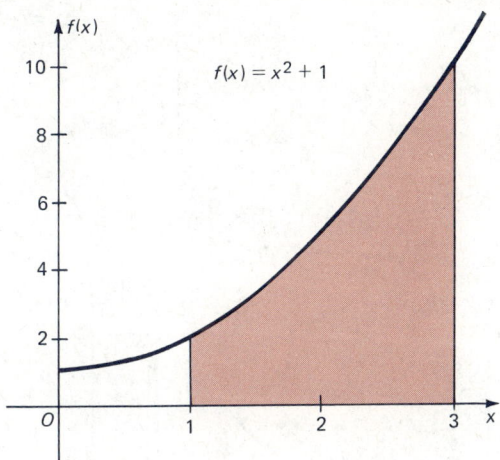

Figure 6.1

To get an approximation of this area, we could use two rectangles as shown in Figure 6.2. We use $f(1) = 2$ as the height of the rectangle on the left and $f(2) = 5$ as the height of the rectangle on the right. The width of each rectangle is 1; thus the total area of the two rectangles is

$$1 \cdot f(1) + 1 \cdot f(2) = 2 + 5 = 7 \text{ square units.}$$

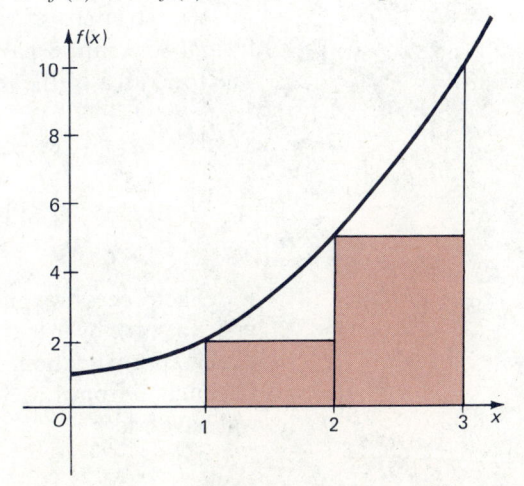

Figure 6.2

As shown in Figure 6.2, this approximation is smaller than the actual area. To improve the accuracy of the approximation, we can divide the interval from 1 to 3 into four equal parts of width 1/2, as shown in Figure 6.3. The total area of the four rectangles is

$$\frac{1}{2} \cdot f(1) + \frac{1}{2} \cdot f\left(1\frac{1}{2}\right) + \frac{1}{2} \cdot f(2) + \frac{1}{2} \cdot f\left(2\frac{1}{2}\right)$$

$$= \frac{1}{2}(2) + \frac{1}{2}\left(\frac{13}{4}\right) + \frac{1}{2}(5) + \frac{1}{2}\left(\frac{29}{4}\right)$$

$$= 1 + \frac{13}{8} + \frac{5}{2} + \frac{29}{8}$$

$$= 8.75 \text{ square units.}$$

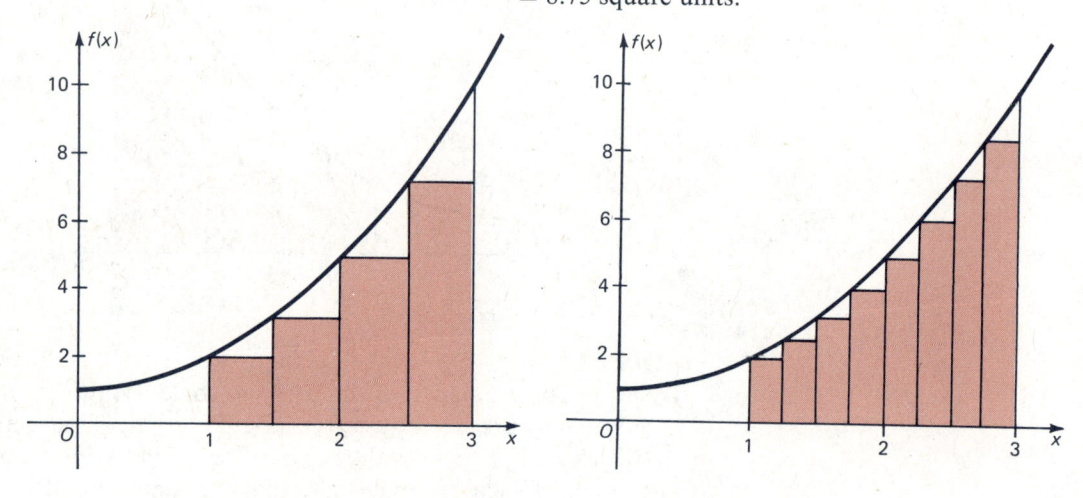

Figure 6.3

Figure 6.4

This approximation is better, but still less than the actual area we seek. To improve the approximation, we could divide the interval from 1 to 3 into 8 parts with equal widths of $\frac{1}{4}$. (See Figure 6.4.) The total area of the rectangles formed would then be

$$\frac{1}{4} \cdot f(1) + \frac{1}{4} \cdot f\left(1\frac{1}{4}\right) + \frac{1}{4} \cdot f\left(1\frac{1}{2}\right) + \frac{1}{4} \cdot f\left(1\frac{3}{4}\right)$$

$$+ \frac{1}{4} \cdot f(2) + \frac{1}{4} \cdot f\left(2\frac{1}{4}\right) + \frac{1}{4} \cdot f\left(2\frac{1}{2}\right) + \frac{1}{4} \cdot f\left(2\frac{3}{4}\right)$$

$$= 9.69 \text{ square units.}$$

The process we have been using here, approximating the area under a curve by using more and more rectangles to get a better and better approximation, can be generalized. To do this, we divide the interval from 1 to 3 into n equal parts. Each of these n intervals will have width

$$\frac{3-1}{n} = \frac{2}{n}.$$

The heights and areas of the n rectangles are shown in the following chart.

Rectangle	Height	Area
1	$f(1)$	$\dfrac{2}{n} \cdot f(1)$
2	$f\left(1 + \dfrac{2}{n}\right)$	$\dfrac{2}{n} \cdot f\left(1 + \dfrac{2}{n}\right)$
3	$f\left(1 + 2 \cdot \dfrac{2}{n}\right)$	$\dfrac{2}{n} \cdot f\left(1 + 2 \cdot \dfrac{2}{n}\right)$
4	$f\left(1 + 3 \cdot \dfrac{2}{n}\right)$	$\dfrac{2}{n} \cdot f\left(1 + 3 \cdot \dfrac{2}{n}\right)$
⋮	⋮	⋮
n	$f\left(1 + (n-1)\dfrac{2}{n}\right)$	$\dfrac{2}{n} \cdot f\left(1 + (n-1)\dfrac{2}{n}\right)$

The total of the areas listed in the last column approximates the area under the curve, above the x-axis, and between the lines $x = 1$ and $x = 3$. As n becomes larger and larger the approximation is better and better.

The simplification of the expression for the total area in the rectangles is not easy, and fortunately it is not necessary for us to go through all the steps. It is sufficient to say that the sum of the areas of the n rectangles simplifies to

$$\frac{32}{3} - \frac{8}{n} + \frac{4}{3n^2}.$$

Since n is the number of rectangles, this result can be used to evaluate the sum of the areas of any number of rectangles. For example, if $n = 20$, the total area of the rectangles is

$$\frac{32}{3} - \frac{8}{20} + \frac{4}{3(20)^2} = 10.27 \text{ square units},$$

a result very close to the actual area under the curve which we are trying to find.

1. Find the value of the sum of the areas of 100 rectangles, using the expression given above.

Answer: 10.59 square units

Work Problem 1 at the side.

If we find the limit of the expression
$$\frac{32}{3} - \frac{8}{n} + \frac{4}{3n^2}$$
as the number of rectangles is increased without bound, that is, as $n \to \infty$, we have
$$\lim_{n \to \infty} \left(\frac{32}{3} - \frac{8}{n} + \frac{4}{3n^2} \right) = \frac{32}{3} = 10\frac{2}{3}.$$
This limit is defined to be the area of the region under the graph of $f(x) = x^2 + 1$ and above the x-axis, between the lines $x = 1$ and $x = 3$. This number is called the **definite integral** of $f(x) = x^2 + 1$ from $x = 1$ to $x = 3$, and is written
$$\int_1^3 (x^2 + 1)\,dx = \frac{32}{3}.$$
The symbol $\int$ is called an **integral sign,** 3 is called the **upper limit** of integration and 1 the **lower limit** of integration.

We can generalize from this example. Figure 6.5 shows the area bounded by the curve $y = f(x)$, the x-axis, and the vertical lines $x = a$ and $x = b$. This area is written
$$\int_a^b f(x)\,dx.$$

ten rectangles of equal width
(a)

twenty rectangles of equal width
(b)

n rectangles of equal width
(c)

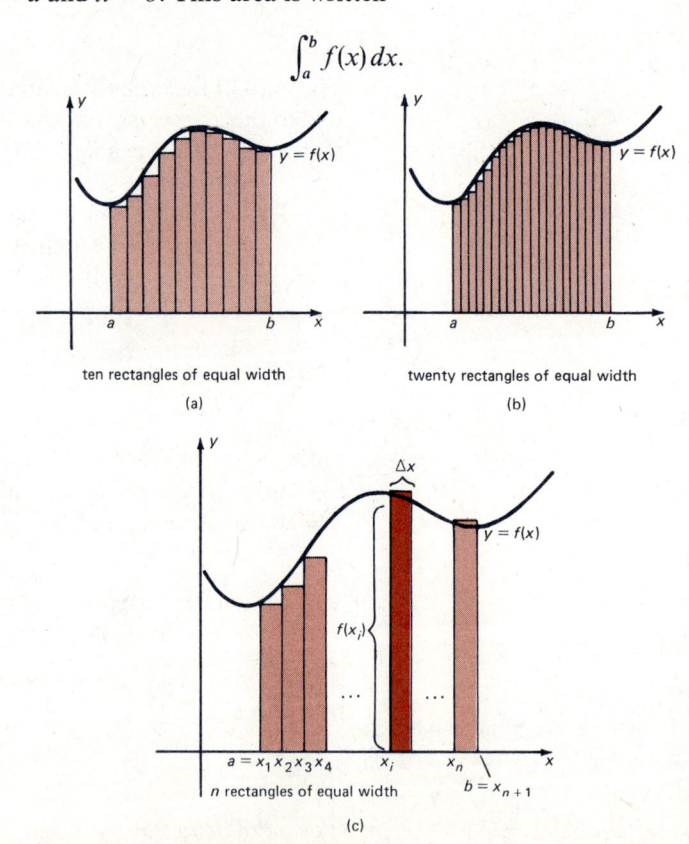

Figure 6.5

To approximate this area, we can divide the area under the curve first into ten rectangles (Figure 6.5(a)) and then into 20 rectangles (Figure 6.5(b)). The sum of the areas of the rectangles gives our approximation to the area under the curve.

To get a number which we can define as the *exact* area, we begin by dividing the interval from a to b into n pieces of equal width and use each of these n pieces as the base of a rectangle. (See Figure 6.5(c).) The sum of the areas of the rectangles gives an approximation to the desired area.

The left endpoints of the n intervals are labeled $x_1, x_2, x_3, \ldots, x_{n+1}$, where $a = x_1$ and $b = x_{n+1}$. In the graph of Figure 6.5(c), the symbol Δx is used to represent the width of each of the intervals. The rectangle in the center is an arbitrary rectangle called the ith rectangle. Its area is given by the product of its length and width. Its width is Δx, and its height is $f(x_i)$:

$$\text{area of } i\text{th rectangle} = f(x_i) \cdot \Delta x.$$

The total area under the curve can be approximated by adding the areas of all n of the rectangles. To represent addition, or summation, the Greek letter sigma, $\sum$, is used. With this symbol, the approximation to the total area becomes

$$\text{area of all } n \text{ rectangles} = \sum_{i=1}^{n} f(x_i) \cdot \Delta x.\text{*}$$

The number that is defined to be the *exact area* is the limit of this sum (if the limit exists) as the number of rectangles increases without bound, or

$$\int_a^b f(x)\,dx = \lim_{n \to \infty} \sum_{i=1}^{n} f(x_i) \cdot \Delta x.$$

The symbol on the left, as we have seen, is the **definite integral** of the function $f(x)$ between the **limits of integration** a and b.

Example 1 Find the following sums.

(a) $\sum_{i=1}^{5} (6i + 2)$

Replace i, in turn, with 1, 2, 3, 4, and 5.

$$\begin{aligned}\sum_{i=1}^{5} (6i + 2) &= (6 \cdot 1 + 2) + (6 \cdot 2 + 2) + (6 \cdot 3 + 2) \\ &\quad + (6 \cdot 4 + 2) + (6 \cdot 5 + 2) \\ &= 8 + 14 + 20 + 26 + 32 \\ &= 100\end{aligned}$$

---

* A computer program for approximating area in this way is given in Margaret L. Lial, *Study Guide with Computer Problems* (Glenview, Ill.: Scott, Foresman, 1979), pp. 75–77.

266 Integration

(b) $\sum_{i=1}^{4} 3^i = 3^1 + 3^2 + 3^3 + 3^4$
$= 3 + 9 + 27 + 81$
$= 120$ ■

2. Find each sum.

(a) $\sum_{i=1}^{6} (3i - 5)$

(b) $\sum_{i=1}^{8} (-11i + 50)$

(c) $\sum_{i=1}^{4} (2i - 1)(3i + 2)$

Answer:
(a) 33
(b) 4
(c) 182

Work Problem 2 at the side.

In the next section we shall see how the antiderivative is used in finding the definite integral and thus the area under a curve.

6.2 EXERCISES

In the following exercises, first approximate the area under the given curve and above the x-axis by using two rectangles. Let the height of the rectangle be given by the value of the function at the left side of the rectangle. Then repeat the process and approximate the area using four rectangles.

1. $f(x) = 3x + 2$ from $x = 1$ to $x = 5$
2. $f(x) = -2x + 1$ from $x = -4$ to $x = 0$
3. $f(x) = x + 5$ from $x = 2$ to $x = 4$
4. $f(x) = 3 + x$ from $x = 1$ to $x = 3$
5. $f(x) = x^2$ from $x = 1$ to $x = 5$
6. $f(x) = x^2$ from $x = 0$ to $x = 4$
7. $f(x) = x^2 + 2$ from $x = -2$ to $x = 2$
8. $f(x) = -x^2 + 4$ from $x = -2$ to $x = 2$
9. $f(x) = e^x - 1$ from $x = 0$ to $x = 4$
10. $f(x) = e^x + 1$ from $x = -2$ to $x = 2$
11. $f(x) = \dfrac{1}{x}$ from $x = 1$ to $x = 5$
12. $f(x) = \dfrac{2}{x}$ from $x = 1$ to $x = 9$
13. Let $f(x) = x^2$. If the region above the x-axis between $x = 1$ and $x = 4$ is divided into n rectangles, as in the text, and the sum of the areas of the n rectangles evaluated, the result simplifies to

$$21 - \frac{45}{2n} + \frac{9}{2n^2}.$$

(a) Approximate the area of the region described above using four rectangles.
(b) Approximate the area of the region using eight rectangles.
(c) Find $\int_1^4 x^2 \, dx$. This definite integral gives the exact area.

14. Let $f(x) = 4 - \tfrac{1}{4}x^2$. If the region from $x = 0$ to $x = 3$ is divided into n rectangles, the sum of the areas of the n rectangles is

$$\frac{39}{4} - \frac{27}{8n} - \frac{9}{8n^2}. \qquad (*)$$

Find $\int_0^3 (4 - \frac{1}{4}x^2)\,dx$ by taking the limit of (*) as $n \to \infty$. This gives the exact area.

In the following exercises, estimate the area under the curve by counting squares.

Sales of cars 15. The graph on the left shows the rate of sales of new cars in a recent year. Estimate the total sales during that year.

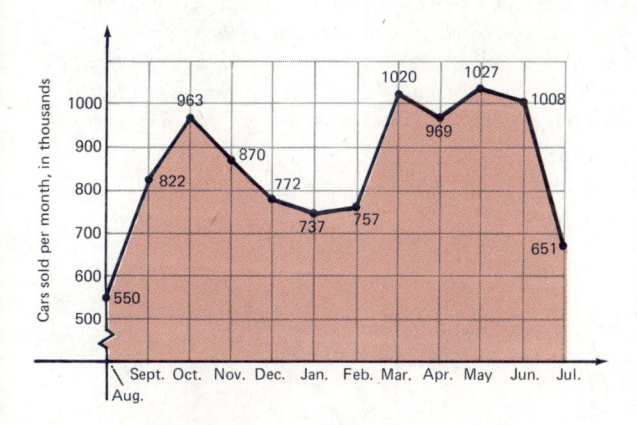

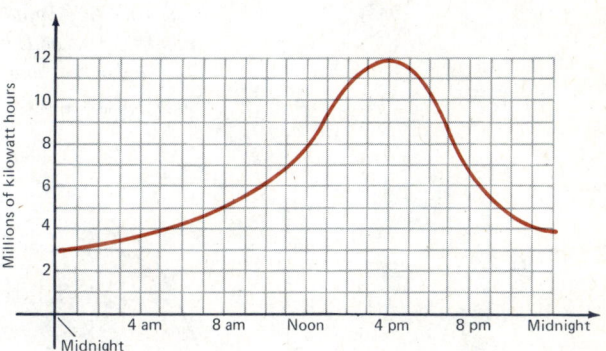

Use of electricity 16. The graph on the right shows the rate of use of electrical energy (in kilowatt hours) in a certain city on a very hot day. Estimate the total usage of electricity on that day.

Concentration of alcohol 17. The graph on the left shows the approximate concentration of alcohol in a person's bloodstream t hours after drinking 2 ounces of alcohol. Estimate the total amount of alcohol in the bloodstream by estimating the area under the curve.

Inhalation of oxygen 18. The graph on the right shows the rate of inhalation of oxygen by a person riding a bicycle very rapidly for 10 minutes. Estimate the total volume of oxygen inhaled in the first 20 minutes after the beginning of the ride.

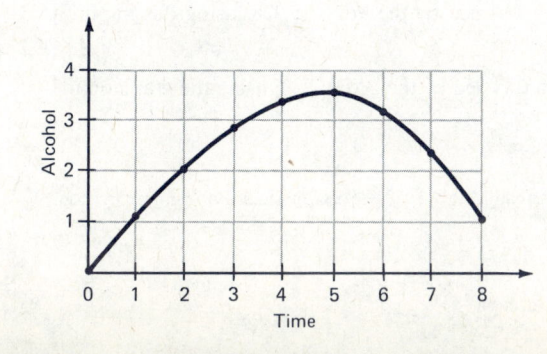

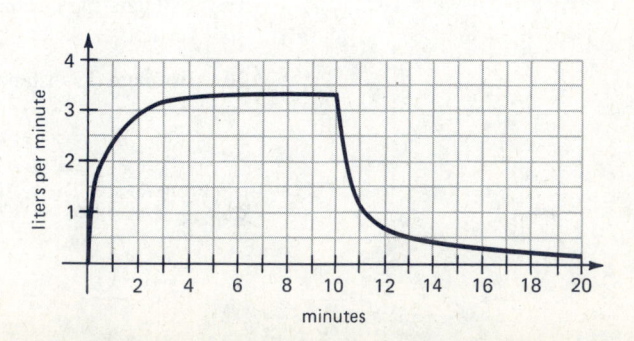

268 Integration

The next two graphs are from Road and Track *magazine.* The curve shows the velocity at time t, in seconds, when the car accelerates from a dead stop. To find the total distance traveled by the car in reaching* 100 *miles per hour, we must estimate the definite integral*

$$\int_0^t v(t)\,dt,$$

where t represents the number of seconds it takes for the car to reach 100 *mph.*

Use the graphs below to estimate this distance. Each square on the graphs represents 10 *miles on the side and* 5 *seconds across the bottom. Estimate the total number of squares in the desired area and multiply this number by* $5 \times 10 = 50$. *To adjust for the different units, divide this result by* 3600 *(the number of seconds in an hour). You then have the number of miles that the car traveled in reaching* 100 *mph. Finally, multiply by* 5280 *feet per mile to convert the answers to feet.*

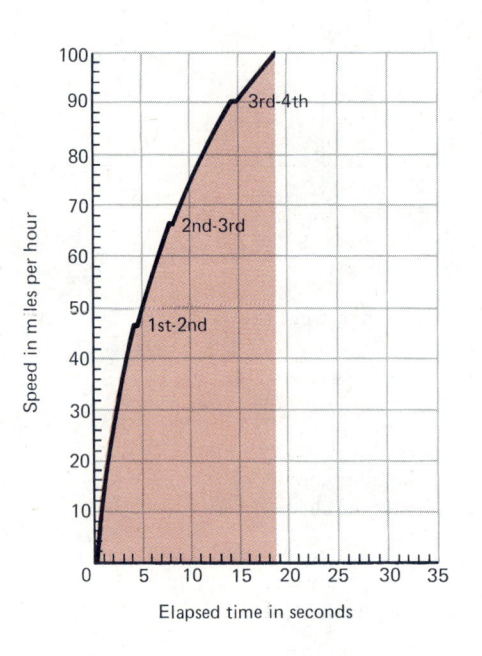

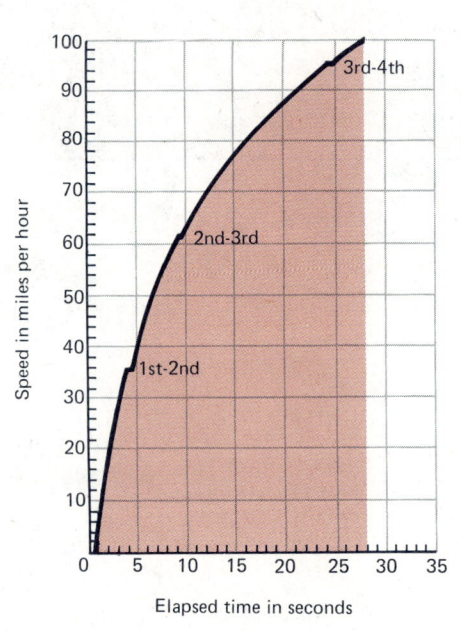

19. Estimate the distance traveled by the Porsche 928, using the graph on the left.

20. Estimate the distance traveled by the BMW 733i using the graph on the right.

---

* From *Road & Track*, April and May, 1978. CBS Consumer Publishing Division. Copyright © 1978.

Evaluate the following sums. See Example 1.

21. $\sum_{i=1}^{4} 3i$ 　　　　　　　　　　22. $\sum_{i=1}^{6} -5i$

23. $\sum_{i=1}^{5} (2i + 7)$ 　　　　　　　24. $\sum_{i=1}^{10} (5i - 8)$

25. $\sum_{i=1}^{7} (2i + 1)(3i - 9)$ 　　　26. $\sum_{i=1}^{8} (4i - 9)(5i - 12)$

27. Let $x_1 = -5$, $x_2 = 8$, $x_3 = 7$, and $x_4 = 10$. Find $\sum_{i=1}^{4} x_i$.

28. Let $x_1 = 10$, $x_2 = 15$, $x_3 = -8$, $x_4 = -12$, and $x_5 = 0$. Find $\sum_{i=1}^{5} x_i$.

6.3 THE FUNDAMENTAL THEOREM OF CALCULUS

In the previous section we found that

$$\int_{1}^{3} (x^2 + 1)\, dx = \frac{32}{3}.$$

This definite integral was found by a long calculation using rectangles and a limit. In this section we develop a simpler procedure for finding this number. An antiderivative of $f(x) = x^2 + 1$ is $F(x) = \frac{1}{3}x^3 + x + C$. If we calculate $F(3) - F(1)$ we get

$$F(3) - F(1) = \left(\frac{27}{3} + 3 + C\right) - \left(\frac{1}{3} + 1 + C\right)$$
$$= \frac{32}{3},$$

which is the value of the definite integral

$$\int_{1}^{3} (x^2 + 1)\, dx,$$

which we found in the last section. This amazing coincidence is expressed as the fundamental theorem of calculus.

Theorem 6.6 *Fundamental theorem of calculus* Let $f(x)$ be the derivative of $F(x)$ and let $f(x)$ be continuous in the interval $[a, b]$. Then

$$\int_{a}^{b} f(x)\, dx = F(b) - F(a).$$

To represent $F(b) - F(a)$ we often use

$$F(x)\Big]_a^b.$$

The fundamental theorem of calculus is really quite remarkable. It deserves its name which sets it apart as the most important theorem of calculus. It is the key connection between the differential calculus and the integral calculus, which originally were developed separately without knowledge of this connection between them.

Because of this relationship between the definite integral and the antiderivative given in the fundamental theorem, it is customary to call an antiderivative an indefinite integral (or sometimes just an integral). Indefinite integrals are indicated with the $\int$ symbol, but with no limits of integration. For example, $F(x) = x^3 - 4x^2 + 5x$ is an antiderivative of $f(x) = 3x^2 - 8x + 5$. This fact can be expressed by writing

$$\int (3x^2 - 8x + 5)\,dx = x^3 - 4x^2 + 5x + C.$$

This process of finding an antiderivative is called **integration;** we integrate $3x^2 - 8x + 5$ to obtain $x^3 - 4x^2 + 5x + C$.

Example 1 First find $\int 4x^3\,dx$ and then find $\int_1^2 4x^3\,dx$.

By the theorems in Section 6.1, we have

$$\int 4x^3\,dx = x^4 + C.$$

Thus, by the fundamental theorem, we have

$$\int_1^2 4x^3\,dx = x^4 + C\Big]_1^2$$
$$= (16 + C) - (1 + C)$$
$$= 15. \blacksquare$$

The number 15 represents the area under the curve $f(x) = 4x^3$ that is above the x-axis and between the lines $x = 1$ and $x = 2$. Since the constant C does not appear in the final answer, it can be omitted when evaluating the definite integral. It must be included, however, when finding an indefinite integral.

Work Problem 1 at the side.

Theorems 6.2 and 6.3 given in Section 6.1 also apply to the definite integral and so are restated here.

---

1. Find the following.

   (a) $\int (2x^3 + 4x)\,dx$

   (b) $\int_{-1}^2 (2x^3 + 4x)\,dx$

   Answer:

   (a) $\dfrac{x^4}{2} + 2x^2 + C$

   (b) $\dfrac{27}{2}$

Theorem 6.7 For any real constant k,
$$\int_a^b k \cdot f(x)\,dx = k \cdot \int_a^b f(x)\,dx.$$

Theorem 6.8 If all indicated definite integrals exist, then
$$\int_a^b [f(x) \pm g(x)]\,dx = \int_a^b f(x)\,dx \pm \int_a^b g(x)\,dx.$$

Example 2 Evaluate
$$\int_1^4 2x^3\,dx.$$

Using Theorems 6.1 and 6.7 and the fundamental theorem of calculus, we have
$$\int_1^4 2x^3\,dx = 2\int_1^4 x^3\,dx = 2\left(\frac{x^4}{4}\right)\Big]_1^4 = 2\left(\frac{256}{4} - \frac{1}{4}\right) = \frac{255}{2}. \blacksquare$$

Work Problem 2 at the side.

Example 3 Evaluate
$$\int_1^4 (\sqrt{x} + 2x^{3/2})\,dx.$$

By using the theorems of this section and Section 6.1 we have
$$\int_1^4 (\sqrt{x} + 2x^{3/2})\,dx = \int_1^4 x^{1/2}\,dx + \int_1^4 2x^{3/2}\,dx$$
$$= \int_1^4 x^{1/2}\,dx + 2\int_1^4 x^{3/2}\,dx$$
$$= \frac{x^{3/2}}{3/2}\Big]_1^4 + 2\left(\frac{x^{5/2}}{5/2}\Big]_1^4\right)$$
$$= \frac{2}{3}\left(x^{3/2}\Big]_1^4\right) + \frac{2\cdot 2}{5}\left(x^{5/2}\Big]_1^4\right)$$
$$= \frac{2}{3}(4^{3/2} - 1^{3/2}) + \frac{4}{5}(4^{5/2} - 1^{5/2})$$
$$= \frac{2}{3}(8-1) + \frac{4}{5}(32-1)$$
$$= \frac{14}{3} + \frac{124}{5}$$
$$= \frac{442}{15}. \blacksquare$$

Work Problem 3 at the side.

---

2. Find the following.

(a) $\int_1^4 5x^4\,dx$

(b) $\int_0^6 -8x^3\,dx$

(c) $\int_{-2}^2 x^2\,dx$

Answer:
(a) 1023
(b) -2592
(c) 16/3

3. Find the following.

(a) $\int_1^8 x^{2/3}\,dx$

(b) $\int_4^9 (x^{1/2} - 4x^{3/2})\,dx$

Answer:

(a) $\dfrac{93}{5}$

(b) $\dfrac{-4874}{15}$

272 Integration

Example 4 $\int_1^2 \frac{dx}{x} = \ln|x|\Big]_1^2$

$= \ln|2| - \ln|1|$
$= \ln 2 - \ln 1$
$= .6931 - 0$
$= .6931$ ∎

Work Problem 4 at the side.

Example 5 Find the area of the region between the x-axis and the graph of $f(x) = \frac{1}{2}x^2 + 3x$ from $x = 1$ to $x = 2$.

The area will be given by the definite integral,

$$\int_1^2 \left(\frac{1}{2}x^2 + 3x\right) dx.$$

Evaluating this integral, we have

$$\int_1^2 \left(\frac{1}{2}x^2 + 3x\right) dx = \frac{x^3}{6} + \frac{3x^2}{2}\Big]_1^2$$

$$= \left(\frac{8}{6} + 6\right) - \left(\frac{1}{6} + \frac{3}{2}\right)$$

$$= \frac{17}{3}.$$

Thus, the required area is 17/3 square units. ∎

Work Problem 5 at the side.

6.3 EXERCISES

Evaluate the following definite integrals. See Examples 1–4.

1. $\int_1^6 x \, dx$
2. $\int_1^5 x \, dx$
3. $\int_0^4 (4x + 3) \, dx$
4. $\int_5^8 (6x - 5) \, dx$
5. $\int_4^8 6 \, dx$
6. $\int_3^9 2 \, dx$
7. $\int_8^{10} (2x + 3) \, dx$
8. $\int_5^{10} (4x - 7) \, dx$
9. $\int_{-1}^1 (x + 6) \, dx$
10. $\int_1^4 (8x + 3) \, dx$
11. $\int_1^5 3x^2 \, dx$
12. $\int_0^3 -9x^2 \, dx$
13. $\int_1^3 (6x^2 - 4x + 5) \, dx$
14. $\int_1^4 (15x^2 - 6x + 3) \, dx$

---

4. Find the following.

(a) $\int_0^4 e^x \, dx$

(b) $\int_3^5 \frac{dx}{x}$

(c) $\int_1^6 e^x \, dx$

(d) $\int_2^8 \frac{4}{x} dx$

Answer:
(a) 53.59815
(b) .5108
(c) 400.71051
(d) 5.5452

5. Find the area of the region between the x-axis and the graph of $f(x) = x^2 + 4$ from $x = 0$ to $x = 3$.

Answer:
21 square units

15. $\int_{1}^{4} (3x^2 - 4x + 1)\,dx$ 16. $\int_{0}^{3} (3x^2 - 2x + 5)\,dx$

17. $\int_{4}^{9} (10x^{3/2} - 6x^{1/2})\,dx$ 18. $\int_{4}^{9} (5x^{3/2} + 3x^{1/2})\,dx$

19. $\int_{1}^{4} \sqrt{x}\,dx$ 20. $\int_{4}^{9} \sqrt{x}\,dx$

21. $\int_{2}^{5} \frac{3}{x}\,dx$ 22. $\int_{1}^{4} \frac{2}{x}\,dx$

23. $\int_{1}^{2} e^x\,dx$ 24. $\int_{0}^{5} e^x\,dx$

25. $\int_{1}^{3} (e^x + x)\,dx$ 26. $\int_{0}^{4} (x^2 - e^x)\,dx$

27. $\int_{1}^{4} (x^{-2} + x^{-1})\,dx$ 28. $\int_{2}^{5} (3x^{-1} - 2x^{-2})\,dx$

29. Let $f(x) = (x^2 + 2x)^2$. Verify that $f'(x) = 2(x^2 + 2x)(2x + 2) = 4(x + 1)(x^2 + 2x)$, and then find

$$\int_{2}^{3} 4(x + 1)(x^2 + 2x)\,dx.$$

30. Let $f(x) = \sqrt{x^2 - 4}$. Verify that $f'(x) = x/\sqrt{x^2 - 4}$, and then find

$$\int_{3}^{4} \frac{x\,dx}{\sqrt{x^2 - 4}}.$$

For Exercises 31–36, see Example 5.

31. Find the area of the region enclosed by the x-axis, the graph of $f(x) = x^2 + 4x + 4$, $x = -1$, and $x = 2$.

32. Find the area of the region enclosed by the x-axis, $y = x$, $x = 0$, and $x = 4$.

33. Find the area of the region enclosed by the x-axis, $y = x^3$, $x = 0$, and $x = 3$.

34. Find the area of the region enclosed by the x-axis, $y = x^2 + x$, $x = 1$, and $x = 4$.

35. Find the area of the region enclosed by the x-axis, $y = \frac{1}{x}$, $x = 1$, and $x = 2$.

36. Find the area of the region enclosed by the x-axis, $y = \frac{1}{x}$, $x = 3$, and $x = 4$.

6.4 SOME APPLICATIONS OF THE DEFINITE INTEGRAL

The definite integral can be used to express *total value over a period of time* as in the following example.

Suppose a leasing company wants to decide on the yearly lease fee for a certain new typewriter. The company expects to lease the typewriter for 5 years and it expects the rate of maintenance, in dollars, to be approximated by

$$M(t) = 10 + 2t + t^2,$$

where t is the number of years the typewriter has been used. Figure 6.6 shows the graph of $M(t)$. The total maintenance charge for the 5-year period will be given by the shaded area of the figure and can be found using the definite integral as follows.

$$\int_0^5 (10 + 2t + t^2)\, dt = 10t + t^2 + \frac{t^3}{3}\Big]_0^5$$

$$= 50 + 25 + \frac{125}{3} - 0$$

$$\approx 116.67$$

The company can expect the total maintenance charge for 5 years to be about $117. Hence, the company should add about

$$\frac{\$117}{5} = \$23.40$$

to its annual lease price to pay for maintenance.

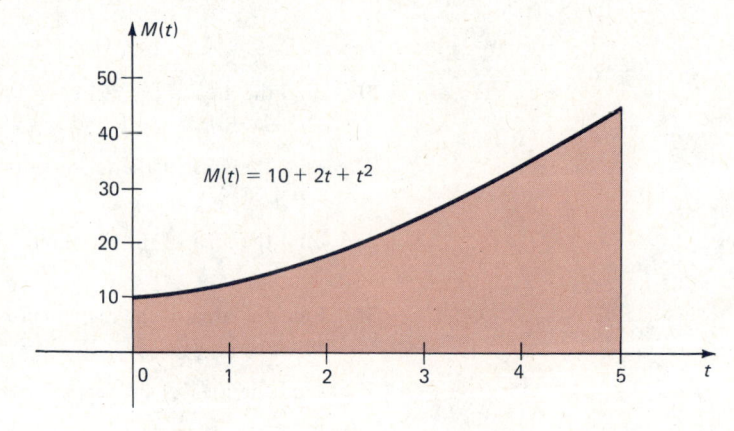

Figure 6.6

Work Problem 1 at the side.

1. Find the total maintenance charge for a lease of
   (a) 1 year;
   (b) 2 years.

   Answer:
   (a) $11.33
   (b) $26.67

Example 1 Elizabeth, who runs a factory that makes signs, has been shown a new machine to staple the signs to the handles. She estimates that the rate of savings from the machine will be approximated by

$$S(x) = 3 + 2x,$$

where x represents the number of years the stapler has been used. If the machine costs $70, would it pay for itself in 5 years?

6.4 Some Applications of the Definite Integral

We need to find the area under the savings curve shown in Figure 6.7, between the lines $x = 0$ and $x = 5$ and the x-axis. Using definite integrals, we have

$$\int_0^5 (3 + 2x)\,dx = 3x + x^2 \Big]_0^5 = 40.$$

The total savings in five years is $40, so the machine will not pay for itself in this time period. ∎

2. Find the amount of savings over the first 6 years for the machine in Example 1.

Answer: $54

Work Problem 2 at the side.

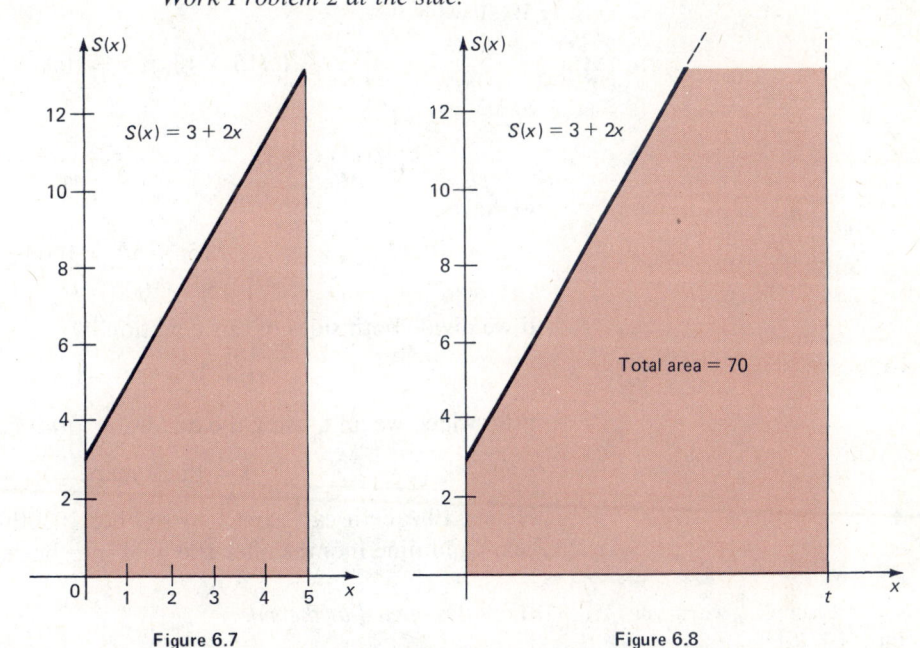

Figure 6.7

Figure 6.8

To find the number of years it will take for the machine of Example 1 to pay for itself, note that, since the machine costs a total of $70, it will pay for itself when the area under the savings curve of Figure 6.8 equals 70, or at a time t such that

$$\int_0^t (3 + 2x)\,dx = 70.$$

Evaluating the definite integral we have

$$3x + x^2 \Big]_0^t = (3t + t^2) - (3 \cdot 0 + 0^2) = 3t + t^2.$$

Since we want the total savings to equal 70,

$$3t + t^2 = 70.$$

3. Find the number of years in which the machine in Example 1 will pay for itself if the rate of savings is $S(x) = 5 + 4x$.

Answer: 4.8 years

Solve this quadratic equation to verify that the machine will pay for itself in seven years.

Work Problem 3 at the side.

Example 2 Elizabeth, of the previous example, believes that the rate of sales of her signs is given by

$$T(x) = 15 + 10x,$$

where x is the number of years she has been in business. When can she expect to sell her 1000th sign?

Let t be the time at which Elizabeth will sell her 1000th sign. The area under the sales curve between $x = 0$ and $x = t$ gives the total sales during that period. Since we want the total sales to be 1000, we have

$$\int_0^t (15 + 10x)\,dx = 1000.$$

Since

$$\int_0^t (15 + 10x)\,dx = 15x + 5x^2 \Big]_0^t = 15t + 5t^2,$$

we have

$$15t + 5t^2 = 1000$$
$$5t^2 + 15t - 1000 = 0.$$

If we divide both sides of this equation by 5, we have

$$t^2 + 3t - 200 = 0,$$

from which we find, using the quadratic formula,

$$t = 12.72 \text{ years}.$$

Hence, Elizabeth can expect to sell her 1000th sign about twelve years and nine months after she goes into business. ∎

Work Problem 4 at the side.

4. In Example 2, when will Elizabeth sell
(a) the 500th sign;
(b) the 1200th sign?

Answer:
(a) 8.61 years
(b) 14.06 years

Example 3 The rate of reaction to a given dose of a drug at time t hours after administration is given by

$$R(t) = \frac{1}{t},$$

where $R(t)$ is measured in appropriate units. Evaluate the total reaction to the given dose of the drug from $t = 1$ to $t = 24$ hours.

The total reaction will be given by

$$\int_1^{24} \frac{1}{t}\,dt = \ln|t|\Big]_1^{24}$$
$$= \ln 24 - \ln 1$$
$$= \ln(4)(6) - 0$$
$$= \ln 4 + \ln 6$$
$$\approx 1.3863 + 1.7918$$
$$= 3.1781. \quad \blacksquare$$

5. The rate of reaction to a given dose of a drug at time t measured in hours is

$$R(t) = \frac{9}{t^2}.$$

(a) Find the total reaction from time $t = 1$ to $t = 12$.
(b) Find the total reaction on the second day.

Answer:
(a) 8.25
(b) 9/48, or .1875

Work Problem 5 at the side.

Example 4 How much is added to the total reaction in Example 3 during the next hour?

We want to find the total reaction from $t = 24$ to $t = 25$.

$$\int_{24}^{25} \frac{1}{t}\,dt = \ln|t|\Big]_{24}^{25}$$

$$= \ln 25 - \ln 24$$

$$\approx 3.2189 - 3.1781$$

$$= .0408$$

We found ln 25 in the table, and ln 24 came from Example 3. There is very little change in total reaction after 24 hours. ∎

6. In Example 4, find the reaction to the drug from time $t = 6$ to $t = 10$.

Answer: .5108

Work Problem 6 at the side.

6.4 APPLIED PROBLEMS

Maintenance charges

1. A car-leasing firm must decide how much to charge for maintenance on the cars it leases. After careful study, it is decided that the rate of maintenance on a new car is approximated by

$$M(x) = 60(1 + x^2),$$

where x is the number of years the car has been in use. What total maintenance charge can the company expect for a two-year lease? What amount should it add to the monthly lease payments to pay for maintenance?

2. Using the function of Exercise 1, find the maintenance charge the company can expect during the third year. Find the total charge during the first three years. What monthly charge should be added to take care of a three-year lease which includes maintenance?

Savings

3. A company is considering a new manufacturing process. It knows that the rate of savings from the process will be about

$$S(t) = 2t + 30,$$

where t is the number of years of use of the process. Find the total savings during the first year. Find the total savings during the first six years. See Example 1.

4. Assume that the new process in Exercise 3 costs $1000. About when will it pay for itself?

Production

5. A company is introducing a new product. Production is expected to grow slowly because of difficulties in the start-up process. It is expected that the rate of production will be approximated by

$$P(x) = 600x^{3/2},$$

where x is the number of years since the introduction of the product. Will the company be able to supply 10,000 units during the first four years? See Example 2.

Pollution

6. Pollution from a factory is entering a lake. The rate of concentration of the pollutant at time t is given by

$$P(t) = 140t^{5/2},$$

where t is the number of years since the factory started introducing pollutants into the lake. Ecologists estimate that the lake can accept a total level of pollution of 4850 units before all the fish life in the lake ends. Can the factory operate for 4 years without killing all the fish in the lake?

Expenditure

7. De Win Enterprises has found that its expenditure rate per day on a certain type of job is given by

$$E(x) = 4x + 2,$$

where x is the number of days since the start of the job. Find the total expenditure if the job takes 10 days.

8. In Exercise 7, how much will be spent on the job from the 10th to the 20th day?

9. In Exercise 7, if the company wants to spend no more than $5000 on the job, in how many days must they complete it?

Income

10. De Win Enterprises also knows that the rate of income per day for the same job is

$$I(x) = 100 - x,$$

where x is the number of days since the job was started. Find the total income for the first 10 days.

11. In Exercise 10, find the income from the 10th to the 20th day.

12. In Exercise 10, how many days must the job last for the total income to be at least $5000?

Profit

13. After a new firm starts in business, it finds that its rate of profits (in hundreds of dollars) after t years of operation, is given by

$$P(t) = 6t^2 + 4t + 5.$$

Find the total profits in the first three years.

14. Find the profit in Exercise 13 in the fourth year of operation.

Oil leakage

15. An oil tanker is leaking oil at the rate of $20t + 50$ barrels per hour, where t is time in hours after the tanker hits a hidden rock. Find the total number of barrels that the ship will leak on the first day.

16. Find the number of barrels that the ship of Exercise 15 will leak on the second day.

Tree growth

17. After long study, tree scientists conclude that a eucalyptus tree will grow at the rate of $.2 + 4t^{-4}$ feet per year, when t is time in years. Find the number of feet that the tree will grow in the second year.

18. Find the number of feet the tree in Exercise 17 will grow in the third year.

Worker efficiency

19. A worker new to a job will improve his efficiency with time so that it takes him fewer hours to produce an item with each day on the job up to a certain point. Suppose the rate of change of the number of hours it takes a worker in a certain factory to produce the xth item is given by
$$H(x) = 20 - 2x.$$
(a) What is the total number of hours required to produce the first 5 items?
(b) What is the total number of hours required to produce the first 10 items?

Growth rate

20. The rate at which a substance grows is given by
$$R(x) = 200e^x,$$
where x is the time in days. What is the total accumulated growth after 2.5 days?

Drug reaction

21. For a certain drug, the rate of reaction in appropriate units is given by
$$R(t) = \frac{5}{t} + \frac{2}{t^2},$$
where t is measured in hours after the drug is administered. See Example 3. Find the total reaction to the drug
(a) from $t = 1$ to $t = 12$;
(b) from $t = 12$ to $t = 24$.

22. For another drug, the rate of reaction in appropriate units is
$$R(t) = \frac{4}{t^2} + 1,$$
where t is measured in hours after the drug is administered. Find the total reaction to the drug
(a) from $t = \frac{1}{4}$ to $t = 1$;
(b) from $t = 12$ to $t = 24$.

6.5 THE AREA BETWEEN TWO CURVES

In Section 6.3 we saw that the definite integral $\int_a^b f(x)\,dx$ can be used to find the area below the graph of the function $y = f(x)$, above the x-axis, and between the lines $x = a$ and $x = b$. In this section we extend this idea to find other areas.

Example 1 Find the area between the x-axis and the graph of $f(x) = x^2 - 4$ from $x = 0$ to $x = 4$.

As shown in Figure 6.9, part of the area lies above the x-axis and part lies below the x-axis. If we were to use the definite integral to evaluate the area below the x-axis, the result would be a negative number, since the function values there are all negative. Since area is a nonnegative quantity, the correct value of that part of the area is given by the negative of the appropriate definite integral. In cases like this, the two parts of the area must be calculated separately and the two positive numbers then added to give the total area.

To find the area between the x-axis and the graph of $f(x) = x^2 - 4$ from $x = 0$ to $x = 4$, we must first find the point where the graph crosses the x-axis. This is done by solving the equation

$$x^2 - 4 = 0.$$

The solutions of this equation are 2 and -2. We are only interested in the values of x in the interval $[0, 4]$ so we discard the solution -2. Thus, the total area will be given by

$$-\int_0^2 (x^2 - 4)\,dx + \int_2^4 (x^2 - 4)\,dx$$
$$= -\left(\frac{1}{3}x^3 - 4x\right)\Big]_0^2 + \left(\frac{1}{3}x^3 - 4x\right)\Big]_2^4$$
$$= -\left(\frac{8}{3} - 8\right) + \left(\frac{64}{3} - 16\right) - \left(\frac{8}{3} - 8\right)$$
$$= 16 \text{ square units.} \blacksquare$$

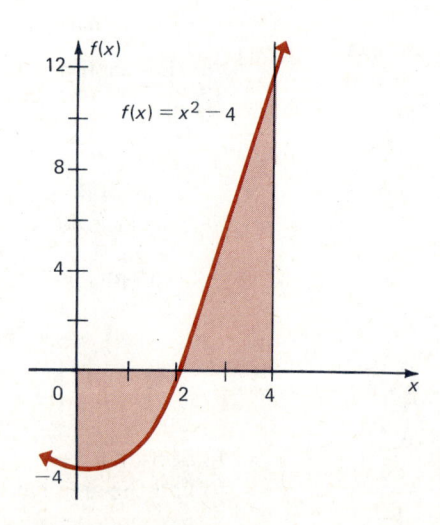

Figure 6.9

1. Find the area between the x-axis and the graph of $f(x) = x - 4$ from $x = 1$ to $x = 5$.

Answer: 5

Work Problem 1 at the side.

Example 2 Find the area between the curves $f(x) = x^{1/2}$ and $g(x) = x^3$ from $x = 0$ to $x = 1$.

To find the area bounded by these two curves and the lines $x = 0$ and $x = 1$, we use definite integrals. The area between $f(x) = x^{1/2}$ and the x-axis from $x = 0$ and $x = 1$ is

$$\int_0^1 x^{1/2} \, dx,$$

while the area between $g(x) = x^3$ and the x-axis from $x = 0$ to $x = 1$ is

$$\int_0^1 x^3 \, dx.$$

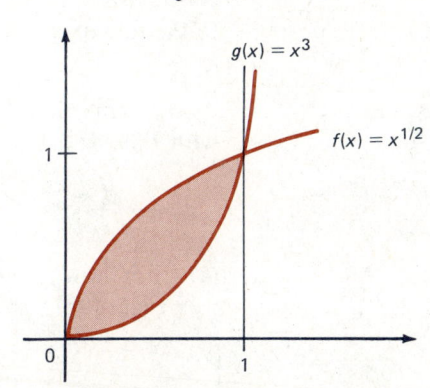

Figure 6.10

As shown in Figure 6.10, the area between these two curves is given by the difference between these two integrals, or

$$\int_0^1 x^{1/2} \, dx - \int_0^1 x^3 \, dx,$$

which, by Theorem 6.6, can be written as

$$\int_0^1 (x^{1/2} - x^3) \, dx.$$

Using the fundamental theorem of calculus, we have

$$\int_0^1 (x^{1/2} - x^3) \, dx = \frac{x^{3/2}}{3/2} - \frac{x^4}{4} \Big]_0^1$$

$$= \frac{2}{3} x^{3/2} - \frac{x^4}{4} \Big]_0^1$$

$$= \frac{2}{3} \cdot 1 - \frac{1}{4}$$

$$= \frac{5}{12} \text{ square units.} \blacksquare$$

2. Find the area between $y = x$ and $y = x^2$ from $x = 0$ to $x = 1$.

Answer: 1/6 square units

Work Problem 2 at the side.

282 Integration

The difference between two integrals can be used to find the area between the graphs of two functions even if one graph lies below the x-axis or if both graphs lie below the x-axis. In general, if $f(x) \le g(x)$ for all values of x in the interval $[a, b]$, then the area between the two graphs is

$$\int_a^b [g(x) - f(x)]\, dx.$$

Work Problem 3 at the side.

3. Find the area between $f(x) = -x^2 + 4$ and $g(x) = x^2 - 4$.

Answer: 64/3 square units

Example 3 A company is considering introducing a new manufacturing process in one of its plants. The new process provides substantial savings, with the savings declining with time x according to the rate of savings function

$$S(x) = 100 - x^2.$$

At the same time, the cost of operating the new process increases with time x, according to the rate of cost function

$$C(x) = x^2 + \frac{14}{3}x.$$

(a) For how many years will the company realize savings?

Figure 6.11 shows the graphs of the rate of savings and the

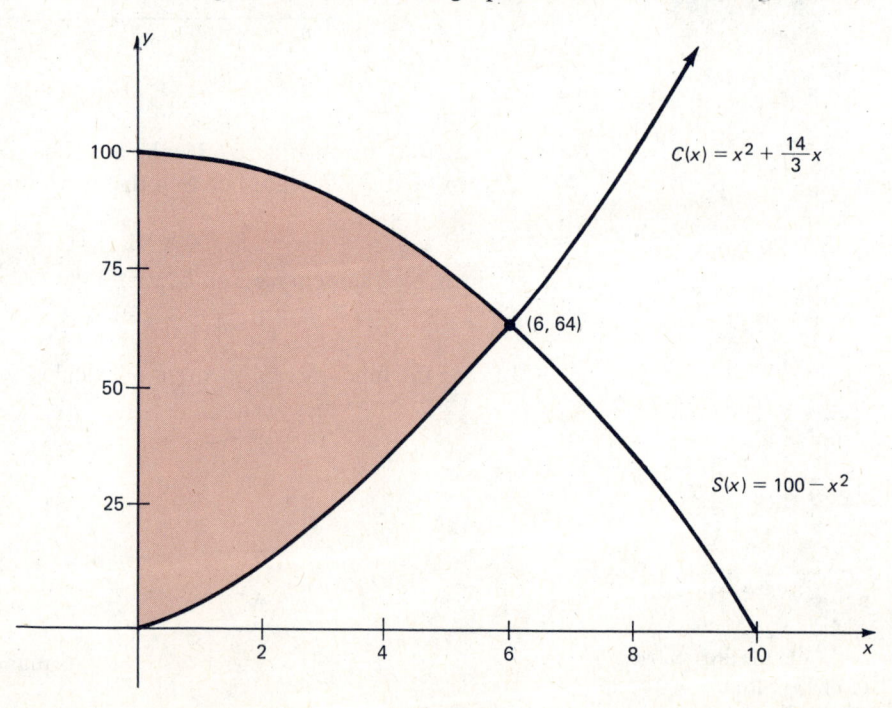

Figure 6.11

rate of cost functions. The company should use this new process until the time at which these functions intersect. That is, the company should find the value of x such that

$$C(x) = S(x),$$

or such that

$$100 - x^2 = x^2 + \frac{14}{3}x.$$

Solving this equation, we find that the only valid solution is $x = 6$. Thus, the company should use the new process for 6 years.

(b) What will the total savings be during this period?

The total savings over the 6-year period are given by the area between the rate of cost and the rate of savings curves and the lines $x = 0$ and $x = 6$, which we can evaluate with definite integrals as follows:

$$\begin{aligned} \text{Total savings} &= \int_0^6 (100 - x^2)\, dx - \int_0^6 \left(x^2 + \frac{14}{3}x\right) dx \\ &= \int_0^6 \left[(100 - x^2) - \left(x^2 + \frac{14}{3}x\right)\right] dx \\ &= \int_0^6 \left(100 - \frac{14}{3}x - 2x^2\right) dx \\ &= 100x - \frac{7}{3}x^2 - \frac{2}{3}x^3 \Big]_0^6 \\ &= 100(6) - \frac{7}{3}(36) - \frac{2}{3}(216) \\ &= 372. \end{aligned}$$

Thus, the company will save a total of $372 over the 6-year period. ■

Work Problem 4 at the side.

4. In Example 3, find the total savings if pollution-control regulations permit the new process for only 4 years.

Answer: $320

Example 4 A farmer has been using a new fertilizer that gives him a better yield, but because it exhausts the soil of other nutrients he must use other fertilizers in greater and greater amounts, so that his costs increase each year. The new fertilizer produces a rate of increase in revenue (in hundreds of dollars) given by

$$R(t) = -0.4t^2 + 8t + 10,$$

where t is measured in years. The rate of yearly costs due to use of the fertilizer is given by

$$C(t) = 2t + 5.$$

How long can the farmer profitably use the fertilizer? What will be his increase in revenue over this period?

The farmer should use the new fertilizer until the additional costs equal the increase in revenue. Thus, we need to solve the equation $R(t) = C(t)$ as follows.

$$-0.4t^2 + 8t + 10 = 2t + 5$$
$$-4t^2 + 80t + 100 = 20t + 50$$
$$-4t^2 + 60t + 50 = 0$$
$$t = 15.8$$

The new fertilizer will be profitable for about 15.8 years.

To find the total amount of additional revenue over the 15.8-year period, we must find the area between the graphs of the revenue and the cost functions, as shown in Figure 6.12. We have

$$\text{total savings} = \int_0^{15.8} [R(t) - C(t)]\, dt$$
$$= \int_0^{15.8} [(-0.4t^2 + 8t + 10) - (2t + 5)]\, dt$$
$$= \int_0^{15.8} (-0.4t^2 + 6t + 5)\, dt$$
$$= \frac{-0.4t^3}{3} + \frac{6t^2}{2} + 5t \Big]_0^{15.8}$$
$$= 302.01.$$

The total savings will amount to about $30,000 over the 15.8-year period.

It is probably not realistic to say that the farmer will need to use the new process for 15.8 years—he will probably have to use it for 15 years or for 16 years. In this case, when the mathematical model produces results that are not in the domain of the function, it will be necessary to find the total savings after 15 years and after 16 years and then select the best result. ∎

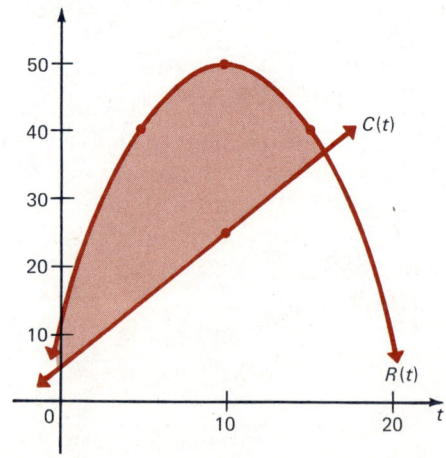

Figure 6.12

5. In Example 4, suppose
$C(x) = 3t + 5$.
(a) How long should the farmer use the fertilizer?
(b) What is his increase in revenue over that period?

Answer:
(a) about 13.4 years
(b) about $19,500

Work Problem 5 at the side.

Example 5 Suppose the price, in cents, for a certain product is

$$P(x) = 900 - 20x - x^2,$$

when the demand for the product is x units. Also, suppose the function

$$P(x) = x^2 + 10x$$

gives the price, in cents, when the supply is x units. The graphs of both functions are shown in Figure 6.13 along with the equilibrium point at which supply and demand are equal. To find the equilibrium supply or demand, x^*, we solve the equation

$$900 - 20x^* - (x^*)^2 = (x^*)^2 + 10x^*$$
$$0 = 2(x^*)^2 + 30x^* - 900$$
$$0 = (x^*)^2 + 15x^* - 450.$$

The only positive solution of the equation is $x^* = 15$.

At the equilibrium point where supply = demand = 15 units, the price is

$$P(15) = 900 - 20(15) - 15^2 = 375,$$

or $3.75.

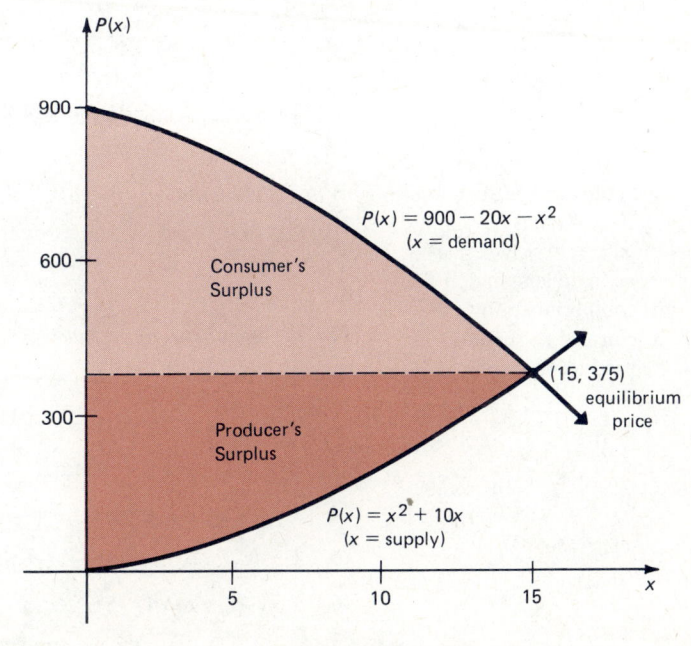

Figure 6.13

As the demand graph shows, there are consumers who are willing to pay more than the equilibrium price, so they benefit from the equilibrium price. Some benefit a lot, some less, and so on. Their total benefit is called the *consumer's surplus* and is represented by an area as shown in Figure 6.13. The consumer's surplus is evaluated by the definite integral

$$\left[\int_0^{15}(900-20x-x^2)\,dx\right]-(15)(375)$$

$$=\left(900x-10x^2-\frac{x^3}{3}\right]_0^{15}\right)-5625$$

$$=\left(900(15)-10(15)^2-\frac{(15)^3}{3}\right)-5625$$

$$=4500.$$

Here, the consumer's surplus is 4500 cents, or $45.00.

On the other hand, some suppliers would have offered the product at a price below the equilibrium price, so they too gain from the equilibrium price. The total of the supplier's gains is called the *producer's surplus* and is represented by the area shown in Figure 6.13. Producer's surplus is given by the definite integral

$$(15)(375)-\int_0^{15}(x^2+10x)\,dx = 5625-\left(\frac{x^3}{3}+5x^2\right)\Big]_0^{15}$$

$$=5625-\left(\frac{15^3}{3}+5(15)^2\right)$$

$$=3375.$$

The producer's surplus is $33.75. ∎

Work Problem 6 at the side.

6.5 EXERCISES

Find the areas between the following curves. See Examples 1 and 2.

1. $x=-2$, $x=1$, $f(x)=x^2+4$, $y=0$
2. $x=1$, $x=2$, $f(x)=x^3$, $y=0$
3. $x=-3$, $x=1$, $f(x)=x+1$, $y=0$
4. $x=-2$, $x=0$, $f(x)=1-x^2$, $y=0$
5. $x=0$, $x=5$, $f(x)=\frac{7}{5}x$, $g(x)=\frac{3}{5}x+10$
6. $x=-1$, $x=1$, $f(x)=x$, $S(x)=x^2-3$
7. $x=0$, $x=2$, $f(x)=150-x^2$, $S(x)=x^2+\frac{11}{4}x$
8. $x=0$, $x=8$, $f(x)=150-x^2$, $g(x)=x^2+\frac{11}{4}x$

6. Given the demand function $p(x)=12-.07x$ and the supply function $p(x)=.05x$, where $p(x)$ is the price in dollars, find
(a) the equilibrium point;
(b) the consumer's surplus;
(c) the producer's surplus.

Answer:
(a) $x=100$
(b) $350
(c) $250

9. $x = 0$, $x = 2$, $f(x) = x^2$, $g(x) = \frac{1}{2}x$
10. $x = 1$, $x = 4$, $f(x) = x^2$, $g(x) = x^3$

APPLIED PROBLEMS

Savings vs. costs

11. Suppose a company wants to introduce a new machine which will produce a rate of annual savings given by

$$S(x) = 150 - x^2,$$

where x is the number of years of operation of the machine, while producing a rate of annual costs of

$$C(x) = x^2 + \frac{11}{4}x.$$

See Examples 3 and 4.
(a) For how many years will it be profitable to use this new machine?
(b) What are the total savings during the first year of the use of the machine?
(c) What are the total savings over the entire period of use of the machine?

Pollution control

12. A new smog-control device will reduce the output of oxides of sulfur from automobile exhausts. It is estimated that the rate of savings to the community from the use of this device will be approximated by

$$S(x) = -x^2 + 4x + 8,$$

where $S(x)$ is the rate of savings in millions of dollars after x years of use of the device. The new device cuts down on the production of oxides of sulfur, but it causes an increase in the production of oxides of nitrogen. The rate of additional costs in millions to the community after x years is approximated by

$$C(x) = \frac{3}{25}x^2.$$

(a) For how many years will it pay to use the new device?
(b) What will be the total savings over this period of time?

Profit

In the exercises for Section 6.4, De Win Enterprises had an expenditure rate (in hundreds of dollars) of $E(x) = 4x + 2$, and an income rate (in hundreds of dollars) of $I(x) = 100 - x$ on a particular job, where x was the number of days from the start of the job. Their profit on that job will equal total income less total expenditure. Profit will be maximized if the job ends at the optimum time, which is the point where the two curves meet. Find the following.

13. The optimum number of days for the job to last.
14. The total income for the optimum number of days.
15. The total expenditure for the optimum number of days.
16. The maximum profit for the job.

For Exercises 17–22 refer to Example 5.

Producer's and consumer's surplus

17. Find the producer's surplus if the price of some item is given by
$$p(x) = x^2 + 2x + 50,$$
where x is the supply, assuming supply and demand are in equilibrium at $x^* = 20$.

18. Suppose a price for a certain commodity is given by
$$p(x) = 100 + 3x + x^2,$$
where x is the supply. Suppose that supply and demand are in equilibrium at $x^* = 3$. Find the producer's surplus.

19. Find the consumer's surplus if the price for an item is given by
$$p(x) = 50 - x^2$$
at a demand of x, assuming supply and demand are in equilibrium at $x^* = 5$.

20. Find the consumer's surplus if the price for an item is given by
$$p(x) = -(x + 4)^2 + 66,$$
where x is the demand and supply and demand are in equilibrium at $x^* = 3$.

21. Suppose the price of a certain item is given by
$$p(x) = \frac{7}{5}x,$$
when the supply is x items, and the price is given by
$$p(x) = -\frac{3}{5}x + 10,$$
when the demand is for x items.
   (a) Graph the supply and demand curves.
   (b) Find the point at which supply and demand are in equilibrium.
   (c) Find the consumer's surplus.
   (d) Find the producer's surplus.

22. Repeat the four steps in Exercise 21 for the supply function
$$p(x) = x^2 + \frac{11}{4}x$$
and the demand function
$$p(x) = 150 - x^2.$$

Income distribution

Suppose that all the people in a country are ranked according to their incomes, starting at the bottom. Let x represent the fraction of the community making the lowest income ($0 \le x \le 1$); $x = .4$, therefore, represents the lower 40% of all income producers. Let $I(x)$ represent the proportion of the total income earned by the lowest x of all people. Thus, $I(.4)$ represents the fraction of total income earned by the lowest 40% of the population. Suppose

$$I(x) = .9x^2 + .1x.$$

Find and interpret the following.

23. $I(.1)$ 24. $I(.4)$ 25. $I(.6)$ 26. $I(.9)$

If income were distributed uniformly, we would have $I(x) = x$. The area between the curves $I(x) = x$ and the particular function $I(x)$ for a given country is called the **coefficient of inequality** for that country.

27. Graph $I(x) = x$ and $I(x) = .9x^2 + .1x$ for $0 \le x \le 1$ on the same axes.

28. Find the area between the curves.

6.6 INTEGRATION BY SUBSTITUTION

In this section we discuss a technique for integrating functions which is related to the chain rule for derivatives. This technique depends on the idea of a differential, which was discussed in Chapter 4. Recall that

$$dy = f'(x) \cdot dx.$$

For example, if $y = 6x^4$, then $dy = 24x^3 \, dx$, and so on.

Work Problem 1 at the side.

We can use the idea of a differential to help us find the indefinite integral

$$\int 2x(x^2 - 1)^4 \, dx.$$

Substitution is used to change the function to one which can be integrated using the methods of Section 1 of this chapter. Let $u = x^2 - 1$. By the definition of the differential, if $u = x^2 - 1$, then $du = 2x \, dx$. Now substitute into the integral above.

$$\int 2x(x^2 - 1)^4 \, dx = \int (x^2 - 1)^4 (2x \, dx) = \int u^4 \, du$$

This last integral can now be found.

$$\int u^4 \, du = \frac{1}{5}u^5 + C$$

Finally, substitute $x^2 - 1$ for u.

$$\int 2x(x^2 - 1)^4 \, dx = \frac{1}{5}(x^2 - 1)^5 + C$$

Work Problem 2 at the side.

---

1. Find dy for the following.
(a) $y = 9x$
(b) $y = 5x^3 + 2x^2$
(c) $y = e^{-2x}$

Answer:
(a) $dy = 9dx$
(b) $dy = (15x^2 + 4x)\,dx$
(c) $dy = -2e^{-2x}\,dx$

2. Find the following
(a) $\int 8x(4x^2 - 9)^3 \, dx$
(b) $\int 12x^3(3x^4 + 8)^5 \, dx$

Answer:
(a) $\frac{1}{4}(4x^2 - 9)^4 + C$
(b) $\frac{1}{6}(3x^4 + 8)^6 + C$

Example 1 Find $\int 6x(3x^2 + 4)^4 dx$.

Trial and error must be used to decide on the expression to set equal to u. Try $u = 3x^2 + 4$, so that $du = 6x\,dx$. Then substitute.

$$\int 6x(3x^2 + 4)^4 dx = \int (3x^2 + 4)^4 (6x\,dx) = \int u^4\,du$$

Since

$$\int u^4\,du = \frac{u^5}{5} + C,$$

and since $u = 3x^2 + 4$, we then have

$$\int 6x(3x^2 + 4)^4 dx = \frac{u^5}{5} + C = \frac{(3x^2 + 4)^5}{5} + C.$$

We can verify this by using the chain rule to find the derivative of $(3x^2 + 4)^5/5 + C$. ∎

Work Problem 3 at the side.

Example 2 Find $\int x^2 \sqrt{x^3 + 1}\,dx$.

If we let $u = x^3 + 1$, then $du = 3x^2\,dx$. The integral does not contain the constant 3, which is needed for du. To take care of this, multiply by 3 inside the integral sign and $\frac{1}{3}$ outside. (This is permissible with constants, but *not* with variables.)

$$\int x^2 \sqrt{x^3 + 1}\,dx = \frac{1}{3} \cdot \int 3x^2 \sqrt{x^3 + 1}\,dx$$

Now substitute u for $x^3 + 1$ and du for $3x^2\,dx$.

$$\frac{1}{3}\int 3x^2\sqrt{x^3+1}\,dx = \frac{1}{3}\int \sqrt{u}\,du$$

$$= \frac{1}{3}\int u^{1/2}\,du$$

$$= \frac{1}{3} \cdot \frac{u^{3/2}}{3/2} + C$$

$$= \frac{2}{9} u^{3/2} + C$$

Since $u = x^3 + 1$,

$$\int x^2 \sqrt{x^3 + 1}\,dx = \frac{2}{9}(x^3 + 1)^{3/2} + C. \quad ∎$$

Work Problem 4 at the side.

---

3. Find the following.
   (a) $\int 8x(4x^2 - 1)^5 dx$
   (b) $\int 18x^2(6x^3 - 5)^{3/2} dx$

   Answer:
   (a) $\dfrac{(4x^2 - 1)^6}{6} + C$
   (b) $\dfrac{2(6x^3 - 5)^{5/2}}{5} + C$

4. Find the following.
   (a) $\int x(5x^2 + 6)^4 dx$
   (b) $\int x^2(x^3 - 2)^5 dx$
   (c) $\int x\sqrt{x^2 + 16}\,dx$

   Answer:
   (a) $\frac{1}{50}(5x^2 + 6)^5 + C$
   (b) $\frac{1}{18}(x^3 - 2)^6 + C$
   (c) $\frac{1}{3}(x^2 + 16)^{3/2} + C$

6.6 Integration by Substitution

The substitution method given in the examples above will not always work. For example, we might try to find

$$\int x^3 \sqrt{x^3 + 1}\, dx$$

by substituting $u = x^3 + 1$, so that $du = 3x^2\, dx$. There is no constant which can be inserted inside the integral sign to give $3x^2$. This integral, and a great many others, cannot be evaluated by the methods of this course. In fact, a large class of functions do not have antiderivatives at all.

Work Problem 5 at the side.

5. Find the following.
(a) $\int x(6x + 2)^{1/2}\, dx$
(b) $\int 5x^2(8x^4 + 2)^3\, dx$

Answer: Neither can be found by the substitution method shown above.

Example 3 Find the value of

$$\int_0^4 (x^2 + 5x)^{3/2}(2x + 5)\, dx.$$

To evaluate this definite integral, we first find

$$\int (x^2 + 5x)^{3/2}(2x + 5)\, dx.$$

Let $u = (x^2 + 5x)$, so that $du = (2x + 5)\, dx$. This gives

$$\int (x^2 + 5x)^{3/2}(2x + 5)\, dx = \int u^{3/2}\, du$$

$$= \frac{2}{5} u^{5/2} + C.$$

We cannot use the limits of integration given in the original problem, 0 and 4, since they refer to x and this antiderivative contains u. Thus, we must substitute $x^2 + 5x$ for u, obtaining

$$\frac{2}{5} u^{5/2} + C = \frac{2}{5}(x^2 + 5x)^{5/2} + C.$$

This antiderivative can now be used to evaluate the given definite integral. Using the fundamental theorem of calculus, we have

$$\int_0^4 (x^2 + 5x)^{3/2}(2x + 5)\, dx = \frac{2}{5}(x^2 + 5x)^{5/2} \Big]_0^4$$

$$= \frac{2}{5}(4^2 + 5 \cdot 4)^{5/2} - \frac{2}{5}(0^2 + 5 \cdot 0)^{5/2}$$

$$= \frac{2}{5}(36)^{5/2} - 0$$

$$= \frac{2}{5}(6)^5. \blacksquare$$

6. Find the following.
(a) $\int_0^4 \sqrt{x^2 + 4x}\,(2x + 4)\, dx$
(b) $\int_5^8 (x^2 - 5x + 1)^{1/2}(2x - 5)\, dx$

Answer:
(a) $\frac{2}{3}(32)^{3/2}$
(b) $\frac{248}{3}$

Recall that it is not necessary to consider the constant C when evaluating a definite integral.

Work Problem 6 at the side.

Example 4 Find $\int e^{4x} \, dx$.

Let $u = 4x$, so that $du = 4 \, dx$. Place 4 inside the integral sign and $\frac{1}{4}$ outside.

$$\int e^{4x} \, dx = \frac{1}{4} \int 4e^{4x} \, dx$$

Now substitute.

$$\frac{1}{4} \int 4e^{4x} \, dx = \frac{1}{4} \int e^u \, du$$

$$= \frac{1}{4} e^u + C$$

$$= \frac{1}{4} e^{4x} + C \quad \blacksquare$$

Based on results similar to this one, we can state a more general form of Theorem 6.4 as follows.

Theorem 6.4A For any real number $a \neq 0$, we have

$$\int e^{ax} \, dx = \frac{1}{a} e^{ax} + C.$$

Work Problem 7 at the side.

The following example shows further how substitution can be used in integrals involving exponentials.

Example 5 Evaluate $\int xe^{x^2} \, dx$.

If we let $u = x^2$, then $du = 2x \, dx$.

$$\int xe^{x^2} \, dx = \frac{1}{2} \int 2xe^{x^2} \, dx$$

$$= \frac{1}{2} \int e^u \, du$$

$$= \frac{1}{2} e^u + C$$

$$= \frac{1}{2} e^{x^2} + C \quad \blacksquare$$

Work Problem 8 at the side.

---

7. Find the following.
(a) $\int 6e^{3x} \, dx$
(b) $\int -8e^{-x/2} \, dx$
(c) $\int e^{-.01x} \, dx$

Answer:
(a) $2e^{3x} + C$
(b) $16e^{-x/2} + C$
(c) $-e^{-.01x}/.01 + C$ or $-100e^{-.01x} + C$

8. Find the following.
(a) $\int 8xe^{3x^2} \, dx$
(b) $\int x^2 e^{x^3} \, dx$

Answer:
(a) $\frac{4}{3} e^{3x^2} + C$
(b) $\frac{1}{3} e^{x^3} + C$

6.6 Integration by Substitution

Recall that the antiderivative of $f(x) = 1/x$ is $\ln|x|$. The next example shows how substitution can be used to extend the use of this antiderivative.

Example 6 Find $\displaystyle\int \frac{(2x - 3)\,dx}{x^2 - 3x}$.

Let $u = x^2 - 3x$, so that $du = (2x - 3)\,dx$. Then,

$$\int \frac{(2x - 3)\,dx}{x^2 - 3x} = \int \frac{du}{u}$$
$$= \ln|u| + C$$
$$= \ln|x^2 - 3x| + C. \quad \blacksquare$$

Work Problem 9 at the side.

9. Find the following.

(a) $\displaystyle\int \frac{4\,dx}{x - 3}$

(b) $\displaystyle\int \frac{(2x - 9)\,dx}{x^2 - 9x}$

(c) $\displaystyle\int \frac{(3x^2 + 8)\,dx}{x^3 + 8x + 5}$

(a) $4\ln|x - 3| + C$
(b) $\ln|x^2 - 9x| + C$
(c) $\ln|x^3 + 8x + 5| + C$

6.6 EXERCISES

Find the following indefinite integrals. See Examples 1, 2, and 4–6.

1. $\displaystyle\int 4(2x + 3)^4\,dx$

2. $\displaystyle\int (-4x + 1)^3\,dx$

3. $\displaystyle\int \frac{4}{(x - 2)^3}\,dx$

4. $\displaystyle\int \frac{-3}{(x + 1)^4}\,dx$

5. $\displaystyle\int \sqrt{x - 4}\,dx$

6. $\displaystyle\int (\sqrt{x + 5})^3\,dx$

7. $\displaystyle\int 2x(x^2 + 1)^3\,dx$

8. $\displaystyle\int 3x^2(x^3 - 4)^3\,dx$

9. $\displaystyle\int (\sqrt{x^2 + 12x})(2x + 12)\,dx$

10. $\displaystyle\int (\sqrt{x^2 - 6x})(2x - 6)\,dx$

11. $\displaystyle\int e^{2x}\,dx$

12. $\displaystyle\int 3e^{-2x}\,dx$

13. $\displaystyle\int (-4e^{2x})\,dx$

14. $\displaystyle\int 5e^{-0.3x}\,dx$

15. $\displaystyle\int 3x^2 e^{2x^3}\,dx$

16. $\displaystyle\int xe^{-x^2}\,dx$

17. $\displaystyle\int \frac{-8}{1 + x}\,dx$

18. $\displaystyle\int \frac{9}{2 + x}\,dx$

19. $\displaystyle\int \frac{dx}{2x + 1}$

20. $\displaystyle\int \frac{dx}{5x - 2}$

21. $\displaystyle\int \frac{2x + 4}{x^2 + 4x}\,dx$

22. $\displaystyle\int \frac{x}{3x^2 - 2}\,dx$

23. $\displaystyle\int \frac{6x\,dx}{(3x^2 + 2)^4}$

24. $\displaystyle\int \frac{4x\,dx}{(2x^2 - 5)^3}$

25. $\int \dfrac{x-1}{(2x^2-4x)^2}\,dx$ 26. $\int \dfrac{2x+1}{(x^2+x)^3}\,dx$

27. $\int \left(\dfrac{1}{x}+x\right)\left(1-\dfrac{1}{x^2}\right)dx$ 28. $\int \left(\dfrac{2}{x}-x\right)\left(\dfrac{-2}{x^2}-1\right)dx$

Find the following definite integrals. See Example 3.

29. $\int_1^4 \sqrt{x}\,dx$ 30. $\int_4^9 2x^{1/2}\,dx$

31. $\int_0^1 2x(x^2+1)^3\,dx$ 32. $\int_0^1 3x^2(x^3-4)^3\,dx$

33. $\int_0^4 (\sqrt{x^2+12x})(2x+12)\,dx$ 34. $\int_0^8 (\sqrt{x^2-6x})(2x-6)\,dx$

35. $\int_{-1}^{e-2} \dfrac{1}{x+2}\,dx$ (Hint: $\ln e = 1$.)

36. $\int_{-2}^{0} \dfrac{-4}{x+3}\,dx$ 37. $\int_0^1 e^{2x}\,dx$

38. $\int_1^2 e^{-x}\,dx$ 39. $\int_2^4 e^{-0.2x}\,dx$

40. $\int_1^4 e^{-0.3x}\,dx$

APPLIED PROBLEMS

Cost

When a new technological development comes along, the cost to manufacture the first few units can be very high, but the cost can decline as more experience in manufacturing is obtained. (This effect is called the **learning curve**.) Suppose that the rate of change of cost to manufacture item number x is given by $C'(x)$ dollars, where

$$C'(x) = 5e^{x/5}.$$

41. Find the total cost of manufacturing the first 20 items by finding $\int_0^{20} C'(x)\,dx$.

42. Find the average cost of each of the first 20 items.

Oil consumption

(Most of the numbers in this exercise come from an article in *The Wall Street Journal*.) Suppose that the rate of consumption of a natural resource is

$$C(t) = k \cdot e^{rt},$$

where t is time in years, r is a constant representing the rate of growth in use, and k is the consumption in the year when $t = 0$. In 1977, Texaco sold 1.2 billion barrels of oil. Assume that oil sales increase at the rate of 4% per year, so that $r = .04$.

43. Write $C(t)$ for Texaco, letting $t = 0$ represent 1977.

44. Set up a definite integral for the amount of oil that Texaco will sell in the ten years from 1977 to 1987.

45. Evaluate the definite integral of Exercise 44.

46. Texaco has about 20 billion barrels of oil in reserve. To find the number of years that this amount will last, we must solve the equation

$$\int_0^t 1.2 e^{.04t}\, dt = 20.$$

On the left, show that we get $30 e^{.04t} - 30$.

47. We now have $30e^{.04t} - 30 = 20$, or $30e^{.04t} = 50$, and, finally, $e^{.04t} = 50/30 = 1.67$. Look down the e^x column of Table 4 until you find the number closest to 1.67. What is the corresponding value of x?

48. You should now have $e^{.04t} = e^{.50}$, or $.04t = .50$. Find t.

49. Rework the last three problems assuming a rate of increase of only 2% per year.

50. Repeat Exercise 49 for 1% per year.

6.7 TABLES OF INTEGRALS

It is a fairly straightforward task to find the derivative of most of the useful functions. The chain rule, together with other formulas of Chapter 3, can be used to find derivatives of almost all functions that are useful in practical applications. However, this is not true for integration. There are many useful functions whose integrals cannot be found by the methods we have discussed. For example, $f(x) = e^{-x^2}$, used to obtain the normal curve of statistics, is one of many such functions. Some calculus courses spend a considerable amount of time on techniques of integration. These techniques often depend on trigonometry and complicated algebraic substitution and manipulation.

Another less time-consuming approach is possible. This is to list all commonly needed integrals in a table of integrals. This table can then be referred to as needed. One such table is included inside the back cover of this text. Larger and more complete tables are listed in the books in the bibliography. The remainder of this section will be devoted to examples showing how to use the table in this book.

Example 1 Find $\int \dfrac{1}{\sqrt{x^2 + 16}}\, dx$.

By inspecting the table, we see that if $a = 4$, this antiderivative is the same as entry 5 of the table. Entry 5 of the table is

$$\int \frac{1}{\sqrt{x^2 + a^2}}\, dx = \ln\left|\frac{x + \sqrt{x^2 + a^2}}{a}\right| + C.$$

1. Find the following.

(a) $\int \dfrac{4}{\sqrt{x^2+100}}\,dx$

(b) $\int \dfrac{-9}{\sqrt{x^2-4}}\,dx$

(c) $\int -6\ln|x|\,dx$

Answer:

(a) $4\ln\left|\dfrac{x+\sqrt{x^2+100}}{10}\right| + C$

(b) $-9\ln\left|\dfrac{x+\sqrt{x^2-4}}{2}\right| + C$

(c) $-6x(\ln|x| - 1) + C$

2. Find the following.

(a) $\int \dfrac{1}{x^2-4}\,dx$

(b) $\int \dfrac{-6}{x\sqrt{25-x^2}}\,dx$

(c) $\int \dfrac{5}{x\sqrt{36+x^2}}\,dx$

Answer:

(a) $\dfrac{1}{4}\ln\left(\dfrac{x-2}{x+2}\right) + C \quad (x^2 > 4)$

(b) $\dfrac{6}{5}\ln\left(\dfrac{5+\sqrt{25-x^2}}{x}\right) + C$

$(0 < x < 5)$

(c) $-\dfrac{5}{6}\ln\left|\dfrac{6+\sqrt{36+x^2}}{x}\right| + C$

By substituting 4 for a in this entry, we get

$$\int \dfrac{1}{\sqrt{x^2+16}}\,dx = \ln\left|\dfrac{x+\sqrt{x^2+16}}{4}\right| + C.$$

This result could be verified by taking the derivative of the right-hand side of this last equation. ∎

Work Problem 1 at the side.

Example 2 Find $\int \dfrac{8}{16-x^2}\,dx$.

We can convert this antiderivative into the one given in entry 7 of the table by writing the 8 in front of the integral sign (permissible only with constants) and by letting $a = 4$. Doing this gives

$$8\int \dfrac{1}{16-x^2}\,dx = 8\left[\dfrac{1}{2\cdot 4}\ln\left(\dfrac{4+x}{4-x}\right)\right] + C$$

$$= \ln\left(\dfrac{4+x}{4-x}\right) + C.$$

In entry 7 of the table, the condition $x^2 < a^2$ is given. Here $a = 4$. Hence, the result given above is valid only for $x^2 < 16$, so that the final answer should be written as

$$\int \dfrac{8}{16-x^2}\,dx = \ln\left(\dfrac{4+x}{4-x}\right) + C, \quad \text{for } x^2 < 16.$$

Because of the condition $x^2 < 16$, the expression in parentheses is always positive, so that absolute value bars are not needed. ∎

Work Problem 2 at the side.

Example 3 Evaluate $\int \dfrac{12x}{2x+1}\,dx$.

For $a = 2$ and $b = 1$, this antiderivative can be rewritten to match entry 11 of this table. Thus

$$\int \dfrac{12x}{2x+1}\,dx = 12\int \dfrac{x}{2x+1}\,dx$$

$$= 12\left[\dfrac{x}{2} - \dfrac{1}{4}\ln|2x+1|\right] + C$$

$$= 6x - 3\ln|2x+1| + C. \quad ∎$$

3. Find the following.

(a) $\int \dfrac{3x}{5x-2}\, dx$

(b) $\int \dfrac{-6x}{(9x+1)^2}\, dx$

(c) $\int \dfrac{2}{x(2x-3)}\, dx$

Answer:

(a) $\dfrac{3x}{5} + \dfrac{6}{25}\ln|5x-2| + C$

(b) $\dfrac{-2}{27(9x+1)} - \dfrac{2}{27}\ln|9x+1| + C$

(c) $\dfrac{-2}{3}\ln\left|\dfrac{x}{2x-3}\right| + C$

4. Find the following.

(a) $\int \dfrac{3}{16x^2-1}\, dx$

(b) $\int \dfrac{-1}{100x^2-1}\, dx$

Answer:

(a) $\dfrac{3}{8}\ln\left|\dfrac{x-1/4}{x+1/4}\right| + C$, $x^2 > 1/16$

(b) $-\dfrac{1}{20}\ln\left|\dfrac{x-1/10}{x+1/10}\right| + C$

$x^2 > 1/100$

5. Find the following.
(a) $\int x^4 \ln|x|\, dx$
(b) $\int xe^{2x}\, dx$

Answer:

(a) $x^5\left(\dfrac{\ln|x|}{5} - \dfrac{1}{25}\right) + C$

(b) $\dfrac{xe^{2x}}{2} - \dfrac{1}{4}e^{2x} + C$

Work Problem 3 at the side.

Example 4 Find $\int \sqrt{9x^2+1}\, dx$.

This antiderivative seems most similar to entry 15 of the table. However, entry 15 requires that the coefficient of the x^2 term be 1. We can satisfy that requirement here by factoring out the 9.

$$\int \sqrt{9x^2+1}\, dx = \int \sqrt{9\left(x^2+\dfrac{1}{9}\right)}\, dx$$

$$= \int 3\sqrt{x^2+\dfrac{1}{9}}\, dx$$

$$= 3\int \sqrt{x^2+\dfrac{1}{9}}\, dx$$

Now, using entry 15 with $a = 1/3$, we have

$$\int \sqrt{9x^2+1}\, dx = 3\left[\dfrac{x}{2}\sqrt{x^2+\dfrac{1}{9}} + \dfrac{(1/3)^2}{2}\cdot \ln\left|x+\sqrt{x^2+\dfrac{1}{9}}\right|\right] + C$$

$$= \dfrac{3x}{2}\sqrt{x^2+\dfrac{1}{9}} + \dfrac{1}{6}\ln\left|x+\sqrt{x^2+\dfrac{1}{9}}\right| + C. \blacksquare$$

Work Problem 4 at the side.

Example 5 Find $\int x^2 e^x\, dx$.

Using entry 17 with $n = 2$ and $a = 1$, we have

$$\int x^2 e^x\, dx = \dfrac{x^2 e^x}{1} - \dfrac{2}{1}\int xe^x\, dx + C. \qquad (1)$$

We must now use entry 17 again, this time to find $\int xe^x\, dx$. Here $n = 1$ and $a = 1$. Thus,

$$\int xe^x\, dx = \dfrac{xe^x}{1} - \dfrac{1}{1}\int x^0 e^x\, dx + K$$

$$= xe^x - \int e^x\, dx + K \qquad \text{Recall: } x^0 = 1$$

$$= xe^x - e^x + K.$$

Substituting this result back into equation (1) gives

$$\int x^2 e^x\, dx = x^2 e^x - 2(xe^x - e^x + K) + C$$

$$= x^2 e^x - 2xe^x + 2e^x + M$$

$$= e^x(x^2 - 2x + 2) + M,$$

where M is an arbitrary constant. $\blacksquare$

Work Problem 5 at the side.

6.7 EXERCISES

Find the following antiderivatives, using the table inside the back cover. See Examples 1–5.

1. $\int \ln|4x|\, dx$

2. $\int \ln\left|\dfrac{3}{5}x\right| dx$

3. $\int \dfrac{-4}{\sqrt{x^2 + 36}}\, dx$

4. $\int \dfrac{9}{\sqrt{x^2 + 9}}\, dx$

5. $\int \dfrac{6}{x^2 - 9}\, dx$

6. $\int \dfrac{-12}{x^2 - 16}\, dx$

7. $\int \dfrac{-4}{x\sqrt{9 - x^2}}\, dx$

8. $\int \dfrac{3}{x\sqrt{121 - x^2}}\, dx$

9. $\int \dfrac{-2x}{3x + 1}\, dx$

10. $\int \dfrac{6x}{4x - 5}\, dx$

11. $\int \dfrac{2}{3x(3x - 5)}\, dx$

12. $\int \dfrac{-4}{3x(2x + 7)}\, dx$

13. $\int \dfrac{4}{4x^2 - 1}\, dx$

14. $\int \dfrac{-6}{9x^2 - 1}\, dx$

15. $\int \dfrac{3}{x\sqrt{1 - 9x^2}}\, dx$

16. $\int \dfrac{-2}{x\sqrt{1 - 16x^2}}\, dx$

17. $\int x^4 \ln|x|\, dx$

18. $\int 4x^2 \ln|x|\, dx$

19. $\int \dfrac{\ln|x|}{x^2}\, dx$

20. $\int \dfrac{-2\ln|x|}{x^3}\, dx$

21. $\int xe^{-2x}\, dx$

22. $\int xe^{3x}\, dx$

23. $\int 2x^2 e^{-2x}\, dx$

24. $\int -3x^2 e^{-4x}\, dx$

25. $\int x^3 e^x\, dx$

26. $\int x^3 e^{2x}\, dx$

CHAPTER 6 SUMMARY

Key Words

antiderivative
definite integral
integral sign
limits of integration
integration

Things to Remember

Theorem 6.1 If $n \neq -1$, then an antiderivative of $f(x) = x^n$ is
$$F(x) = \frac{1}{n+1}x^{n+1} + C.$$

Theorem 6.2 If $F(x)$ is an antiderivative of $f(x)$ and if k is any constant, then the antiderivative of $k \cdot f(x)$ is $k \cdot F(x) + C$, where C is a constant.

Theorem 6.3 If $F(x)$ is an antiderivative of $f(x)$ and $G(x)$ is an antiderivative of $g(x)$, then an antiderivative of $f(x) \pm g(x)$ is $F(x) \pm G(x) + C$.

Theorem 6.4 An antiderivative of $f(x) = e^x$ is
$$F(x) = e^x + C.$$

Theorem 6.4A For any real number $a \neq 0$, we have
$$\int e^{ax}\, dx = \frac{1}{a}e^{ax} + C.$$

Theorem 6.5 An antiderivative of $f(x) = \frac{1}{x} = x^{-1}$ is
$$F(x) = \ln|x| + C.$$

Theorem 6.6 *Fundamental theorem of calculus* Let $f(x)$ be the derivative of $F(x)$ and let $f(x)$ be continuous in the interval $[a, b]$. Then
$$\int_a^b f(x)\, dx = F(b) - F(a).$$

Theorem 6.7 For any real constant k,
$$\int_a^b k \cdot f(x)\, dx = k \cdot \int_a^b f(x)\, dx.$$

Theorem 6.8 If all indicated definite integrals exist, then
$$\int_a^b [f(x) \pm g(x)]\, dx = \int_a^b f(x)\, dx \pm \int_a^b g(x)\, dx.$$

CHAPTER 6 TEST

[6.1] *Find an antiderivative of each of the following.*

1. $f(x) = 2x^3$
2. $f(x) = x^2 + 5x$
3. $f(x) = (\sqrt{x})^3$
4. $f(x) = \dfrac{1}{x^4}$
5. $f(x) = 3e^x$
6. $f(x) = \dfrac{5}{x}$

[6.2] 7. Approximate the area under the graph of $f(x) = 2x + 3$ and above the x-axis from $x = 0$ to $x = 4$ using four rectangles. Let the height of each rectangle be given by the value of the function at the left side.

[6.3] 8. Find the exact area under the graph of $f(x) = 2x + 3$ and above the x-axis from $x = 0$ to $x = 4$.

Evaluate the following definite integrals.

9. $\int_{-1}^{2} (3x^2 + 5) dx$

10. $\int_{1}^{3} x^{-1} dx$

11. $\int_{0}^{4} -2e^x dx$

[6.4]

12. The rate of change of sales of a new brand of tomato soup, in thousands, is given by
$$S(x) = \sqrt{x} + 2,$$
where x is the time in months that the new product has been on the market. Find the total sales after 9 months.

13. The rate of change of a population of prairie dogs, in terms of the number of coyotes, x, which prey on them, is given by
$$P(x) = 25 - .1x.$$
Find the total number of prairie dogs as the coyote population grows from 100 to 200.

[6.5]

14. Find the area between the graphs of $f(x) = 5 - x^2$ and $g(x) = x^2 - 3$.

15. A company has installed new machinery which will produce a savings rate (in thousands of dollars) of
$$S(x) = 225 - x^2,$$
where x is the number of years the machinery is to be used. The rate of additional costs to the company due to the new machinery is expected to be
$$C(x) = x^2 + 25x + 150.$$
For how many years should the company use the new machinery? What will the net savings in thousands of dollars be over this period?

[6.6] *Use substitution to find the following.*

16. $\int x\sqrt{5x^2 + 6}\, dx$

17. $\int (x^2 - 5x)^4 (2x - 5)\, dx$

18. $\int e^{-4x}\, dx$

19. $\int \frac{12(2x + 9)}{x^2 + 9x + 1}\, dx$

20. $\int_{2}^{3} (3x^2 - 6x)^2 (6x - 6)\, dx$

[6.7] *Use the table of integrals to find the following.*

21. $\int \frac{1}{\sqrt{x^2 - 64}}\, dx$

22. $\int \frac{5}{x\sqrt{25 + x^2}}\, dx$

23. $\int \frac{-2}{x(3x + 4)}\, dx$

24. $\int \sqrt{x^2 + 49}\, dx$

25. $\int 3x^2 \ln|x|\, dx$

CASE 7 ESTIMATING DEPLETION DATES FOR MINERALS

It is becoming more and more obvious that the earth contains only a finite quantity of minerals. The "easy and cheap" sources of minerals are being used up, forcing an ever more expensive search for new sources. For example, oil from the North Slope of Alaska would never have been used in the United States during the 1930's since there was so much Texas and California oil readily available.

We said in Chapter 5 that population tends to follow an exponential growth curve. Mineral usage also follows such a curve. Thus, if q represents the rate of consumption of a certain mineral at time t, while q_0 represents consumption when $t = 0$, then

$$q = q_0 e^{kt},$$

where k is the annual rate of increase in use of the mineral. For example, the world consumption of petroleum in a recent year was about 19,600 million barrels, with the annual rate of increase running about 6%. If we let $t = 0$ correspond to this base year, then $q_0 = 19,600$, $k = 0.06$, and

$$q = 19,600 e^{0.06t}$$

is the rate of consumption at time t, assuming that all present trends continue.

Based on estimates of the National Academy of Science, we shall use 2,000,000 as the number of millions of barrels of oil that are now in provable reserves or that are likely to be discovered in the future. At the present rate of consumption, how many years would be necessary to deplete these estimated reserves? We can use the integral calculus of this chapter to find out.

To begin, we need to know the total quantity of petroleum that would be used between time $t = 0$ and some future time $t = t_1$. Figure 1 shows a typical graph of the function $q = q_0 e^{kt}$.

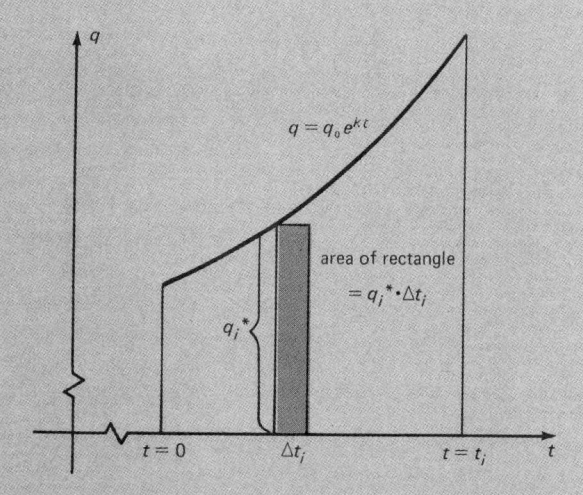

Figure 1

Following the work we did in Section 6.2, divide the time interval from $t = 0$ to $t = t_1$ into n subintervals. Let the ith subinterval have width Δt_i. Let the rate of consumption for the ith subinterval be approximated by q_i^*. Thus, the approximate total consumption for the subinterval is given by

$$q_i^* \cdot \Delta t_i,$$

and the total consumption over the interval from time $t = 0$ to $t = t_1$ is approximated by

$$\sum_{i=1}^{n} q_i^* \cdot \Delta t_i.$$

The limit of this sum as each of the Δt_i's approaches 0 gives the total consumption from time $t = 0$ to $t = t_1$. That is,

$$\text{total consumption} = \lim_{\Delta t_i \to 0} \sum q_i^* \cdot \Delta t_i.$$

However, we have seen that this limit is the definite integral of the function $q = q_0 e^{kt}$ from $t = 0$ to $t = t_1$, or

$$\text{total consumption} = \int_0^{t_1} q_0 e^{kt} \, dt.$$

We can now evaluate this definite integral.

$$\int_0^{t_1} q_0 e^{kt} \, dt = q_0 \int_0^{t_1} e^{kt} \, dt$$

$$= q_0 \left(\frac{1}{k} e^{kt} \right) \Big]_0^{t_1}$$

$$= \frac{q_0}{k} e^{kt} \Big]_0^{t_1}$$

$$= \frac{q_0}{k} e^{kt_1} - \frac{q_0}{k} e^0$$

$$= \frac{q_0}{k} e^{kt_1} - \frac{q_0}{k}(1)$$

$$= \frac{q_0}{k}(e^{kt_1} - 1) \tag{1}$$

Now let us return to the numbers we gave for petroleum. We said that $q_0 = 19,600$ million barrels where q_0 represents consumption in the base year. We have $k = 0.06$, with total petroleum reserves estimated as 2,000,000 million barrels. Thus, using equation (1) we have

$$2,000,000 = \frac{19,600}{0.06}(e^{0.06t_1} - 1).$$

Multiply both sides of the equation by 0.06:

$$120,000 = 19,600(e^{0.06t_1} - 1).$$

Divide both sides of the equation by 19,600.

$$6.1 = e^{0.06t_1} - 1$$

Add 1 to both sides.

$$7.1 = e^{0.06t_1}$$

Take natural logarithms of both sides:

$$\ln 7.1 = \ln e^{0.06t_1}$$
$$\ln 7.1 = 0.06t_1 \ln e$$
$$\ln 7.1 = 0.06t_1 \quad \text{(since } \ln e = 1\text{).}$$

Finally,

$$t_1 = \frac{\ln 7.1}{0.06}.$$

From Table 5, estimate $\ln 7.1$ as about 1.96. Thus,

$$t_1 = \frac{1.96}{0.06}$$
$$= 33.$$

By this result, petroleum reserves will last the world for thirty-three years.

The results of mathematical analyses such as this must be used with great caution. By the analysis above, the world would use all the petroleum that it wants in the thirty-second year after the base year, but there would be none at all in thirty-four years. This is not at all realistic. As petroleum reserves decline, the price will increase, causing demand to decline and supplies to increase.

EXERCISES

1. Find the number of years that the estimated petroleum reserves would last if used at the same rate as in the base year.

2. How long would the estimated petroleum reserves last if the rate of growth of use was only 2% instead of 6%?

Estimate the length of time until depletion for each of the following minerals.

3. Bauxite (the ore from which aluminum is obtained), estimated reserves in base year 15,000,000 thousand tons, rate of consumption 63,000 thousand tons, annual rate of increase of consumption 6%.

4. Bituminous coal, estimated world reserves 2,000,000 million tons, rate of consumption 2200 million tons, annual rate of increase of consumption 4%.

7 Further Techniques and Applications of Integration

In this chapter we discuss methods of integrating more complicated functions. We also look at numerical methods of integration, which are often used with practical data or when a function cannot be integrated at all. Then we see how to use integration to find the volume of a solid. The chapter ends with a discussion of an application of integration to probability.

7.1 INTEGRATION BY PARTS

In this section we introduce a technique of integration which often makes it possible to reduce a complicated integral to a simpler integral. If u and v are both differentiable functions, then uv is also differentiable and, by the product rule for derivatives,

$$\frac{d(uv)}{dx} = u\frac{dv}{dx} + v\frac{du}{dx}.$$

We can rewrite this expression, using differentials, as

$$d(uv) = u\,dv + v\,du.$$

Now, integrating both sides of this last equation, we have

$$\int d(uv) = \int u\,dv + \int v\,du,$$
$$uv = \int u\,dv + \int v\,du.$$

By rearranging terms, we obtain the formula:

$$\int u\,dv = uv - \int v\,du.$$

7.1 Integration by Parts

The technique using this formula to find integrals is called **integrating by parts** and is illustrated in the following examples.

Example 1 Find $\int x\sqrt{1-x}\,dx$.

To use integration by parts, first write the expression $x\sqrt{1-x}\,dx$ as a product of functions u and dv in such a way that v can be found. Let us select $u = x$ and $dv = \sqrt{1-x}\,dx$. Then $du = dx$. We find v by integrating dv. Therefore,

$$v = \int \sqrt{1-x}\,dx = -\frac{2}{3}(1-x)^{3/2} + C.$$

(Verify this.) For simplicity, we ignore the constant C and just add it at the end. Thus, $v = -\frac{2}{3}(1-x)^{3/2}$. Now substitute this into the formula for integration by parts as follows.

$$\int u\,dv = uv - \int v\,du,$$

$$\int x\sqrt{1-x}\,dx = x\left[-\frac{2}{3}(1-x)^{3/2}\right] - \int \left[-\frac{2}{3}(1-x)^{3/2}\right]dx$$

$$= -\frac{2}{3}x(1-x)^{3/2} - \frac{2}{3}\int -(1-x)^{3/2}\,dx$$

$$= -\frac{2}{3}x(1-x)^{3/2} - \frac{2}{3}\left[\frac{2}{5}(1-x)^{5/2}\right]$$

$$\int x\sqrt{1-x}\,dx = -\frac{2}{3}x(1-x)^{3/2} - \frac{4}{15}(1-x)^{5/2} + C$$

We added a constant C in the last step. ■

Work Problem 1 at the side.

Example 2 Find $\int xe^{5x}\,dx$.

Let $dv = e^{5x}\,dx$ and let $u = x$. Then $v = \frac{1}{5}e^{5x}$ and $du = dx$. Substitute into the formula for integration by parts.

$$\int u\,dv = uv - \int v\,du$$

$$\int xe^{5x}\,dx = x\left(\frac{1}{5}e^{5x}\right) - \int \frac{1}{5}e^{5x}\,dx$$

$$= \frac{1}{5}xe^{5x} - \frac{1}{25}e^{5x} + C \quad ■$$

Work Problem 2 at the side.

---

1. Find the following.
   (a) $\int x(x+1)^{-2}\,dx$
   (b) $\int x\sqrt{x+1}\,dx$

   Answer:
   (a) $-x(x+1)^{-1} + \ln|x+1| + C$
   (b) $\frac{2}{3}x(x+1)^{3/2} - \frac{4}{15}(x+1)^{5/2} + C$

2. Find the following.
   (a) $\int 8xe^{-2x}\,dx$
   (b) $\int -3(x+1)e^{2x}\,dx$

   Answer:
   (a) $-4xe^{-2x} - 2e^{-2x} + C$
   (b) $-\frac{3}{2}(x+1)e^{2x} + \frac{3}{4}e^{2x} + C$

Sometimes it is necessary to use the technique of integrating by parts more than once as in the following example.

Example 3 Find $\int 2x^2 e^x \, dx$.

Since $\int e^x \, dx$ can be found, let us choose $dv = e^x \, dx$, so that $v = e^x$. Then $u = 2x^2$ and $du = 4x \, dx$. Now we substitute into the formula for integrating by parts.

$$\int u \, dv = uv - \int v \, du$$
$$\int 2x^2 e^x \, dx = 2x^2 e^x - \int e^x(4x \, dx) \tag{1}$$

We must find $\int e^x(4x \, dx)$ by parts. Again, we choose $dv = e^x \, dx$, which gives $v = e^x$, $u = 4x$, and $du = 4 \, dx$. Substituting again into the formula for integrating by parts, we have the following.

$$\int e^x(4x \, dx) = 4xe^x - \int e^x(4 \, dx)$$
$$= 4xe^x - 4e^x$$

Now we must substitute back into our first result, equation (1), to get the final answer.

$$\int 2x^2 e^x \, dx = 2x^2 e^x - (4xe^x - 4e^x)$$
$$= 2x^2 e^x - 4xe^x + 4e^x + C,$$

where a constant C was added at the last step. ∎

Work Problem 3 at the side.

The method of integration by parts requires choosing the factor dv so that $\int dv$ can be found. If this is not possible, or if the remaining factor, which becomes u, does not have a differential du such that $v \, du$ can be integrated, the technique cannot be used. For example, to integrate

$$\int \frac{1}{4 - x^2} \, dx$$

we might choose $dv = dx$ and $u = (4 - x^2)^{-1}$. Then $v = x$ and $du = 2x \, dx/(4 - x^2)^2$. Then we have

$$\int \frac{1}{4 - x^2} \, dx = \frac{x}{4 - x^2} - \int \frac{2x^2 \, dx}{(4 - x^2)^2},$$

where the integral on the right is more complicated than the original integral. A second use of integration by parts on the new integral would make matters even worse. Since we cannot choose $dv = (4 - x^2)^{-1}$ because we cannot integrate it, integration by parts is

---

3. Find $\int 3x^2 e^{-5x} \, dx$.

Answer:
$-\frac{3}{5}x^2 e^{-5x} - \frac{6}{25}xe^{-5x} - \frac{6}{125}e^{-5x} + C$

not suitable for this problem. In fact, there are many functions whose integrals cannot be found by any of the methods we have described. Many of these can be found by more advanced methods and are available in tables of integrals, discussed in Section 6.7. However, there are functions which cannot be integrated at all, for example, $f(x) = e^{-x^2}$, a function which is important in statistical theory.

7.1 EXERCISES

Use integration by parts to find the following. See Examples 1–3.

1. $\int x(x+1)^5 \, dx$
2. $\int x\sqrt{5x-1} \, dx$
3. $\int x^3(1+x^2)^{1/4} \, dx$
4. $\int 2x^5(1-3x^3)^{1/2} \, dx$
5. $\int \dfrac{x}{\sqrt{x-1}} \, dx$
6. $\int \dfrac{4x}{\sqrt{8-x}} \, dx$
7. $\int \dfrac{2x}{(x+5)^6} \, dx$
8. $\int \dfrac{3x^3}{(x^2-5)^5} \, dx$
9. $\int \dfrac{x^3 \, dx}{\sqrt{3-x^2}}$
10. $\int \dfrac{x^7 \, dx}{(x^4+3)^{2/3}}$
11. $\int xe^x \, dx$
12. $\int (x+1)e^x \, dx$
13. $\int (5x-9)e^{-3x} \, dx$
14. $\int (6x+3)e^{-2x} \, dx$
15. $\int (x^2+1)e^x \, dx$
16. $\int 3x^2 e^{-x} \, dx$
17. $\int (1-x^2)e^{2x} \, dx$
18. $\int x^2 e^{2x} \, dx$

19. Use integration by parts to find $\int \ln|x| \, dx$. Let $u = \ln|x|$, so that $du = \dfrac{1}{x}$.

20. Use integration by parts to find $\int \ln|5x-1| \, dx$. $\left(\text{Hint: Write } \dfrac{5x}{5x-1} \text{ as } 1 + \dfrac{1}{5x-1}.\right)$

Use the results of Exercise 19 and integration by parts to find the following.

21. $\int x \ln|x| \, dx$
22. $\int x^2 \ln|x| \, dx$
23. $\int (2x-1)\ln|3x| \, dx$
24. $\int (8x+7)\ln|5x| \, dx$

25. Find the area between $y = (x-2)e^x$ and the x-axis from $x = 2$ to $x = 4$.
26. Find the area between $y = xe^x$ and the x-axis from $x = 0$ to $x = 1$.

7.2 NUMERICAL INTEGRATION

As mentioned in Section 7.1, some integrals cannot be evaluated by any technique or found in any table. Since $\int_a^b f(x)\,dx$ represents an area, any approximation of that area also approximates the definite integral. Many methods of approximating definite integrals by areas, made feasible by the availability of pocket calculators and the high-speed computer, are in use today. These methods are referred to as **numerical integration.** We shall discuss two methods of numerical integration, *the trapezoidal rule* and *Simpson's rule.**

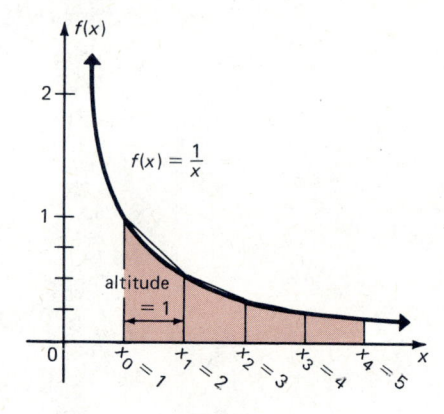

Figure 7.1

As an example, let us consider

$$\int_1^5 \frac{1}{x}\,dx.$$

The shaded region in Figure 7.1 shows the area under the graph of $f(x) = 1/x$ above the x-axis and between the lines $x = 1$ and $x = 5$. Since $\int (1/x)\,dx = \ln|x| + C$, we have

$$\int_1^5 \frac{1}{x}\,dx = \ln|x| \Big]_1^5$$
$$= \ln 5 - \ln 1$$
$$= \ln 5 - 0$$
$$= \ln 5,$$

a number which can be found in Table 5. Using the table, we find that $\ln 5 \approx 1.6094$. How was this number in the table obtained? One way to find this number is to approximate the area under the curve in Figure 7.1. This can be done by the trapezoidal rule. To

---

* For a computer method which uses the trapezoidal rule, see Margaret L. Lial, *Study Guide with Computer Problems* (Glenview, Ill.: Scott, Foresman, 1979), pp. 77–80.

7.2 Numerical Integration

get a first approximation to $\ln 5$ by the trapezoidal rule, we might first find the sum of the areas of the four trapezoids indicated in Figure 7.1. The area of a trapezoid is one-half the product of the sum of the bases and the altitude. Each of the trapezoids in Fig. 7.1 has altitude 1. Thus, we have

$$\ln 5 = \int_1^5 \frac{1}{x} dx \approx \frac{1}{2}\left(\frac{1}{1}+\frac{1}{2}\right)(1) + \frac{1}{2}\left(\frac{1}{2}+\frac{1}{3}\right)(1) + \frac{1}{2}\left(\frac{1}{3}+\frac{1}{4}\right)(1) + \frac{1}{2}\left(\frac{1}{4}+\frac{1}{5}\right)(1)$$

$$= \frac{1}{2}\left(\frac{3}{2}+\frac{5}{6}+\frac{7}{12}+\frac{9}{20}\right)$$

$$\approx 1.68.$$

To get a better approximation, we would divide the interval $1 \le x \le 5$ into more subintervals. The larger the number of subintervals, the better the approximation will be. In general, suppose f is a continuous function on an interval $a \le x \le b$. Divide the interval from a to b into n equal subintervals by the points $a = x_0$, $x_1, x_2, \ldots, x_n = b$. Use the subintervals to make trapezoids which approximate the area under the curve (as in Figure 7.1). The sum of the areas of the trapezoids is given by

$$\frac{1}{2}\left[f(x_0) + f(x_1)\right]\left(\frac{b-a}{n}\right) + \frac{1}{2}\left[f(x_1) + f(x_2)\right]\left(\frac{b-a}{n}\right) + \cdots$$
$$+ \frac{1}{2}\left[f(x_{n-1}) + f(x_n)\right]\left(\frac{b-a}{n}\right)$$

$$= \left(\frac{b-a}{n}\right)\left[\frac{1}{2}f(x_0) + \frac{1}{2}f(x_1) + \frac{1}{2}f(x_1) + \frac{1}{2}f(x_2) + \frac{1}{2}f(x_2) + \cdots \right.$$
$$\left. + \frac{1}{2}f(x_{n-1}) + \frac{1}{2}f(x_n)\right]$$

$$= \left(\frac{b-a}{n}\right)\left[\frac{1}{2}f(x_0) + f(x_1) + f(x_2) + \cdots + f(x_{n-1}) + \frac{1}{2}f(x_n)\right].$$

A general statement of the trapezoidal rule follows.

Trapezoidal rule Let f be a continuous function in $a \le x \le b$. Let $a \le x \le b$ be divided into n equal subintervals by the points $a = x_0, x_1, x_2, \ldots, x_n = b$. Then

$$\int_a^b f(x) dx \approx \left(\frac{b-a}{n}\right)\left[\frac{1}{2}f(x_0) + f(x_1) + \cdots + f(x_{n-1}) + \frac{1}{2}f(x_n)\right].$$

Work Problem 1 at the side.

---

1. (a) Approximate $\int_2^6 (1+x)\,dx$ by using the trapezoidal rule with $n = 4$.
   (b) Evaluate the definite integral and compare your answers.

Answer:
(a) 20
(b) 20

310 Further Techniques and Applications of Integration

Another numerical method, called Simpson's rule, approximates consecutive portions of the curve with portions of parabolas rather than with line segments as in the trapezoidal rule. As shown in Figure 7.2, a parabola is fitted through points A, B, and C, another through C, D, and E, and so on. (It is necessary to have an even number of intervals for this process to come out right.) Then the sum of the areas under these parabolas will approximate the area under the graph of the function.

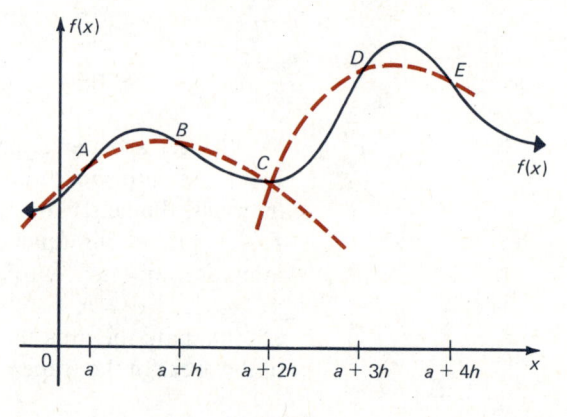

Figure 7.2

We do not derive Simpson's rule here; see one of the books listed in the bibliography under "Calculus Books for Proofs."

Simpson's rule Let f be a continuous function on $a \leq x \leq b$. Let $a \leq x \leq b$ be divided into an even number n of equal subintervals by the points $a = x_0, x_1, x_2, \ldots, x_n = b$. Then

$$\int_a^b f(x)\,dx \approx \frac{b-a}{3n}[f(x_0) + 4f(x_1) + 2f(x_2) + 4f(x_3) + \cdots + 2f(x_{n-2}) + 4f(x_{n-1}) + f(x_n)].$$

Note that it is necessary for n to be an even number. For example, let us use the same function, $1/x$, and divide the interval $1 \leq x \leq 5$ into 4 subintervals in the same way as in the example of the trapezoidal rule. By Simpson's rule,

$$\int_1^5 \frac{1}{x}\,dx \approx \frac{4}{12}\left[\frac{1}{1} + 4\left(\frac{1}{2}\right) + 2\left(\frac{1}{3}\right) + 4\left(\frac{1}{4}\right) + \frac{1}{5}\right]$$

$$= \frac{1}{3}\left(1 + 2 + \frac{2}{3} + 1 + \frac{1}{5}\right)$$

$$\approx 1.62.$$

Simpson's rule gives a better approximation than the trapezoidal rule. However, as n is increased, the two approximations would differ by less and less.

2. (a) Approximate
$$\int_2^6 (x^2 + 1)\,dx$$
using Simpson's rule with $n = 4$.
(b) Evaluate the integral and compare your answers.

Answer:
(a) $73\frac{1}{3}$
(b) $73\frac{1}{3}$

Work Problem 2 at the side.

7.2 EXERCISES

In Exercises 1–10, use $n = 4$ to approximate the value of each of the given integrals (a) by the trapezoidal rule; (b) by Simpson's rule.

1. $\int_0^2 x^2\,dx$
2. $\int_0^2 (2x + 1)\,dx$
3. $\int_0^4 \sqrt{x+1}\,dx$
4. $\int_1^5 \dfrac{1}{x+1}\,dx$
5. $\int_{-2}^2 (2x^2 + 1)\,dx$
6. $\int_0^3 (2x^2 + 1)\,dx$
7. $\int_1^5 \dfrac{1}{x^2}\,dx$
8. $\int_{-1}^3 \dfrac{1}{4-x}\,dx$
9. $\int_2^4 \dfrac{1}{x^3}\,dx$
10. $\int_0^4 x\sqrt{2x-1}\,dx$

APPLIED PROBLEMS

In the following applications, values of $f(x)$ are given.

Chemical experiment

11. The table below shows the results from a chemical experiment.

Concentration of chemical A, x	1	2	3	4	5	6	7
Rate of formation of chemical B, $f(x)$	12	16	18	21	24	27	32

(a) Plot these points. Connect the points with line segments.
(b) Use the trapezoidal rule to find the area bounded by the broken line of part (a), the x-axis, the line $x = 1$, and the line $x = 7$.
(c) Find the same area using Simpson's rule.

Psychology

12. The results from a research study in psychology were as follows.

Number of hours of study, x	1	2	3	4	5	6	7
Number of extra points earned on a test, $f(x)$	4	7	11	9	15	16	23

Sales 13. A sales manager presented the following results at a sales meeting.

Year, x	1	2	3	4	5	6	7
Rate of sales, $f(x)$	.4	.6	.9	1.1	1.3	1.4	1.6

Repeat steps (a)–(c) of Exercise 11 for this data to find the total sales over the 7-year period.

Costs 14. A company found that its marginal costs in hundreds of dollars were as follows over a 7-year period.

Year, x	1	2	3	4	5	6	7
Marginal cost, $f(x)$	9.0	9.2	9.5	9.4	9.8	10.1	10.5

Repeat steps (a)–(c) of Exercise 11 for this data to find the total sales

Bioavailability *In the study of* bioavailability *in pharmacy, a drug is given to a patient. The level of concentration of the drug is then measured periodically, producing* blood level *curves such as the ones shown below. The areas under the curves give the total amount of the drug available to the patient. Use Simpson's rule to find the following areas. Break the problem into two parts: find the area from 0 to 4 using 1-hour intervals, and find the area from 4 to 20 using 2-hour intervals. You will have to estimate the height for many of the values of time. Simpson's rule requires an even number of equal subintervals. (These graphs are taken from* Basics of Bioavailability, *by D. J. Chodos and A. R. DeSantos. The Upjohn Company. Copyright © 1978.)*

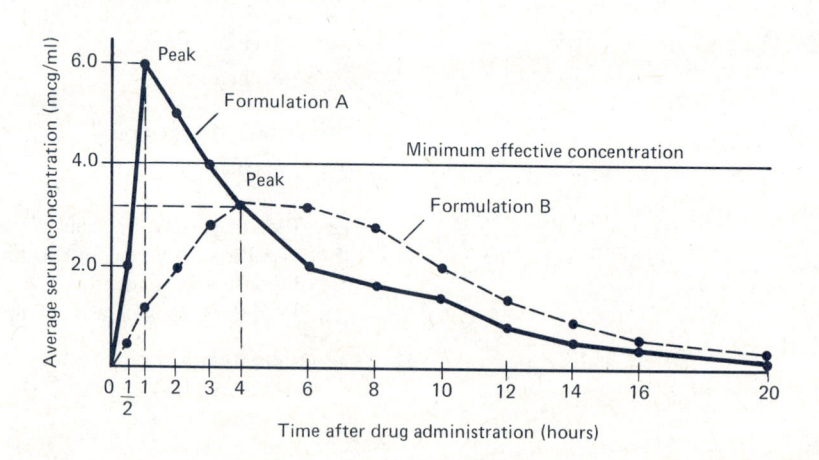

15. Find the area under the curve for Formulation A.

16. Find the area under the curve for Formulation B.

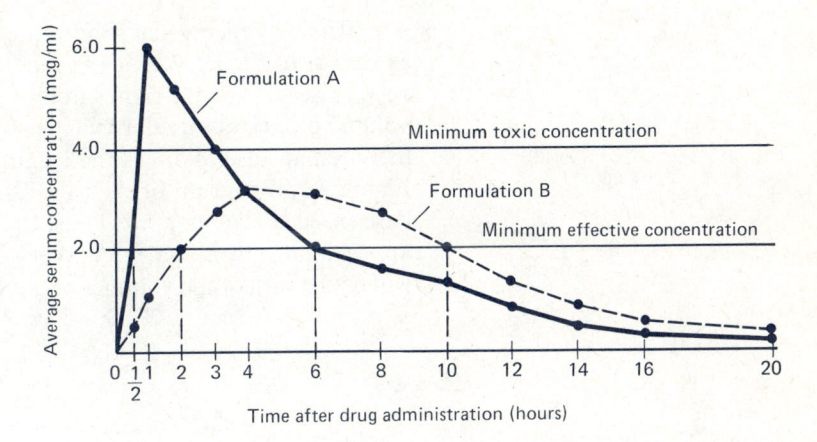

17. Find the area under the curve between the minimum toxic concentration line and the minimum effective concentration line for Formulation A.

18. Find the area under the curve and above the minimum effective concentration line for Formulation B.

7.3 FINDING VOLUMES BY INTEGRATION

Figure 7.3 shows the region below the graph of some function $y = f(x)$, above the x-axis, and between $x = a$ and $x = b$. We have already used integrals to find the area of such a region. Now, suppose this region is revolved about the x-axis as shown in Figure 7.4. The resulting figure is called a **solid of revolution.** The volume of a solid of revolution can also be found by integrals.

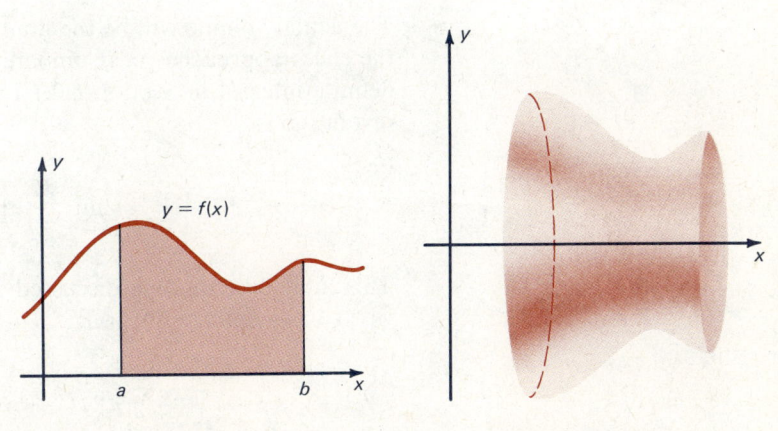

Figure 7.3

Figure 7.4

Think of slicing the solid into n slices of equal thickness Δx, as shown in Figure 7.5. If the slices are thin enough, each slice is very close to being a right circular cylinder. The formula for the volume of a right circular cylinder is $\pi r^2 h$, where r is the radius of the circular base and h is the height of the cylinder. As shown in Figure 7.6, the height of each slice is Δx. The radius of the circular base of each slice is $f(x_i)$. Thus, the volume of the slice is closely approximated by $\pi [f(x_i)]^2 \Delta x$. The volume of the solid of revolution will be the sum of the volumes of the slices. We can write this sum as

$$V \approx \sum_{i=1}^{n} \pi [f(x_i)]^2 \Delta x.$$

(a)

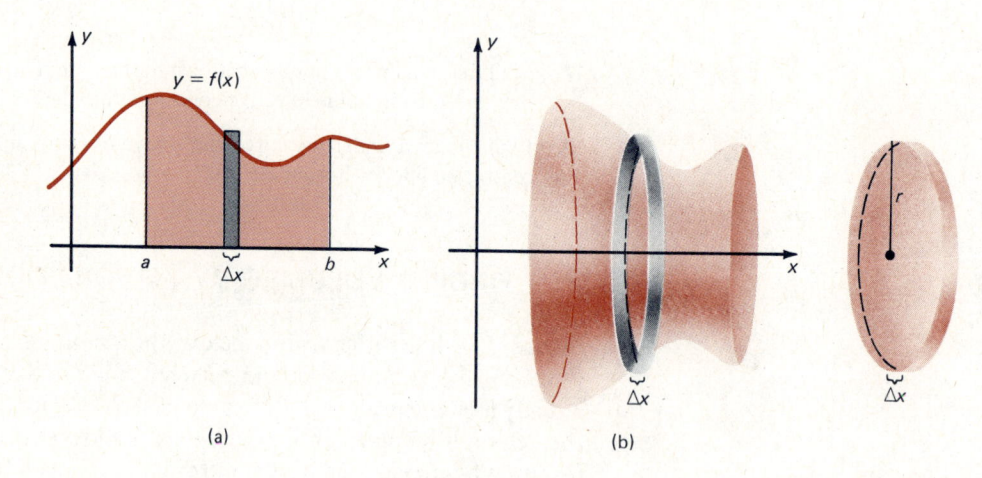

Figure 7.5

(b)

Figure 7.6

The actual volume will be the limit of this sum as the thickness of the slices approaches 0. (Compare this with the definition of the definite integral in Section 6.2.) Thus, the volume of the solid of revolution is

$$V = \lim_{\Delta x \to 0} \sum_{i=1}^{n} \pi [f(x_i)]^2 \Delta x.$$

This limit, like the one discussed in Section 6.2 for area, can be written as a definite integral.

$$V = \lim_{\Delta x \to 0} \sum_{i=1}^{n} \pi [f(x_i)]^2 \Delta x = \int_{a}^{b} \pi [f(x)]^2 \, dx.$$

7.3 Finding Volumes by Integration

Example 1 Find the volume of the solid of revolution formed by rotating about the x-axis the region bounded by $y = x + 1$, $y = 0$, $x = 1$, and $x = 4$. (See Figure 7.7(a).)

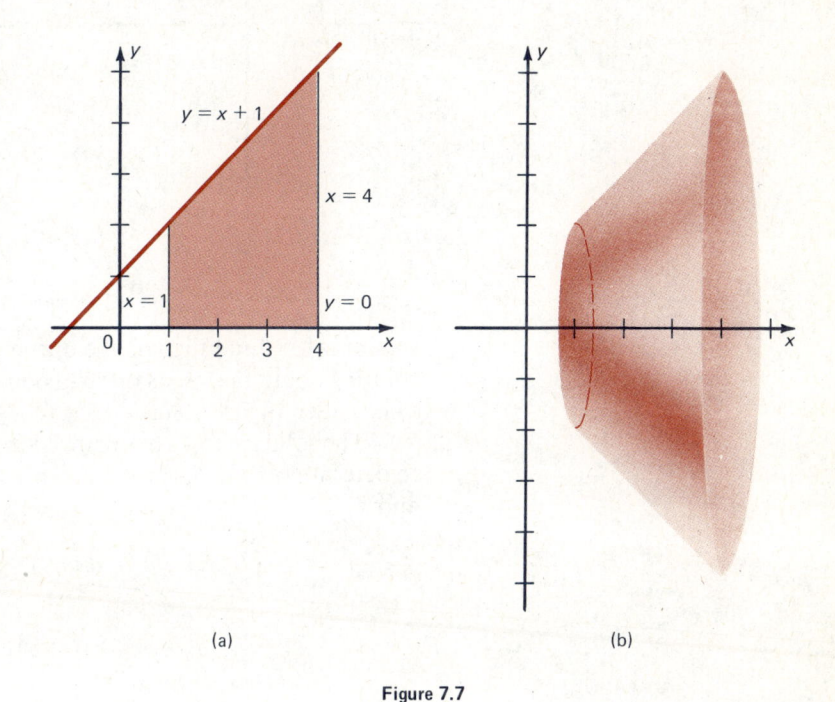

(a)

(b)

Figure 7.7

The solid is shown in Figure 7.7(b). Use the formula given above for the volume. Here $a = 1$, $b = 4$, and $f(x) = x + 1$.

$$V = \int_1^4 \pi(x + 1)^2 \, dx = \pi \left(\frac{(x + 1)^3}{3} \right) \Big]_1^4$$

$$= \frac{\pi}{3}(5^3 - 2^3)$$

$$= \frac{117\pi}{3} \quad \blacksquare$$

1. Find the volume of the solid of revolution formed by rotating about the x-axis the region bounded by $y = 2x$, $y = 0$, $x = 0$, and $x = 3$.

   Answer: 36π

Work Problem 1 at the side.

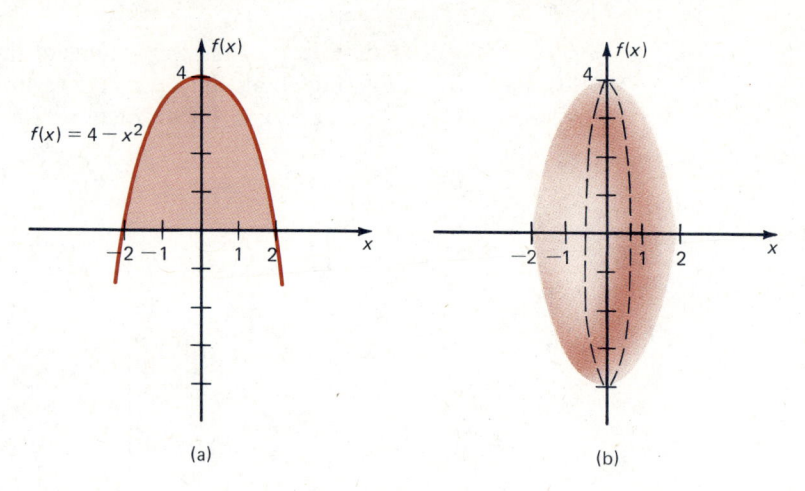

Figure 7.8

Example 2 Find the volume of the solid of revolution formed by rotating about the x-axis the area bounded by $f(x) = 4 - x^2$ and the x-axis. (See Figure 7.8(a).)

The solid is shown in Figure 7.8(b). We must use the x-intercepts to determine a and b. If $y = 0$, then $x = 2$ or $x = -2$. Thus, $a = -2$ and $b = 2$. The volume is

$$V = \int_{-2}^{2} \pi(4 - x^2)^2 \, dx = \int_{-2}^{2} \pi(16 - 8x^2 + x^4) \, dx$$

$$= \pi \left(16x - \frac{8x^3}{3} + \frac{x^5}{5} \right) \Big]_{-2}^{2}$$

$$= \frac{512\pi}{15}. \blacksquare$$

7.3 EXERCISES

Find the volume of the solid of revolution formed by rotating about the x-axis the regions bounded by the following graphs. See Examples 1 and 2.

1. $f(x) = x$, $y = 0$, $x = 0$, $x = 2$
2. $f(x) = 2x$, $y = 0$, $x = 0$, $x = 3$
3. $f(x) = 2x + 1$, $y = 0$, $x = 0$, $x = 4$
4. $f(x) = x - 4$, $y = 0$, $x = 4$, $x = 10$
5. $f(x) = \frac{1}{3}x + 2$, $y = 0$, $x = 1$, $x = 3$
6. $f(x) = \frac{1}{2}x + 4$, $y = 0$, $x = 0$, $x = 5$
7. $f(x) = \sqrt{x}$, $y = 0$, $x = 1$, $x = 2$

8. $f(x) = \sqrt{x+1}$, $y = 0$, $x = 0$, $x = 3$
9. $f(x) = \sqrt{2x+1}$, $y = 0$, $x = 1$, $x = 4$
10. $f(x) = \sqrt{3x+2}$, $y = 0$, $x = 1$, $x = 2$
11. $f(x) = e^x$, $y = 0$, $x = 0$, $x = 2$
12. $f(x) = 2e^x$, $y = 0$, $x = -2$, $x = 1$
13. $f(x) = \dfrac{1}{\sqrt{x}}$, $y = 0$, $x = 1$, $x = 4$
14. $f(x) = \dfrac{1}{\sqrt{x+1}}$, $y = 0$, $x = 0$, $x = 2$
15. $f(x) = x^2$, $y = 0$, $x = 1$, $x = 5$
16. $f(x) = \dfrac{x^2}{2}$, $y = 0$, $x = 0$, $x = 4$
17. $f(x) = 1 - x^2$, $y = 0$
18. $f(x) = 2 - x^2$, $y = 0$

The function $y = \sqrt{r^2 - x^2}$ has a graph which is a semicircle of radius r with center at $(0, 0)$. (See the figure.) In each of the following exercises, find the volume that results when this semicircle is rotated about the x-axis.

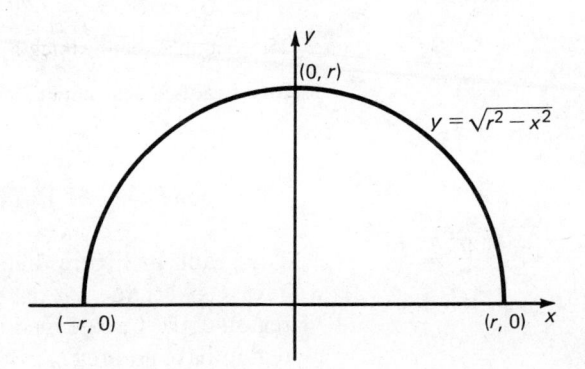

19. $f(x) = \sqrt{1 - x^2}$, $y = 0$
20. $f(x) = \sqrt{9 - x^2}$, $y = 0$
21. $f(x) = \sqrt{4 - x^2}$, $y = 0$
22. $f(x) = \sqrt{16 - x^2}$, $y = 0$
23. $f(x) = \sqrt{r^2 - x^2}$, $y = 0$
24. Use the result of Exercise 23 to find a formula for the volume of a sphere.

APPLIED PROBLEMS

Blood flow

The figure shows the blood flow in a small artery of the body. The flow of blood is laminar, which means that the velocity is very low near the artery walls and highest in the center of the artery. To calculate the total flow in the artery we think of the flow as being made up of many layers of concentric tubes sliding one on the other.

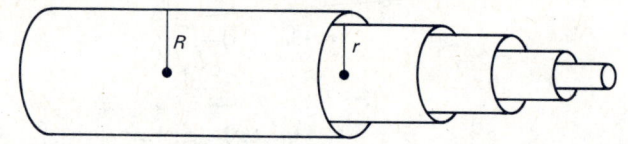

Suppose R is the radius of an artery and r is the distance from a given layer to the center. Then the velocity of blood in a given layer can be shown to equal

$$v(r) = k(R^2 - r^2),$$

where k is a numerical constant.

The area of the cross section of one of the layers can be found by differentials to be approximately

$$a(r) \approx 2\pi r \, \Delta r,$$

where Δr is the thickness of the layer. The total flow in the layer is given by the product of volume and cross-section area, or

$$F(r) = 2\pi r k(R^2 - r^2) \, \Delta r.$$

25. Set up a definite integral to find the total flow in the artery.

26. Evaluate this definite integral.

7.4 IMPROPER INTEGRALS

The graph in Figure 7.9(a) shows the area bounded by the curve $f(x) = x^{-3/2}$, the x-axis, and the vertical line $x = 1$. The shaded region could be extended indefinitely to the right. Does this shaded region have an area?

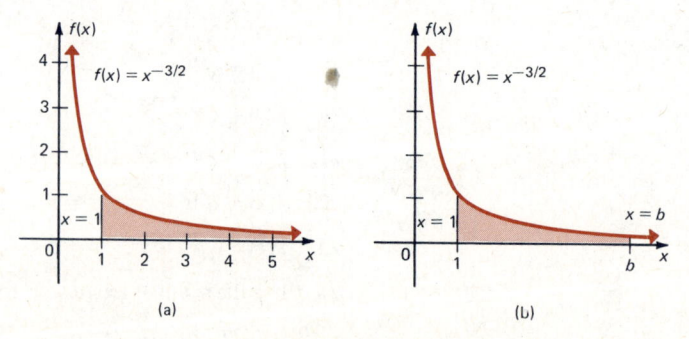

Figure 7.9

To see if we can find the area of this region, introduce a vertical line at $x = b$, as shown in Figure 7.9(b). With this vertical line we have a region with both upper and lower limits of integration. This region is given by the definite integral

$$\int_1^b x^{-3/2} \, dx.$$

By the fundamental theorem of calculus,

$$\int_1^b x^{-3/2} \, dx = -2x^{-1/2} \Big]_1^b$$
$$= -2b^{-1/2} + (2 \cdot 1^{-1/2})$$
$$= -2b^{-1/2} + 2$$
$$= 2 - \frac{2}{b^{1/2}}.$$

Suppose we now let the vertical line $x = b$ of Figure 7.9(b) move further to the right. That is, suppose we let $b \to \infty$. The expression $-2/b^{1/2}$ would then approach 0, and

$$\lim_{b \to \infty} \left(2 - \frac{2}{b^{1/2}} \right) = 2 - 0 = 2.$$

This limit is defined as the *area* shown in Figure 7.9(a), so that

$$\int_1^\infty x^{-2/3} \, dx = 2.$$

In general, an integral of the form

$$\int_a^\infty f(x) \, dx, \quad \int_{-\infty}^b f(x) \, dx, \quad \text{or} \quad \int_{-\infty}^\infty f(x) \, dx$$

is called an improper integral. These **improper integrals** are defined as follows.

$$\int_a^\infty f(x) \, dx = \lim_{b \to \infty} \int_a^b f(x) \, dx$$

$$\int_{-\infty}^b f(x) \, dx = \lim_{a \to -\infty} \int_a^b f(x) \, dx$$

$$\int_{-\infty}^\infty f(x) \, dx = \int_{-\infty}^c f(x) \, dx + \int_c^\infty f(x) \, dx$$

If the expression on the right side exists, the integrals are **convergent**; otherwise, they are **divergent**.

Work Problem 1 at the side.

1. Find the following.

(a) $\int_4^\infty x^{-3/2} \, dx$

(b) $\int_9^\infty x^{-3/2} \, dx$

Answer:
(a) 1
(b) $\frac{2}{3}$

Example 1 Find the following.

(a) $\int_{1}^{\infty} \frac{dx}{x}$

A graph of this region is shown in Figure 7.10. Use the definition of an improper integral:

$$\int_{1}^{\infty} \frac{dx}{x} = \lim_{b \to \infty} \int_{1}^{b} \frac{dx}{x}.$$

Use the fundamental theorem of calculus to find $\int_{1}^{b} \frac{dx}{x}$.

$$\int_{1}^{b} \frac{dx}{x} = \ln|x|\Big]_{1}^{b} = \ln|b| - \ln|1| = \ln|b| - 0 = \ln|b|$$

Take the limit as $b \to \infty$. We know that $\ln|b|$ gets larger and larger as b gets larger and larger. Therefore,

$$\lim_{b \to \infty} \ln|b| \text{ does not exist.}$$

Since the limit does not exist, $\int_{1}^{\infty} \frac{dx}{x}$ is divergent.

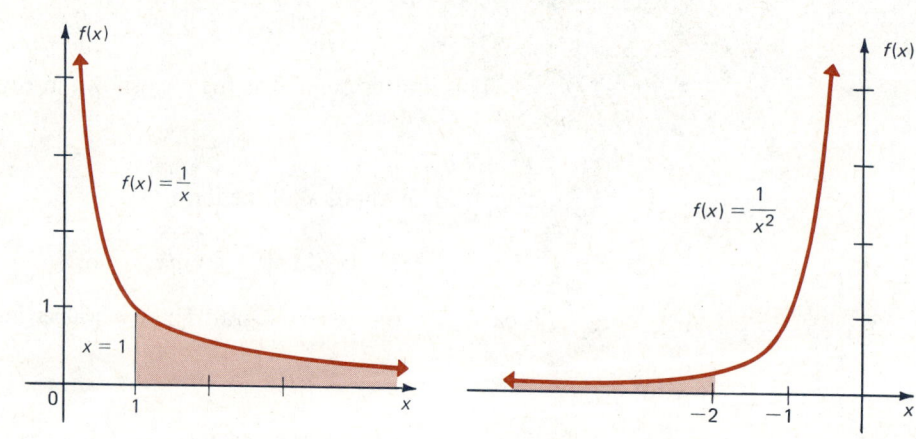

Figure 7.10

Figure 7.11

(b) $\int_{-\infty}^{-2} \frac{1}{x^2} dx = \lim_{a \to -\infty} \int_{a}^{-2} \frac{1}{x^2} dx$

$$= \lim_{a \to -\infty} \frac{-1}{x}\Big]_{a}^{-2}$$

$$= \lim_{a \to -\infty} \left(\frac{1}{2} + \frac{1}{a}\right)$$

$$= \frac{1}{2}$$

A graph of this region is shown in Figure 7.11. This integral converges. ∎

2. Find the following.

(a) $\int_{-\infty}^{-2} \frac{1}{x} dx$

(b) $\int_{-\infty}^{-3} -4x^{-3} dx$

(c) $\int_{1}^{\infty} \sqrt{x}\, dx$

Answer:
(a) divergent
(b) 2/9
(c) divergent

3. Find the following.

(a) $\int_{-\infty}^{0} e^{2x} dx$

(b) $\int_{0}^{\infty} 3e^{-3x} dx$

Answer:
(a) $\frac{1}{2}$
(b) 1

4. Let R represent the fixed annual rent on a property. The capitalized value of the property is given by $\int_{0}^{\infty} Re^{-kt} dt$, where k is the average rate of interest in the economy. Find the capitalized value of property when rent is paid perpetually if
(a) $k = .04$ and $R = 5000$;
(b) $k = .08$ and $R = 12{,}500$.

Answer:
(a) $125{,}000
(b) $156{,}250

Work Problem 2 at the side.

Example 2 Find $\int_{0}^{\infty} 4e^{-3x} dx$
By definition,

$$\int_{0}^{\infty} 4e^{-3x} dx = \lim_{b \to \infty} \int_{0}^{b} 4e^{-3x} dx$$

$$= \lim_{b \to \infty} -\frac{4}{3} e^{-3x} \Big]_{0}^{b}$$

$$= \lim_{b \to \infty} \left[-\frac{4}{3} e^{-3b} - \left(-\frac{4}{3} e^{-0} \right) \right]$$

$$= \lim_{b \to \infty} \left[\frac{-4}{3e^{3b}} + \frac{4}{3} \right]$$

$$= 0 + \frac{4}{3} = \frac{4}{3}. \quad \blacksquare$$

Work Problem 3 at the side.

Example 3 A chemical is being released into a small stream. The rate at which the chemical is being released into the stream at time t is given by Pe^{-kt}, where P is the amount of chemical released into the stream initially. Suppose $P = 1000$ and $k = .06$. Find the total amount of the chemical that will be released into the stream into the indefinite future.

We need to find

$$\int_{0}^{\infty} Pe^{-kt} dt = \int_{0}^{\infty} 1000 e^{-.06t} dt.$$

Work as above.

$$\int_{0}^{\infty} 1000 e^{-.06t} dt = \lim_{b \to \infty} \int_{0}^{b} 1000 e^{-.06t} dt$$

$$= \lim_{b \to \infty} \left(\frac{1000}{-.06} e^{-.06t} \right) \Big]_{0}^{b}$$

$$= \lim_{b \to \infty} \left[\frac{1000}{-.06 e^{.06b}} - \frac{1000}{-.06} e^{0} \right]$$

$$= \frac{-1000}{-.06} = 16{,}667$$

total units of the chemical. $\blacksquare$

Work Problem 4 at the side.

7.4 EXERCISES

Find the value of the following which exist. See Examples 1 and 2.

1. $\int_{2}^{\infty} \frac{1}{x^2} dx$
2. $\int_{5}^{\infty} \frac{1}{x^2} dx$
3. $\int_{1}^{\infty} \frac{1}{\sqrt{x}} dx$
4. $\int_{16}^{\infty} \frac{-3}{\sqrt{x}} dx$
5. $\int_{-\infty}^{-1} \frac{2}{x^3} dx$
6. $\int_{-\infty}^{-4} \frac{3}{x^4} dx$
7. $\int_{1}^{\infty} \frac{1}{x^{1.001}} dx$
8. $\int_{1}^{\infty} \frac{1}{x^{.999}} dx$
9. $\int_{-\infty}^{-1} x^{-2} dx$
10. $\int_{-\infty}^{-4} x^{-2} dx$
11. $\int_{-\infty}^{-1} x^{-8/3} dx$
12. $\int_{-\infty}^{27} x^{-5/3} dx$
13. $\int_{0}^{\infty} 4e^{-4x} dx$
14. $\int_{0}^{\infty} 10e^{-10x} dx$
15. $\int_{-\infty}^{0} 4e^{x} dx$
16. $\int_{-\infty}^{0} 3e^{4x} dx$
17. $\int_{-\infty}^{-1} \ln|x| dx$
18. $\int_{1}^{\infty} \ln|x| dx$
19. $\int_{0}^{\infty} \frac{dx}{(x+1)^2}$
20. $\int_{0}^{\infty} \frac{dx}{(2x+1)^3}$
21. $\int_{-\infty}^{-1} \frac{2x-1}{x^2-x} dx$
22. $\int_{0}^{\infty} \frac{2x+3}{x^2+3x} dx$

Use the table inside the back cover as necessary for the following.

23. $\int_{-\infty}^{1} \frac{2}{3x(2x-7)} dx$
24. $\int_{1}^{\infty} \frac{7}{2x(5x+1)} dx$
25. $\int_{1}^{\infty} \frac{4}{9x(x+1)^2} dx$
26. $\int_{1}^{5} \frac{5}{4x(x+2)^2} dx$
27. $\int_{0}^{\infty} xe^{2x} dx$
28. $\int_{-\infty}^{0} xe^{3x} dx$

APPLIED PROBLEMS

Capitalized values

Find the capitalized values of the following properties. See Problem 4 in the margin at the end of the text above.

29. A castle for which annual rent of $60,000 will be paid in perpetuity; the interest rate is 8%

30. A fort on a strategic peninsula in the North Sea, annual rent $500,000. paid in perpetuity; the interest rate is 6%

Nuclear wastes

Radioactive waste is entering the atmosphere over an area at a decreasing rate. Use the improper integral

$$\int_{0}^{\infty} Pe^{-kt} dt$$

with $P = 50$ *to find the total amount of the waste that will enter the atmosphere for each of the following values of k. See Example 3.*

31. $k = .04$ **32.** $k = .06$

7.5 PROBABILITY DENSITY FUNCTIONS

A bank is interested in improving its services to the public. The manager decides to begin by finding out how much time its tellers spend on each transaction. She decides to time the transactions to the nearest minute. The time for a complete transaction t will then be one of the numbers 1, 2, 3, ... Because the value of the variable t will occur randomly, t is called a **random variable.**

A table of the results of 75 transactions, where $f(t)$ is the relative frequency of a transaction, is shown below. The shortest transaction time was 1 minute, and there were 3 transactions of 1-minute duration. The longest time was 10 minutes. Only one transaction took that long.

t	1	2	3	4	5	6	7	8	9	10
Number of occurrences	3	5	9	12	15	11	10	6	3	1
$f(t)$	$\frac{3}{75}$	$\frac{5}{75}$	$\frac{9}{75}$	$\frac{12}{75}$	$\frac{15}{75}$	$\frac{11}{75}$	$\frac{10}{75}$	$\frac{6}{75}$	$\frac{3}{75}$	$\frac{1}{75}$

A graph of the relative frequency with which the various values of t occurred is shown in Figure 7.12.

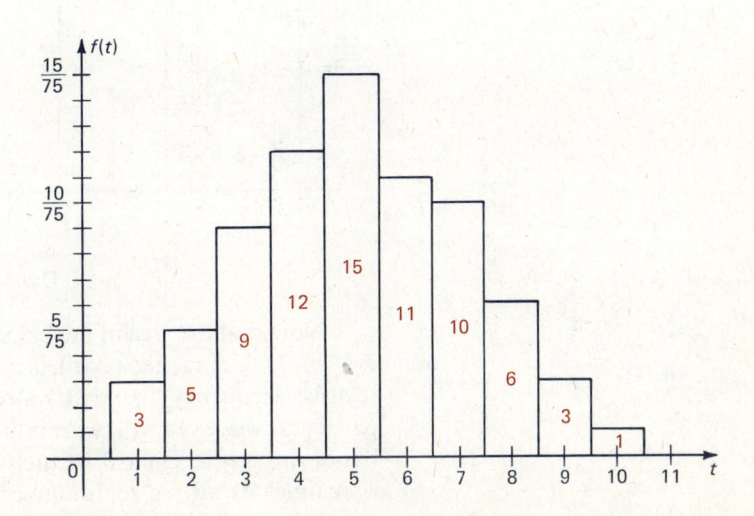

Figure 7.12

324 Further Techniques and Applications of Integration

The manager could use the results of her survey to get the probability that a transaction would last t minutes. The **probability** of an event can be defined as the relative frequency of occurrence of the event. Thus, the probability that $t = 1$, written $P(t = 1)$, is 3/75. Similarly, $P(t = 2)$ is 5/75 and $P(t = 3)$ is 9/75. To find the probability that the transaction time is no more than 3 minutes, we can add the three probabilities found above. That is,

$$P(t \leq 3) = P(t = 1) + P(t = 2) + P(t = 3) = 17/75.$$

Work Problem 1 at the side.

1. Find the following using the table of values given above.
   (a) $P(t = 5)$
   (b) $P(t = 6)$
   (c) $P(5 \leq t)$

Answer:
(a) 1/5
(b) 11/75
(c) 46/75

It would have been possible to time the transactions more accurately—to the nearest tenth of a minute or even to the nearest second (or 1/60th of a minute) if desired. Theoretically, at least, the transaction times t could take on any positive real-number value (between, say, 0 and 11 minutes). This allows us to think of the graph of the relative frequency of transaction times as the continuous curve shown in Figure 7.13. As indicated in Figure 7.13 the curve was derived from the graph of Figure 7.12 by connecting the points at the top of the bars and smoothing the resulting polygon into a curve.

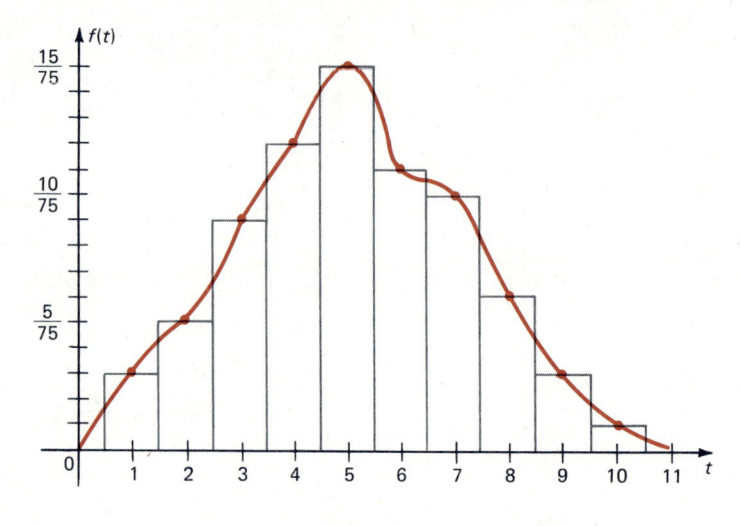

Figure 7.13

Note that the area of the bar above $t = 1$ in Figure 7.9 is 1 times 3/75 or 3/75. Since each bar has a width of 1, its area is equal to the relative frequency for that value of t. The probability that a particular value of t will occur is given by the area of the appropriate bar of the graph. Thus, if we prefer to think of the possible transaction times as all the real numbers between 0 and 11, we can use the area under the curve of Figure 7.13 between any two values of t to find the probability that a transaction time will be between those numbers. For example, the shaded region in Figure 7.14 corresponds to the probability that t is between a and b, written $P(a \leq t \leq b)$.

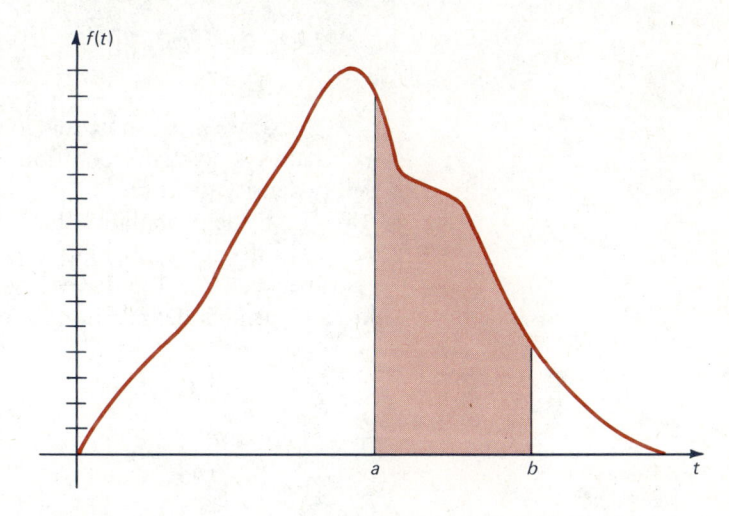

Figure 7.14

In Chapter 6, we saw that the definite integral can be used to find the area under the graph of $f(x)$ from $x = a$ to $x = b$. If we can find a function $f(x)$ to describe a relative frequency curve, we can use the definite integral to find the area under the curve from a to b. Since that area represents the probability that x will be between a and b, we have

$$P(a \leq x \leq b) = \int_a^b f(x)\, dx.$$

A function $f(x)$ which can be used to describe a relative frequency curve is called a **probability density function.** Such a function must satisfy certain conditions. If x is a random variable over the interval $[a, b]$, then $f(x)$ is a probability density function if

1. $\int_a^b f(x)\, dx = 1$, and

2. $f(x) \geq 0$ for all x in the interval $[a, b]$.

Intuitively, property (1) says that the total probability must be 1; *something* must happen. Property (2) says that the probability of a particular event can never be negative.

Example 1 (a) Show that $f(x) = \frac{3}{26}x^2$ is a probability density function for the interval $[1, 3]$.

First, we show that condition 1 holds.

$$\int_1^3 \frac{3}{26}x^2\,dx = \frac{3}{26}\left(\frac{x^3}{3}\right)\Big]_1^3$$
$$= \frac{3}{26}\left(9 - \frac{1}{3}\right)$$
$$= 1$$

Next, we must show that $f(x) \geq 0$ for the interval $[1, 3]$. Since x^2 is always positive, condition 2 also holds. Thus, $f(x)$ is a probability density function.

(b) Find the probability that x will be between 1 and 2.

The desired probability is given by the area under the graph of $f(x)$ between $x = 1$ and $x = 2$ as shown in Figure 7.15. We find the area by using a definite integral.

$$P(1 \leq x \leq 2) = \int_1^2 \frac{3}{26}x^2\,dx = \frac{3}{26}\left(\frac{x^3}{3}\right)\Big]_1^2 = \frac{7}{26} \quad \blacksquare$$

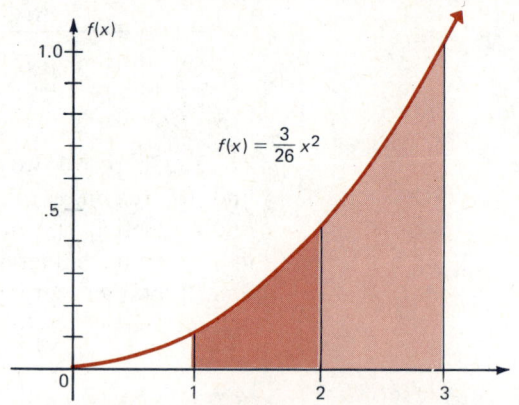

Figure 7.15

Example 2 Is $f(x) = 3x^2$ a probability density function for the interval $[0, 4]$? If not, convert it to one.

We have

$$\int_0^4 3x^2\,dx = x^3\Big]_0^4 = 64.$$

Since the integral is not equal to 1, the function is not a probability density function. To convert it to one, we simply multiply $f(x)$ by $\frac{1}{64}$. The function $P(x) = \frac{3}{64}x^2$ for $[0, 4]$ will be a probability density function since

$$\int_0^4 \frac{3}{64}x^2 = 1,$$

and $P(x) \geq 0$ for all x in $[0, 4]$. $\blacksquare$

2. (a) Is $f(x) = 2x$ a probability density function for the interval $[2, 6]$?
(b) If your answer to (a) is no, convert $f(x)$ to a probability density function.

Answer:
(a) no
(b) let $f(x) = \dfrac{x}{16}$

3. A certain species of plant normally grows to a height between 12 cm and 25 cm. Find the probability that a particular plant, selected at random, has a height between 18 and 20 cm. Assume a uniform distribution.

Answer: 2/13

Work Problem 2 at the side.

We now discuss some common probability density functions.

Uniform distribution This probability density function is given by

$$f(x) = \frac{1}{b-a} \quad \text{for } x \text{ in } [a, b],$$

where a and b are real numbers. The graph is shown in Figure 7.16. The uniform distribution requires the assumption that the probability of any value of x between a and b is equally likely.

Uniform distribution

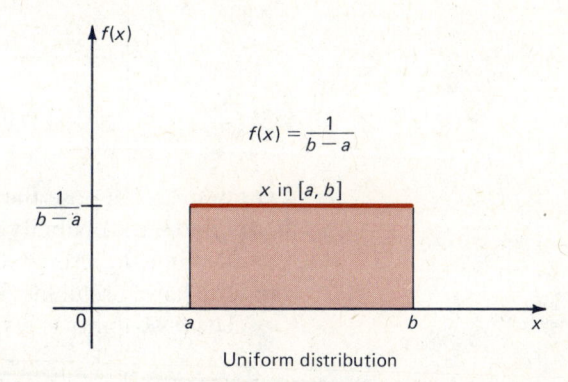

Figure 7.16

Example 3 A couple is planning to vacation in San Francisco. They have been told that the maximum daily temperature during the time they plan to be there ranges from 15°C to 27°C. What is the probability that the maximum temperature on the day they arrive will be greater than 24°C?

If we let t be the maximum temperature on a given day, then $f(t) = \tfrac{1}{12}$ for the interval $[15, 27]$ is the uniform probability density function for t. We have

$$P(t > 24) = \int_{24}^{27} \frac{1}{12} dt = \frac{1}{12} t \Big]_{24}^{27} = \frac{1}{4}.$$

By using the uniform probability density function here, we are assuming that the probability of any temperature between 15°C and 27°C is equally likely for any given day during the specified time period. ∎

Work Problem 3 at the side.

Exponential distribution The exponential distribution function is given by

$$f(x) = ae^{-ax} \quad \text{for } 0 \leq x < \infty,$$

where a is a positive real number. The graph is shown in Figure 7.17.

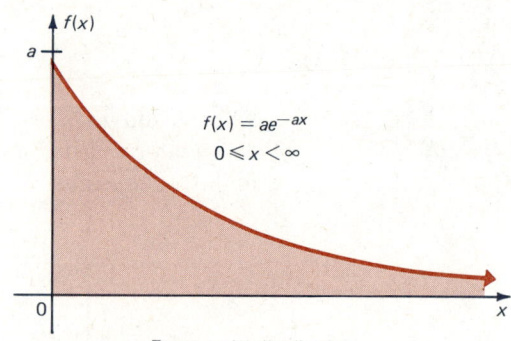

$f(x) = ae^{-ax}$
$0 \leq x < \infty$

Exponential distribution

Figure 7.17

Example 4 Suppose the useful life of a flashlight battery is t, in hours, and the probability density function is given by $f(t) = \frac{1}{20}e^{-t/20}$ ($t \geq 0$). Find the probability that a particular battery, selected at random, has a useful life of less than 100 hours.

The probability is given by

$$\int_0^{100} \frac{1}{20} e^{-t/20} \, dt = \frac{1}{20}\left(-20 e^{-t/20}\right)\Big]_0^{100}$$
$$= -(e^{-100/20} - e^0)$$
$$= -(e^{-5} - 1)$$
$$\approx 1 - .0067$$
$$= .9933. \quad \blacksquare$$

Work Problem 4 at the side.

4. The life span (in days) of a certain insect is a continuous random variable whose probability density function is given by

$$f(x) = .04 e^{-.04x}.$$

Find the probability that a particular one of these insects, selected randomly, will die within 10 days of birth.

Answer: .32968

Normal distribution The normal bell-shaped distribution is undoubtedly the most important probability density function. It is widely used in various applications of statistics. The normal distribution function is given by

$$f(x) = \frac{1}{\sigma\sqrt{2\pi}} e^{-(x-\mu)^2/2\sigma^2} \quad \text{for } -\infty < x < \infty$$

where μ and σ are real numbers, with $\sigma \geq 0$. Using very advanced techniques, it can be shown that

$$\int_{-\infty}^{\infty} \frac{1}{\sigma\sqrt{2\pi}} e^{-(x-\mu)^2/2\sigma^2} = 1.$$

It would be far too much work to calculate values for this probability distribution for various values of μ and σ. Instead, values are calculated for the **standard normal distribution,** having $\mu = 0$ and $\sigma = 1$. The graph of the standard normal distribution is the bell-shaped curve shown in Figure 7.18.

$$f(x) = \frac{1}{\sqrt{2\pi}} e^{-x^2/2}$$
$$-\infty < x < \infty$$

Standard normal distribution

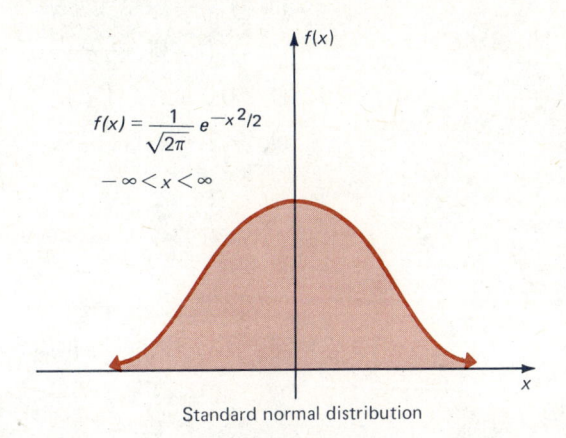

Figure 7.18

To find probabilities for the standard normal distribution we need to evaluate the definite integral

$$\int_a^b \frac{1}{\sqrt{2\pi}} e^{-x^2/2}.$$

As mentioned earlier, $e^{-x^2/2}$ is a function that does not have an antiderivative, so numerical methods must be used to find values of the definite integral.

7.5 EXERCISES

Decide if the following functions are probability density functions. If not, tell why. See Example 1.

1. $f(x) = \dfrac{1}{9}x - \dfrac{1}{18}, \quad 2 \leq x \leq 5$

2. $f(x) = \dfrac{1}{3}x - \dfrac{1}{6}, \quad 3 \leq x \leq 4$

3. $f(x) = \dfrac{3}{63}x^2, \quad 1 \leq x \leq 4$

4. $f(x) = \dfrac{3}{98}x^2, \quad 3 \leq x \leq 5$

5. $f(x) = 4x^3, \quad 0 \leq x \leq 3$

6. $f(x) = \dfrac{x^3}{81}, \quad 0 \leq x \leq 3$

7. $f(x) = \dfrac{x^2}{16}, \quad -2 \leq x \leq 2$

8. $f(x) = 2x^2, \quad -1 \leq x \leq 1$

9. $f(x) = 2x$, $-2 \leq x \leq \sqrt{5}$ 10. $f(x) = 4x^3$, $-1 \leq x \leq \sqrt[4]{2}$

For each of the following, find a value of k which will make f(x) a probability density function. See Example 2.

11. $f(x) = kx^{1/2}$, $1 \leq x \leq 4$ 12. $f(x) = kx^{3/2}$, $4 \leq x \leq 9$
13. $f(x) = kx^2$, $0 \leq x \leq 5$ 14. $f(x) = kx^2$, $-1 \leq x \leq 2$
15. $f(x) = kx$, $0 \leq x \leq 3$ 16. $f(x) = kx$, $2 \leq x \leq 3$
17. $f(x) = kx$, $1 \leq x \leq 5$ 18. $f(x) = kx$, $0 \leq x \leq 4$
19. $f(x) = kx^2$, $1 \leq x \leq 3$ 20. $f(x) = kx^3$, $2 \leq x \leq 4$

APPLIED PROBLEMS

Solve the following problems. See Examples 3 and 4.

21. The probability density function of a random variable x is

$$f(x) = 1 - \frac{1}{\sqrt{x}} \quad \text{for } 1 \leq x \leq 4.$$

Find the following probabilities.
(a) $P(x \geq 3)$
(b) $P(x \leq 2)$
(c) $P(2 \leq x \leq 3)$

22. The probability density function of a random variable x is

$$f(x) = \frac{1}{11}\left(1 + \frac{3}{\sqrt{x}}\right) \quad \text{for } 4 \leq x \leq 9.$$

Find the following probabilities.
(a) $P(x \geq 6)$
(b) $P(x \leq 5)$
(c) $P(4 \leq x \leq 7)$

23. The probability density function of a random variable x is

$$f(x) = \frac{1}{15}\left(x + \frac{1}{2}\right) \quad \text{for } 0 \leq x \leq 5.$$

Find the following probabilities.
(a) $P(x \geq 3)$
(b) $P(x \leq 2)$
(c) $P(1 \leq x \leq 4)$

24. The probability density function of a random variable x is

$$f(x) = \frac{1}{2} \quad \text{for } 1 \leq x \leq 3.$$

Find the following probabilities.
(a) $P(x \geq 2)$
(b) $P(x \leq \frac{3}{2})$
(c) $P(1 \leq x \leq \frac{5}{2})$

25. The length (in centimeters) of the leaf of a certain plant is a continuous random variable with a probability density function of

$$f(x) = \frac{5}{4} \quad \text{for } 4 \leq x \leq 4.8.$$

Find the probability that a particular leaf from one of these plants will be between 4.5 and 4.8 cm in length.

26. The price of an item in dollars is a continuous random variable with a probability density function of

$$f(x) = 2 \quad \text{for } 1.25 \leq x \leq 1.75.$$

Find the probability that the price will be less than $1.35.

27. The length of time, t, in years until a particular radioactive particle decays is a random variable with a probability density function of

$$f(t) = .03e^{-.03t}, \quad 0 \leq t < \infty.$$

Find the probability that the particle decays within 20 years.

28. The length of time, t, in years that a seedling tree survives is a random variable with a probability density function of

$$f(t) = .05e^{-.05t}, \quad 0 \leq t < \infty.$$

Find the probability that a particular tree survives fewer than 10 years.

29. The length of time, t, in days required to learn a certain task is a random variable with a probability density function of

$$f(t) = e^{-t}, \quad 0 \leq t < \infty.$$

Find the probability that a certain individual learns the task in 2 to 4 days.

30. The distance, x, in meters that seeds are dispersed from a certain kind of plant is a random variable with a probability density function of

$$f(x) = .1e^{-.1x}, \quad 0 \leq x < \infty.$$

Find the probability that the seeds of a particular plant are dispersed from 1 to 2 meters away.

CHAPTER 7 SUMMARY

Key Words

integrating by parts
numerical integration
trapezoidal rule
Simpson's rule
solid of revolution

improper integral
convergent
divergent
random variable
probability density function

Things to Remember

Formula for integrating by parts

$$\int u \, dv = uv - \int v \, du$$

Trapezoidal rule Let f be a continuous function on $a \leq x \leq b$. Let $a \leq x \leq b$ be divided into n equal subintervals by the points $a = x_0, x_1, x_2, \ldots, x_n = b$. Then

$$\int_a^b f(x) \, dx \approx \frac{b-a}{n} \left[\frac{1}{2} f(x_0) + f(x_1) + \cdots + f(x_{n-1}) + \frac{1}{2} f(x_n) \right].$$

Simpson's rule Let f be a continuous function on $a \leq x \leq b$. Let $a \leq x \leq b$ be divided into an even number n of equal subintervals by the points $a = x_0, x_1, x_2, \ldots, x_n = b$. Then

$$\int_a^b f(x) \, dx \approx \frac{b-a}{3n} \left[f(x_0) + 4f(x_1) + 2f(x_2) + 4f(x_3) + \cdots \right.$$
$$\left. + 2f(x_{n-2}) + 4f(x_{n-1}) + f(x_n) \right]$$

Definition If x is a random variable over $[a, b]$ then $f(x)$ is a probability density function if

1. $\int_a^b f(x) \, dx = 1$, and

2. $f(x) \geq 0$ for all x in the interval from a to b.

Probability Density Functions

Uniform distribution: $f(x) = \dfrac{1}{b-a}$ for $a \leq x \leq b$

Exponential distribution: $f(x) = ae^{-ax}$ for $0 \leq x < \infty$

Normal distribution: $f(x) = \dfrac{1}{\sigma \sqrt{2\pi}} e^{-(x-\mu)^2/2\sigma^2}$ for $-\infty < x < \infty$

Standard normal distribution: $f(x) = \dfrac{1}{\sqrt{2\pi}} e^{-x^2/2}$ for $-\infty < x < \infty$

CHAPTER 7 TEST

Use integration by parts to integrate the following.

[7.1]   1. $\int x(8+x)^{3/2}\,dx$ 2. $\int x e^x\,dx$

         3. $\int (x+2)e^{-3x}\,dx$ 4. $\int (3+x^2)e^{2x}\,dx$

[7.2]   5. Find $\int_2^6 \dfrac{dx}{x^2-1}$ by using the trapezoidal rule with $n=4$.

         6. Find $\int_2^6 \dfrac{dx}{x^2-1}$ by using Simpson's rule with $n=4$.

[7.3]   7. Find the volume of the solid generated by revolving about the x-axis the region bounded by $f(x)=x+2$, $y=0$, $x=1$, and $x=3$.

         8. Find the volume of the solid generated by revolving about the x-axis the region bounded by $f(x)=\sqrt{x+4}$, $y=0$, $x=-4$, and $x=4$.

[7.4]   *Find the following improper integrals which are convergent.*

         9. $\int_{-\infty}^{-2} x^{-2}\,dx$ 10. $\int_1^{\infty} x^{-1}\,dx$

        11. $\int_0^{\infty} 6e^{-8x}\,dx$ 12. $\int_0^{\infty} \dfrac{dx}{(5x+2)^2}$

[7.5]   13. Is $f(x)=2x+4$ for $1 \le x \le 4$ a probability density function? If not, tell why.

       14. Find k such that $f(x)=kx+1$ for $1 \le x \le 3$ is a probability density function.

       15. The time t in years until a certain machine requires repairs is a random variable whose density function is

$$f(x) = \frac{5}{112}(1-x^{-3/2}) \quad \text{for} \quad 1 \le x \le 25.$$

Find the probability that no repairs are required in the first four years by finding the probability that a repair will be needed in years 4 through 25.

       16. The distance in meters that a certain animal moves away from a release point is a random variable x whose probability density function is

$$f(x) = .01 e^{-.01x} \quad \text{for} \quad 0 \le x < \infty.$$

Find the probability that the animal will move no farther than 100 meters away.

CASE 8   A CROP-PLANTING MODEL*

Many firms in food processing, seed production, and similar industries face a problem every year deciding how many acres of land to plant in each of various crops. Demand for the crop is unknown, as is the actual yield per acre. In this case, we set up a mathematical model for determining the optimum number of acres to plant in a crop.

This model is designed to tell the company the number of acres of seed that it should plant. The model uses the following variables (a simplified version of this model was given in Case 1).

D = number of tons of seed demanded
$f(D)$ = continuous probability density function for the quantity of seed demanded, D
$F(D)$ = cumulative probability distribution for the quantity of seed demanded, D
X = quantity of seed produced per acre of land
Q = quantity of seed carried over in inventory from previous years
S = selling price per ton of seed
C_p = variable costs of production, marketing, etc., per ton of seed
C_c = cost to carry over a ton of seed from previous years
A = number of acres of land to be planted
C_A = variable cost per acre of land contracted
T = total number of tons of seed available for sale
a = lower limit of the domain of $f(D)$
b = upper limit of the domain of $f(D)$

To decide on the optimum number of acres to plant, it is necessary to calculate the *expected value of the profit* from the planting of D acres. Expected value, an idea from probability and statistics, is defined as the product of the profit (or loss) from a certain outcome and the probability of that outcome happening.

Based on the definition of the variables above, the total number of tons of seed that will be available for sale is given by the product of the number of acres planted, A, and the yield per acre, X, added to the carryover, Q. If T represents this total, then

$$T = AX + Q.$$

The variable here is A; we assume X and Q are known and fixed.

The expected profit can be broken down into several parts. The first portion comes from multiplying the profit per ton and the average number of tons demanded. The profit per ton is found by subtracting the variable cost per ton, C_p, from the selling price per ton, S:

$$\text{profit per ton} = S - C_p.$$

---

*Based on work by David P. Rutten, Senior Mathematician, The Upjohn Company, Kalamazoo, Michigan.

It is shown in more advanced courses that the average number of tons demanded for our interval of concern is given by

$$\int_a^T D \cdot f(D)\,dD.$$

Thus, this portion of the expected profit is

$$(S - C_p) \cdot \int_a^T D \cdot f(D)\,dD. \tag{1}$$

A second portion of expected profit is found by multiplying the profit per ton, $S - C_p$, the total number of tons available, T (recall that this is a variable), and the probability that T or more tons will be demanded by the marketplace.

$$(S - C_p)(T) \int_T^b f(D)\,dD \tag{2}$$

If T is greater than D, there will be costs associated with carrying over the excess seeds. The expected value of these costs is given by the product of the carrying cost per ton, C_c, and the number of tons to be carried over, or

$$-C_c \int_a^T (T - D) f(D)\,dD. \tag{3}$$

The minus sign shows that these costs reduce profit. If $T < D$, this term would be omitted.

Finally, the total cost of producing the seeds is given by the product of the variable cost per acre and the number of acres:

$$-C_A \cdot A. \tag{4}$$

The expected profit is the sum of the expressions in (1)–(4), or

$$\text{expected profit} = (S - C_p) \cdot \int_a^T D \cdot f(D)\,dD + (S - C_p)(T) \int_T^b f(D)\,dD$$
$$- C_c \int_a^T (T - D) f(D)\,dD - C_A \cdot A. \tag{5}$$

As an example, suppose that

$$\text{demand density function} = f(D) = \frac{1}{1000} \quad \text{for } 500 \leq D \leq 1500 \text{ tons}$$

$$a = 500$$
$$b = 1500$$
$$\text{selling price} = S = \$10{,}000 \text{ per ton}$$
$$\text{variable cost} = C_p = \$5000 \text{ per ton}$$
$$\text{carrying cost} = C_c = \$3000 \text{ per ton}$$
$$\text{variable cost per acre} = C_A = \$100$$
$$\text{inventory carryover} = Q = 200 \text{ tons}$$
$$\text{yield per acre} = X = .1 \text{ ton}$$
$$T = AX + Q = .1A + 200.$$

Substitute all this into equation (5).

$$\text{expected profit} = (10{,}000 - 5000)\int_{500}^{.1A+200} D \cdot \frac{1}{1000} dD$$

$$+ (10{,}000 - 5000)(.1A + 200)\int_{.1A+200}^{1500} \frac{1}{1000} dD$$

$$- 3000\int_{500}^{.1A+200} (.1A + 200 - D) \cdot \frac{1}{1000} dD - 100A$$

Simplify all this.

$$\text{expected profit} = \frac{5000}{1000} \cdot \frac{D^2}{2}\bigg]_{500}^{.1A+200} + (5000)(.1A + 200)\frac{D}{1000}\bigg]_{.1A+200}^{1500}$$

$$- 3000\left(\frac{.1AD}{1000} + \frac{200D}{1000} - \frac{D^2}{2000}\right)\bigg]_{500}^{.1A+200}$$

$$= \frac{5}{2}(.1A + 200)^2 - \frac{5}{2} \cdot 500^2 + 5(.1A + 200)1500$$

$$- 5(.1A + 200)(.1A + 200) - .3A(.1A + 200)$$

$$- 600(.1A + 200) + \frac{3}{2}(.1A + 200)^2 + 150A$$

$$+ 600(500) - \frac{3}{2}(500)^2 - 100A$$

$$= -(.1A + 200)^2 + 6900(.1A + 200) - .3A(.1A + 200)$$

$$+ 50A - 700{,}000$$

To find the maximum expected profit, take the derivative of this function with respect to A and then place it equal to 0.

$$D[\text{expected profit}] = -2(.1A + 200)(.1) + 690 - .06A - 60 + 50$$

$$= -.02A - 40 + 690 - .06A - 10$$

$$= -.08A + 640$$

Place this derivative equal to 0.

$$-.08A + 640 = 0$$

$$640 = .08A$$

$$\frac{640}{.08} = A$$

$$8000 = A$$

If 8000 acres are planted, the maximum profit will be obtained. Compare this with the answer to Case 1.

8
Differential Equations

Many practical problems involve the rate of change of one variable with respect to another. In Chapter 3, we expressed the rate of change of a variable y with respect to another variable x as a derivative. Given an equation for y, we were able to find dy/dx. In this chapter, we see how to solve an equation involving a derivative, such as dy/dx, for y. This kind of equation is frequently used as a model in problems of growth and decay of populations, Gross National Product, investments, profits, and radioactive substances, for example.

8.1 GENERAL AND PARTICULAR SOLUTIONS

The rate of change of many natural growth processes with respect to time can be expressed as a function of time in the form

$$\frac{dy}{dx} = kx \quad \text{or} \quad \frac{dy}{dx} = ky,$$

where x represents time and k is some constant. Such equations containing a derivative or differential are called **differential equations**. In most differential equations, it is convenient to represent the derivative as dy/dx, rather than y' or $f'(x)$. For a second derivative, the notation d^2y/dx^2 is used. Some examples of differential equations include

$$\frac{dy}{dx} = x^{1/2}, \quad \frac{dy}{dx} = 2y^2, \quad \text{and} \quad \frac{d^2y}{dx^2} = 3x^2 + 4.$$

The time-dating of dairy products depends on the solution of a differential equation. The rate of growth of bacteria in such products increases with time. If y is the number of bacteria in millions present at a time t in days, then the rate of growth of bacteria can be expressed as dy/dt, and we can write

$$\frac{dy}{dt} = kt,$$

where k is an appropriate constant. For simplicity, let us assume that $k = 10$ for a particular product, so that

$$\frac{dy}{dt} = 10t. \tag{1}$$

In Section 4.8, we defined the differential dy as $f'(x)\,dx$, and since $f'(x) = dy/dx$,

$$dy = \frac{dy}{dx}\,dx.$$

Using this idea, we have $dy = (dy/dt)\,dt$, and we can rewrite equation (1) as

$$\frac{dy}{dt}\,dt = 10t\,dt$$

or

$$dy = 10t\,dt.$$

Here we treat dy/dt as a quotient.

If we now integrate both sides of this equation, we get

$$\int dy = \int 10t\,dt.$$

The differential dy on the left indicates that we should integrate there with respect to y, while the dt on the right indicates that we should integrate with respect to t on that side. Performing the two integrations gives

$$y + C_1 = 5t^2 + C_2$$
$$y = 5t^2 + C_2 - C_1.$$

We can replace $C_2 - C_1$ with the single constant C to get

$$y = 5t^2 + C. \tag{2}$$

Equation (2) is called the **general solution** of differential equation (1). (This process of replacing two or more arbitrary constants with one can be skipped from now on—we shall just include one arbitrary constant and let it go at that.)

Suppose there is a known number of bacteria present at time $t = 0$, say $y = 50$ in millions. Such a condition, called an **initial**

8.1 General and Particular Solutions

condition or **boundary condition,** can be used in the general solution (2) to find C as follows.

$$y = 5t^2 + C$$
$$50 = 5(0)^2 + C$$
$$C = 50$$

With this value of C, equation (2) becomes

$$y = 5t^2 + 50. \tag{3}$$

Since this solution to the differential equation depends on the particular values of t and y that were given, it is called a **particular solution** of differential equation (1).

After the maximum acceptable value of y is found, the number of days, t, the product is usable can be determined. For example, if the maximum value of y is to be 550 million, from equation (3) we have

$$y = 5t^2 + 50$$
$$550 = 5t^2 + 50$$
$$t^2 = 100$$
$$t = 10.$$

Thus, the product should be dated for 10 days from the date when $t = 0$.

To check the solution we found for equation (1) we differentiate $y = 5t^2 + 50$.

$$y' = \frac{dy}{dt} = 10t$$

Since the result agrees with the given equation we have the correct solution.

Example 1 Find the particular solution to

$$\frac{dy}{dx} = 3x^2 + 4x + 5,$$

given that $y = -1$ when $x = 0$.

We first find the general solution.

$$\frac{dy}{dx} = 3x^2 + 4x + 5$$
$$dy = (3x^2 + 4x + 5)\,dx$$
$$\int dy = \int (3x^2 + 4x + 5)\,dx$$
$$y = x^3 + 2x^2 + 5x + C$$

340 Differential Equations

Then, substituting the given values for x and y, we can find C.
$$-1 = 0^3 + 2(0)^2 + 5(0) + C$$
$$C = -1$$

Thus, this particular solution is
$$y = x^3 + 2x^2 + 5x - 1,$$
as can be verified by differentiating. ■

Work Problem 1 at the side.

Example 2 Find the particular solution to
$$\frac{d^2y}{dx^2} = 2x + 4, \qquad (4)$$
given that $y = 2$ when $x = 0$ and $y = -4$ when $x = 3$.

Integrating both sides with respect to x, we have
$$\int \frac{d^2y}{dx^2}\,dx = \int (2x + 4)\,dx.$$

The integral of the second derivative becomes
$$\int \frac{d^2y}{dx^2}\,dx = \frac{dy}{dx} + C.$$

Thus, the differential equation can be written as
$$\frac{dy}{dx} = x^2 + 4x + C_1,$$
for some constant C_1. Integrating again, we have
$$y = \frac{x^3}{3} + 2x^2 + C_1 x + C_2 \qquad (5)$$
for some constant C_2. Equation (5) is a general solution of equation (4). To find values of C_1 and C_2, we use the given pairs of values. If $x = 0$, then $y = 2$, so that
$$2 = \frac{0^3}{3} + 2 \cdot 0^2 + C_1 \cdot 0 + C_2$$
$$C_2 = 2.$$

Using $C_2 = 2$, $x = 3$, and $y = -4$, we have
$$-4 = \frac{3^3}{3} + 2 \cdot 3^2 + C_1 \cdot 3 + 2$$
$$-4 = 29 + 3C_1$$
$$C_1 = \frac{-33}{3} = -11.$$

1. Find and check the particular solution to
$$\frac{dy}{dx} = 4x^3 - 3x^2 + 2x$$
given that $y = 2$ when $x = 1$.

Answer: $y = x^4 - x^3 + x^2 + 1$

The particular solution is thus

$$y = \frac{x^3}{3} + 2x^2 - 11x + 2.$$

To check this solution, differentiate twice. ■

Work Problem 2 at the side.

2. Find and check the particular solution to

$$\frac{d^2y}{dx^2} = x^2 - 4x,$$

given that $y = 5$ when $x = 0$ and $y = 1$ when $x = -1$.

Answer: $y = \dfrac{x^4}{12} - \dfrac{2x^3}{3} + \dfrac{57x}{12} + 5$

8.1 EXERCISES

Find general solutions for the following differential equations.

1. $\dfrac{dy}{dx} = x^2$
2. $\dfrac{dy}{dx} = -x + 2$
3. $\dfrac{dy}{dx} = -2x + 3x^2$
4. $\dfrac{dy}{dx} = 6x^2 - 4x$
5. $\dfrac{dy}{dx} = -4 + 3x^3$
6. $\dfrac{dy}{dx} = 4x^3 + 3$
7. $\dfrac{dy}{dx} = e^x$
8. $\dfrac{dy}{dx} = e^{3x}$
9. $\dfrac{dy}{dx} = 2e^{-x}$
10. $\dfrac{dy}{dx} = 3e^{-2x}$
11. $3\dfrac{dy}{dx} = -4x^2$
12. $3x^3 - 2\dfrac{dy}{dx} = 0$
13. $4\dfrac{dy}{dx} - x = 0$
14. $3x^2 - 3\dfrac{dy}{dx} = 2$
15. $4x - 4 + \dfrac{dy}{dx} = 6x^2$
16. $-8x^3 - 3x^2 + \dfrac{1}{2} \cdot \dfrac{dy}{dx} = 3$
17. $\dfrac{d^2y}{dx^2} = -8x$
18. $\dfrac{d^2y}{dx^2} = 4x$
19. $\dfrac{d^2y}{dx^2} = 5 - 4x$
20. $\dfrac{d^2y}{dx^2} = 2 + 8x$
21. $\dfrac{d^2y}{dx^2} = 4x^2 - 2x$
22. $x^2 - \dfrac{d^2y}{dx^2} = 3x$
23. $5x - 3 + \dfrac{d^2y}{dx^2} = 0$
24. $\dfrac{d^2y}{dx^2} = 6 - x$
25. $\dfrac{d^2y}{dx^2} = e^x$
26. $\dfrac{d^2y}{dx^2} = 6e^{-x}$

Find particular solutions for the following differential equations. See Examples 1 and 2.

27. $\dfrac{dy}{dx} + 2x = 3x^2; \quad y = 2 \text{ when } x = 0$

28. $\dfrac{dy}{dx} = 4x + 3; \quad y = -4 \text{ when } x = 0$

29. $\dfrac{dy}{dx} = 5x + 2; \quad y = -3 \text{ when } x = 0$

30. $\dfrac{dy}{dx} = 6 - 5x; \quad y = 6 \text{ when } x = 0$

31. $\dfrac{dy}{dx} = 3x^2 - 4x + 2; \quad y = 3 \text{ when } x = -1$

32. $\dfrac{dy}{dx} = 4x^3 - 3x^2 + x; \quad y = 0 \text{ when } x = 1$

33. $\dfrac{d^2y}{dx^2} = 2x + 1; \quad y = 2 \text{ when } x = 0; \quad y = 3 \text{ when } x = -2$

34. $\dfrac{d^2y}{dx^2} = -3x + 2; \quad y = -3 \text{ when } x = 0; \quad y = -19 \text{ when } x = 3$

35. $\dfrac{d^2y}{dx^2} = e^x + 1; \quad y = 2 \text{ when } x = 0; \quad y = 3/2 \text{ when } x = 1$

36. $\dfrac{d^2y}{dx^2} = -e^x - 2; \quad y = -2 \text{ when } x = 0; \quad y = 3/2 \text{ when } x = 1$

37. $3x^2 - \dfrac{d^2y}{dx^2} = 2; \quad y = 4 \text{ when } x = 0; \quad y = 6 \text{ when } x = 2$

38. $\dfrac{d^2y}{dx^2} + 4x^2 = 1; \quad y = 1/6 \text{ when } x = 1; \quad y = -1 \text{ when } x = 0$

APPLIED PROBLEMS

Time-dating

39. In the time-dating of dairy products example discussed in the text, suppose the number of bacteria present at time $t = 0$ was 250.
    (a) Find a particular solution of the differential equation.
    (b) Find the number of days the product can be sold if the maximum value of y is 970.

Social Science

40. Suppose the rate at which a rumor spreads—that is, the number of people who have heard the rumor over a period of time—increases with the number of days. If y is the number of people who have heard the rumor, then

$$\dfrac{dy}{dt} = kt,$$

where t is the time in days.
(a) If y is 0 when $t = 0$, and y is 100 when $t = 2$, find k.
(b) Using the value of k from part (a), find y when $t = 3; 5; 10$.

8.2 SEPARATION OF VARIABLES

The differential equations we have discussed up to this point have been of the form

$$\frac{dy}{dx} = f(x) \quad \text{or} \quad \frac{d^2y}{dx^2} = f(x).$$

Another common type of differential equation can be written in the form

$$h(y)\,dy = f(x)\,dx,$$

where all terms involving y (including dy) are on one side of the equation, and all terms involving x (and dx) are on the other side. The process of obtaining a differential equation in this form is called **separating the variables.** When we have a differential equation in which the variables are separated, we can solve the equation by integrating both sides.

Example 1 Find the general solution of

$$y\frac{dy}{dx} = x^2.$$

We begin by separating the variables to get

$$y\,dy = x^2\,dx.$$

The general solution can be found by integrating both sides.

$$\int y\,dy = \int x^2\,dx$$

$$\frac{y^2}{2} = \frac{x^3}{3} + C$$

$$y^2 = \frac{2}{3}x^3 + 2C$$

$$y^2 = \frac{2}{3}x^3 + K \quad \blacksquare$$

When the solution of a differential equation leads to a result with y raised to a power, as in Example 1, it is customary to leave the solution in that form, rather than solve explicitly for y.

Work Problem 1 at the side.

1. Find the general solution of

$$3y^2\frac{dy}{dx} = x + 1.$$

Answer: $y^3 = \dfrac{x^2}{2} + x + C$

Example 2 Find the general solution of

$$2xy\frac{dy}{dx} = 4.$$

We separate the variables as follows.

$$2xy\frac{dy}{dx} = 4$$

$$2y\frac{dy}{dx} = \frac{4}{x} \quad \text{(assume } x \neq 0\text{)}$$

$$2y\,dy = \frac{4}{x}dx$$

Integrating both sides gives the general solution

$$y^2 = 4\ln|x| + C.$$

Verify that this is the correct solution by differentiating implicitly and substituting in the original equation. ∎

Work Problem 2 at the side.

Example 3 Find the general solution of $\frac{dy}{dx} = ky$, where k is a constant.

By separating variables we get

$$\frac{1}{y}dy = k \cdot dx$$

To solve this equation, we integrate both sides.

$$\int \frac{1}{y}dy = \int k \cdot dx$$

$$\ln|y| = kx + C.$$

This general solution can be rewritten, using the definition of natural logarithms, as

$$y = e^{kx+C}$$
$$= e^{kx}e^C$$
$$= Me^{kx},$$

where we replaced the constant e^C with the constant M. Since Me^{kx} is never negative, we don't need the absolute value bars around y. ∎

In general, in the equation

$$y = Me^{kx},$$

the constant k gives the rate of change of a quantity of size M at time $x = 0$. The equation represents growth when k is positive and decay

---

2. Find the general solution of

$$3x^2\frac{dy}{dx} = 4.$$

Answer: $y = -\dfrac{4}{3x} + C$

8.2 Separation of Variables

when k is negative. The function $y = Me^{kx}$ is extremely important because of its many applications in management and the social and behavioral sciences, which we will see in the next section.

Example 4 Find the general solution of

$$x^3 \frac{d^2y}{dx^2} = 3.$$

Since $y' = \frac{dy}{dx}$, we can write $\frac{d^2y}{dx^2}$ as $\frac{dy'}{dx}$.

$$x^3 \frac{dy'}{dx} = 3$$

Now, separate the variables.

$$dy' = \frac{3}{x^3} dx$$

If we integrate both sides, we have

$$y' = -\frac{3}{2} x^{-2} + C_1.$$

Substitute $\frac{dy}{dx}$ for y' and integrate again to get the general solution.

$$\frac{dy}{dx} = -\frac{3}{2} x^{-2} + C_1$$

$$dy = \left(-\frac{3}{2} x^{-2} + C_1\right) dx$$

$$y = \frac{3}{2} x^{-1} + C_1 x + C_2$$

$$y = \frac{3}{2x} + C_1 x + C_2 \quad \blacksquare$$

Work Problem 3 at the side.

In Sections 8.1 and 8.2, we have discussed only the simplest techniques for solving differential equations. It is not always possible to separate the variables in a differential equation. For example, try to separate the variables in

$$\frac{dy}{dx} = x^2 + y^2.$$

For this equation and other types which arise in applications, more advanced methods are required.

3. Find the general solution of

$$x^2 \frac{d^2y}{dx^2} = 12x^4 + 2x^2.$$

Answer: $y = x^4 + x^2 + C_1 x + C_2$.

8.2 EXERCISES

Find general solutions for the following differential equations. See Examples 1–4.

1. $y \dfrac{dy}{dx} = x$

2. $y^2 \cdot \dfrac{dy}{dx} = x^2 - 1$

3. $y \dfrac{dy}{dx} = x^2 - 1$

4. $2x + x^2 - y \dfrac{dy}{dx} = 0$

5. $\dfrac{dy}{dx} = 2xy$

6. $\dfrac{dy}{dx} = x^2 y$

7. $\dfrac{dy}{dx} = 3x^2 y - 2xy$

8. $(y^2 - y) \dfrac{dy}{dx} = x$

9. $\dfrac{d^2 y}{dx^2} - 4x = 0$

10. $\dfrac{d^2 y}{dx^2} - 2x + 3x^2 = 0$

11. $2 \dfrac{d^2 y}{dx^2} - 3x^2 = 0$

12. $2 \dfrac{d^2 y}{dx^2} = x^2$

Find particular solutions for the following.

13. $\dfrac{dy}{dx} = \dfrac{x^2}{y}; \quad y = 3$ when $x = 0$

14. $x^2 \dfrac{dy}{dx} = y; \quad y = -1$ when $x = 1$

15. $(2x + 3)y = \dfrac{dy}{dx}; \quad y = 1$ when $x = 0$

16. $x \dfrac{dy}{dx} - y\sqrt{x} = 0; \quad y = 1$ when $x = 0$

17. $\dfrac{d^2 y}{dx^2} + x = 2; \quad y = 2$ when $x = 0$ and $y = -2$ when $x = 1$

18. $\dfrac{4}{x} \cdot \dfrac{dy^2}{dx^2} = 8x; \quad y = -1$ when $x = 0$ and $y = 3$ when $x = -1$

APPLIED PROBLEMS

Radioactive decay

19. A radioactive substance decays at a rate given by

$$\dfrac{dy}{dx} = -0.05y$$

where y represents the amount, in grams, present at time x, in months.
(a) Find the general solution.
(b) Find a particular solution if $y = 90$ when $x = 0$.
(c) Find the amount left when $x = 10$.

Sales   20. Extensive experiments have shown that under relatively constant market conditions, sales of a product, in the absence of promotional activities such as advertising, decrease at a constant yearly rate. This rate of sales decline varies considerably from product to product, but it seems to be relatively constant for a particular product. Suppose the yearly rate of sales decrease for a certain company is given by

$$\frac{dy}{dt} = -0.25y,$$

where t is the time in years and y is sales in thousands of dollars.
(a) Find a particular solution if $y = 80$ when $t = 0$.
(b) Find y when $t = 2; 4$.

8.3 APPLICATIONS OF DIFFERENTIAL EQUATIONS

Differential equations have many practical applications, especially in biology and economics. For example, marginal productivity is the rate at which production changes (increases or decreases) for a unit change in capitalization. Thus, marginal productivity can be expressed as the first derivative of the function which gives production in terms of capitalization. Suppose the marginal productivity of an operation is given by

$$P'(x) = 3x^2 - 10, \qquad (6)$$

where x is the amount of capitalization in hundred thousands. If the operation produces 100 units per month with its present capitalization of $300,000 ($x = 3$), how much would production increase if the capitalization is increased to $500,000?

To obtain an equation for production, we can take antiderivatives on both sides of equation (6) to get

$$P(x) = x^3 - 10x + C.$$

Using the given initial values of $P(x) = 100$ when $x = 3$, we can find C.

$$100 = 3^3 - (10)(3) + C = 27 - 30 + C$$
$$C = 103$$

Thus, production is given by

$$P(x) = x^3 - 10x + 103,$$

and if capitalization is increased to $500,000, production becomes

$$P(5) = 5^3 - 10(5) + 103 = 178.$$

An increase to $500,000 in capitalization will increase production from 100 units to 178 units.

Work Problem 1 at the side.

Example 1 *Unlimited growth model* It is common for a population (of people, bacteria, or insects, for example) to increase at a rate proportional to the number of individuals present, at least until food supply and the accumulation of waste products affects the rate. That is, when only a few individuals are present, the rate of growth of the population is small. When the number of individuals is large, the rate of growth of the population is high. If y is the population at a time x, then such a situation can be expressed as

$$\frac{dy}{dx} = ky,$$

for some constant k. As we saw in the last section, the general solution of this differential equation is

$$y = Me^{kx}.$$

To obtain a particular solution it is necessary to know some boundary conditions for y and x. For example, suppose 1000 individuals are present when $x = 0$. (In other words, $y = 1000$ when $x = 0$.) Then

$$1000 = Me^{k \cdot 0}$$
$$= M \cdot 1$$
$$= M,$$

so that

$$y = 1000e^{kx}.$$

Now suppose $y = 1500$ when $x = 10$. Then

$$1500 = 1000e^{10k},$$
$$1.5 = e^{10k}.$$

Taking natural logarithms of both sides, we have

$$\ln 1.5 = \ln e^{10k}$$
$$= 10k \ln e$$
$$= 10k,$$

from which

$$k = \frac{\ln 1.5}{10}.$$

---

1. The marginal cost of an item in terms of x, the number of items produced, is given by

$$C'(x) = \frac{1}{x+1}.$$

Suppose the fixed cost is $100.
(a) Write an equation for cost.
(b) Find the cost when $x = 94$.

Answer:
(a) $C(x) = \ln(x+1) + 100$
(b) about $105

Using Table 5, we have

$$k \approx \frac{0.41}{10} = 0.041.$$

Hence, $y \approx 1000e^{0.041x}$. As mentioned in the last section, k (.041 here) gives the rate of growth of the population. ∎

Work Problem 2 at the side.

Example 2 *Limited growth model* Suppose a population is limited (perhaps by a fixed food supply or competition with other species) and cannot exceed some fixed value, such as M. It is plausible to assume that the rate of change of the population is proportional both to the number of individuals present, y, and to the difference $M - y$. That is,

$$\frac{dy}{dx} = ky(M - y) \tag{7}$$

for some constant k.

We can rewrite equation (7) as

$$\frac{1}{y(M-y)} dy = k \, dx$$

from which

$$\int \frac{1}{y(M-y)} dy = \int k \, dx. \tag{8}$$

Now we need a little algebra to verify that

$$\frac{1}{y(M-y)} = \frac{1}{M}\left(\frac{1}{y} + \frac{1}{M-y}\right).$$

Thus,

$$\int \frac{1}{y(M-y)} dy = \int \frac{1}{M}\left(\frac{1}{y} + \frac{1}{M-y}\right) dy$$

$$= \frac{1}{M}\left(\int \frac{1}{y} dy + \int \frac{1}{M-y} dy\right)$$

$$= \frac{1}{M}[\ln y - \ln(M - y)].$$

(Both y and $M - y$ are positive, so we do not need absolute values.) Using this result, and the fact that $\int k \, dx = kx + C$, equation (8) becomes

$$\frac{1}{M}[\ln y - \ln(M - y)] = kx + C.$$

---

2. The rate of change of a chemical substance depends on the amount of substance present. At the beginning of an experiment, there were 10 moles of the substance present. After 5 minutes, there were 8 moles left. Write an equation for the number of moles, y, present at any time t.

Answer: $y = 10e^{-.04t}$

If we multiply both sides by M and use properties of logarithms, we have

$$\ln \frac{y}{M-y} = Mkx + C_1,$$

where $C_1 = MC$. By the definition of logarithm, this yields

$$\frac{y}{M-y} = C_2 e^{Mkx},$$

where C_2 is e^{C_1}. If we let y_0 represent the population at time $x = 0$, we get

$$\frac{y_0}{M-y_0} = C_2 e^{Mk(0)} = C_2(1) = C_2.$$

Hence,

$$\frac{y}{M-y} = \frac{y_0}{M-y_0} e^{Mkx}.$$

After considerable algebraic manipulation, this last result can be solved for y. The result is

$$y = \frac{My_0}{y_0 + (M-y_0)e^{-Mkx}}. \tag{9}$$

As time increases without bound, that is, *as* $x \to \infty$, the expression e^{-Mkx} approaches 0. Thus,

$$\lim_{x \to \infty} \frac{My_0}{y_0 + (M-y_0)e^{-Mkx}} = \frac{My_0}{y_0 + (M-y_0)(0)} = \frac{My_0}{y_0} = M,$$

so that the population does tend to approach M, the limiting value, as required by the statement of the problem. ∎

Work Problem 3 at the side.

According to the result obtained above, population will tend to increase right up to the absolute limit that the land can support, even if that support is only at a sustenance level. When another variable, such as the quality of life, is introduced, the situation may change. The individuals within a society may seek to limit the growth of that society so that the quality of life is enhanced. For example, the number M, the upper limit on population, will soon be reached in some underdeveloped countries. On the other hand, people in many countries now seem to believe that a population substantially less than M provides a better overall quality of life. The results of a society's permitting population to try to reach the upper limit M are explored in Dennis L. Meadows and Donella H. Meadows, *Limits to Growth*, (New York: New American Library). A good discussion of the flaws in *Limits to Growth* is given in

3. In equation (9) above, let the initial population $y_0 = 100$, the limit of the population $M = 500$, and $k = .002$.
 (a) Write the equation for y.
 (b) Find y when $x = 3$ years.

Answer:

(a) $y = \dfrac{50{,}000}{100 + 400e^{-x}}$

(b) about 420

Carl Kaysen, "The Computer That Printed Out W*O*L*F*" *Foreign Affairs*, July 1972, page 660.

Example 3 A small lake will sustain no more than 150 fish. One year, a naturalist determined that there were 80 fish in the lake. Five years later, the naturalist found that the population had increased to 115. Let the initial year correspond to $x = 0$.
(a) Write an equation for the size of the population y in any year x using the given information and equation (9).
We set $y_0 = 80$ and $M = 150$. Then,

$$y = \frac{80(150)}{80 + (150 - 80)e^{-150kx}}$$

$$y = \frac{12000}{80 + 70e^{-150kx}}.$$

To find k, we use the fact that $y = 115$ when $x = 5$.

$$115 = \frac{12000}{80 + 70e^{(-150)(5)k}}.$$

Multiply by the denominator on both sides.

$$9200 + 8050e^{-750k} = 12000$$
$$8050e^{-750k} = 2800$$
$$e^{-750k} = .3478$$

Now take the natural logarithm on both sides.

$$\ln e^{-750k} = \ln .3478$$
$$(-750k)(\ln e) = \ln .3478$$
$$-750k = -1.0561$$
$$k = .0014$$

Thus, the required equation is

$$y = \frac{12000}{80 + 70e^{-.21x}}.$$

(b) Find the fish population in ten years. The tenth year corresponds to $x = 10$.

$$y = \frac{12000}{80 + 70e^{-(.21)(10)}}$$
$$y = \frac{12000}{80 + 70e^{-2.1}}$$
$$y = 135.$$

We would expect about 135 fish in the lake ten years later. ∎

Work Problem 4 at the side.

---

4. Rework Example 3 with $y_0 = 50$.

Answer:

(a) $y = \dfrac{7500}{50 + 100e^{-.38x}}$

(b) about 144 fish

8.3 APPLIED PROBLEMS

Solve the following problems.

Product demand

1. The rate of change of demand for a certain product is given by
$$\frac{dy}{dx} = -4x + 40,$$
where x represents the price of the item. Find the demand at the following price levels if $y = 6$ when $x = 0$.
   (a) $x = 5$
   (b) $x = 8$
   (c) $x = 10$
   (d) $x = 20$

Profit

2. The marginal profit of a certain company is given by
$$\frac{dy}{dx} = 32 - 4x,$$
where x represents the amount in thousands of dollars spent on advertising. Find the profit for each of the following advertising expenditures if the profit is 1000 when nothing is spent for advertising.
   (a) $x = 3$
   (b) $x = 5$
   (c) $x = 7$
   (d) $x = 10$

Bacteria

3. The rate at which the number of bacteria (in thousands) in a culture is changing after the introduction of a bactericide is given by
$$\frac{dy}{dx} = 50 - 10x,$$
where y is the number of bacteria (in thousands) present at time x. Find the number of bacteria present at each of the following times if there were 1000 thousand bacteria present at time $x = 0$.
   (a) $x = 2$
   (b) $x = 5$
   (c) $x = 10$
   (d) $x = 15$

Cost

4. Suppose the marginal cost of producing x copies of a book is given by
$$\frac{dy}{dx} = \frac{50}{\sqrt{x}}.$$
If $y = 25{,}000$ when $x = 0$, find the cost of producing the following number of books.
   (a) 100 books
   (b) 400 books
   (c) 625 books

Radioactive decay

5. Suppose a radioactive sample decays according to the relationship
$$\frac{dy}{dx} = -0.08y,$$

where x is measured in years. Find the amount left after 5 years if $y = 20$ grams when $x = 0$.

Production   6. A company has found that the rate at which a person new to the assembly line produces items is

$$\frac{dy}{dx} = 7.5e^{-0.3x},$$

where x is the number of days the person has worked on the line. How many items can a new worker be expected to produce on the 8th day if he produces none when $x = 0$?

Population   7. Assume the rate of change of the population of a certain city is given by

$$\frac{dy}{dt} = 6000e^{0.06t},$$

where y is the population at time t, measured in years. If the population was 100,000 in 1960 (assume $t = 0$ in 1960), predict the population in 1980.

Sales   8. A company has found that the rate of change of sales during an advertising campaign is

$$\frac{dy}{dx} = 30e^{.03x},$$

where y is the sales in thousands at time x in months. If the sales were \$200,000 when $x = 0$, find the sales at the end of 6 months.

For each of the following, use the unlimited growth model $y = Me^{bx}$ to write an equation and then solve the problem. See Example 1.

Inflation   9. If inflation grows at the rate of 6% per year, how long will it take for \$1 to lose half its value?

10. The Gross National Product (GNP) of a country increases at the rate of 2% per year. Ten years ago, the GNP was 10^5 dollars. What will the GNP amount to in 5 years?

Tracer dye   11. The amount of a tracer dye injected into the bloodstream decreases at the rate of 3% per minute. If 6 cc are present initially, how many cc are present after 10 minutes? (Here b will be negative.)

Population   12. A population of mites increases at the rate of 5% per week. At the beginning of an observation period there were 30 mites. How many are present 4 weeks later?

13. A colony of bacteria grows at a rate proportional to the number present. Initially, 4000 bacteria were present, while 40,000 were present at time $t = 10$.
   (a) Find an equation showing the number of bacteria present at time t.
   (b) How many bacteria will be present at time $t = 20$? (Hint: $e^{4.6} \approx 99.5$.)

Algae growth

14. The phosphate compounds found in many detergents are highly water soluble and are excellent fertilizers for algae. Assume that the rate of growth of algae, in the presence of sufficient phosphates, is proportional to the number present. Assume that there are 3000 algae present at time $t = 0$ and 60,000 present at time $t = 10$.
    (a) Find a function showing the number of algae present at time t.
    (b) How many algae will be present at time $t = 20$?

Cell damage

15. If long-chain polymers in a cell, such as a protein or nucleic acid, are hit by a beam of ionized particles, some of the polymers may be damaged. Let n_0 be the number of undamaged polymers of a specific protein present in a cell. Let D be the number of ionizing particles penetrating the cell. Let n be the number of chains undamaged after exposure to the radiation. The higher the value of D, the more damage to the chains of the protein, so that dn/dD must be negative. If dn/dD is proportional to n, we have

$$\frac{dn}{dD} = -kn,$$

for some positive constant k. Solve this equation for n.

For each of the following, use the limited growth model. See Examples 2 and 3.

Population

16. Suppose the growth of a population of insects is given by

$$y = \frac{My_0}{y_0 + (M - y_0)e^{-0.001Mx}},$$

where x represents time in weeks, y_0 represents the initial number of insects, and M represents the upper limit of the population. Find the number present at each of the following times if $M = 1000$ and $y_0 = 100$.
    (a) $x = 1$
    (b) $x = 3$
    (c) $x = 10$

17. An isolated fish population is limited by the amount of food available. If $M = 5000$, $y_0 = 150$, $k = 0.0001$, and x is in years, use the equation of Example 2 to find the number present at the end of each of the following times.
    (a) $x = 1$
    (b) $x = 2$
    (c) $x = 5$

Bankruptcies

18. Let B represent the number of all small business firms threatened by bankruptcy. If y is the number who are bankrupt by time t, then $B - y$ is the number who are not yet bankrupt by time t. The rate of change of y depends on both y and $B - y$. Thus,

$$\frac{dy}{dt} = ky(B - y).$$

This differential equation is the same as the limited growth differential equation discussed in Example 2. The solution is

$$y = \frac{By_0}{y_0 + (B - y_0)e^{-Bkt}},$$

where y_0 is the number of firms which are bankrupt at $t = 0$. Let 1970 correspond to $t = 0$. Suppose $B = 1500$, 50 of the 1500 firms were bankrupt at time $t = 0$, and 75 were bankrupt by 1975. How many firms will be bankrupt by 1980?

Newton's law of cooling

Newton's law of cooling states that the rate of change of temperature of an object is proportional to the difference in temperature between the object and surrounding medium. Thus, if T is the temperature of the object after t hours, and T_F is the (constant) temperature of the surrounding medium,

$$\frac{dT}{dt} = -k(T - T_F),$$

where k is a constant.

19. Show that the solution of this equation is

$$T = ce^{-kt} + T_F,$$

where c is a constant.

20. Suppose a container of soup with a temperature of $180°F$ is put into a refrigerator where the temperature is $40°F$. When checked 1/2 hour later, the temperature of the soup is $110°F$. Find an equation for the temperature of the soup in terms of the number of hours it has been in the refrigerator.

21. Find the temperature of the soup after one hour;

22. after 5 hours.

23. A ceramic jar is taken from a kiln at a temperature of $50°C$ and set to cool in a room where the temperature is $22°C$. In one hour, the temperature of the jar is $35°C$. Find an equation for the temperature of the jar at time t.

24. Find the temperature of the jar after 2 hours.

CHAPTER 8 SUMMARY

Key Words

differential equation
general solution
initial condition
boundary condition

particular solution
separating the variables
unlimited growth model
limited growth model

CHAPTER 8 TEST

Find general solutions for the following.

[8.1]   1. $\dfrac{dy}{dx} = -3 + 3x^2 + 5x^4$

2. $\dfrac{dy}{dx} = 4e^{-2x}$

3. $\dfrac{d^2y}{dx^2} - e^x = 3x^2$

[8.2]   4. $(y + y^2)\dfrac{dy}{dx} = x^2 y$

Find particular solutions for the following.

[8.1]   5. $\sqrt{x} + \dfrac{dy}{dx} = x$; $y = 2$ when $x = 0$

[8.2]   6. $\dfrac{dy}{dx} - 4xy = 0$; $y = -1$ when $x = 0$.

[8.3]   7. The marginal profit of a small grocery chain is given by

$$\dfrac{dy}{dx} = 400 - 30x,$$

where x represents the amount in thousands of dollars spent on advertising. If nothing is spent on advertising, the profit is 700. Find the profit when $x = 1$.

8. The rate of change of the population of a rare species of Australian spiders is given by

$$\dfrac{dy}{dx} = -.02y,$$

where y is the number of spiders present at time x, measured in years. Find y when x is 5, if there were 10,000 spiders at time $x = 0$.

9. A radioactive substance decays at a rate proportional to the amount present at any given time.
   (a) Write an equation to represent the rate of decay, dy/dx, in terms of the time x in hours.
   (b) If there are 200 grams present at time $x = 0$ and 100 grams left after 4 hours, how much is remaining after 10 hours?

10. An isolated population is limited by space to a maximum of 1000 individuals. The present population is 60 and the rate of change of the population, k, is .001.
    (a) Give an equation for the population, y, in terms of time, x, in years.
    (b) What will the population be in 1 year?

CASE 9 A MARINE FOOD CHAIN*

In the oceans there is a relatively simple food chain from nutrients in the water to phytoplankton, to zooplankton, and then to a series of carnivores. As part of a model of this food chain, we consider the role of the phytoplankton, which are microscopic plant organisms.

The growth of phytoplankton, like that of other plants, depends upon photosynthesis, the formation of carbohydrates from water and carbon dioxide, using sunlight as the source of energy. On the other hand, growth is retarded by respiration, the giving off (or exhaling) of the by-products of photosynthesis. The population growth of phytoplankton is also retarded by the grazing (feeding) of zooplankton on phytoplankton.

The rate of change in the phytoplankton population can be described by the differential equation

$$\frac{dx}{dt} = x(P - R - G),$$

where

$x =$ the phytoplankton population
$t =$ time
$P =$ rate of photosynthesis per unit of population
$R =$ rate of respiration
$G =$ rate of grazing loss.

The rate of photosynthesis, P, when ample nutrients are available, depends on the amount of light and is given by

$$P = pI, \tag{1}$$

where I is the amount of sunlight, in appropriate units, and p is a constant (about 2.5). The amount of sunlight at any depth z can be expressed as

$$I_z = I_0 e^{-kz}. \tag{2}$$

I_0 is the amount of sunlight at the surface of the sea and k is the extinction coefficient.

Substituting from equation (2) into equation (1), we have

$$P_z = pI_0 e^{-kz},$$

where P_z is the rate of photosynthesis at depth z. To get the average rate of photosynthesis, we integrate the expression for P_z from 0 to z_1, the maximum depth of photosynthesis, and divide by $z_1 - 0 = z_1$. This gives

---

* "Factors controlling phytoplankton populations on Georges Bank" by Gordon A. Riley in *Readings in Marine Ecology*, edited by James W. Nybakken. Copyright © 1971 by James W. Nybakken. Reprinted by permission of Harper & Row, Publishers, Inc.

$$P = \frac{1}{z_1}\int_0^{z_1} pI_0 e^{-kz}\,dz = \frac{pI_0}{z_1}\int_0^{z_1} e^{-kz}\,dz \qquad (p, I_0, \text{ and } k \text{ are constants})$$

$$= \frac{pI_0}{-kz_1}\left(e^{-kz}\Big]_0^{z_1}\right)$$

$$= -\frac{pI_0}{kz_1}e^{-kz_1} + \frac{pI_0}{kz_1}$$

$$P = \frac{pI_0}{kz_1}(1 - e^{-kz_1}).$$

This function describes the rate of photosynthesis as a function of incoming light and the clearness of the water.

The model can be improved by considering the effect of nutrient shortages which occur in the summer. This is done by multiplying the equation above by the fraction N, where

$$N = \frac{\text{observed nutrient concentration}}{\text{limiting nutrient concentration}},$$

The model now becomes

$$P = \frac{pI_0}{kz_1}(1 - e^{-kz_1})(N).$$

Another factor can be introduced to account for vertical mixing losses. During the winter, turbulence moves the phytoplankton down from the surface of the sea to deep areas which receive no light. To allow for this factor, the equation above is multiplied by the fraction

$$V = \frac{\text{depth of photosynthesis}}{\text{depth of mixed layer}},$$

which gives

$$P = \frac{pI_0}{kz_1}(1 - e^{-kz_1})(N)(V).$$

The loss due to respiration, which is assumed to be due to temperature alone, is

$$R = R_0 e^{rT}$$

where

R = respiration rate at temperature T

R_0 = respiration rate at 0°C

r = a constant.

Finally, the loss due to grazing can be given by

$$G = gZ$$

where

G = rate of grazing

g = a constant

Z = quantity of zooplankton.

Returning now to our original equation for rate of change of phytoplankton population, we have

$$\frac{dx}{dt} = x(P - R - G)$$

$$= x\left[\frac{pI_0}{kz_1}(1 - e^{-kz_1})(N)(V) - R_0 e^{rT} - gZ\right].$$

EXERCISES

1. Let $p = 2.5$, $I_0 = 210$, $N = .7$, $V = 1$, $k = .0001$, $z_1 = 20{,}000$, $R_0 = .05$, $r = .2$, $T = 10$, $g = .1$, and $Z = 178$, all in appropriate units. Assume an initial population of $x_0 = 100{,}000$. (When $t = 0$, $x_0 = 100{,}000$.) Find an equation for x, the change in the phytoplankton population. Find x when $t = 10$. Is the population increasing or decreasing at that point?

2. If $I_0 = 170$, $N = 1$, $V = .8$, $T = 4$, and all other values are the same as in Exercise 1, find an equation for x. Find x when $t = 10$. Is the population increasing or decreasing at that point?

CASE 10 DIFFERENTIAL EQUATIONS IN ECOLOGY

Consider a closed area containing only two animal or plant species. Assume that there is effective interaction between the two species. For example, the first species may be a source of food for the second species. A species of trees might reduce the light falling on another plant species. One species might provide shelter for another. A species of insects might pollinate a species of plant. One species may poison the soil for another species. There are many applications of such interaction.

Let $N_1 = N_1(t)$ and $N_2 = N_2(t)$ represent the number of individuals of the two species at time t. Let ΔN_1 represent the change in the population of species 1 during a given time period, while ΔN_2 is the change in the population of species 2 during the same time period. During a given time interval of length Δt, we may describe the populations of the species as follows:

Population of species 1:

$$\Delta N_1 = \begin{pmatrix} \text{any change in the} \\ \text{absence of interaction} \end{pmatrix} + \begin{pmatrix} \text{any change due to} \\ \text{interaction with species 2} \end{pmatrix}.$$

Population of species 2:

$$\Delta N_2 = \begin{pmatrix} \text{any change in the} \\ \text{absence of interaction} \end{pmatrix} + \begin{pmatrix} \text{any change due to} \\ \text{interaction with species 1} \end{pmatrix}.$$

Divide both sides of each of these equations by Δt, and assume that each term approaches a limit as $\Delta t \to 0$. Then, by the definition of derivative, we have

$$\frac{dN_1}{dt} = \begin{pmatrix} \text{rate of change in the} \\ \text{absence of interaction} \end{pmatrix} + \begin{pmatrix} \text{rate of change due to} \\ \text{interaction with species 2} \end{pmatrix}$$

$$\frac{dN_2}{dt} = \begin{pmatrix} \text{rate of change in the} \\ \text{absence of interaction} \end{pmatrix} + \begin{pmatrix} \text{rate of change due to} \\ \text{interaction with species 1} \end{pmatrix}.$$

Let us assume, for simplicity, that the birth and death rates in both populations are constant, so that the rate of change of a population is proportional to the size of the population. We can also assume that the rate of change due to interaction is proportional to the size of the interacting population. These assumptions lead to differential equations of the form

$$\frac{dN_1}{dt} = a \cdot N_1 + b \cdot N_2, \quad \frac{dN_2}{dt} = c \cdot N_2 + d \cdot N_1$$

for appropriate constants a, b, c, and d.

The common solution of these two differential equations requires a knowledge of the values of a, b, c, and d. In more advanced courses, it is shown that the solutions will be of the form

$$N_1 = p_1 \cdot e^{qt} + k \cdot p_2 \cdot e^{rt}, \quad N_2 = k \cdot p_1 \cdot e^{qt} + p_2 \cdot e^{rt},$$

where q and r are solutions of the quadratic equation

$$m^2 - (a+c)m + (ac + bd) = 0, \qquad (1)$$

and where p_1, p_2, and k are constants depending on a, b, c, and d.

If either q or r (or both) is positive, then both populations will grow with time and increase without bound. If both q and r are negative, the populations will both tend to decrease in size. If there are no real-number solutions of the quadratic equation (1) above, then the populations will oscillate from very large to very small, and so on.

If specific values of a, b, c, and d are known, the changes in the populations of the two species can be found from the solutions of the differential equations given above. For example, if the two species are in a predator-prey relationship, then the solutions of the quadratic equation (1) above will not be real numbers, and the populations will oscillate. To see how this works in practice, consider populations of hares and lynx in some limited area. If the hare population starts to increase, the number of lynx will also increase because of the additional food supply. The lynx will compete for this additional food, causing the number of hares to decline; this causes the number of lynx to decline. As the number of lynx goes down, the hares have more chance to increase their numbers, starting the whole cycle over again. An idealized graph of this situation is shown in Figure 1.

To test the accuracy of this theory, we shall use the records on lynx and hare trappings kept by the Hudson's Bay Company in Canada. For many years, starting in 1847, the company caught as many lynx and hares annually as possible and kept records of this catch. We shall assume that the company catches are roughly proportional to the total population of each species, although there are some outside influences on the size of the catch (for

Case 10 Differential Equations in Ecology 361

example, notice the fall-off in the catches during the years 1861–65, the years of the American Civil War). The company data has been graphed in Figure 2, with a smooth curve for each animal drawn through the data points. From the graph we see that if the hare population begins to increase, then the lynx population will also increase about a year later. As the hare population declines, lynx population also starts to decline, again with a time lag of one year. Thus, the populations oscillate on about a ten-year cycle. Using the data from Figure 2, a graph very similar to the one of Figure 1 could be constructed.

Figure 1

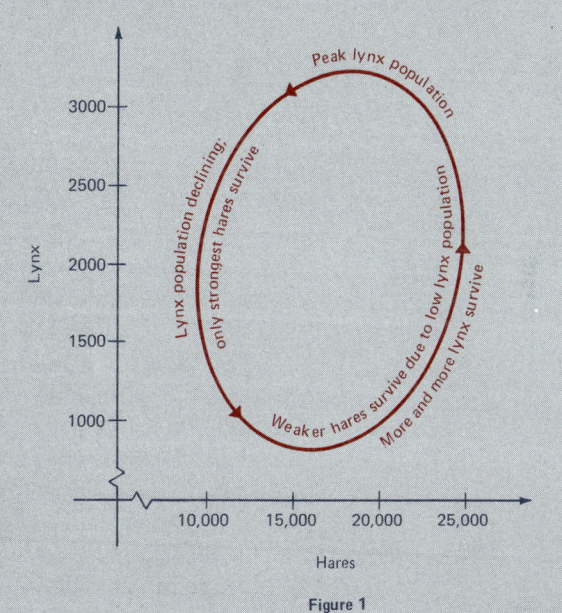

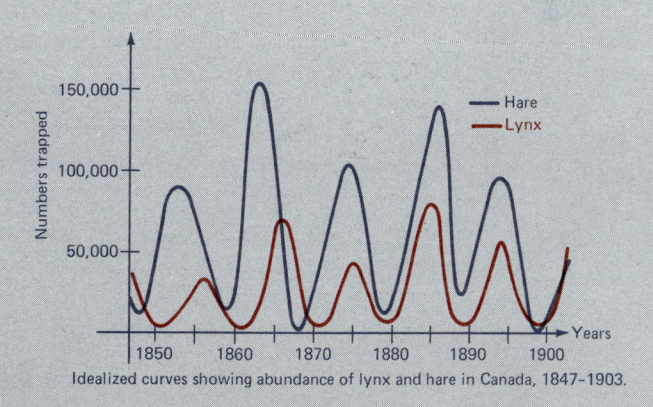

Figure 2*

Idealized curves showing abundance of lynx and hare in Canada, 1847–1903.

---

* Hudson Bay Company records on lynx and hare 1847–1903 from *Fluctuations in the Numbers of the Varying Hare* by D. A. MacLulich. Toronto: University of Toronto Press, 1937, reprinted 1974.

9
Functions of Several Variables

So far in this book all the functions that we have discussed have used only one independent variable. We have seen functions that give sales in terms of price or advertising, and functions that give populations in terms of time. In many cases, however, a valid mathematical model must include more than one independent variable.

For example, a mathematical model for the sales of air conditioners would include several variables—the price, the weather, the availability and price of electricity, among other possible variables. A mathematical model for the population of foxes in an area might involve such variables as the population of rabbits, squirrels, gophers, and other food resources.

In this chapter, we shall look at the mathematics of functions of more than one independent variable. We shall see that many of the ideas developed for functions of one variable also apply to functions of more than one variable. In particular, the fundamental idea of derivative generalizes in a very natural way to functions of more than one variable.

9.1 FUNCTIONS OF SEVERAL VARIABLES

Suppose the value of the variable z depends on the value of x and y according to the relationship

$$z = 4x^2 + 2xy + 3y.$$

For every pair of values of x and y that we might choose we can find exactly one value of z. For example, if $x = 1$ and $y = -4$, we have

$$z = 4x^2 + 2xy + 3y$$
$$z = 4(1)^2 + 2(1)(-4) + 3(-4)$$
$$= 4 - 8 - 12$$
$$z = -16.$$

9.1 Functions of Several Variables

Therefore, if $x = 1$ and $y = -4$, then $z = -16$. To abbreviate this, we let $z = f(x, y)$, so that

$$f(x, y) = 4x^2 + 2xy + 3y.$$

"If $x = 1$ and $y = -4$, then $z = -16$" can now be written as just

$$f(1, -4) = -16.$$

To find $f(3, -2)$, let $x = 3$ and $y = -2$.

$$f(x, y) = 4x^2 + 2xy + 3y$$
$$f(3, -2) = 4(3)^2 + 2(3)(-2) + 3(-2)$$
$$= 36 - 12 - 6$$
$$f(3, -2) = 18$$

Also, $f(0, 0) = 0$ and $f(2, 5) = 51$.

Work Problem 1 at the side.

In general, $z = f(x, y)$ is a **function of two variables** whenever each pair of values for x and y leads to just one value of z. The variables x and y are called **independent variables**; z is the **dependent variable**.

Example 1 Let $f(x, y) = 4x\sqrt{x^2 + y^2}$. Find the following.
(a) $f(0, 0)$
   Let $x = 0$ and $y = 0$. $f(0, 0) = 4(0)\sqrt{0^2 + 0^2} = 0$
(b) $f(3, -4)$
   $f(3, -4) = 4(3)\sqrt{3^2 + (-4)^2} = 12\sqrt{9 + 16} = 12\sqrt{25}$
   $= 12(5) = 60$
(c) $f(-5, -7)$
   $f(-5, -7) = 4(-5)\sqrt{(-5)^2 + (-7)^2} = -20\sqrt{74}$ ∎

Work Problem 2 at the side.

Example 2 The amount of money, $M(x, y)$, in dollars, that a twelve-year-old girl will spend in a week on rock-and-roll records depends on her weekly allowance, x, in dollars, and the amount, y, in dollars, that she spends weekly on cosmetics. In a suburb of Carmichael, it was found that $M(x, y)$ can be closely approximated by the mathematical model

$$M(x, y) = 4x - 2y - 9.$$

Find $M(4, 2.5)$.
   Let $x = 4$ and $y = 2.5$.

$$M(4, 2.5) = 4(4) - 2(2.5) - 9 = 16 - 5 - 9 = 2$$

---

1. Let $f(x, y) = 8x + 9xy - 4y^2 + 1$. Find the following.
(a) $f(2, 1)$
(b) $f(-3, 4)$
(c) $f(0, 0)$

Answer:
(a) 31
(b) −195
(c) 1

2. Let $g(x, y) = -4x/\sqrt{x^2 + y^2}$. Find the following, where possible.
(a) $g(3, 0)$
(b) $g(-4, -3)$
(c) $g(0, 0)$

Answer:
(a) −4
(b) 16/5
(c) function not defined at $(0, 0)$—can't have a 0 denominator

364 Functions of Several Variables

3. In Example 2, how much will be spent on records by a girl whose allowance is $3 per week and who spends $1.25 per week on cosmetics?

Answer: 50¢

4. In Example 3, find C when $x = 275$ and $y = 5$.

Answer: 5500 cc or 5.5 liters

5. Let $g(x, y, z) = 5x^2 - 4xz + 2yz - 3$. Find the following.
   (a) $g(3, 0, 1)$
   (b) $g(-2, 1, 4)$
   (c) $g(6, -1, 0)$

Answer:
(a) 30
(b) 57
(c) 177

The girl spends $2 on records. This amount, together with the $2.50 spent on cosmetics, exceeds her allowance by 50¢. ■

Work Problem 3 at the side.

Example 3 Let x represent the amount of carbon dioxide released by the lungs in one minute; x is measured in cc of carbon dioxide. Let y be the change in the carbon dioxide content of the blood as it leaves the lungs (y is measured in cc of carbon dioxide per 100 cc of blood). Let C be the total output of blood from the heart in one minute (measured in cc). Then

$$C = \frac{100x}{y}.$$

Let $y = 6$ cc of carbon dioxide per 100 cc of blood per minute and find C when $x = 320$ cc of carbon dioxide per minute.

We have

$$C = \frac{100(320)}{6} \approx 5333 \text{ cc of blood per minute}$$

$$\approx 5.3 \text{ liters of blood per minute.} \blacksquare$$

Work Problem 4 at the side.

While we have defined only a function of two independent variables, similar definitions could be given for functions of three, four, or more independent variables.

Example 4 Let $f(x, y, z) = 4xz - 3yx^2 + 2z^2$. Find the following.
(a) $f(2, -3, 1)$

$$f(2, -3, 1) = 4(2)(1) - 3(-3)(2)^2 + 2(1)^2$$
$$= 8 + 36 + 2$$
$$= 46$$

(b) $f(-4, 3, -2) = 4(-4)(-2) - 3(3)(-4)^2 + 2(-2)^2$
$$= 32 - 144 + 8$$
$$= -104 \blacksquare$$

Work Problem 5 at the side.

9.1 EXERCISES

Let $f(x, y) = 4x + 5y + 3$. Find the following. See Example 1.

1. $f(2, -1)$ **2.** $f(-4, 1)$ **3.** $f(-2, -3)$ **4.** $f(0, 8)$

Let $g(x, y) = -x^2 - 4xy + y^3$. Find the following.

5. $g(-2, 4)$ **6.** $g(-1, -2)$ **7.** $g(-2, 3)$ **8.** $g(5, 1)$

Let $h(x, y) = \sqrt{x^2 + 2y^2}$. Find the following.

9. $h(5, 3)$ **10.** $h(2, 4)$ **11.** $h(-1, -3)$ **12.** $h(-3, -1)$

APPLIED PROBLEMS

Population

The population of cats on a certain farm is approximated by the mathematical model

$$C(x, y) = x^2 + 200y - 1200,$$

where x is the population, in hundreds, of small mice and y is the population, in tens, of large rats. Find the following. See Examples 2 and 3.

13. $C(50, 0)$ **14.** $C(30, 4)$

15. How many cats will be present if there are 1400 small mice and 150 large rats?

16. If the farm has 3000 small mice and 200 large rats, how many cats will be present?

Labor costs

The labor charge for assembling a precision camera is given by

$$L(x, y) = 12x + 6y + 2xy + 40,$$

where x is the number of work hours required by a skilled craftsperson and y is the number of hours required by a semiskilled person. Find the following.

17. $L(3, 5)$

18. $L(5, 2)$

19. If a skilled craftsperson requires 7 hours and a semiskilled person needs 9 hours, find the total labor charge.

20. Find the total labor charge if a skilled worker needs 12 hours and a semiskilled worker requires 4 hours.

Use a calculator with an x^y key, or logarithms, to work the following problems.*

---

* From *Mathematics in Biology* by Duane J. Clow and N. Scott Urquhart. Copyright © 1974 by W. W. Norton & Company, Inc. Used by permission.

Oxygen consumption

The oxygen consumption of a well-insulated mammal which is not sweating is approximated by

$$m = \frac{2.5(T - F)}{w^{.67}},$$

where T is the internal body temperature of the animal (in °C), F is the temperature of the outside of the animal's fur (in °C), and w is the animal's weight in kilograms. Find m for the following data. See Example 4.

21. $T = 38°$, $F = 6°$, $w = 32$ kg

22. $T = 40°$, $F = 20°$, $w = 43$ kg

Human surface area

The surface area of a human (in square meters) is approximated by

$$A = 2.02 W^{.425} H^{.725},$$

where W is the weight of the person in kilograms and H is the height in meters. Find A for the following data.

23. $W = 72$, $H = 1.78$

24. $W = 65$, $H = 1.40$

25. $W = 70$, $H = 1.60$

26. Find your own surface area.

9.2 GRAPHING FUNCTIONS OF TWO VARIABLES

When we graph functions of one independent variable, we use ordered pairs. We use a similar method to graph functions of two independent variables. For example, if $z = f(x, y) = 4x + 2y$, we might let $x = 2$ and $y = -6$:

$$z = f(2, -6) = 4(2) + 2(-6)$$
$$= -4.$$

If $x = 2$ and $y = -6$, then $z = -4$. This result is written as the **ordered triple** $(2, -6, -4)$. Other ordered triples for the function $f(x, y) = 4x + 2y$ are $(-3, 5, -2)$, $(6, -4, 16)$, $(0, 0, 0)$, and so on. In the ordered triple (x, y, z), we assume that $z = f(x, y)$.

1. Let $z = f(x, y) = -9x + 2y + 5$. Complete the following ordered triples.
(a) $(0, 0,)$
(b) $(-3, 2,)$
(c) $(1, 7,)$

Answer:
(a) $(0, 0, 5)$
(b) $(-3, 2, 36)$
(c) $(1, 7, 10)$

Work Problem 1 at the side.

Ordered pairs are graphed with two axes, an x-axis and a y-axis. Ordered triples are graphed with three axes, an x-axis, a y-axis, and a z-axis. Each of these three axes is perpendicular to the other two. Figure 9.1 shows one possible way to draw the graph of these three axes. In this figure, the plane containing the y-axis and the z-axis is in the plane of the page, while the x-axis is perpendicular to the plane of the page.

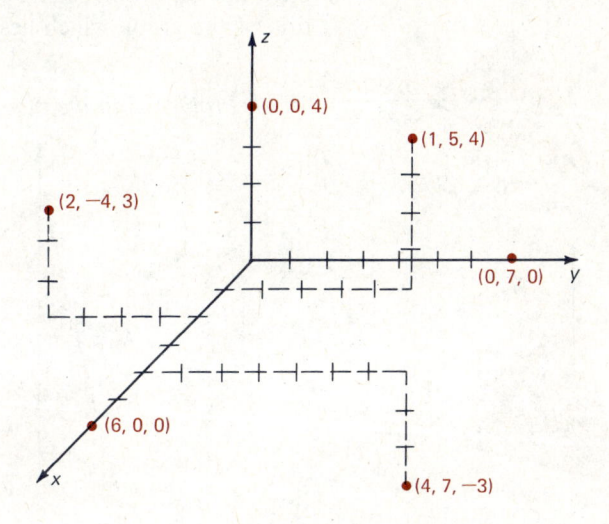

Figure 9.1

To locate the point corresponding to the ordered triple $(2, -4, 3)$, start at the origin and go 2 units along the positive x-axis. Then go 4 units in a negative direction, parallel to the y-axis. Finally, go up 3 units, parallel to the z-axis. The point representing $(2, -4, 3)$ and several other sample points are shown in Figure 9.1. The region of three-dimensional space where all coordinates are positive is called the **first octant**.

Work Problem 2 at the side.

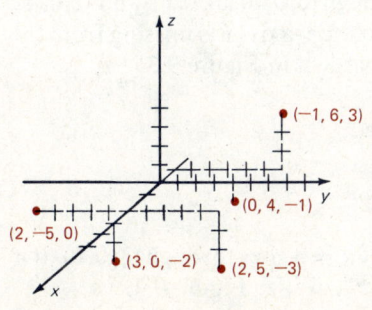

In the rest of this section, we look at various types of functions and the graphs they produce in three-dimensional space.

In two-dimensional space the graph of $ax + by = c$ is a straight line, if not both a and b are 0. In three-dimensional space the graph of

$$ax + by + cz = d,$$

is a **plane** if a, b, c, and d are real numbers, and not all of a, b, and c are 0.

2. Graph the following ordered triples.
(a) $(2, 5, -3)$
(b) $(-1, 6, 3)$
(c) $(0, 4, -1)$
(d) $(3, 0, -2)$
(e) $(2, -5, 0)$

Answer:

368 Functions of Several Variables

3. Graph the following planes.
   (a) $2x + 3y + 4z = 12$
   (b) $5x + 2y + 3z = 15$

Answer:
(a) [Graph showing plane with intercepts (0, 0, 3), (0, 4, 0), (6, 0, 0)]

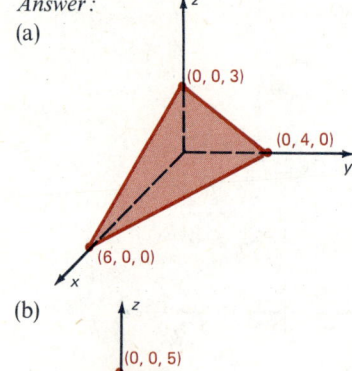

(b) [Graph showing plane with intercepts (0, 0, 5), $(0, 7\frac{1}{2}, 0)$, (3, 0, 0)]

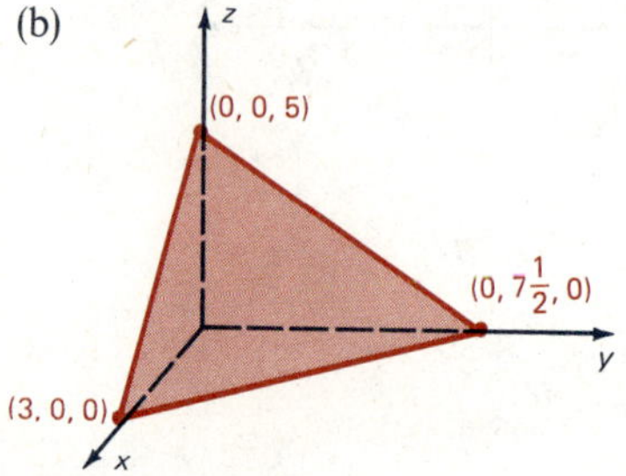

4. Graph the following.
   (a) $y + z = 7$
   (b) $x + 2y = 8$

Answer:
(a) [Graph showing plane through (0, 0, 7) and (0, 7, 0), parallel to x-axis]

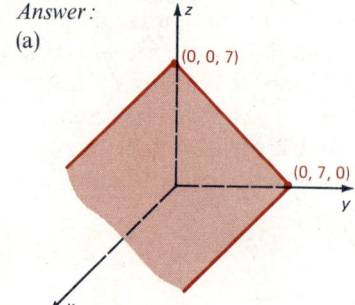

(b) [Graph showing plane through (0, 4, 0) and (8, 0, 0), parallel to z-axis]

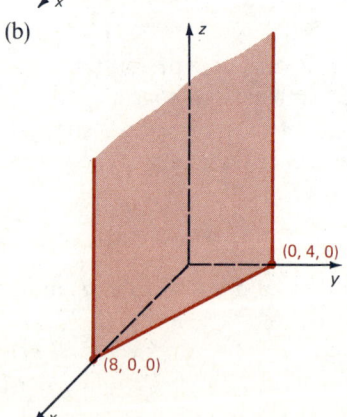

Example 1 Graph $2x + y + z = 6$.

This equation has a graph which is a plane. To graph the plane, find some typical ordered triples. Some of these ordered triples are $(0, 0, 6)$, $(3, 0, 0)$, and $(0, 6, 0)$. If we plot them we can get the graph shown in Figure 9.2. It is common to show only the portion of the graph which lies in the first octant. ∎

Work Problem 3 at the side.

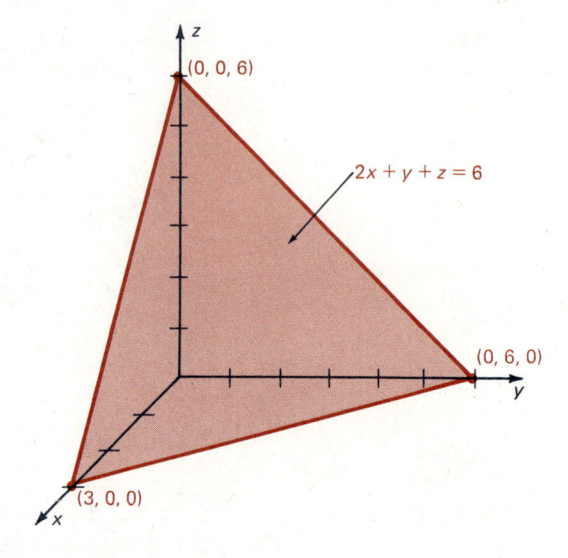

[Figure showing the plane $2x + y + z = 6$ with intercepts (0, 0, 6), (3, 0, 0), (0, 6, 0)]

Figure 9.2

Recall from Chapter 2 that the graph of $x = 3$ is a vertical line going through the point $(3, 0)$. The graph is parallel to the y-axis, with y not appearing in the equation $x = 3$. Similar results hold with planes, as shown in the next examples.

Example 2 Graph $x + z = 6$.

The graph is a plane going through $(6, 0, 0)$, $(0, 0, 6)$, and $(2, 0, 4)$, for example. The plane is parallel to the y-axis (y is missing from the equation $x + z = 6$). The graph is shown in Figure 9.3. ∎

Work Problem 4 at the side.

Example 3 Graph $x = 3$.

The plane which is the graph of $x = 3$ goes through $(3, 0, 0)$. It is parallel to both the y-axis and the z-axis. See Figure 9.4. ∎

9.2 Graphing Functions of Two Variables 369

5. Graph the following.
(a) $y = 2$
(b) $z = 3$
(c) $y = -4$

Answer:
(a)

(b)

(c)

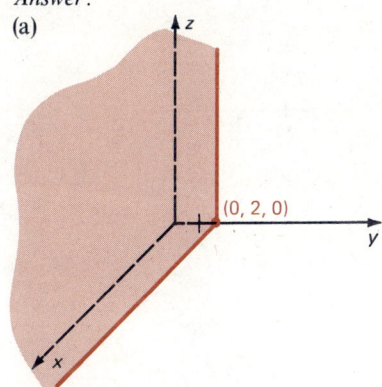

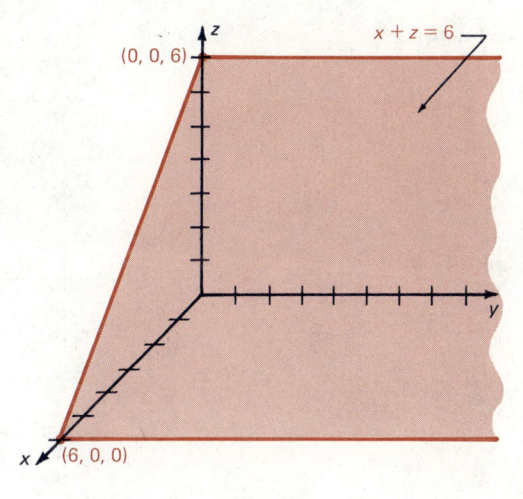

Figure 9.3

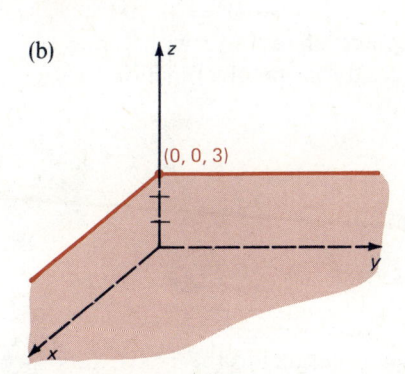

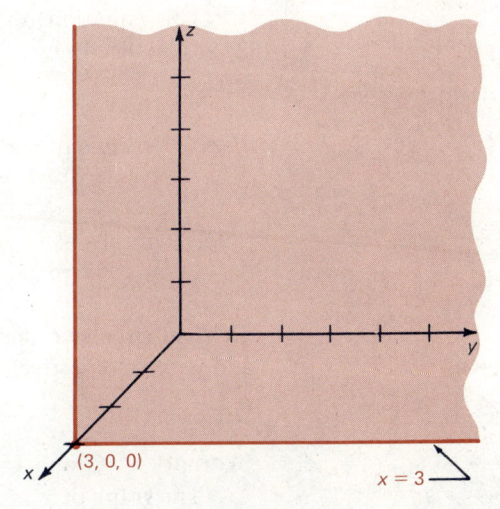

Figure 9.4

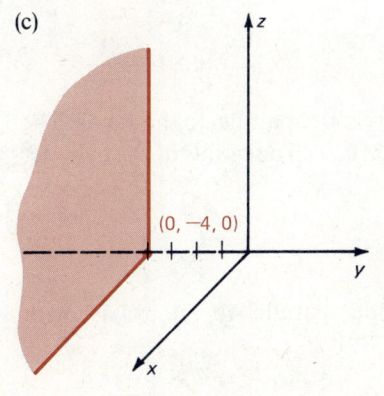

Work Problem 5 at the side.

The graph of
$$(x - h)^2 + (y - k)^2 + (z - j)^2 = r^2$$
is a **sphere** with center at (h, k, j) and radius r.

370 Functions of Several Variables

6. Graph the following.
(a) $x^2 + y^2 + z^2 = 9$
(b) $(x-3)^2 + (y-4)^2 + (z-1)^2 = 16$

Answer:
(a) [graph showing sphere with intercepts (0,0,3), (0,3,0), (3,0,0)]

(b) [graph showing sphere centered at (3, 4, 1)]

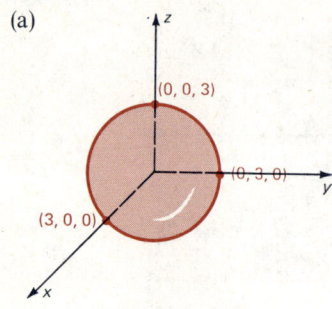

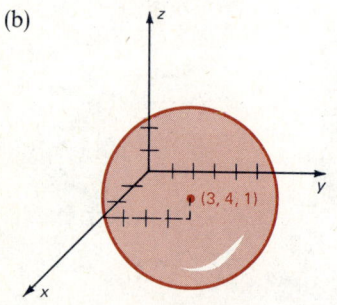

7. Graph the following.
(a) $y^2 + z^2 = 9$
(b) $x^2 + z^2 = 100$

Answer:
(a) [cylinder parallel to x-axis through (0, 0, 3) and (0, 3, 0)]

(b) [cylinder parallel to y-axis through (0, 0, 10) and (10, 0, 0)]

$(x-2)^2 + (y+3)^2 + (z-4)^2 = 4$

[graph showing sphere centered at (2, −3, 4)]

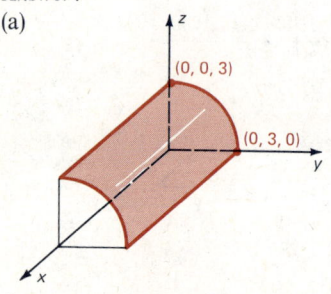

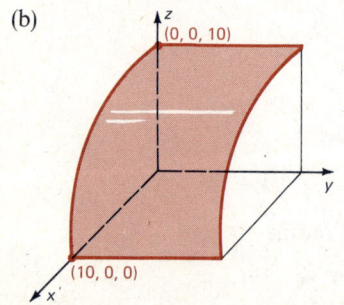

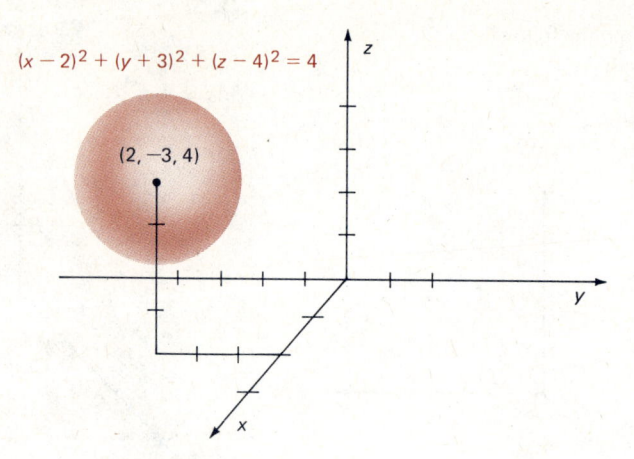

Figure 9.5

Example 4 Graph $(x-2)^2 + (y+3)^2 + (z-4)^2 = 4$.

This equation represents a sphere of radius 2 with center at $(2, -3, 4)$. See Figure 9.5. Is a sphere the graph of a function? ∎

Work Problem 6 at the side.

The graphs of

$$x^2 + y^2 = r^2, \qquad y^2 + z^2 = r^2, \qquad \text{and} \qquad x^2 + z^2 = r^2$$

are **right circular cylinders** of radius r, parallel to the z-axis, x-axis, and y-axis, respectively.

Example 5 Graph $x^2 + y^2 = 4$.

The graph of $x^2 + y^2 = 4$ is a cylinder parallel to the z-axis. The cylinder goes through $(0, 2, 0)$ and $(2, 0, 0)$. The radius of the cylinder is 2. See Figure 9.6. ∎

Example 6 Graph $y^2 + z^2 = 25$.

This graph represents a cylinder parallel to the x-axis, with radius 5. See Figure 9.7. ∎

Work Problem 7 at the side.

9.2 Graphing Functions of Two Variables 371

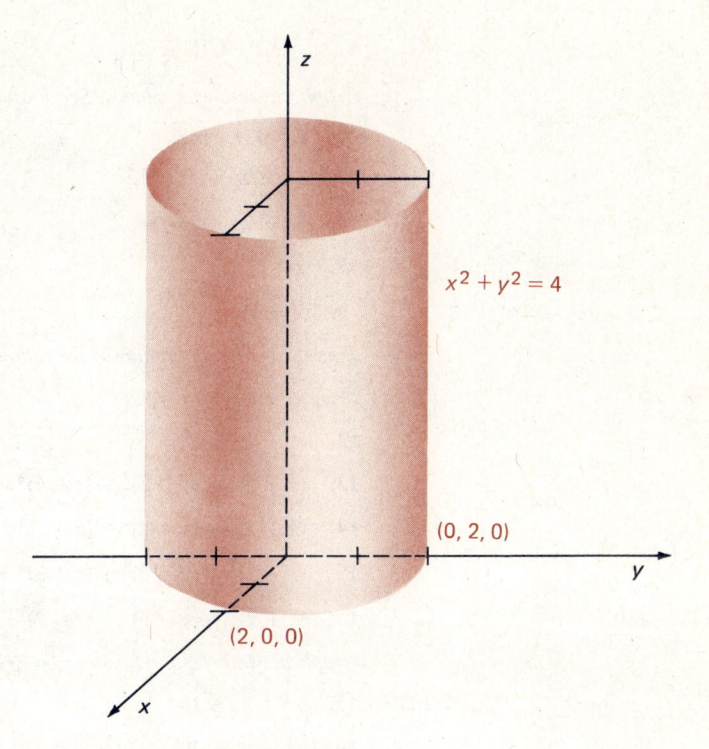

Figure 9.6

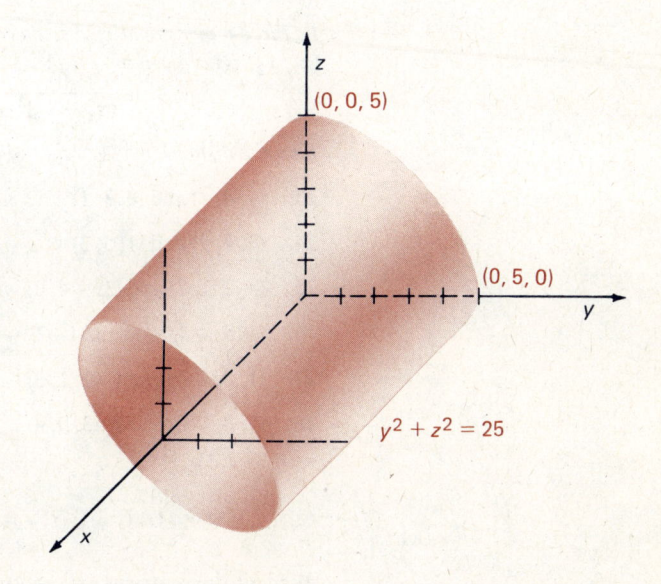

Figure 9.7

9.2 EXERCISES

Graph the following planes. See Examples 1–3.

1. $x + y + z = 6$
2. $x + y + z = 12$
3. $2x + 3y + 4z = 12$
4. $4x + 2y + 3z = 24$
5. $3x - 2y + z = 18$
6. $4x - 3y - z = 12$
7. $x + y = 6$
8. $y + z = 4$
9. $x = 2$
10. $z = -3$

Find the center and radius of the following spheres. Graph each. See Example 4.

11. $x^2 + y^2 + z^2 = 49$
12. $x^2 + y^2 + z^2 = 16$
13. $(x - 5)^2 + (y - 3)^2 + (z + 4)^2 = 9$
14. $(x + 3)^2 + (y + 2)^2 + (z - 1)^2 = 16$
15. $(x - 2)^2 + (y - 5)^2 + (z + 3)^2 = 25$
16. $(x + 2)^2 + (y - 3)^2 + (z - 2)^2 = 1$

Graph the following cylinders. See Examples 5 and 6.

17. $x^2 + z^2 = 16$
18. $x^2 + y^2 = 25$
19. $z^2 + y^2 = 49$
20. $x^2 + z^2 = 36$
21. $x^2 + y^2 = 121$
22. $y^2 + z^2 = 100$

In three-dimensional space, the distance between the points (x_1, y_1, z_1) and (x_2, y_2, z_2) is given by

$$\sqrt{(x_1 - x_2)^2 + (y_1 - y_2)^2 + (z_1 - z_2)^2}.$$

Find the distance between the following pairs of points.

23. $(1, 4, 2)$ and $(0, 3, 1)$
24. $(3, 7, 9)$ and $(1, 5, 3)$
25. $(-2, 1, 7)$ and $(3, -4, 0)$
26. $(-5, -3, -8)$ and $(-2, -4, -6)$
27. $(1, 0, -4)$ and $(-2, 1, 3)$
28. $(5, 8, -6)$ and $(-3, 0, 4)$

9.3 PARTIAL DERIVATIVES

Just as derivatives are useful in the study of functions of a single variable, partial derivatives are useful with functions of several variables. A *partial derivative* is found by taking the derivative of a function of several variables with respect to one variable at a time.

9.3 Partial Derivatives

As an example of this process, suppose that a small firm makes only two products, radios and cassette recorders. The profit of the firm is given by

$$P(x, y) = 40x^2 - 10xy + 5y^2 - 80,$$

where x is the number of radios sold and y is the number of cassette recorders sold. Lately, sales of radios have been steady at 10 units; only the sales of recorders vary. The management would like to know the marginal profit when $y = 12$.

We know that marginal profit is given by the derivative of the profit function. We also know that in our example x is fixed at 10. Using this information, we begin by finding a new function, $f = P(10, y)$. We let $x = 10$ and get

$$f = P(10, y) = 40(10)^2 - 10(10)y + 5y^2 - 80$$
$$= 3920 - 100y + 5y^2.$$

The function f shows the profit from the sale of y recorders, assuming that x is fixed at 10. To find the marginal profit when $y = 12$, we first find the derivative df/dy.

$$\frac{df}{dy} = -100 + 10y$$

When $y = 12$, we have $-100 + 10(12) = -100 + 120 = 20$, so the marginal profit is 20 when $y = 12$.

Work Problem 1 at the side.

The derivative of the function f that we found above was with respect to y only; we assumed that x was fixed. In general, let $z = f(x, y)$.

The **partial derivative of f with respect to x** is the derivative of f obtained by treating x as a variable and y as a constant.

The **partial derivative of f with respect to y** is the derivative of f obtained by treating y as a variable and x as a constant.

The symbols f_x (no prime is used) and $\partial f/\partial x$ are used to represent the partial derivative of f with respect to x.

Example 1 Let $f(x, y) = 4x^2 - 9xy + 6y^3$. Find f_x and f_y.

To find f_x, treat y as a constant and x as a variable. This gives

$$f_x = 8x - 9y.$$

If we treat y as a variable and x as a constant, we can find f_y.

$$f_y = -9x + 18y^2 \quad \blacksquare$$

Work Problem 2 at the side.

---

1. Assume that x increases to a constant 12.
   (a) Find $f = P(12, y)$.
   (b) Find the marginal profit under this assumption when $y = 8$.

Answer:
(a) $f = P(12, y) = 5680 - 120y + 5y^2$
(b) -40

2. Let $f(x, y) = -2x^4 + 6x^3y^3 + 5y^2 - 8$. Find f_x and f_y.

Answer: $f_x = -8x^3 + 18x^2y^3$;
$f_y = 18x^3y^2 + 10y$

374 Functions of Several Variables

Example 2 Let $f(x, y) = \ln(x^2 + y)$. Find f_x and f_y.

Recall the formula for the derivative of a natural logarithm function. [If $g(x) = \ln x$, then $g'(x) = 1/x$.] Using this formula and the chain rule, we have

$$f_x = \frac{2x}{x^2 + y} \quad \text{and} \quad f_y = \frac{1}{x^2 + y}. \quad \blacksquare$$

Work Problem 3 at the side.

3. Find f_x and f_y for the following functions.
  (a) $f(x, y) = \ln(4x + 9y^2)$
  (b) $f(x, y) = xe^y$

Answer:
(a) $f_x = 4/(4x + 9y^2)$;
$f_y = 18y/(4x + 9y^2)$
(b) $f_x = e^y$; $f_y = xe^y$

The notation $f_x(a, b)$ or $(\partial f/\partial x)_{(a, b)}$ represents the value of a partial derivative when $x = a$ and $y = b$, as shown in the next example.

Example 3 Let $f(x, y) = 2x^2 + 3xy^3 + 2y + 5$. Find the following.
(a) $f_x(-1, 2)$
First, find f_x by holding y constant.

$$f_x = 4x + 3y^3$$

Now let $x = -1$ and $y = 2$.

$$f_x(-1, 2) = 4(-1) + 3(2)^3 = -4 + 24 = 20$$

(b) $f_y(-4, -3)$

$$f_y = 9xy^2 + 2$$
$$f_y(-4, -3) = 9(-4)(-3)^2 + 2$$
$$= 9(-36) + 2 = -322 \quad \blacksquare$$

Work Problem 4 at the side.

4. Let $f(x, y) = 8x^2 + 5xy^2 - 9y + 4$. Find the following.
  (a) $f_x(0, 4)$
  (b) $f_x(-1, 2)$
  (c) $f_y(-2, 5)$
  (d) $f_y(0, 0)$

Answer:
(a) 80
(b) 4
(c) −109
(d) −9

The derivative of $y = f(x)$ gives the rate of change of y with respect to x. In the same way, if $z = f(x, y)$, then f_x gives the rate of change of z with respect to x, provided that y is held constant.

Example 4 Suppose that the temperature of the water at the point on a river where a nuclear power plant discharges its hot waste water is approximated by

$$T(x, y) = 2x + 5y + xy - 40,$$

where x represents the temperature of the river water in °C before it reaches the power plant and y is the number of megawatts (in hundreds) of electricity being produced annually by the plant.
(a) Find the temperature of the discharge water if the water reaching the plant has a temperature of 8°C and if 300 megawatts of electricity are being produced.

Here $x = 8$ and $y = 3$ (since 300 megawatts of electricity are being produced). Thus,

$$T(8, 3) = 2(8) + 5(3) + 8(3) - 40 = 15.$$

The water at the outlet of the plant is at a temperature of 15°C.

Work Problem 5 at the side.

(b) For the function $T(x, y) = 2x + 5y + xy - 40$, find and interpret $T_x(9, 5)$.

$$T_x = 2 + y$$

This partial derivative gives the rate of change of T with respect to x.

$$T_x(9, 5) = 2 + 5 = 7$$

This result, 7, is the approximate change in temperature of the output water if input water temperature changes from $x = 9$ to $x = 9 + 1 = 10$ while y remains constant.

(c) Use the same function to find and interpret $T_y(9, 5)$.

$$T_y = 5 + x$$

This partial derivative gives the rate of change of T with respect to y.

$$T_y(9, 5) = 5 + 9 = 14$$

This result, 14, is the approximate change in temperature resulting from an increase in production of electricity from $y = 5$ to $y = 6$ (600 megawatts) while x remains constant. ■

Work Problem 6 at the side.

When working with functions of one variable, we found second derivatives by taking the derivative of the first derivative. In much the same way, we can define **second partial derivatives.**

$$\frac{\partial}{\partial x}\left(\frac{\partial z}{\partial x}\right) = \frac{\partial^2 z}{\partial x^2} = f_{xx} \qquad \frac{\partial}{\partial y}\left(\frac{\partial z}{\partial y}\right) = \frac{\partial^2 z}{\partial y^2} = f_{yy}$$

$$\frac{\partial}{\partial y}\left(\frac{\partial z}{\partial x}\right) = \frac{\partial^2 z}{\partial y\, \partial x} = f_{xy} \qquad \frac{\partial}{\partial x}\left(\frac{\partial z}{\partial y}\right) = \frac{\partial^2 z}{\partial x\, \partial y} = f_{yx}$$

Be careful with these last two symbols. If we first find the partial derivative of z with respect to x, and then with respect to y, we have found $(f_x)_y$, or f_{xy}. We have also found

---

5. Find the temperature of the water at the outlet if the water has an initial temperature of 10°C and if 400 megawatts of electricity are being produced.

Answer: 40°C

6. Use the function of Example 4 to find and interpret the following.
(a) $T_x(5, 4)$
(b) $T_y(8, 3)$

Answer:
(a) $T_x(5, 4) = 6$; the approximate increase in temperature if the input temperature increases from 5 to 6 degrees.
(b) $T_y(8, 3) = 13$; the approximate increase in temperature if the production of electricity increases from 300 to 400 megawatts.

$$\frac{\partial}{\partial y}\left(\frac{\partial z}{\partial x}\right) = \frac{\partial^2 z}{\partial y\, \partial x}, \quad \text{so that} \quad \frac{\partial^2 z}{\partial y\, \partial x} = f_{xy}.$$

Note that the order of the symbols x and y is reversed in the last equation.

Example 5 Find all second partial derivatives for
$$f(x, y) = -4x^3 - 3x^2 y^3 + 2y^2.$$
First find f_x and f_y.
$$f_x = -12x^2 - 6xy^3 \quad \text{and} \quad f_y = -9x^2 y^2 + 4y$$
To find f_{xx}, take the partial derivative of f_x with respect to x.
$$f_{xx} = -24x - 6y^3$$
Take the partial derivative of f_y with respect to y; this gives f_{yy}.
$$f_{yy} = -18x^2 y + 4$$
We find f_{xy} by starting with f_x. Take the partial derivative of f_x with respect to y.
$$f_{xy} = -18xy^2$$
Finally, find f_{yx} by starting with f_y; take its partial derivative with respect to x.
$$f_{yx} = -18xy^2 \quad \blacksquare$$

Work Problem 7 at the side.

Example 6 Let $f(x, y) = 2e^x - 8x^3 y^2$. Find all second partial derivatives.

Here $f_x = 2e^x - 24x^2 y^2$ and $f_y = -16x^3 y$. [Recall: If $g(x) = e^x$, then $g'(x) = e^x$.] Now find the second partial derivatives.
$$f_{xx} = 2e^x - 48xy^2 \qquad f_{xy} = -48x^2 y$$
$$f_{yy} = -16x^3 \qquad f_{yx} = -48x^2 y \quad \blacksquare$$

Work Problem 8 at the side.

In all our examples of second partial derivatives, we found that $f_{xy} = f_{yx}$. This happens for many functions, and certainly for all the ones we shall use in this book.

---

7. Let $f(x, y) = 4x^2 y^2 - 9xy + 8x^2 - 3y^4$. Find all second partial derivatives.

Answer: $f_{xx} = 8y^2 + 16$;
$f_{yy} = 8x^2 - 36y^2$;
$f_{xy} = 16xy - 9$;
$f_{yx} = 16xy - 9$

8. Let $f(x, y) = 4e^{x+y} + 2x^3 y$. Find all second partial derivatives

Answer: $f_{xx} = 4e^{x+y} + 12xy$;
$f_{yy} = 4e^{x+y}$;
$f_{xy} = 4e^{x+y} + 6x^2$;
$f_{yx} = 4e^{x+y} + 6x^2$

9.3 EXERCISES

Find f_x and f_y for the following. Then find $f_x(2, -1)$ and $f_y(-4, 3)$. Leave the answers in terms of e in Exercises 7–10. See Examples 1–4.

1. $f(x, y) = 12x^2 + 4y^2$
2. $f(x, y) = -6x^2 + 7y^2$
3. $f(x, y) = -2xy + 6y^3 + 2$
4. $f(x, y) = 4x^2y - 9y^2$
5. $f(x, y) = 3x^3y^2$
6. $f(x, y) = -2x^2y^4$
7. $f(x, y) = e^{x+y}$
8. $f(x, y) = 3e^{2x+y}$
9. $f(x, y) = -5e^{3x-4y}$
10. $f(x, y) = 8e^{7x-y}$

Find all second partial derivatives for the following. See Examples 5 and 6.

11. $f(x, y) = 6x^3y - 9y^2 + 2x$
12. $g(x, y) = 5xy^4 + 8x^3 - 3y$
13. $R(x, y) = 4x^2 - 5xy^3 + 12y^2x^2$
14. $h(x, y) = 30y + 5x^2y + 12xy^2$
15. $r(x, y) = \dfrac{4x}{x + y}$
16. $k(x, y) = \dfrac{-5y}{x + 2y}$
17. $z = 4xe^y$
18. $z = -3ye^x$
19. $r = \ln|x + y|$
20. $k = \ln|5x - 7y|$

APPLIED PROBLEMS

Cost   Suppose that the manufacturing cost of a precision electronic calculator is approximated by
$$M(x, y) = 40x^2 + 30y^2 - 10xy + 30,$$
where x is the cost of the necessary electronic chips and y is the cost of labor. Find the following.

21. $M_y(4, 2)$
22. $M_x(3, 6)$
23. $(\partial M/\partial x)_{(2, 5)}$
24. $(\partial M/\partial y)_{(6, 7)}$

Mating   The total number of matings per day between individuals of a certain species of grasshoppers is approximated by
$$M(x, y) = 2xy + 10xy^2 + 30y^2 + 20,$$
where x represents the temperature in °C and y represents the number of days since the last rain. Find the following.

25. $(\partial M/\partial x)_{(20,4)}$ 26. $(\partial M/\partial y)_{(24,10)}$

27. $M_x(17, 3)$ 28. $M_y(21, 8)$

Revenue

The revenue from the sale of x units of a tranquilizer and y units of an antibiotic is given by
$$R(x, y) = 5x^2 + 9y^2 - 4xy.$$

29. Suppose $x = 9$ and $y = 5$. What is the approximate effect on revenue if x is increased to 10, while y is fixed? (See Example 4.)

30. Suppose $x = 9$ and $y = 5$. What is the approximate effect on revenue if y is increased to 6, while x is fixed?

Production function

A production function $z = f(x, y)$ is a function which gives the quantity of an item produced, z, as a function of two other variables, x and y.

31. The production function z for the United States was once estimated as
$$z = x^{2.1}y^{.3},$$
where x stands for the amount of labor and y stands for the amount of capital. Find the marginal productivity of labor (find $\partial z/\partial x$) and of capital.

32. A similar production function for Canada is
$$z = x^{.4}y^{.6},$$
with x, y, and z as in Exercise 31. Find the marginal productivity of labor and of capital.

Oxygen consumption

(The remaining exercises in this section were first discussed in Section 1 of this chapter.) The oxygen consumption of a well-insulated mammal which is not sweating is approximated by
$$m = m(T, F, w) = \frac{2.5(T - F)}{w^{.67}} = 2.5(T - F)w^{-.67},$$
where T is the internal body temperature of the animal (in °C), F is the temperature of the outside of the animal's fur (in °C), and w is the animal's weight in kilograms.

33. Find m_T.

34. Suppose $T = 38°$, $F = 12°$, and $w = 30$ kg. Find $m_T(38, 12, 30)$.

35. Find m_w.

36. Suppose $T = 40°$, $F = 20°$, and $w = 40$ kg. Find $m_w(40, 20, 40)$.

37. Find m_F.

38. Suppose $T = 36°$, $F = 14°$, and $w = 25$ kg. Find $m_F(36, 14, 25)$.

Human surface area

The surface area of a human, in square meters, is approximated by
$$A(W, H) = 2.02W^{.425}H^{.725},$$
where W is the weight of the person in kilograms and H is the height in meters.

39. Find $\partial A/\partial W$.

40. Suppose $W = 72$ and $H = 1.8$. Find $(\partial A/\partial W)_{(72, 1.8)}$.

41. Find $\partial A/\partial H$.

42. Suppose $W = 70$ and $H = 1.6$. Find $(\partial A/\partial H)_{(70, 1.6)}$.

Physiology

In one method of computing the quantity of blood pumped through the lungs in one minute, a researcher first finds each of the following (in milliliters).

$b =$ quantity of oxygen used by body in one minute

$a =$ quantity of oxygen per liter of blood that has just gone through the lungs

$v =$ quantity of oxygen per liter of blood that is about to enter the lungs

In one minute,

 amount of oxygen used = amount of oxygen per liter
 × number of liters of blood pumped

If C is the number of liters pumped through the blood in one minute, then

$$b = (a - v) \cdot C$$

or

$$C = \frac{b}{a - v}$$

43. Find C if $a = 160$, $b = 200$, and $v = 125$.

44. Find C if $a = 180$, $b = 260$, and $v = 142$.

Find the following partial derivatives.

45. $\partial C/\partial b$ 46. $\partial C/\partial v$

9.4 MAXIMUMS AND MINIMUMS

One of the most important applications of calculus is finding maximums and minimums. Partial derivatives are used to find maximums and minimums of functions of two variables.

To define relative maximums and minimums for a function of two variables, let $z = f(x, y)$ be a function defined for each point in some region of the plane containing the x-axis and y-axis. The function $z = f(x, y)$ has a **relative maximum** at the point (a, b) if there exists a circular region with center at (a, b) such that

$$f(x, y) \leq f(a, b)$$

for every point (x, y) of that region. An example of a relative maximum is shown in Figure 9.8. **Relative minimums** are defined in a similar way. A relative minimum for $z = f(x, y)$ is shown in Figure 9.9.

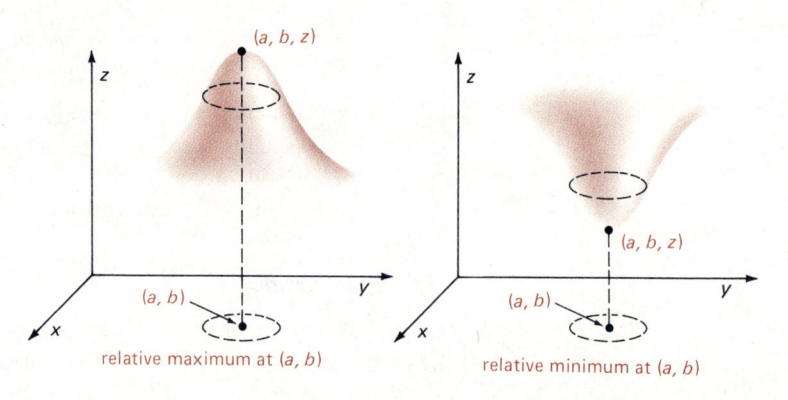

Figure 9.8 — relative maximum at (a, b)

Figure 9.9 — relative minimum at (a, b)

A function of one variable, such as $y = f(x)$, can have a relative maximum or relative minimum at a point $x = a$ that makes the derivative equal 0. (But the function does not necessarily have a relative maximum or minimum at $x = a$ just because $f'(a) = 0$.) A similar result holds for functions of two variables, as shown by the next theorem.

Theorem 9.1 Let a function $z = f(x, y)$ have a relative maximum or relative minimum at the point (a, b). Let $f_x(a, b)$ and $f_y(a, b)$ both exist. Then

$$f_x(a, b) = 0 \quad \text{and} \quad f_y(a, b) = 0.$$

Once again, the fact that $f_x(a, b) = 0$ and $f_y(a, b) = 0$ is no guarantee that the function has a relative maximum or minimum at (a, b). For example, Figure 9.10 shows the graph of $f(x, y) = z = x^2 - y^2$. Both $f_x(0, 0) = 0$ and $f_y(0, 0) = 0$, and yet $(0, 0)$ leads to neither a relative maximum nor a relative minimum for the function. Here the point $(0, 0, 0)$ is called a **saddle point**. A saddle point is a minimum when approached from one direction but a maximum when approached from another direction. Thus, a saddle point is neither a maximum nor a minimum.

9.4 Maximums and Minimums 381

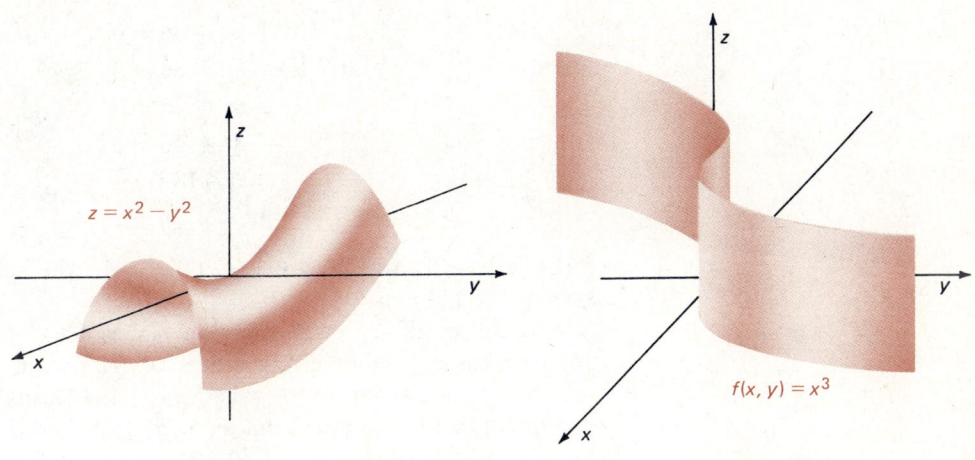

Figure 9.10

Figure 9.11

Figure 9.11 shows the graph of $f(x, y) = x^3$. Check that $f_x(0, 0) = 0$ and $f_y(0, 0) = 0$, but $(0, 0)$ does not lead to a relative maximum, a relative minimum, or a saddle point.

Example 1 At what points might the function

$$f(x, y) = 6x^2 + 6y^2 + 6xy + 36x - 5$$

have any relative maximums or minimums?

We begin by finding all points (a, b) such that $f_x(a, b) = 0$ and $f_y(a, b) = 0$; first find f_x and f_y.

$$f_x = 12x + 6y + 36 \quad \text{and} \quad f_y = 12y + 6x$$

Place each of these two partial derivatives equal to 0.

$$12x + 6y + 36 = 0 \quad \text{and} \quad 12y + 6x = 0$$

These two equations make up a system of linear equations. We need to find all values of x and y that make both of the equations true at the same time. To find these values, we rewrite the simplest equation as follows.

$$12y + 6x = 0$$
$$6x = -12y$$
$$x = -2y$$

Now we substitute $-2y$ for x in the remaining equation.

$$12x + 6y + 36 = 0$$
$$12(-2y) + 6y + 36 = 0$$
$$-24y + 6y + 36 = 0$$
$$-18y + 36 = 0$$
$$-18y = -36$$
$$y = 2$$

Since $x = -2y$, we have $x = -2(2) = -4$. The solution of the system of equations is $(-4, 2)$.

Therefore, if the given function $f(x, y) = 6x^2 + 6y^2 + 6xy + 36x - 5$ has any relative maximums or relative minimums, they occur at $(-4, 2)$. To decide whether $(-4, 2)$ leads to a relative maximum, a relative minimum, or neither, we need the results of the next theorem. ∎

1. At what point might the function $f(x, y) = x^2 + y^2 + 4x - 6y + 2$ have any relative maximums or minimums?

Answer: $(-2, 3)$

Work Problem 1 at the side.

Theorem 9.2 *M-test for relative maximums and minimums* Let all the following partial derivatives exist for a function $z = f(x, y)$. Let (a, b) be a point for which

$$f_x(a, b) = 0 \quad \text{and} \quad f_y(a, b) = 0.$$

Define the number M as

$$M = f_{xx}(a, b) \cdot f_{yy}(a, b) - [f_{xy}(a, b)]^2.$$

Then
- **(a)** $f(a, b)$ is a relative maximum if $M > 0$ and $f_{xx}(a, b) < 0$.
- **(b)** $f(a, b)$ is a relative minimum if $M > 0$ and $f_{xx}(a, b) > 0$.
- **(c)** $f(a, b)$ is a saddle point (neither a maximum nor a minimum) if $M < 0$.
- **(d)** If $M = 0$, the test gives no information.

Example 2 In Example 1 above, we found that if the function

$$f(x, y) = 6x^2 + 6y^2 + 6xy + 36x - 5$$

has any relative maximums or relative minimums, they occur at the point $(-4, 2)$. Does $(-4, 2)$ lead to a relative maximum, a relative minimum, or neither?

We can find out by using the M-test above. We already know that

$$f_x(-4, 2) = 0 \quad \text{and} \quad f_y(-4, 2) = 0.$$

We now must find the various second partial derivatives used in finding M. From $f_x = 12x + 6y + 36$ and $f_y = 12y + 6x$, we get

$$f_{xx} = 12, \quad f_{yy} = 12, \quad \text{and} \quad f_{xy} = 6.$$

(If these second partial derivatives had not all been equal to constants, we would have had to evaluate them at the point $(-4, 2)$.) Now we can find M.

$$M = f_{xx}(-4, 2) \cdot f_{yy}(-4, 2) - [f_{xy}(-4, 2)]^2$$
$$= 12 \cdot 12 - 6^2$$
$$M = 108$$

Since $M > 0$ and $f_{xx}(-4, 2) = 12 > 0$, part (b) of Theorem 9.2 applies. Thus, $f(x, y) = 6x^2 + 6y^2 + 6xy + 36x - 5$ has a relative minimum at $(-4, 2)$. This relative minimum is given by $f(-4, 2) = -77$. ∎

Example 3 Find all points where the function

$$f(x, y) = 50 + 4x - 5y + x^2 + y^2 + xy$$

has any relative maximums or relative minimums.

Here we have

$$f_x = 4 + 2x + y \quad \text{and} \quad f_y = -5 + 2y + x.$$

If we place these partial derivatives equal to 0 and rearrange terms, we get

$$2x + y = -4 \quad \text{and} \quad x + 2y = 5.$$

One way to solve this system of equations is to multiply both sides of the first equation by -2 and then add the two equations.

$$\begin{aligned} -4x - 2y &= 8 \\ x + 2y &= 5 \\ \hline -3x &= 13 \end{aligned}$$

From this last equation, we get $x = -13/3$. Substitute $-13/3$ for x in either equation to find y. You should find that $y = 14/3$. The only possible point leading to a relative maximum or minimum is $(-13/3, 14/3)$.

In order to use the M-test, we find the following second partial derivatives.

$$f_{xx} = 2, \quad f_{yy} = 2, \quad \text{and} \quad f_{xy} = 1$$

Thus, $$M = 2 \cdot 2 - 1^2 = 3.$$

Since $M > 0$ and $f_{xx}(-13/3, 14/3) = 2$, we see that $f(-13/3, 14/3)$ is a relative minimum. ∎

Work Problem 2 at the side.

---

2. Find all points where the following functions have any relative maximums or relative minimums.
(a) $f(x, y) = 4x^2 + 3xy + 2y^2 + 7x - 6y - 6$
(b) $f(x, y) = 4x^2 + 6xy + y^2 + 34x + 8y + 5$

Answer:
(a) $M = 23$, relative minimum at $(-2, 3)$
(b) $M = -20$, saddle point (neither a maximum nor a minimum) at $(1, -7)$

Example 4 Find all points where the function
$$f(x, y) = 9xy - x^3 - y^3 - 6$$
has any relative maximums or relative minimums.

Here we have
$$f_x = 9y - 3x^2 \quad \text{and} \quad f_y = 9x - 3y^2.$$
Place each of these partial derivatives equal to 0.

$$\begin{aligned} f_x &= 0 & f_y &= 0 \\ 9y - 3x^2 &= 0 & 9x - 3y^2 &= 0 \\ 9y &= 3x^2 & 9x &= 3y^2 \\ 3y &= x^2 & 3x &= y^2 \end{aligned}$$

The final equation on the left, $3y = x^2$, can be rewritten as $y = x^2/3$. Substitute this into the final equation on the right.

$$3x = y^2$$
$$3x = \left(\frac{x^2}{3}\right)^2$$
$$3x = \frac{x^4}{9}$$
$$27x = x^4$$

Rewrite this equation as

$$\begin{aligned} x^4 - 27x &= 0 & \\ x(x^3 - 27) &= 0 & \text{Factor} \\ x = 0 \quad \text{or} \quad x^3 - 27 &= 0 \\ x = 0 \quad \text{or} \quad x^3 &= 27 \\ x = 0 \quad \text{or} \quad x &= 3. \end{aligned}$$

We can use these values of x, along with the equation $3x = y^2$, to find y.

If $x = 0$, $\qquad\qquad$ If $x = 3$,
$3x = y^2$ $\qquad\qquad$ $3x = y^2$
$3(0) = y^2$ $\qquad\qquad$ $3(3) = y^2$
$0 = y^2$ $\qquad\qquad$ $9 = y^2$
$0 = y$ $\quad$ or $\quad$ $3 = y$ $\quad$ or $\quad$ $-3 = y$.

Since $f_x(3, -3) \neq 0$ and $f_y(3, -3) \neq 0$, the only possible relative maximums or minimums for $f(x, y) = 9xy - x^3 - y^3 - 6$ occur at $(0, 0)$ or $(3, 3)$. To find out which, we use the *M*-test. Here

$$f_{xx} = -6x, \quad f_{yy} = -6y, \quad \text{and} \quad f_{xy} = 9.$$

Test each of the possible points.

For (0, 0):
$f_{xx}(0, 0) = -6(0) = 0$
$f_{yy}(0, 0) = -6(0) = 0$
$f_{xy}(0, 0) = 9$
$M = 0 \cdot 0 - 9^2 = -81.$
Since $M < 0$, we have a saddle point at (0, 0).

For (3, 3):
$f_{xx}(3, 3) = -6(3) = -18$
$f_{yy}(3, 3) = -6(3) = -18$
$f_{xy}(3, 3) = 9$
$M = -18(-18) - 9^2 = 243.$
Here $M > 0$ and $f_{xx}(3, 3) = -18 < 0$; there is a relative maximum at (3, 3). ∎

3. Find all points where the function $f(x, y) = x^3 + y^2 - 6xy + 5$ has any relative maximums or relative minimums.

Answer: at (0, 0), we have $M = -36 < 0$, a saddle point; at (6, 18) we have $M = 36 > 0$ and $f_{xx}(6, 18) = 36 > 0$, a relative minimum

Work Problem 3 at the side.

Example 5 A company is developing a new soft drink. The cost in dollars to produce a batch of the drink is approximated by

$$C(x, y) = 2200 + 27x^3 - 72xy + 8y^2,$$

where x is the number of kilograms of sugar per batch and y is the number of grams of flavoring per batch.

(a) Find the amounts of sugar and flavoring that result in minimum cost for a batch.

We need the following partial derivatives.

$$C_x = 81x^2 - 72y \quad \text{and} \quad C_y = -72x + 16y$$

Set each of these equal to 0 and solve for y.

$$81x^2 - 72y = 0 \qquad\qquad -72x + 16y = 0$$
$$-72y = -81x^2 \qquad\qquad 16y = 72x$$
$$y = \frac{9}{8}x^2 \qquad\qquad y = \frac{9}{2}x$$

Since $(9/8)x^2$ and $(9/2)x$ both equal y they are equal. Solve the resulting equation for x.

$$\frac{9}{8}x^2 = \frac{9}{2}x$$
$$9x^2 = 36x$$
$$9x^2 - 36x = 0$$
$$9x(x - 4) = 0$$
$$9x = 0 \quad \text{or} \quad x - 4 = 0$$

The equation $9x = 0$ leads to $x = 0$ which is not a useful answer for our problem. Substitute $x = 4$ into $y = (9/2)x$ to find y.

$$y = \frac{9}{2}x = \frac{9}{2}(4) = 18$$

Now we must check to see if the point (4, 18) leads to a relative minimum. For (4, 18) we have

$$C_{xx} = 162x = 162(4) = 648, \quad C_{yy} = 16, \quad \text{and} \quad C_{xy} = -72.$$

Also, $\quad M = (648)(16) - (-72)^2 = 5184.$

Since $M > 0$ and $C_{xx}(4, 18) > 0$, the cost at (4, 18) is a minimum.

(b) What is the minimum cost?

To find the minimum cost, we go back to the cost function and evaluate $C(4, 18)$.

$$C(x, y) = 2200 + 27x^3 - 72xy + 8y^2$$
$$C(4, 18) = 2200 + 27(4)^3 - 72(4)(18) + 8(18)^2 = 1336$$

The minimum cost for a batch is $1336.00. ∎

9.4 EXERCISES

Find all points where the following functions have any relative maximums or relative minimums. Identify any saddle points. See Examples 2–4.

1. $f(x, y) = xy + x - y$
2. $f(x, y) = 4xy + 8x - 9y$
3. $f(x, y) = x^2 - 2xy + 2y^2 + x - 5$
4. $f(x, y) = x^2 + xy + y^2 - 6x - 3$
5. $f(x, y) = x^2 - xy + y^2 + 2x + 2y + 6$
6. $f(x, y) = x^2 + xy + y^2 + 3x - 3y$
7. $f(x, y) = x^2 + 3xy + 3y^2 - 6x + 3y$
8. $f(x, y) = 5xy - 7x^2 - y^2 + 3x - 6y - 4$
9. $f(x, y) = 4xy - 10x^2 - 4y^2 + 8x + 8y + 9$
10. $f(x, y) = x^2 + xy + 3x + 2y - 6$
11. $f(x, y) = x^2 + xy - 2x - 2y + 2$
12. $f(x, y) = x^2 + xy + y^2 - 3x - 5$
13. $f(x, y) = x^2 - y^2 - 2x + 4y - 7$
14. $f(x, y) = 4x + 2y - x^2 + xy - y^2 + 3$
15. $f(x, y) = 2x^3 + 3y^2 - 12xy + 4$
16. $f(x, y) = 5x^3 + 2y^2 - 60xy - 3$
17. $f(x, y) = x^2 + 4y^3 - 6xy - 1$
18. $f(x, y) = 3x^2 + 7y^3 - 42xy + 5$

APPLIED PROBLEMS

Profit   **19.** Suppose that the profit of a certain firm is approximated by

$$P(x, y) = 1000 + 24x - x^2 + 80y - y^2,$$

where x is the cost of a unit of labor and y is the cost of a unit of goods. Find values of x and y that maximize profit. Find the maximum profit.

Labor cost   **20.** The labor cost for manufacturing a precision camera can be approximated by

$$L(x, y) = \frac{3}{2}x^2 + y^2 - 2x - 2y - 2xy + 68,$$

where x is the number of hours required by a skilled craftsperson and y is the number of hours required by a semiskilled person. Find values of x and y that minimize the labor charge. Find the minimum labor charge.

Rooster production   **21.** The number of roosters that can be fed from x pounds of Super-Hen chicken feed and y pounds of Super-Rooster feed is given by

$$R(x, y) = 800 - 2x^3 + 12xy - y^2.$$

Find the number of pounds of each kind of feed that produces the maximum number of roosters.

Profit   **22.** The total profit from one acre of a certain crop depends on the amount spent on fertilizer, x, and hybrid seed, y, according to the model

$$P(x, y) = -x^2 + 3xy + 160x - 5y^2 + 200y + 2{,}600{,}000.$$

Find values of x and y that lead to maximum profit. Find the maximum profit.

Cost   **23.** The total cost to produce x units of electrical tape and y units of packing tape is given by

$$C(x, y) = 2x^2 + 3y^2 - 2xy + 2x - 126y + 3800.$$

Find the number of units of each kind of tape that should be produced so that the total cost is a minimum. Find the minimum total cost.

Revenue   **24.** The total revenue from the sale of x hot tubs and y solar heaters is approximated by

$$R(x, y) = 12 + 74x + 85y - 3x^2 - 5y^2 - 5xy.$$

Find the number of each that should be sold to produce maximum revenue. Find the maximum revenue.

9.5   LAGRANGE MULTIPLIERS

In the previous section we saw how to find any relative maximums or relative minimums for functions of two variables. In practice, many such functions are given with a secondary condition or **constraint**.

388 Functions of Several Variables

For example, we might be asked to find the maximum profit from the sale of x units of small boxes and y units of large boxes, subject to the constraint that the total number of units must be 12 (that is, $x + y = 12$).

Problems with constraints can be worked using the method of **Lagrange multipliers.** This method looks fairly complicated when written down as a theorem, so before we do this we shall look at several examples of the method in use.

Example 1 Find the minimum value of

$$f(x, y) = 5x^2 + 6y^2 - xy,$$

subject to the constraint $x + 2y = 24$.

Go through the following steps.

1. Rewrite the constraint so that it equals 0. Let $g(x, y)$ equal this constraint. In this example,

$$x + 2y = 24 \quad \text{becomes} \quad x + 2y - 24 = 0,$$

and finally,

$$g(x, y) = x + 2y - 24.$$

2. Form the **Lagrange function** $F(x, y, \lambda)$. (The symbol λ is the Greek letter lambda.) The Lagrange function is defined as the sum of the function $f(x, y)$ and the product of λ and $g(x, y)$. In this example we have

$$\begin{aligned} F(x, y, \lambda) &= f(x, y) + \lambda \cdot g(x, y) \\ &= 5x^2 + 6y^2 - xy + \lambda(x + 2y - 24) \\ &= 5x^2 + 6y^2 - xy + \lambda x + 2\lambda y - 24\lambda. \end{aligned}$$

3. Find F_x, F_y, and F_λ. Form the system of equations $F_x = 0$, $F_y = 0$, and $F_\lambda = 0$. In our example, we have

$$F_x = 10x - y + \lambda = 0 \qquad (1)$$

$$F_y = 12y - x + 2\lambda = 0 \qquad (2)$$

$$F_\lambda = x + 2y - 24 = 0. \qquad (3)$$

4. Solve the system of equations from Step 3 for x, y, and λ.

To solve the system, multiply both sides of equation (1) by 12 and add the result to equation (2). Doing this, and rearranging terms, we have

$$\begin{aligned} 120x - 12y + 12\lambda &= 0 \\ -x + 12y + 2\lambda &= 0 \\ \hline 119x + 14\lambda &= 0. \end{aligned} \qquad (4)$$

9.5 Lagrange Multipliers **389**

To eliminate y from equations (2) and (3), multiply both sides of equation (3) by -6 and add the result to equation (2). After rearranging terms, we have

$$-x + 12y + 2\lambda = 0$$
$$-6x - 12y = -144$$
$$\overline{-7x + 2\lambda = -144.} \qquad (5)$$

We can now find x by solving equations (4) and (5) together. Multiply both sides of equation (5) by -7, and add to equation (4).

$$119x + 14\lambda = 0$$
$$49x - 14\lambda = 1008$$
$$\overline{168x = 1008,}$$

from which $x = 6$. Substituting 6 for x in equation (3), we have $y = 9$. Also, although we don't use this number, $\lambda = -51$.

Finally, it can be verified that the minimum value for $f(x, y) = 5x^2 + 6y^2 - xy$, subject to the constraint $x + 2y = 24$, is at the point $(6, 9)$. This minimum value is $f(6, 9) = 612$. ∎

Work Problem 1 at the side.

Example 2 Find two numbers whose sum is 50 and whose product is a maximum.

If we let x and y represent the two numbers, then we wish to maximize the product

$$f(x, y) = xy,$$

subject to the constraint $x + y = 50$. Go through the four steps presented above.

1. $g(x, y) = x + y - 50$
2. $F(x, y, \lambda) = xy + \lambda(x + y - 50)$
3. $F_x = y + \lambda = 0$ \hfill (6)
   $F_y = x + \lambda = 0$ \hfill (7)
   $F_\lambda = x + y - 50 = 0$ \hfill (8)
4. From equation (6), we have $\lambda = -y$. Substituting $-y$ for λ in equation (7), we have

$$x + \lambda = 0$$
$$x + (-y) = 0$$
$$x = y.$$

---

1. Go through the following steps to find a minimum value of $f(x, y) = x^2 + y^2$, subject to the constraint $2x + 3y = 9$.
   (a) Put the constraint equal to 0; let $g(x, y)$ equal the constraint
   (b) Form the Lagrange function $F(x, y, \lambda)$.
   (c) Form the system of equations $F_x = 0, F_y = 0, F_\lambda = 0$.
   (d) Solve the system.

Answer:
(a) $2x + 3y - 9 = 0$;
    $g(x, y) = 2x + 3y - 9$
(b) $F(x, y, \lambda) = x^2 + y^2 + \lambda(2x + 3y - 9)$
(c) $F_x = 2x + 2\lambda$;
    $F_y = 2y + 3\lambda$;
    $F_\lambda = 2x + 3y - 9$
(d) $x = 18/13, y = 27/13,$
    $\lambda = -18/13$

Substituting y for x in equation (8) gives

$$x + y - 50 = 0$$
$$y + y - 50 = 0$$
$$2y = 50$$
$$y = 25.$$

Since $x = y$, we have $x = 25$.

Thus, it can be verified that 25 and 25 are the two numbers whose sum is 50 and whose product is a maximum. The maximum product is given by $25 \cdot 25 = 625$. ∎

Work Problem 2 at the side.

2. Find two numbers x and y whose sum is 12 and such that x^2y is maximized.

Answer: $x = 8$, $y = 4$ leads to the maximum product of $8^2 \cdot 4 = 256$ ($x = 0$, $y = 12$ leads to the *minimum* product of 0.)

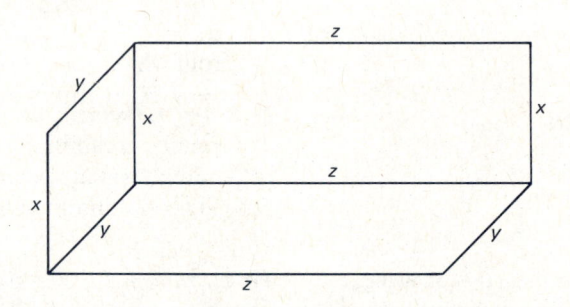

Figure 9.12

Example 3 Find the dimensions of the rectangular box of maximum volume that can be produced from 6 square feet of material.

Let x, y, and z represent the dimensions of the box, as shown in Figure 9.12. The volume of the box is given by

$$f(x, y, z) = xyz.$$

As shown in Figure 9.12, the total amount of material required for the two ends of the box is $2xy$, the total needed for the sides is $2xz$, and the total needed for the top and bottom is $2yz$. Since 6 square feet of material is available, we have

$$2xy + 2xz + 2yz = 6 \quad \text{or} \quad xy + xz + yz = 3.$$

In summary, we must maximize $f(x, y, z) = xyz$ subject to the constraint $xy + xz + yz = 3$. This problem involves functions of *three* variables instead of two, but the method of Lagrange multipliers can be used in the same way as before.

1. $g(x, y, z) = xy + xz + yz - 3$
2. $F(x, y, z, \lambda) = xyz + \lambda(xy + xz + yz - 3)$

3. $F_x = yz + \lambda y + \lambda z = 0$
   $F_y = xz + \lambda x + \lambda z = 0$
   $F_z = xy + \lambda x + \lambda y = 0$
   $F_\lambda = xy + xz + yz - 3 = 0$

4. The solution of the system of equations of Step 3 is $x = 1$, $y = 1$, $z = 1$. In other words, the box we want is a cube, 1 foot on a side. ∎

3. A can is to be made that will hold 128π cubic inches of canned salmon. Go through the four steps of the example above to find the dimensions of the can that will produce minimum surface area. The surface area of a can of radius r and height h is given by $S = 2\pi r^2 + 2\pi rh$. The volume is $V = \pi r^2 h$. Remember to treat π as a constant.

Answer:
(Step 1)
$g(r, h) = \pi r^2 h - 128\pi$
(Step 2)
$F(r, h, \lambda) = 2\pi r^2 + 2\pi rh + \lambda(\pi r^2 h - 128\pi)$
(Step 3)
$F_r = 4\pi r + 2\pi h + 2\lambda \pi rh = 0$;
$F_h = 2\pi r + \lambda \pi r^2 = 0$;
$F_\lambda = \pi r^2 h - 128\pi = 0$
(Step 4) From the third equation, $F_\lambda = 0$, we get $\pi r^2 h - 128\pi = 0$, or $\pi r^2 h = 128\pi$, or $r^2 h = 128$, and $h = 128/r^2$. From the second equation, $F_h = 0$, we get $2 + \lambda r = 0$, or $\lambda = -2/r$. From the first equation, we get $2r + h + \lambda rh = 0$. Substitute for h and λ, obtaining $2r + 128/r^2 - 256/r^2 = 0$. Finally, $r = 4$ and $h = 8$.

Work Problem 3 at the side.

Now we can state the method of Lagrange multipliers as a theorem. As we said above, the method works with two independent variables, or three, or four, or more. For simplicity, we state the theorem only for two independent variables.

Theorem 9.3 *Lagrange multipliers* Any relative maximum or relative minimum values of the function $z = f(x, y)$, subject to a constraint $g(x, y) = 0$, will be found among those points (x, y) for which there exists a value of λ such that

$$F_x(x, y, \lambda) = 0$$
$$F_y(x, y, \lambda) = 0$$
$$F_\lambda(x, y, \lambda) = 0,$$

where $\quad F(x, y, \lambda) = f(x, y) + \lambda \cdot g(x, y)$.

We assume that all indicated partial derivatives exist.

One problem with the method of Lagrange multipliers is that it is not easy to tell whether an answer is a relative maximum or a relative minimum—the *M*-test of the previous section does not apply. (Details on deciding whether a maximum or a minimum has been found are given in Jean E. Draper and Jane S. Klingman, *Mathematical Analysis*, 2nd ed. (New York: Harper and Row, 1972), pp. 376 and 602.) However, we can usually tell from the problem itself—a cost function would probably lead to a minimum, while a problem about profit should lead to a maximum.

9.5 EXERCISES

Find the following relative maximums or relative minimums. See Example 1.

1. Maximum of $f(x, y) = 2xy$, subject to $x + y = 12$.

2. Maximum of $f(x, y) = 4xy + 2$, subject to $x + y = 24$.

3. Maximum of $f(x, y) = x^2y$, subject to $2x + y = 4$.

4. Maximum of $f(x, y) = 4xy^2$, subject to $3x - 2y = 5$.

5. Minimum of $f(x, y) = x^2 + 2y^2 - xy$, subject to $x + y = 8$.

6. Minimum of $f(x, y) = 3x^2 + 4y^2 - xy - 2$, subject to $2x + y = 21$.

7. Maximum of $f(x, y) = x^2 - 10y^2$, subject to $x - y = 18$.

8. Maximum of $f(x, y) = 12xy - x^2 - 3y^2$, subject to $x + y = 16$.

9. Find two numbers whose sum is 20 and whose product is a maximum. See Example 2.

10. Find two numbers whose sum is 100 and whose product is a maximum.

11. Find two numbers x and y such that $x + y = 18$ and xy^2 is maximized.

12. Find two numbers x and y such that $x + y = 36$ and x^2y is maximized.

13. Find three numbers whose sum is 90 and whose product is a maximum.

14. Find three numbers whose sum is 240 and whose product is a maximum.

APPLIED PROBLEMS

Area

15. A farmer has 200 meters of fencing. Find the dimensions of the rectangular field of maximum area that can be enclosed by this amount of fencing.

16. Because of terrain difficulties, two sides of a fence can be built for $6 a foot, while the other two sides cost $4 per foot. (See the sketch.) Find the field of maximum area that can be enclosed for $1200.

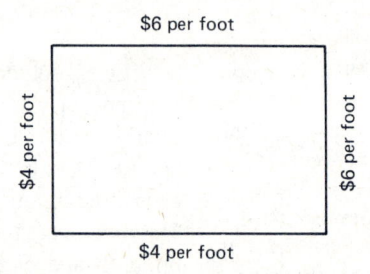

17. Find the area of the largest rectangular field that can be enclosed with 600 meters of fencing. Assume that no fencing is needed along one side of the field.

18. A fence is built against a large building, so that no fencing material is used on that one side. Material for the ends costs $8 a foot; the side can be built for $6 per foot. Find the dimensions of the field of maximum area that can be enclosed for $1200.

Surface area

19. A cylindrical can is to be made that will hold 250π cubic inches of candy. Find the dimensions of the can with minimum surface area.

20. An ordinary 12-ounce beer or soda pop can holds about 25 cubic inches. Use a calculator and find the dimensions of a can with minimum surface area. Measure a can and see how close its dimensions are to the results you found.

Volume **21.** A rectangular box with no top is to be built from 500 square meters of material. Find the dimensions of such a box that will enclose the maximum volume.

22. A one-pound soda cracker box has a volume of 185 cubic inches. The end of the box is square. Find the dimensions of such a box that has minimum surface area.

Cost **23.** The total cost to produce x large needlepoint kits and y small ones is given by

$$C(x, y) = 2x^2 + 6y^2 + 4xy + 10.$$

If a total of ten kits must be made, how should production be allocated so that total cost is minimized?

Profit **24.** The profit from the sale of x units of radiators for automobiles and y units of radiators for generators is given by

$$P(x, y) = -x^2 - y^2 + 4x + 8y.$$

Find values of x and y that lead to a maximum profit if the firm must produce a total of 6 units of radiators.

Production **25.** The production of nails depends on the cost of steel and the price of labor. Suppose that the number of units of nails manufactured is given by

$$f(x, y) = -60x + 100y - x^2 - 2y^2 - 3xy + 390{,}000$$

where x is the number of units of steel and y is the number of units of labor. Assume that steel costs $8 per unit and labor is $10 per unit. Find the maximum number of units of nails that can be made if the total budget for steel and labor is $4600.

26. There are two basic ingredients in soft drinks: sugar and carbonated water. If x units of sugar and y units of carbonated water are used, then the total units of soft drinks which can be produced are given by

$$f(x, y) = 120x - 160y - 3x^2 - 2y^2 - 2xy.$$

One unit of sugar costs $4 and one unit of carbonated water costs $8. Find the maximum number of units of soft drinks that can be produced with a budget of $1800 for sugar and carbonated water.

9.6 AN APPLICATION—THE LEAST SQUARES LINE

In trying to predict the future sales of a product or the total number of matings between two species of insects, it is common to gather as much data as possible and then draw a graph showing the data.

394 Functions of Several Variables

This graph, called a **scatter diagram,** can then be inspected to see if a reasonably simple mathematical curve can be found which fits fairly well through all the given data points.

As an example, suppose a firm gathers data showing the relationship between the price in dollars of an item, y, and the number of units of the item that are sold, x, with results as follows.

units sold, x	10	15	20
price, y	80	68	62

A graph of this data is shown in Figure 9.13.

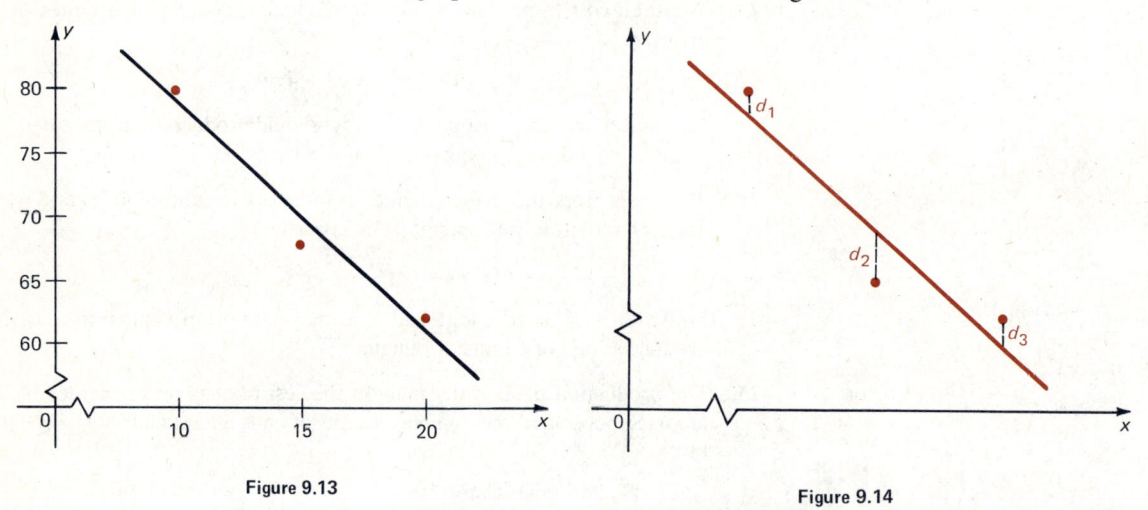

Figure 9.13

Figure 9.14

As shown above, a straight line fits reasonably well through these data points. If all the data points were to lie on the straight line, we could use the point-slope form of the equation of a line from Chapter 2 to find the equation of the line through the points.

In practice, the points on a scatter diagram never fit a straight line exactly. Then we must decide on the "best" straight line through the points. One way to define the "best" line is as follows. Figure 9.14 shows an enlargement of the scatter diagram of Figure 9.13. This time we have added vertical line segments showing the distances of the three points from the "best" line. The "best" straight line is defined as the one that minimizes the sum of the squares of the lengths of these segments, $d_1^2 + d_2^2 + d_3^2$.

Let $y = mx + b$ be the equation of the line that best fits through the three points. Then d_1, in Figure 9.14, is given by

$$d_1 = 80 - (m \cdot 10 + b).$$

Also,
$$d_2 = 68 - (m \cdot 15 + b)$$

and
$$d_3 = 62 - (m \cdot 20 + b).$$

9.6 An Application—The Least Squares Line

We want to find values of m and b that will minimize $d_1^2 + d_2^2 + d_3^2$; that is, we want to minimize

$$S = [80 - (10m + b)]^2 + [68 - (15m + b)]^2 + [62 - (20m + b)]^2.$$

To minimize S we need $\partial S/\partial m$ and $\partial S/\partial b$. Check that

$$\frac{\partial S}{\partial m} = -20[80 - (10m + b)] - 30[68 - (15m + b)] - 40[62 - (20m + b)]$$

$$\frac{\partial S}{\partial b} = -2[80 - (10m + b)] - 2[68 - (15m + b)] - 2[62 - (20m + b)].$$

Simplify each of these, obtaining

$$\frac{\partial S}{\partial m} = -6120 + 1450m + 90b$$

$$\frac{\partial S}{\partial b} = -420 + 90m + 6b.$$

Place $\partial S/\partial m = 0$ and $\partial S/\partial b = 0$ and simplify.

$$-6120 + 1450m + 90b = 0 \quad \text{or} \quad 145m + 9b = 612$$
$$-420 + 90m + 6b = 0 \quad \text{or} \quad 15m + b = 70$$

Solve the system on the right by multiplying both sides of the second equation by -9. The solution is $m = -1.8$, $b = 97$. The equation that best fits through the points of Figure 9.13 is $y = mx + b$, or

$$y = -1.8x + 97.$$

This **least squares regression equation** can be used to predict values of y from values of x. For example, if $x = 16$, then we could predict that

$$y = -1.8(16) + 97$$
$$y = 68.2.$$

Thus, 16 units of the product can be sold if the price is $68.20 per unit.

Work Problem 1 at the side.

Example 1 Find the best straight line through the data points $(x_1, y_1), (x_2, y_2), (x_3, y_3), \ldots, (x_n, y_n)$.

Let $y = mx + b$ be the equation of the line. Once again, we want to minimize the sum of the squares of the vertical distances

---

1. Predict y if
(a) $x = 18$
(b) $x = 22$.
Predict x if
(c) $y = 74$
(d) $y = 56$.

Answer:
(a) 64.6
(b) 57.4
(c) 12.8
(d) 22.8

from the points to the line. The distance of the first point from the line is given by

$$y_1 - (mx_1 + b);$$

in general, the distance from point (x_i, y_i) to the line is

$$y_i - (mx_i + b).$$

We want to minimize the sum

$$\sum_{i=1}^{n} [y_i - (mx_i + b)]^2.$$

Here we are using *sigma notation* ($\sum$ is the Greek letter sigma) to represent the sum.

Since the x_i and y_i values represent known data points, the unknowns in this sum are the numbers m and b. To emphasize this fact, we can write the sum as a function:

$$\begin{aligned} f(m, b) &= \sum_{i=1}^{n} [y_i - (mx_i + b)]^2 \\ &= \sum_{i=1}^{n} [y_i - mx_i - b]^2 \\ &= (y_1 - mx_1 - b)^2 + (y_2 - mx_2 - b)^2 \\ &\quad + \cdots + (y_n - mx_n - b)^2. \end{aligned}$$

To find the minimum value of this function, find the partial derivatives with respect to m and to b and place each equal to 0.

$$\frac{\partial f}{\partial m} = -2x_1(y_1 - mx_1 - b) - 2x_2(y_2 - mx_2 - b) - \cdots$$
$$- 2x_n(y_n - mx_n - b) = 0$$

$$\frac{\partial f}{\partial b} = -2(y_1 - mx_1 - b) - 2(y_2 - mx_2 - b) - \cdots$$
$$- 2(y_n - mx_n - b) = 0.$$

Using some algebra and rearranging terms, these two equations become

$$(x_1^2 + x_2^2 + \cdots + x_n^2)m + (x_1 + x_2 + \cdots + x_n)b$$
$$= x_1 y_1 + x_2 y_2 + \cdots + x_n y_n$$
$$(x_1 + x_2 + \cdots + x_n)m + nb = y_1 + y_2 + \cdots + y_n.$$

We can rewrite these last two equations using sigma notation as follows.

$$(\sum x^2)m + (\sum x)b = \sum xy \tag{1}$$

$$(\sum x)m + nb = \sum y \tag{2}$$

9.6 An Application—The Least Squares Line

To solve this system of equations, multiply the first on both sides by $-n$ and the second on both sides by $\sum x$. This gives

$$-n(\sum x^2)m - n(\sum x)b = -n(\sum xy)$$
$$(\sum x)(\sum x)m + (\sum x)nb = (\sum x)(\sum y)$$
$$\overline{(\sum x)(\sum x)m - n(\sum x^2)m = (\sum x)(\sum y) - n(\sum xy)}$$

Write the product $(\sum x)(\sum x)$ as $(\sum x)^2$. If we use this and solve the last equation for m, we get

$$m = \frac{(\sum x)(\sum y) - n(\sum xy)}{(\sum x)^2 - n(\sum x^2)}. \qquad (3)$$

The easiest way to find b is to solve equation (2) for b. ■

2. Solve equation (2) for b.

Answer: $b = \dfrac{\sum y - m(\sum x)}{n}$

Work Problem 2 at the side.

In practice, find m first, using equation (3). Next, use the result to find b from Problem 2 at the side. Then complete the equation $y = mx + b$.

Example 2 Find the least squares regression equation for the data shown in the scatter diagram of Figure 9.15.

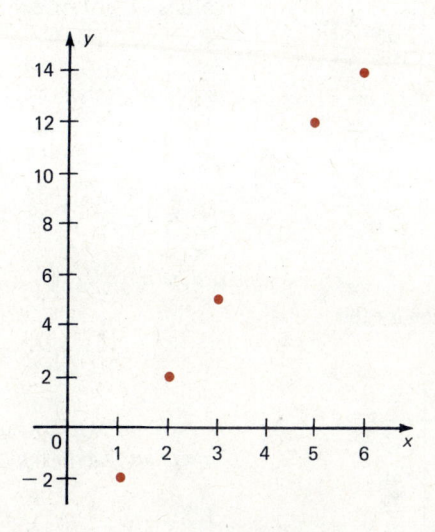

Figure 9.15

From the figure, we find the data values shown in the following chart. The chart also shows the numerical values needed for finding m and b.

x	y	x^2	xy
1	-2	1	-2
2	2	4	4
3	5	9	15
5	12	25	60
6	14	36	84
Totals 17	31	75	161

We have $\sum x = 17$, $\sum y = 31$, $\sum x^2 = 75$, $\sum xy = 161$, and $n = 5$. Using equation (3), we have

$$m = \frac{17(31) - 5(161)}{17^2 - 5(75)} = \frac{527 - 805}{289 - 375} = \frac{-278}{-86} = 3.2.$$

Now we can find b:

$$b = \frac{\sum y - m(\sum x)}{n} = \frac{31 - 3.2(17)}{5} = -4.7.$$

The least squares regression equation is $y = mx + b$, or

$$y = 3.2x - 4.7. \blacksquare$$

This equation can be used to predict values of y for given values of x. For example, if $x = 4$, we have

$$y = 3.2x - 4.7$$
$$= 3.2(4) - 4.7$$
$$= 12.8 - 4.7$$
$$y = 8.1.$$

Also, if $x = 12$, we have $y = 3.2(12) - 4.7 = 38.4 - 4.7 = 33.7$.

Work Problem 3 at the side.

9.6 EXERCISES

In each of the following exercises, draw a scatter diagram for the given set of data points and then find the least squares regression equation. See Example 2.

1.

x	1	2	3	5	9
y	9	13	18	25	41

2.

x	2	3	5	6	8
y	8	13	23	28	38

3. Find the least squares regression equation for the following data.

x	y
8	15
10	21
12	22
14	26
16	32
18	35

Answer: $y = 1.96x - .31$

3.
x	4	5	8	12	14
y	3	7	17	28	35

4.
x	3	4	5	6	8
y	8	12	16	18	28

APPLIED PROBLEMS

Education

5. The ACT test scores of eight students were compared to their GPA's after one year in college, with the following results.

ACT score, x	19	20	22	24	25	26	27	29
GPA, y	2.2	2.4	2.7	2.6	3.0	3.5	3.4	3.8

(a) Plot the eight points on a scatter diagram.
(b) Find the least squares regression equation and graph it on the scatter diagram of part (a).
(c) Using the results of (b), predict the GPA of a student with an ACT score of 28.

Sales

6. Records show that the annual sales for the Sweet Palms Life Insurance Company in 5-year periods for the last 20 years were as follows.

Year	Year in coded form, x	Sales in millions, y
1959	1	1.0
1964	2	1.3
1969	3	1.7
1974	4	1.9
1979	5	2.1

(a) Plot the five points on a scatter diagram.
(b) Find the least squares regression equation and graph it on the scatter diagram of part (a).
(c) Predict the company's sales for 1984.

Profit

7. A fast-food chain wishes to find the relationship between annual store sales and pretax profit in order to estimate increases in profit due to increased sales volume. The data shown below was obtained from a sample of stores across the country.

Annual store sales in $1000, x	250	300	375	425	450	475	500	575	600	650
Pretax percent profit, y	9.3	10.8	14.8	14.0	14.2	15.3	15.9	19.1	19.2	21.0

400 Functions of Several Variables

(a) Plot the ten pairs of values on a scatter diagram.
(b) Find the equation of the least squares regression line and draw it on the scatter diagram of part (a).
(c) Using the equation of part (b), predict the pretax percent profit for annual sales of $700,000; of $750,000.

CHAPTER 9 SUMMARY

Key Words

function of two variables
independent variable
dependent variable
ordered triple
first octant
plane
sphere
cylinder
partial derivative

second partial derivative
relative maximum or minimum
saddle point
constraint
Lagrange multipliers
Lagrange function
scatter diagram
least squares regression equation
sigma notation

Things to Remember

Theorem 9.1 Let a function $z = f(x, y)$ have a relative maximum or relative minimum at the point (a, b). Let $f_x(a, b)$ and $f_y(a, b)$ both exist. Then

$$f_x(a, b) = 0 \quad \text{and} \quad f_y(a, b) = 0.$$

Theorem 9.2 (*M-test for relative maximums and minimums*) Let all the following partial derivatives exist for a function $z = f(x, y)$. Let (a, b) be a point for which

$$f_x(a, b) = 0 \quad \text{and} \quad f_y(a, b) = 0.$$

Define the number M as

$$M = f_{xx}(a, b) \cdot f_{yy}(a, b) - [f_{xy}(a, b)]^2.$$

Then
(a) $f(a, b)$ is a relative maximum if $M > 0$ and $f_{xx}(a, b) < 0$.
(b) $f(a, b)$ is a relative minimum if $M > 0$ and $f_{xx}(a, b) > 0$.
(c) $f(a, b)$ is a saddle point (neither a maximum nor a minimum) if $M < 0$.
(d) If $M = 0$, the test gives no information.

Theorem 9.3 (*Lagrange multipliers*) Any relative maximum or relative minimum values of the function $z = f(x, y)$, subject to a constraint $g(x, y) = 0$, will be found among those points (x, y) for which there exists a value of λ such that

$$F_x(x, y, \lambda) = 0$$
$$F_y(x, y, \lambda) = 0$$
$$F_\lambda(x, y, \lambda) = 0,$$

where

$$F(x, y, \lambda) = f(x, y) + \lambda \cdot g(x, y).$$

We assume that all indicated partial derivatives exist.

CHAPTER 9 TEST

[9.1] Let $f(x, y) = 9x^2 - 2xy + 4y^2 - 8$. Find the following.

1. $f(-4, 8)$
2. $f(1, -6)$

[9.2] Graph the following. Identify each graph.

3. $2x + 3y + 4z = 12$
4. $x + 2z = 10$
5. $(x + 1)^2 + (y - 3)^2 + (z - 4)^2 = 9$
6. $x^2 + z^2 = 49$

[9.3] Find f_x and f_y for the following. Then find $f_x(-2, 5)$ and $f_y(3, 6)$.

7. $f(x, y) = 9x^3 + 2y^2 - 4xy$
8. $f(x, y) = \dfrac{x + 1}{y + 1}$

Find f_{xx}, f_{yy}, and f_{xy} for the following.

9. $f(x, y) = x^2 y^3 + 8xy$
10. $f(x, y) = \dfrac{y}{1 + x}$

[9.4] Find all points where the following functions have maximums or minimums.

11. $f(x, y) = x^2 + 5xy - 10x + 3y^2 - 12y$
12. $z = x^3 - 8y^2 + 6xy + 4$

The total cost to manufacture x units of doggy dishes and y units of kitty plates is given by
$$C(x, y) = x^2 + 5y^2 + 4xy - 70x - 164y + 1800.$$

13. Find values of x and y that lead to a minimum cost.

14. What is the minimum cost?

[9.5]

15. Find the dimensions of the rectangular field of maximum area that can be enclosed with 400 meters of fence.

[9.6]

The following data shows the connection between blood-sugar levels, x, and cholesterol levels, y, for eight different patients.

patient	1	2	3	4	5	6	7	8
blood-sugar level, x	130	138	142	159	165	200	210	250
cholesterol level, y	170	160	173	181	201	192	240	290

For this data, $\sum x = 1394$, $\sum y = 1607$, $\sum xy = 291,990$, $\sum x^2 = 255,214$, and $\sum y^2 = 336,155$.

16. Find the equation of the least squares regression line $y = mx + b$.

17. Predict the cholesterol level for a person whose blood-sugar level is 190.

10
The Trigonometric Functions

We have discussed many different types of functions, including linear, quadratic, exponential, and logarithmic, throughout this book. In this chapter we introduce the trigonometric functions, which differ in a fundamental way from those previously studied. The trigonometric functions describe a *periodic* or *repetitive* relationship.

There are many practical applications of the trigonometric functions: they describe many natural phenomena and are important in the study of optics, heat, electronics, business, X ray, acoustics, seismology, fluctuating populations, and so on. We shall also see that many algebraic functions have antiderivatives involving trigonometric functions.

10.1 TRIGONOMETRIC FUNCTIONS

One of the basic ideas of trigonometry is the angle, which we define in this section. Figure 10.1 shows a line through the two points A and B. The portion of line AB that starts at A and continues through B and on past B is called **ray** AB. Point A is the endpoint of the ray.

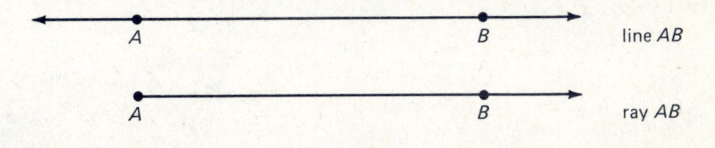

```
●─────────────────────●────────────▶    line AB
A                     B

●─────────────────────●────────────▶    ray AB
A                     B
```

Figure 10.1

An **angle** is formed by rotating a ray about its endpoint. The initial position of the ray is called the **initial side** of the angle and the endpoint of the ray is called the **vertex** of the angle. The location of the ray at the end of its rotation is called the **terminal side** of the angle. Figure 10.2 shows the initial and terminal sides of an angle with vertex A.

If the rotation of an angle is counterclockwise, the angle is *positive*. If the rotation is clockwise, the angle is *negative*. Figure 10.3 shows two angles, one positive and one negative.

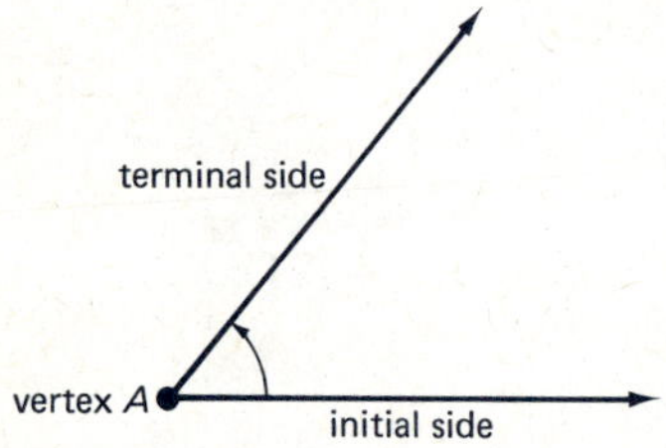

Figure 10.2

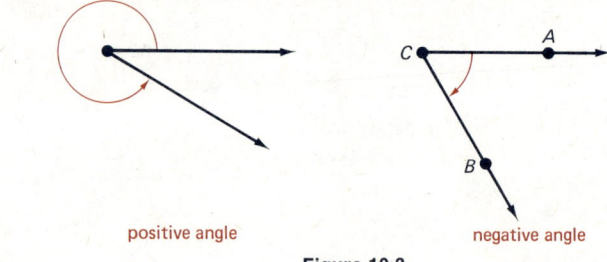

positive angle   negative angle
Figure 10.3

An angle can be named by its vertex. For example, the angle on the right in Figure 10.3 can be called angle C. Also, an angle can be named by using three letters. For example, the angle on the right could also be named angle ACB or angle BCA. (Put the vertex in the middle of the three letters.)

There are two common systems used for measuring the size of angles. The most common unit of measure is the degree. Degree measure has remained unchanged since the Babylonians developed it 4000 years ago. To use degree measure, we assign 360 degrees to the complete rotation of a ray. See Figure 10.4.

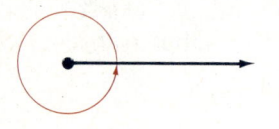

One degree, written $1°$, represents 1/360 of a rotation. Also, $90°$ represents 90/360 or 1/4 of a rotation, and $180°$ represents 180/360 or 1/2 of a rotation. One sixtieth of a degree is called a **minute.** The measure $12°48'$ represents 12 degrees and 48 minutes.

A complete rotation of a ray gives a 360 degree angle.
Figure 10.4

An angle having degree measure between $0°$ and $90°$ is called an **acute angle.** An angle whose measure is $90°$ is a **right angle.** An angle having measure more than $90°$ and less than $180°$ is an **obtuse angle,** while an angle of $180°$ is a **straight angle.**

An angle is in **standard position** if its vertex is at the origin of a coordinate system and its initial side is along the positive x-axis. The two angles of Figure 10.5 are in standard position.

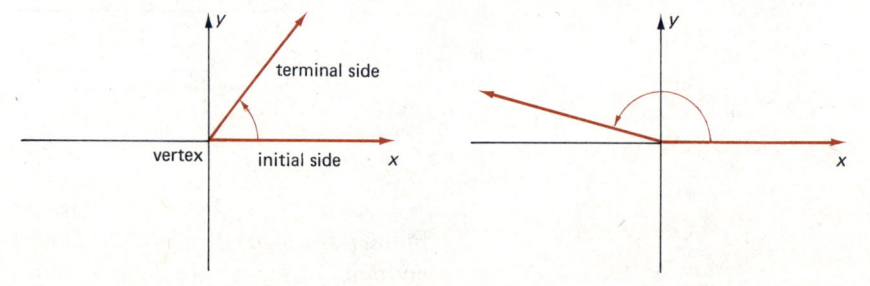

Figure 10.5

An angle in standard position is said to lie in the quadrant where its terminal side lies. For example, an acute angle in standard position is in Quadrant I, while an obtuse angle is in Quadrant II.

Example 1 Find the quadrants for the angles of Figure 10.5.
The angle on the left in Figure 10.5 is in Quadrant I, while the angle on the right is in Quadrant II. ∎

Work Problem 1 at the side.

In most work in applied trigonometry, angles are measured in degrees and minutes. However, in more advanced work in trigonometry, and especially in calculus, the use of degree measure for angles makes many formulas very complicated. These formulas can be simplified if we use another system of angle measure called radian measure.

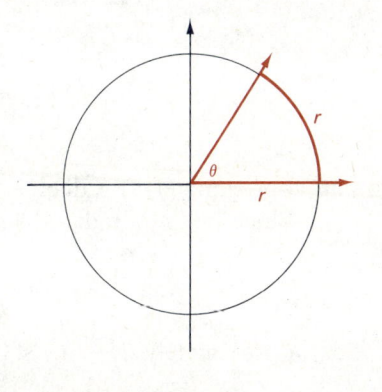

The length of the arc cut by θ is equal to the radius of the circle. Therefore, θ has a measure of 1 radian.

Figure 10.6

Figure 10.6 shows an angle θ (the Greek letter theta). The angle θ is in standard position; a circle of radius r is also shown in Figure 10.6. The vertex of θ is at the center of the circle. Angle θ cuts a piece of the circle called an **arc** of the circle. The length of this arc is equal to the radius of the circle. Because of this, we say that angle θ has a measure of one radian. **One radian** is the measure of an angle whose vertex is at the center of a circle and which cuts an arc on the circle equal in length to the radius of the circle.

To generalize this definition, we say that an angle of measure two radians cuts an arc equal in length to twice the radius of the circle. An angle of measure one-half radian cuts an arc equal in length to half the radius of the circle, and so on.

---

1. Find the quadrants for the angles shown here.
(a) 225°
(b) −47°

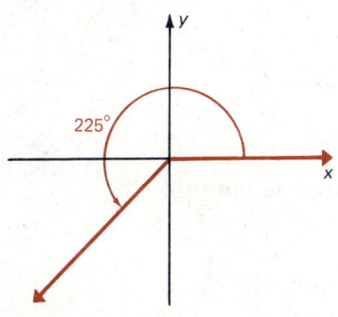

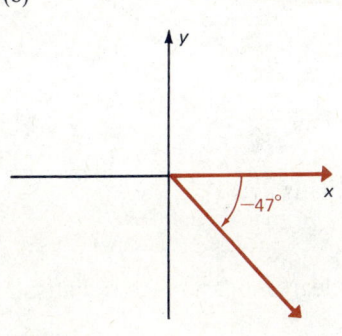

Answer:
(a) III
(b) IV

The circumference of a circle is the distance around the circle. It is given by the formula $C = 2\pi r$, where r is the radius of the circle and π is a constant, the irrational number approximately equal to 3.14159. This formula shows that the radius of the circle can be laid off 2π times around a circle. Therefore, an angle of 360°, that is, a complete circle, cuts off an arc equal in length to 2π times the radius of the circle. Therefore, an angle of 360° has a measure of 2π radians, or

$$360° = 2\pi \text{ radians.}$$

This result gives a basis for comparing degree and radian measure.

We know that an angle of 180° is half the size of an angle of 360°. Therefore, an angle of 180° would have half the radian measure of an angle of 360°, or

$$180° = \frac{1}{2}(2\pi) \text{ radians}$$

180° = π radians.

Since π radians = 180°, we can divide both sides by π to find that

$$1 \text{ radian} = \frac{180°}{\pi}$$

or, approximately, 1 radian = 57°18′.

Since 180° = π radians, we have

$$1° = \frac{\pi}{180} \text{ radians}$$

or, approximately, 1° = .0174533 radians.

Example 2 Convert the following degree measures to radians.
(a) 45°
Since 1° = $\pi/180$ radians, we have

$$45° = 45\left(\frac{\pi}{180}\right) \text{ radians} = \frac{45\pi}{180} \text{ radians} = \frac{\pi}{4} \text{ radians.}$$

The word *radian* is often omitted, so we could write just 45° = $\pi/4$.

(b) $240° = 240\left(\dfrac{\pi}{180}\right) = \dfrac{4\pi}{3}$ ∎

Work Problem 2 at the side.

---

2. Convert the following degree measures to radians.
(a) 60°
(b) 120°
(c) 225°
(d) 360°

Answer:

(a) $\dfrac{\pi}{3}$

(b) $\dfrac{2\pi}{3}$

(c) $\dfrac{5}{4}\pi$

(d) 2π

Example 3 Convert the following radian measures to degrees.

(a) $\dfrac{9\pi}{4}$

We know that 1 radian $= \dfrac{180°}{\pi}$. Thus,

$$\dfrac{9\pi}{4} \text{ radians} = \dfrac{9\pi}{4}\left(\dfrac{180°}{\pi}\right) = 405°$$

(b) $\dfrac{11\pi}{3}$ radians $= \dfrac{11\pi}{3}\left(\dfrac{180°}{\pi}\right) = 660°$ ■

Work Problem 3 at the side.

3. Convert the following radian measures to degrees.
   (a) $3\pi/4$
   (b) $5\pi/6$
   (c) $5\pi/3$
   (d) 7π

   Answer:
   (a) 135°
   (b) 150°
   (c) 300°
   (d) 1260°

We use the idea of an angle as we define the six basic trigonometric functions. To define these functions, assume that we have an angle θ in standard position as shown in Figure 10.7. Choose any point P, having coordinates (x, y), on the terminal side of angle θ. (The point P must not be the vertex of θ.)

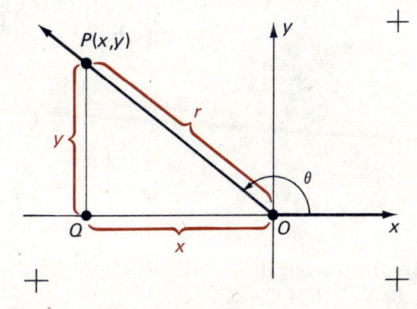

Figure 10.7

A perpendicular from P to the x-axis at point Q sets up a right triangle having vertices at O, P, and Q. The distance from P to O is called the **radius vector**, which we abbreviate as r. Since distance is never negative, we have $r > 0$. The six **trigonometric functions** of angle θ are defined as follows.

$$\text{sine } \theta = \sin \theta = \dfrac{y}{r} \qquad \text{cotangent } \theta = \cot \theta = \dfrac{x}{y}$$

$$\text{cosine } \theta = \cos \theta = \dfrac{x}{r} \qquad \text{secant } \theta = \sec \theta = \dfrac{r}{x}$$

$$\text{tangent } \theta = \tan \theta = \dfrac{y}{x} \qquad \text{cosecant } \theta = \csc \theta = \dfrac{r}{y}$$

408   The Trigonometric Functions

Example 4   The terminal side of an angle α (the Greek letter alpha) goes through the point (8, 15). Find the values of the six trigonometric functions of angle α.

Figure 10.8 shows angle α and the triangle formed by dropping a perpendicular from the point (8, 15). To reach the point (8, 15), we begin at the origin and go to the right 8 units and then turn and go up 15 units. Therefore, $x = 8$ and $y = 15$. To find the radius vector r, we use the **Pythagorean theorem:** In a triangle with a right angle, if the longest side of the triangle is r and the shorter sides are x and y, then

$$r^2 = x^2 + y^2,$$

or

$$r = \sqrt{x^2 + y^2}.$$

[Recall that $\sqrt{a}$ represents the *positive* square root of a.] Now substitute the known values, $x = 8$ and $y = 15$.

$$r = \sqrt{8^2 + 15^2} = \sqrt{64 + 225} = \sqrt{289}$$

From the square root table in the back of the book,

$$r = 17.$$

Finally, $x = 8$, $y = 15$, and $r = 17$. The values of the six trigonometric functions of angle α are now found by using the definitions given above.

$$\sin \alpha = \frac{y}{r} = \frac{15}{17} \qquad \tan \alpha = \frac{y}{x} = \frac{15}{8} \qquad \sec \alpha = \frac{r}{x} = \frac{17}{8}$$

$$\cos \alpha = \frac{x}{r} = \frac{8}{17} \qquad \cot \alpha = \frac{x}{y} = \frac{8}{15} \qquad \csc \alpha = \frac{r}{y} = \frac{17}{15} \qquad ■$$

Work Problem 4 at the side.

To find the six trigonometric functions, we can choose *any* point on the terminal side of the angle except the origin. To see why any point may be used, look at Figure 10.9, which shows an angle θ and two points on its terminal side. Point P has coordinates (x, y) and point P' has coordinates (x', y'). Let r be the length of the *hypotenuse* (longest side) of triangle OPQ, and let r' be the length of the hypotenuse of triangle $OP'Q'$. By similar triangles from geometry, we have

$$\sin \theta = \frac{y}{r} = \frac{y'}{r'},$$

so that $\sin \theta$ is the same no matter which point is used to find it. A similar result holds for the other five functions.

---

4. The terminal side of angle β (beta) goes through $(-3, 4)$. Find the values of the six trigonometric functions of angle β.

Answer: $\sin \beta = \frac{4}{5}$, $\cos \beta = -\frac{3}{5}$, $\tan \beta = -\frac{4}{3}$, $\cot \beta = -\frac{3}{4}$, $\sec \beta = -\frac{5}{3}$, $\csc \beta = \frac{5}{4}$

10.1 Trigonometric Functions

Figure 10.8

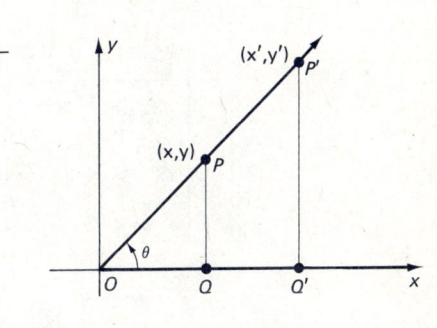

$x = 8$
$y = 15$
$r = 17$

Figure 10.9

Example 5 Find the values of the six trigonometric functions for an angle of 90°.

First, select any point on the terminal side of a 90° angle. Let us select the point (0, 1), as shown in Figure 10.10. Here $x = 0$ and $y = 1$. Check that $r = 1$ also. Then

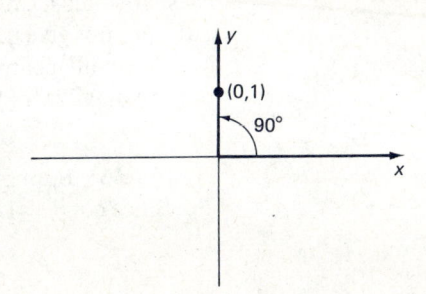

$$\sin 90° = \frac{1}{1} = 1 \qquad \cot 90° = \frac{0}{1} = 0$$

$$\cos 90° = \frac{0}{1} = 0 \qquad \sec 90° = \frac{1}{0} \text{ (undefined)}$$

$$\tan 90° = \frac{1}{0} \text{ (undefined)} \qquad \csc 90° = \frac{1}{1} = 1. \blacksquare$$

Figure 10.10

5. Find the values of the six trigonometric functions for an angle of 180°; of 360°.

Answer: $\sin 180° = 0$, $\cos 180° = -1$, $\tan 180° = 0$, $\cot 180°$ is undefined, $\sec 180° = -1$, $\csc 180°$ is undefined; $\sin 360° = 0$, $\cos 360° = 1$, $\tan 360° = 0$, $\cot 360°$ is undefined, $\sec 360° = 1$, $\csc 360°$ is undefined

Work Problem 5 at the side.

In the same way, we could find the values of the six trigonometric functions for the angles 0° and 270°. These results are summarized in the following table. Note that the results for 360° are the same as for 0°.

θ	$\sin\theta$	$\cos\theta$	$\tan\theta$	$\cot\theta$	$\sec\theta$	$\csc\theta$
0°	0	1	0	undefined	1	undefined
90°	1	0	undefined	0	undefined	1
180°	0	−1	0	undefined	−1	undefined
270°	−1	0	undefined	0	undefined	−1
360°	0	1	0	undefined	1	undefined

Several important properties of the trigonometric functions can be obtained from the definitions of the functions. First recall that the *reciprocal* of a number a is the number $1/a$. (There is no reciprocal for 0.) The numbers y/r and r/y are thus reciprocals of each other. Looking back at the definitions of the trigonometric functions, we see that y/r is the definition of $\sin\theta$ and r/y is the definition of $\csc\theta$. Thus, $\sin\theta$ and $\csc\theta$ are reciprocals of each other, and

$$\sin\theta = \frac{1}{\csc\theta} \qquad \csc\theta = \frac{1}{\sin\theta}$$

Also, $\cos\theta$ and $\sec\theta$ are reciprocals, as are $\tan\theta$ and $\cot\theta$.

$$\cos\theta = \frac{1}{\sec\theta} \qquad \sec\theta = \frac{1}{\cos\theta}$$

$$\tan\theta = \frac{1}{\cot\theta} \qquad \cot\theta = \frac{1}{\tan\theta}$$

These formulas, called the **reciprocal identities,** hold for any angle θ that does not give a zero denominator. (**Identities** are equations that are true for all meaningful values of the variables. We study identities in more detail later.)

Example 6 Suppose $\tan\theta = 3/4$. Find $\cot\theta$.
Since $\cot\theta = 1/\tan\theta$, we have

$$\cot\theta = \frac{1}{\frac{3}{4}} = \frac{4}{3}. \qquad \blacksquare$$

Work Problem 6 at the side.

6. Find the following.
(a) $\cos\alpha$ if $\sec\alpha = \frac{7}{4}$
(b) $\csc\theta$ if $\sin\theta = \frac{6}{11}$

Answer:
(a) $\cos\alpha = \frac{4}{7}$
(b) $\csc\theta = \frac{11}{6}$

10.1 EXERCISES

Convert the following degree measures to radians. Leave answers as multiples of π. See Example 2.

1. 60°
2. 90°
3. 150°
4. 135°
5. 210°
6. 300°
7. 390°
8. 480°

Convert the following radian measures to degrees. See Example 3.

9. $7\pi/4$
10. $2\pi/3$
11. $11\pi/6$
12. $-\pi/4$
13. $8\pi/5$
14. $7\pi/10$
15. $4\pi/15$
16. 5π

Find the values of the six trigonometric functions for the angles in standard position having the following points on their terminal sides. See Example 4.

17. $(-3, 4)$
18. $(-12, -5)$
19. $(6, 8)$
20. $(-7, 24)$
21. $(0, 2)$
22. $(-4, 0)$
23. $(-2\sqrt{3}, -2)$
24. $(8, -8\sqrt{3})$

Evaluate each of the following.

Example Find $\sin 90° + 4 \cot 270° + 3 \sec^2 0°$.

Here $\sec^2 0°$ means $(\sec 0°)^2$. Using the table of values of trigonometric functions given in the text, we have

$$\sin 90° + 4 \cot 270° + 3 \sec^2 0° = 1 + 4(0) + 3(1)^2 = 4. \quad \blacksquare$$

25. $\cos 90° + 3 \sin 270°$
26. $3 \sec 180° - 5 \tan 360°$
27. $\tan 360° + 4 \sin 180° + 5 \cos^2 180°$
28. $2 \sec 0° + 4 \cot^2 90° + \cos 360°$
29. $\sin^2 180° + \cos^2 180°$
30. $\sin^2 360° + \cos^2 360°$
31. $\sec^2 180° - 3 \sin^2 360° + 2 \cos 180°$
32. $5 \sin^2 90° + 2 \cos^2 270° - 7 \tan^2 360°$
33. $-4|\sin 90°| + 3|\cos 180°| + 2|\csc 270°|$
34. $-|\cos 270°| - 2|\sin 90°| + 5|\cos 180°|$

Find the sine of the following angles. See Example 6.

35. $\csc \alpha = 3$
36. $\csc \theta = -4$
37. $\csc \theta = \sqrt{5}$
38. $\csc \beta = \sqrt{12}$

In Quadrant I, x, y, and r are all positive, so that all six trigonometric functions have positive values. In Quadrant II, x is negative and y positive (r is always positive). Thus, in Quadrant II, sine is positive, cosine is negative, and so on. Complete the following table of values for the signs of the trigonometric functions.

	θ in quadrant	$\sin\theta$	$\cos\theta$	$\tan\theta$	$\cot\theta$	$\sec\theta$	$\csc\theta$
39.	I						
40.	II						
41.	III						
42.	IV						

Identify the quadrant or quadrants for the angles satisfying the following conditions.

43. $\sin\alpha > 0, \cos\alpha < 0$
44. $\cos\beta > 0, \tan\beta > 0$
45. $\sec\theta < 0, \csc\theta < 0$
46. $\tan\alpha > 0, \cot\alpha > 0$
47. $\sin\beta < 0, \cos\beta > 0$
48. $\cos\beta > 0, \sin\beta > 0$
49. $\sin\alpha > 0$
50. $\cos\theta < 0$
51. $\tan\theta > 0$
52. $\csc\alpha < 0$

10.2 FINDING VALUES OF TRIGONOMETRIC FUNCTIONS

In this section we shall see how to draw graphs of the basic trigonometric functions. In order to do this, we will need to know how to evaluate the trigonometric functions of common angles, such as 30°, 45°, 60°, and so on. The trigonometric functions of 30° and 60° are found by starting with a 30°–60° right triangle. To obtain a 30°–60° right triangle, we begin with an **equilateral triangle,** a triangle with all sides equal in length. Each angle of such a triangle has a measure of 60°. See Figure 10.11.

(a) equilateral triangle
(b) 30° – 60° right triangles

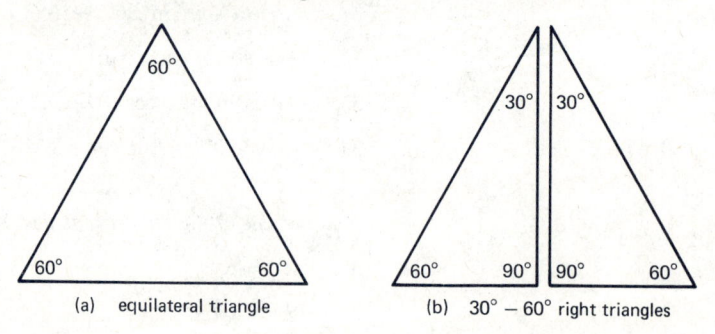

Figure 10.11

If we bisect one angle of an equilateral triangle we obtain two right triangles, each of which has angles of 30°, 60°, and 90°, as shown in Figure 10.11(b). If the hypotenuse of one of these right triangles has a length of 2, then the shortest side will have a length of 1. (Why?) If we use x to represent the length of the medium side, then we can use the Pythagorean theorem to find x.

$$2^2 = 1^2 + x^2$$
$$4 = 1 + x^2$$
$$3 = x^2$$
$$\sqrt{3} = x$$

The length of the medium side is thus $\sqrt{3}$. In summary, in a 30°–60° right triangle, the hypotenuse is always twice as long as the shortest side, and the medium side has a length which is $\sqrt{3}$ times as long as that of the shortest side. Also, the shortest side is opposite the 30° angle.

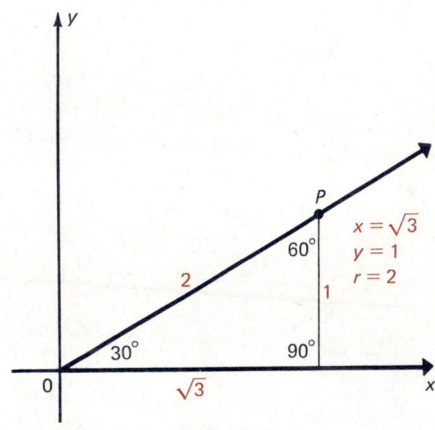

Figure 10.12

Now we can find the trigonometric functions for an angle of 30°. To do so, place a 30° angle in standard position, as shown in Figure 10.12. Choose a point P on the terminal side of the angle so that $r = 2$. By the work above, P will have coordinates $(\sqrt{3}, 1)$. Thus, $x = \sqrt{3}$, $y = 1$, and $r = 2$. By the definitions of the trigonometric functions, we have

$$\sin 30° = \frac{1}{2} \qquad \tan 30° = \frac{\sqrt{3}}{3} \qquad \sec 30° = \frac{2\sqrt{3}}{3}$$

$$\cos 30° = \frac{\sqrt{3}}{2} \qquad \cot 30° = \sqrt{3} \qquad \csc 30° = 2.$$

To find the values of $\tan 30°$ and $\sec 30°$, we rationalized the denominators

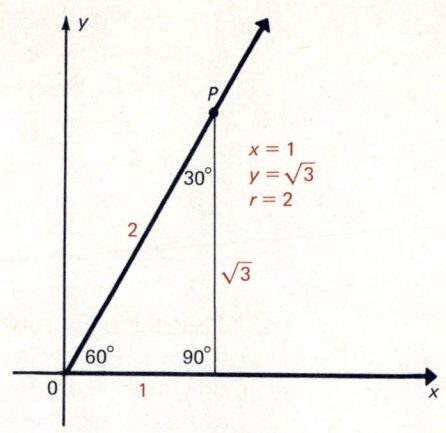

414   The Trigonometric Functions

Figure 10.13

Example 1   Find the values of the trigonometric functions for an angle of 60°.

To find the values of the trigonometric functions for 60°, place a 60° angle in standard position, as in Figure 10.13. Choose point P on the terminal side of the angle so that $r = 2$. Then, by the work above, $x = 1$, $y = \sqrt{3}$, and $r = 2$. Thus

$$\sin 60° = \frac{\sqrt{3}}{2} \qquad \tan 60° = \sqrt{3} \qquad \sec 60° = 2$$

$$\cos 60° = \frac{1}{2} \qquad \cot 60° = \frac{\sqrt{3}}{3} \qquad \csc 60° = \frac{2\sqrt{3}}{3}. \blacksquare$$

Work Problem 1 at the side.

To find the values of the trigonometric functions for 45°, let us start with a 45°–45° right triangle, as shown in Figure 10.14. Such a triangle is **isosceles** and thus has two sides of equal length.

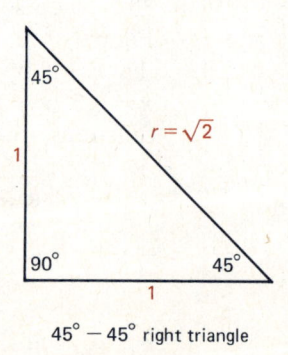

1. Find the values of the trigonometric functions for an angle of 210°. Use the sketch.

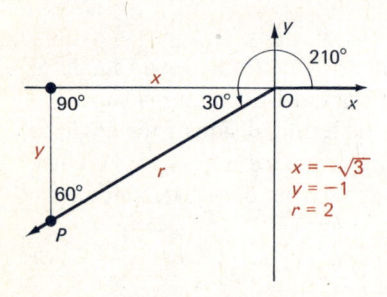

Answer: $\sin 210° = -1/2$, $\cos 210° = -\sqrt{3}/2$, $\tan 210° = \sqrt{3}/3$, $\cot 210° = \sqrt{3}$, $\sec 210° = -2\sqrt{3}/3$, $\csc 210° = -2$

45° – 45° right triangle

Figure 10.14

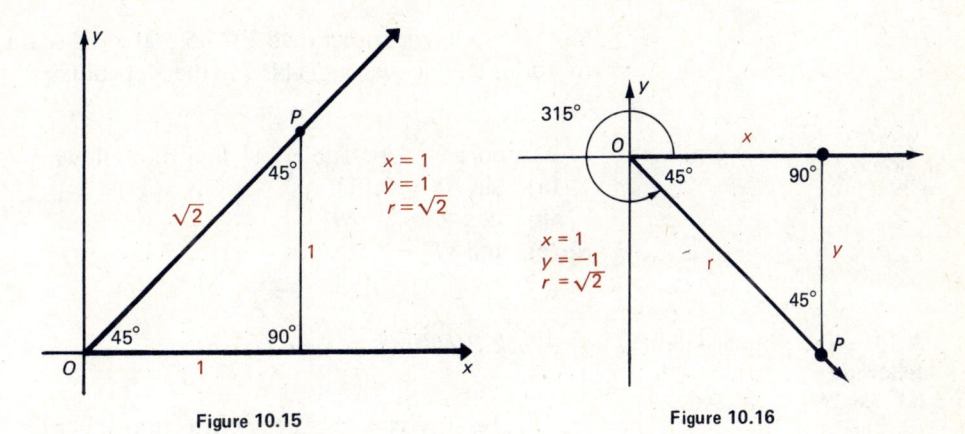

10.2 Finding Values of Trigonometric Functions

Figure 10.15

Figure 10.16

If we let the shorter sides each have length 1 and if r represents the length of the hypotenuse, we have

$$1^2 + 1^2 = r^2$$
$$2 = r^2$$
$$\sqrt{2} = r.$$

In general, in a 45°–45° right triangle, the hypotenuse has a length which is $\sqrt{2}$ times as long as the length of either of the shorter sides. To find the values of the trigonometric functions for 45°, place a 45° angle in standard position, as in Figure 10.15. Choose point P on the terminal side of the angle so that $r = \sqrt{2}$. Then the coordinates of P are $(1, 1)$. Hence, $x = 1$, $y = 1$, and $r = \sqrt{2}$, so that

$$\sin 45° = \frac{\sqrt{2}}{2} \qquad \tan 45° = 1 \qquad \sec 45° = \sqrt{2}$$

$$\cos 45° = \frac{\sqrt{2}}{2} \qquad \cot 45° = 1 \qquad \csc 45° = \sqrt{2}.$$

To find the values of sin 45° and cos 45°, we rationalized the denominators.

Example 2 Find the values of the trigonometric functions for an angle of 315°.

Figure 10.16 shows an angle of measure 315° in standard position. Choose point P on the terminal side of the angle so that $r = \sqrt{2}$. The coordinates of P become $(1, -1)$, and

$$\sin 315° = -\frac{\sqrt{2}}{2} \qquad \tan 315° = -1 \qquad \sec 315° = \sqrt{2}$$

$$\cos 315° = \frac{\sqrt{2}}{2} \qquad \cot 315° = -1 \qquad \csc 315° = -\sqrt{2}. \blacksquare$$

Work Problem 2 at the side.

2. Find the values of the trigonometric functions for an angle of 135°.

Answer: $\sin 135° = \sqrt{2}/2$, $\cos 135° = -\sqrt{2}/2$, $\tan 135° = -1$, $\cot 135° = -1$, $\sec 135° = -\sqrt{2}$, $\csc 135° = \sqrt{2}$

For angles other than 30°, 45°, 60°, and so on, we use a calculator or a table such as Table 7 in the Appendix.

Example 3 Use Table 7 to find the following.
(a) $\sin 10° = 0.1736$
(b) $\cos 48° = 0.6691$
(c) $\tan 82° = 7.1154$ ∎

3. Use Table 7 to find the following.
(a) $\cos 26°$
(b) $\sin 74°$
(c) $\tan 16°$

Answer:
(a) .8988
(b) .9613
(c) .2867

Work Problem 3 at the side.

Table 7 can also be used to find the value of trigonometric functions for angles measured in radians. Just use the "Radians" column of the table.

Example 4 (a) $\sin .2618 = 0.2588$
(b) $\tan 1.2043 = 2.6051$
(c) $\sec .7679 = 0.7193$ ∎

4. Use Table 7 to find the following.
(a) $\cos .2967$
(b) $\sin 1.0647$
(c) $\tan .8029$

Answer:
(a) .9563
(b) .8746
(c) 1.0355

Work Problem 4 at the side.

Now we can draw graphs of the trigonometric functions. These functions have a very special property: they are **periodic,** that is, their graphs repeat indefinitely. In general, a function $y = f(x)$ is periodic if there exists a positive real number a such that

$$f(x) = f(x + a)$$

for all values of x. The smallest possible value of a is called the **period** of the function.

For example, we know that $\sin 0 = 0$ and $\sin 2\pi = 0$, so that

$$\sin 0 = \sin 2\pi = \sin(0 + 2\pi).$$

Also, $$\sin \frac{\pi}{2} = \sin\left(\frac{\pi}{2} + 2\pi\right)$$

and $$\sin \pi = \sin(\pi + 2\pi).$$

In fact, $\sin x = \sin(x + 2\pi)$ for any real number x. Thus, $y = \sin x$ is periodic with period 2π.

The sine function, $y = \sin x$, has the set of all real numbers as its domain, while its range is restricted to $-1 \leq y \leq 1$. Since the sine function has period 2π, we can sketch the graph of the sine function by concentrating on values of x between 0 and 2π. We can then repeat this portion of the graph as many times as necessary.

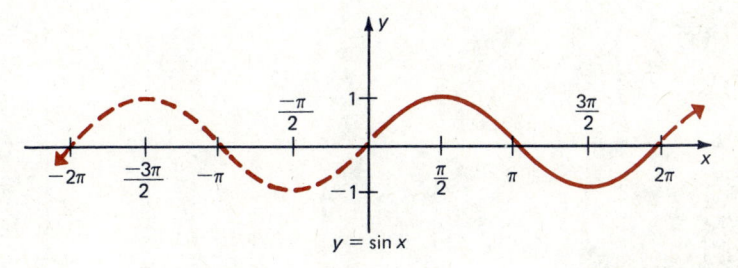

Figure 10.17 $y = \sin x$

From Table 7, we see that sin x gradually increases from 0 to 1 as x increases from 0 to $\pi/2$. The values of sin x then decrease back to 0 as x goes from $\pi/2$ to π. For $\pi < x < 2\pi$, sin x is negative. These facts are summarized in the following table, where decimals have been rounded to the nearest tenth.

x	0	$\pi/4$	$\pi/2$	$3\pi/4$	π	$5\pi/4$	$3\pi/2$	$7\pi/4$	2π
sin x	0	.7	1	.7	0	$-.7$	-1	$-.7$	0

Plotting the points from the table of values and connecting them with a smooth line, we get the solid portion of the graph in Figure 10.17. Since $y = \sin x$ is periodic and has the set of all real numbers as its domain, the graph continues in both directions indefinitely, as shown with dashed lines.

The graph of $y = \cos x$ in Figure 10.18 can be found in much the same way.

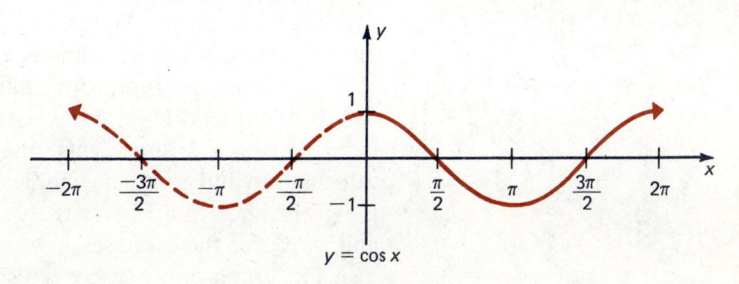

Figure 10.18 $y = \cos x$

The maximum distance of the graph of $y = \sin x$ or $y = \cos x$ from the x-axis is called the **amplitude**. The graphs of sin x and cos x in Figures 10.17 and 10.18 show that each has an amplitude of 1.

The period of $y = \tan x$ is π, so we investigate the tangent function only within an interval of π units. A convenient interval for this purpose is $-\pi/2 < x < \pi/2$. Although the endpoints $-\pi/2$ and $\pi/2$ are not in the domain of $y = \tan x$ (Why?), tan x exists for all other values in the interval.

$y = \tan x$

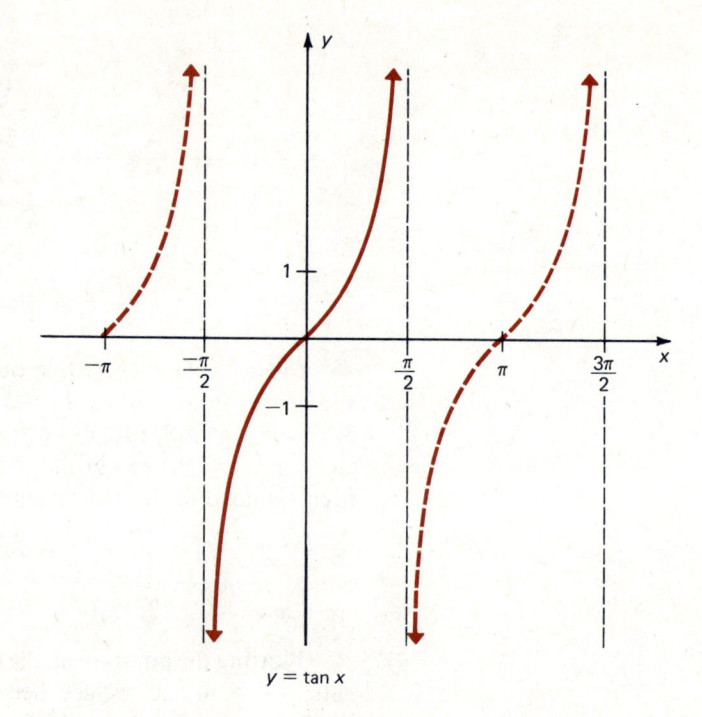

Figure 10.19

There is no point on the graph of $y = \tan x$ at $x = -\pi/2$ or at $x = \pi/2$. However, in the interval $0 < x < \pi/2$, $\tan x$ is positive. As x goes from 0 to $\pi/2$, we see from Table 7 that $\tan x$ gets larger and larger. The graph approaches the vertical line $x = \pi/2$ but never touches it. This line is a vertical asymptote. We show the asymptotes on the graph with a faint line. See Figure 10.19.

In the interval $-\pi/2 < x < 0$, $\tan x$ is negative, and as x goes from 0 to $-\pi/2$, $\tan x$ gets smaller and smaller. If we use this information and complete a table of values for $y = \tan x$, we can get the graph shown in Figure 10.19. The graph of $y = \cot x$ is similar and is left for the exercises.

The graph of $y = \csc x$ is restricted to values of x that are not integer multiples of π. (Why?) Thus, the lines $x = k\pi$ are vertical asymptotes. Since $\csc x = 1/\sin x$, the period is 2π, the same as for $\sin x$. When $\sin x = 1$, we have $\csc x = 1$, and when $0 < \sin x < 1$, then $\csc x > 1$. Thus, the graph takes the shape of the solid line shown in Figure 10.20. To show how the two curves are related, the graph of $y = \sin x$ is also shown, as a dashed curve. The graph of $y = \sec x$, which is related to the graph of $\cos x$ in a similar way, is left for the exercises.

We close this section with some applications of the graphs of trigonometric functions.

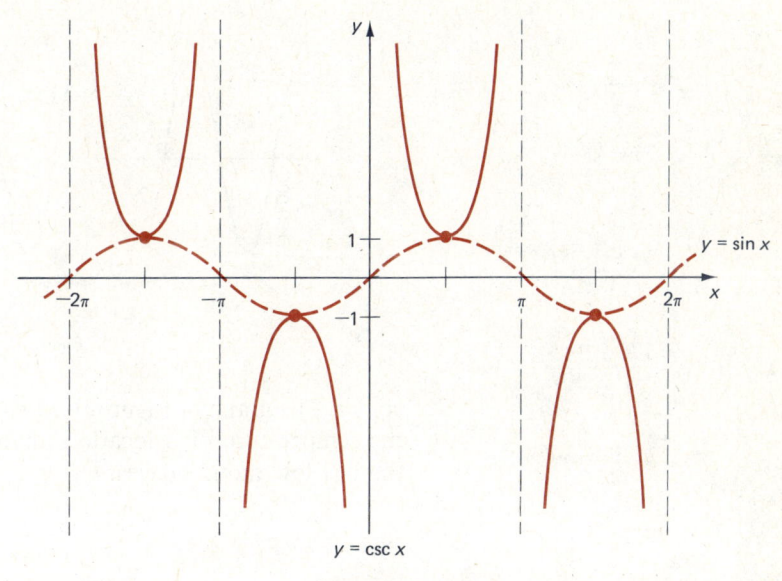

Figure 10.20

Sound waves A sound wave is made up of vibrations. Our eardrum picks up these vibrations and transfers them to our brain as electrical impulses. An *oscilloscope* is an electronic instrument with a televisionlike screen that converts sound waves into electrical impulses. Everyday sounds make very complex patterns on the screen of an oscilloscope. For example, the photograph on the left is an oscilloscope pattern of Helen Reddy singing "I am Woman." On the right is an oscilloscope photograph of Roy Clark singing "Thank God and Greyhound She's Gone." While the sounds appear very complex in the photographs, it is shown in more advanced courses that they are composed of sums of a large number of sine waves.

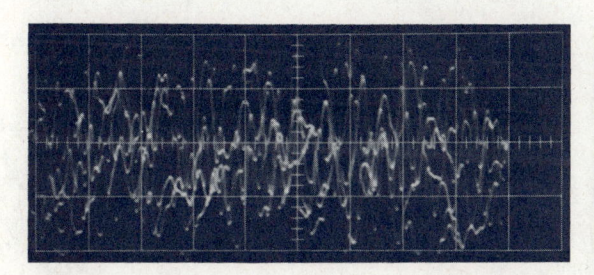

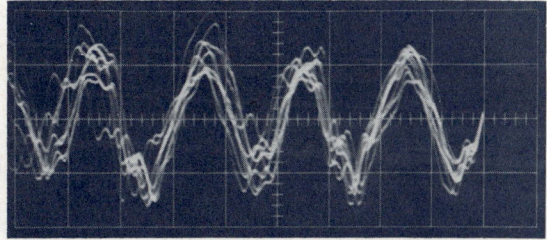

Temperature Sine waves are used to describe many of the periodic relationships that occur in nature. For example, scientists now believe that the average annual temperature in a given location is periodic. The overall temperature at a given place during a given season fluctuates, as time goes on, from colder to warmer and back to

420 The Trigonometric Functions

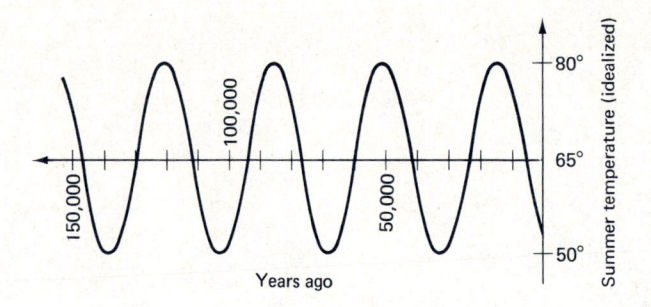

Figure 10.21

colder. The graph of Figure 10.21 shows an idealized description of the temperature of a location at the latitude of Anchorage over the last few thousand years.

10.2 EXERCISES

Find the values of the six trigonometric functions for the following angles. See Examples 1 and 2.

1. 120°	**2.** 135°	**3.** 150°	**4.** 225°
5. 240°	**6.** 300°	**7.** 330°	**8.** 390°
9. 420°	**10.** 495°	**11.** 510°	**12.** 570°

Complete the following table.

	θ	$\sin\theta$	$\cos\theta$	$\tan\theta$	$\cot\theta$	$\sec\theta$	$\csc\theta$
13.	30°	1/2	$\sqrt{3}/2$			$2\sqrt{3}/3$	
14.	45°			1	1		
15.	60°		1/2	$\sqrt{3}$		2	
16.	120°	$\sqrt{3}/2$		$-\sqrt{3}$			$2\sqrt{3}/3$
17.	135°	$\sqrt{2}/2$	$-\sqrt{2}/2$			$-\sqrt{2}$	$\sqrt{2}$
18.	150°		$-\sqrt{3}/2$	$-\sqrt{3}/3$			2
19.	210°	$-1/2$		$\sqrt{3}/3$	$\sqrt{3}$		-2
20.	240°	$-\sqrt{3}/2$	$-1/2$			-2	$-2\sqrt{3}/3$

Use Table 7 to find a value for the following. See Examples 3 and 4.

21. sin 39° **22.** cos 58°

23. tan 82° **24.** tan 54°

25. sin 37° **26.** tan 29°

27. $\tan 41°$ **28.** $\cos 69°$

29. $\sin .4014$ **30.** $\tan 1.0123$

31. $\cos 1.4137$ **32.** $\sin 1.5359$

Graph the following functions over a two-period interval.

33. $y = 2 \sin x$ **34.** $y = 2 \cos x$

35. $y = \cos 2x$ **36.** $y = \sin 2x$

37. $y = \cot x$ **38.** $y = \sec x$

APPLIED PROBLEMS

Sound   Pure sounds produce single sine waves on an oscilloscope. Find the amplitude and period of each sine wave in the following photographs. On the vertical scale each square represents .5, and on the horizontal scale each square represents 30°.

39.

40.

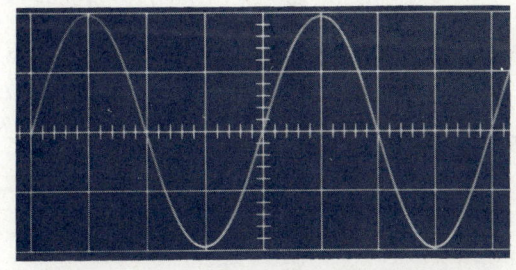

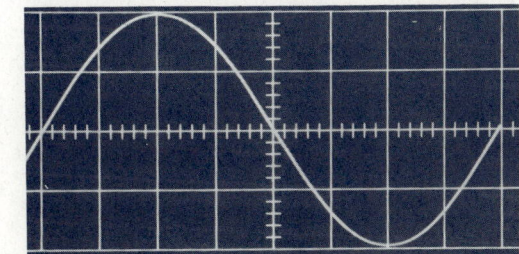

Temperature

41. See Figure 10.21 to find the highest temperature recorded.

42. Find the lowest temperature recorded.

43. Use these two numbers to find the amplitude. (Hint: An alternate definition of the amplitude is half the difference of the highest and lowest points on the graph.)

44. Find the period of the graph.

45. What is the trend of the temperature now?

The maximum afternoon temperature in a given city is approximated by

$$T(x) = 60 - 30\cos\frac{x}{2},$$

where x represents the month, with $x = 0$ representing January, $x = 1$ representing February, and so on. Find the maximum afternoon temperature for the following months.

46. January
47. March
48. October (Hint: $\cos 4.5 \approx -.2$.)
49. June (Hint: $\cos 2.5 \approx -.8$.)
50. August (Hint: $\cos 3.5 \approx -.9$.)

Pollution levels

The amount of pollution in the air fluctuates with the seasons. It is lower after heavy spring rains and higher after periods of little rain. In addition to this seasonal fluctuation, the long-term trend is upward. An idealized graph of this situation is shown in the figure. Trigonometric functions can be used to describe the fluctuating part of the pollution levels. Powers of the number e can be used to show the long-term growth. In fact, the pollution level in a certain area might be given by

$$P(t) = 7(1 - \cos 2\pi t)(t + 10) + 100e^{.2t},$$

where t is time in years, with $t = 0$ representing January 1 of the base year. Thus, July 1 of the same year would be represented by $t = .5$, while October 1 of the following year would be represented by $t = 1.75$. Find the pollution levels on the following dates.

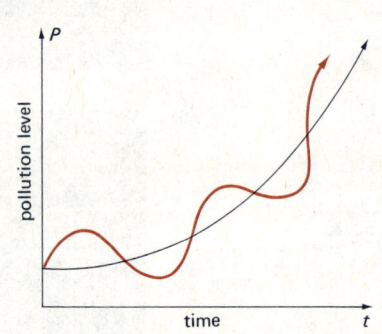

51. January 1, base year
52. July 1, base year
53. January 1, following year
54. July 1, following year

A mathematical model for the temperature in Fairbanks is

$$T(x) = 37 \sin\left[\frac{2\pi}{365}(x - 101)\right] + 25,*$$

where $T(x)$ is the temperature in degrees Celsius on day x, with $x = 0$ corresponding to January 1 and $x = 365$ corresponding to December 31. Use a calculator

---

* Barbara Lando and Clifton Lando, "Is the Graph of Temperature Variation a Sine Curve?" *The Mathematics Teacher* 70 (September, 1977) 534–37.

with trigonometric function keys or Table 7 to estimate the temperature on each of the following days.

55. March 1 (day 60)

56. April 1 (day 91)

57. day 101

58. day 150

59. Find the amplitude for the function $T(x)$.

60. Use the amplitude to find the highest and lowest temperatures in Fairbanks during the year.

10.3 TRIGONOMETRIC IDENTITIES

An **identity** is a statement which is true for every value in its domain. Examples of identities include

$$3x + 2x = 5x \qquad (a + b)^2 = a^2 + 2ab + b^2 \qquad m^0 = 1.$$

In this section we introduce the fundamental trigonometric identities and discuss some of their uses. We do not prove all of these identities; proofs can be found in any standard trigonometry text. We begin by listing the three **reciprocal identities** from Section 1 of this chapter.

$$\sin\theta = \frac{1}{\csc\theta} \qquad \tan\theta = \frac{1}{\cot\theta} \qquad \cos\theta = \frac{1}{\sec\theta}$$

Each of these three identities leads to another. For example, from $\sin\theta = 1/\csc\theta$ we have $\csc\theta = 1/\sin\theta$.

From the definitions of the trigonometric functions of Section 1, we can get

$$\tan\theta = \frac{y}{x} = \frac{\frac{y}{r}}{\frac{x}{r}} = \frac{\sin\theta}{\cos\theta}$$

or

$$\tan\theta = \frac{\sin\theta}{\cos\theta}.$$

In the same way,

$$\cot\theta = \frac{\cos\theta}{\sin\theta}.$$

In the definitions of the trigonometric functions, x, y, and r are the lengths of the sides of a right triangle. Thus, by the Pythagorean theorem,

$$x^2 + y^2 = r^2.$$

Dividing both sides of this equation by r^2 and rearranging terms gives

$$\frac{y^2}{r^2} + \frac{x^2}{r^2} = \frac{r^2}{r^2} \quad \text{or} \quad \left(\frac{y}{r}\right)^2 + \left(\frac{x}{r}\right)^2 = 1.$$

Substituting $\sin\theta$ for y/r and $\cos\theta$ for x/r, we have

$$(\sin\theta)^2 + (\cos\theta)^2 = 1.$$

It is customary to write $\sin^2\theta$ for $(\sin\theta)^2$ and $\cos^2\theta$ for $(\cos\theta)^2$. Making this change, we get the identity

$$\sin^2\theta + \cos^2\theta = 1.$$

This identity can also be written as

$$\sin^2\theta = 1 - \cos^2\theta \quad \text{or} \quad \cos^2\theta = 1 - \sin^2\theta.$$

A little algebraic manipulation of the identity $\sin^2\theta + \cos^2\theta = 1$ gives two more identities. First, divide both sides by $\cos^2\theta$.

$$\frac{\sin^2\theta}{\cos^2\theta} + \frac{\cos^2\theta}{\cos^2\theta} = \frac{1}{\cos^2\theta} \quad (\cos\theta \neq 0)$$

Now substitute $\tan\theta$ and $\sec\theta$.

$$\tan^2\theta + 1 = \sec^2\theta \quad \text{or} \quad \tan^2\theta = \sec^2\theta - 1$$

In the same way, we can divide both sides of $\sin^2\theta + \cos^2\theta = 1$ by $\sin^2\theta$ to get the identities

$$1 + \cot^2\theta = \csc^2\theta \quad \text{or} \quad \cot^2\theta = \csc^2\theta - 1.$$

From the circle in Figure 10.22, we see that for any angle θ with a point (x, y) on its terminal side, there is a corresponding angle $-\theta$ with a point $(x, -y)$ on its terminal side. From the definition of sine, we have

$$\sin(-\theta) = \frac{-y}{r} \quad \text{and} \quad \sin\theta = \frac{y}{r},$$

so that $\sin(-\theta)$ is the negative of $\sin\theta$, or

$$\sin(-\theta) = -\sin\theta.$$

In the same way,

$$\cos(-\theta) = \cos\theta.$$

It is sometimes helpful to rewrite a trigonometric function of the sum of two angles in terms of trigonometric functions of each angle. For example, we can rewrite $\cos(A - B)$ in terms of $\sin A$, $\cos A$, $\sin B$, and $\cos B$. Although we shall not prove it, it turns out that

$$\cos(A - B) = \cos A \cos B + \sin A \sin B.$$

There is a similar expression for $\cos(A + B)$:

$$\cos(A + B) = \cos A \cos B - \sin A \sin B.$$

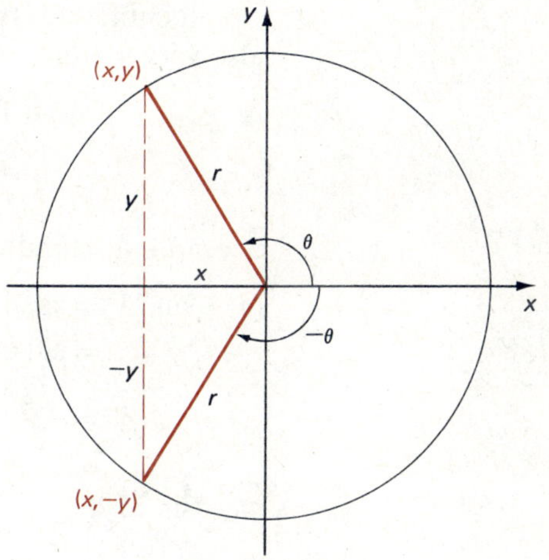

10.3 Trigonometric Identities

$\sin(-\theta) = -y/r = -\sin\theta$

Figure 10.22

These two identities are useful in finding certain values of the trigonometric functions, as shown in the next example.

Example 1 Use identities to find the value of the following.
(a) $\cos 15°$

To find $\cos 15°$, we must write $15°$ as the sum or difference of two angles which have known function values. Since we know the trigonometric function values of both $45°$ and $30°$, we can write $15° = 45° - 30°$. Then we use the identity for the cosine of the difference of two angles.

$$\cos 15° = \cos(45° - 30°)$$
$$= \cos 45° \cos 30° + \sin 45° \sin 30°$$
$$= \frac{\sqrt{2}}{2} \cdot \frac{\sqrt{3}}{2} + \frac{\sqrt{2}}{2} \cdot \frac{1}{2}$$
$$\cos 15° = \frac{\sqrt{6} + \sqrt{2}}{4}$$

(b) $\cos\dfrac{5\pi}{12} = \cos\left(\dfrac{\pi}{6} + \dfrac{\pi}{4}\right)$

$$= \cos\frac{\pi}{6}\cos\frac{\pi}{4} - \sin\frac{\pi}{6}\sin\frac{\pi}{4}$$
$$= \frac{\sqrt{3}}{2} \cdot \frac{\sqrt{2}}{2} - \frac{1}{2} \cdot \frac{\sqrt{2}}{2}$$
$$\cos\frac{5\pi}{12} = \frac{\sqrt{6} - \sqrt{2}}{4} \quad \blacksquare$$

Work Problem 1 at the side.

---

1. Use identities to find the following.
(a) $\cos(45° + 60°)$
(b) $\cos(180° - 45°)$

Answer:
(a) $\dfrac{\sqrt{2} - \sqrt{6}}{4}$

(b) $-\dfrac{\sqrt{2}}{2}$

Identities for the sine of the sum or difference of two angles are also very useful.

$$\sin(A + B) = \sin A \cos B + \cos A \sin B$$

$$\sin(A - B) = \sin A \cos B - \cos A \sin B$$

Example 2 Find the following.

(a) $\sin 75° = \sin(45° + 30°)$
$= \sin 45° \cos 30° + \cos 45° \sin 30°$
$= \dfrac{\sqrt{2}}{2} \cdot \dfrac{\sqrt{3}}{2} + \dfrac{\sqrt{2}}{2} \cdot \dfrac{1}{2}$

$\sin 75° = \dfrac{\sqrt{6} + \sqrt{2}}{4}$

(b) $\sin 105° = \sin(135° - 30°)$
$= \sin 135° \cos 30° - \cos 135° \sin 30°$
$= \dfrac{\sqrt{2}}{2} \cdot \dfrac{\sqrt{3}}{2} - \left(-\dfrac{\sqrt{2}}{2}\right)\left(\dfrac{1}{2}\right)$

$\sin 105° = \dfrac{\sqrt{6} + \sqrt{2}}{4}$ ∎

Work Problem 2 at the side.

Some special cases of the identities for the sum of two angles are used often enough to be expressed as separate identities. These are the identities that result from the addition identities when $A = B$, so that $A + B = A + A = 2A$. These identities are called **double-angle identities**.

In the identity for $\cos(A + B)$, we let $A = B$ to get an identity for $\cos 2A$.

$$\cos 2A = \cos(A + A)$$
$$= \cos A \cos A - \sin A \sin A$$
$$\cos 2A = \cos^2 A - \sin^2 A$$

By substitution from either $\cos^2 A = 1 - \sin^2 A$ or $\sin^2 A = 1 - \cos^2 A$, we get two alternate forms.

$$\cos 2A = 1 - 2\sin^2 A \qquad \cos 2A = 2\cos^2 A - 1$$

We can do the same thing with $\sin(A + B)$.

$$\sin 2A = \sin(A + A)$$
$$= \sin A \cos A + \cos A \sin A$$
$$\sin 2A = 2 \sin A \cos A$$

---

2. Use identities to find the following.
(a) $\sin(120° + 30°)$
(b) $\sin(90° + 45°)$
(c) $\sin(180° - 45°)$
(d) $\sin(150° - 30°)$

Answer:
(a) $1/2$
(b) $\sqrt{2}/2$
(c) $\sqrt{2}/2$
(d) $\sqrt{3}/2$

10.3 Trigonometric Identities

From the alternate forms of the double-angle identity for cosine, we can derive three additional identities. These **half-angle identities** are as follows.

$$\cos\frac{A}{2} = \pm\sqrt{\frac{1 + \cos A}{2}}$$

$$\sin\frac{A}{2} = \pm\sqrt{\frac{1 - \cos A}{2}}$$

$$\tan\frac{A}{2} = \pm\sqrt{\frac{1 - \cos A}{1 + \cos A}}$$

In these identities, the plus or minus sign is chosen according to the quadrant in which $A/2$ terminates. For example, if A represents an angle of $324°$, then $A/2 = 162°$, which lies in Quadrant II. Thus, $\cos A/2$ and $\tan A/2$ would be negative, while $\sin A/2$ would be positive.

Example 3 Find $\sin 22\frac{1}{2}°$.

Since $22\frac{1}{2}° = 45°/2$, we can use the half-angle identity for sine.

$$\sin 22\frac{1}{2}° = \sin\frac{45°}{2}$$
$$= \sqrt{\frac{1 - \cos 45°}{2}}$$
$$= \sqrt{\frac{1 - \frac{\sqrt{2}}{2}}{2}}$$
$$= \sqrt{\frac{2 - \sqrt{2}}{4}}$$
$$\sin 22\frac{1}{2}° = \frac{\sqrt{2 - \sqrt{2}}}{2}$$

We selected the + sign since $22\frac{1}{2}°$ is in Quadrant I. ∎

Work Problem 3 at the side.

10.3 EXERCISES

For each expression in Column I, choose the expression from Column II which completes an identity.

Column I

1. $\dfrac{\cos x}{\sin x}$

Column II

(a) $\sin^2 x + \cos^2 x$

---

3. Find the following.
 (a) $\cos 22\frac{1}{2}°$
 (b) $\tan 22\frac{1}{2}°$
 (c) $\cos 112\frac{1}{2}°$

Answer:

(a) $\dfrac{\sqrt{2 + \sqrt{2}}}{2}$

(b) $\sqrt{\dfrac{2 - \sqrt{2}}{2 + \sqrt{2}}} = \sqrt{2} - 1$

(c) $-\dfrac{\sqrt{2 - \sqrt{2}}}{2}$

2. $\tan x$ (b) $\cot x$

3. $\cos(-x)$ (c) $\sec^2 x$

4. $\tan^2 x + 1$ (d) $\dfrac{\sin x}{\cos x}$

5. 1 (e) $\cos x$

For each expression in Column I, choose the expression from Column II which completes an identity. You will have to rewrite one or both expressions to recognize the matches.

Column I   Column II

6. $-\tan x \cos x$ (a) $\dfrac{\sin^2 x}{\cos^2 x}$

7. $\sec^2 x - 1$ (b) $\dfrac{1}{\sec^2 x}$

8. $\dfrac{\sec x}{\csc x}$ (c) $\sin(-x)$

9. $1 - \sin^2 x$ (d) $\csc^2 x - \cot^2 x - \sin^2 x$

10. $\cos^2 x$ (e) $\tan x$

Identify the following as true or false.

11. $\cos 42° = \cos(30° + 12°)$

12. $\cos(-24°) = \cos 16° - \cos 40°$

13. $\cos 74° = \cos 60° \cos 14° + \sin 60° \sin 14°$

14. $\cos 140° = \cos 60° \cos 80° - \sin 60° \sin 80°$

15. $\sin(100° - 70°) = \sin 100° \cos 70° - \cos 100° \sin 70°$

16. $\sin(35° + 15°) = \sin 35° \cos 15° - \cos 35° \sin 15°$

17. $\cos(80° + 30°) = \cos 80° \cos 30° + \sin 80° \sin 30°$

18. $\cos(110° - 42°) = \cos 110° \cos 42° + \sin 110° \sin 42°$

Use the sum or difference identities to find the following without using tables. See Examples 1 and 2.

19. $\cos(-15°)$

20. $\cos 75°$

21. $\cos \dfrac{7\pi}{12}$

22. $\cos\left(-\dfrac{5\pi}{12}\right)$

23. $\cos 14° \cos 31° - \sin 14° \sin 31°$

24. $\cos 40° \cos 50° - \sin 40° \sin 50°$

25. $\cos 80° \cos 35° + \sin 80° \sin 35°$

26. $\cos(-10°)\cos 35° + \sin(-10°)\sin 35°$

27. $\sin 105°$

28. $\sin 285°$

29. $\sin 76° \cos 31° - \cos 76° \sin 31°$

30. $\sin 80° \cos(-55°) - \cos 80° \sin(-55°)$

For each of the following, determine whether the positive or negative square root should be selected.

31. $\sin 195° = \pm \sqrt{\dfrac{1 - \cos 390°}{2}}$

32. $\cos 58° = \pm \sqrt{\dfrac{1 + \cos 116°}{2}}$

33. $\tan 225° = \pm \sqrt{\dfrac{1 - \cos 450°}{1 + \cos 450°}}$

34. $\sin(-10°) = \pm \sqrt{\dfrac{1 - \cos(-20°)}{2}}$

Use the half-angle identities to find the sine, cosine, and tangent for the following. See Example 3.

35. $\theta = 22\tfrac{1}{2}°$

36. $\theta = 15°$

37. $\theta = 195°$

38. $\theta = 67\tfrac{1}{2}°$

39. $x = -\pi/8$

40. $x = -5\pi/6$

APPLIED PROBLEMS

Speed

An airplane flying faster than sound sends out sound waves that form a cone, as shown in the figure. The cone intersects the ground to form a curve called a hyperbola. As this hyperbola passes over a particular point on the ground, a sonic boom is heard at that point. If α is the angle at the vertex of the cone, then

$$\sin \frac{\alpha}{2} = \frac{1}{m},$$

where m is the mach number of the plane. (We assume $m > 1$.) The mach number is the ratio of the speed of the plane to the speed of sound. Thus, a speed of mach 1.4 means that the plane is flying 1.4 times the speed of sound. Find α or m, as necessary, for each of the following.

41. $m = 3/2$

42. $m = 5/4$

43. $m = 2$
44. $m = 5/2$
45. $\alpha = 30°$
46. $\alpha = 60°$

10.4 INVERSE TRIGONOMETRIC FUNCTIONS

To solve the equation $y = e^x$ for x, we had to introduce a new function, the logarithmic function. We do something similar now. In order to be able to solve the equation $y = \sin x$ for x, we introduce the inverse sine function. The inverse sine function is formed by exchanging the x and y in $y = \sin x$ to get $x = \sin y$. To solve this expression for y, write

$$y = \text{Arcsin } x.*$$

As Figure 10.23 shows, the domain of $y = \text{Arcsin } x$ is $-1 \leq x \leq 1$, while the range is $-\pi/2 \leq y \leq \pi/2$.

For each of the other trigonometric functions, we can define an inverse function by a suitable restriction on the range of the inverse function. The three most commonly used **inverse trigonometric functions** and their ranges are listed below.

$$y = \textbf{Arcsin } x, \quad -\pi/2 \leq y \leq \pi/2$$

$$y = \textbf{Arccos } x, \quad 0 \leq y \leq \pi$$

$$y = \textbf{Arctan } x, \quad -\pi/2 < y < \pi/2$$

$y = \text{Arc sin } x$

Figure 10.23

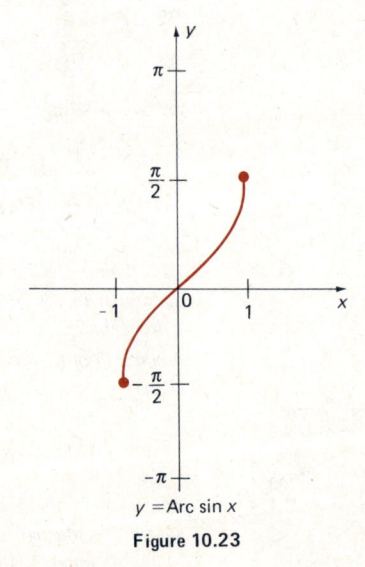

* An alternate notation for Arcsin x is $\text{Sin}^{-1} x$.

10.4 Inverse Trigonometric Functions

The graphs of $y = \text{Arccos } x$ and $y = \text{Arctan } x$ are shown in Figure 10.24 and Figure 10.25.

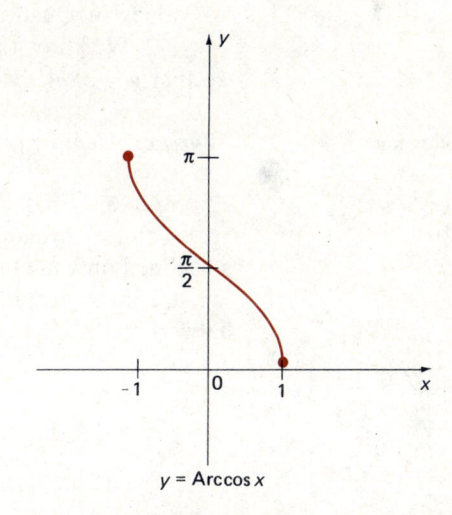

$y = \text{Arccos } x$

Figure 10.24

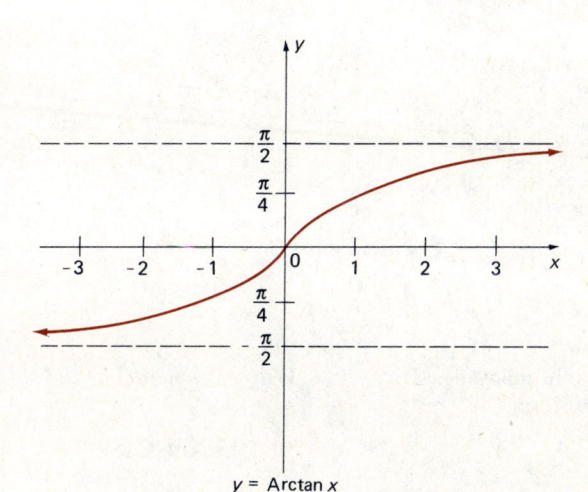

$y = \text{Arctan } x$

Figure 10.25

1. Find the following.
   (a) $\text{Arcsin } \sqrt{3}/2$
   (b) $\text{Arccos } 1$
   (c) $\text{Arctan } \sqrt{3}/3$
   (d) $\text{Arccos } \sqrt{2}/2$

   Answer:

   (a) $\dfrac{\pi}{3}$

   (b) 0

   (c) $\dfrac{\pi}{6}$

   (d) $\dfrac{\pi}{4}$

Example 1 Find $\text{Arcsin } 1/2$.

Let $y = \text{Arcsin } 1/2$. Then $\sin y = 1/2$. Since $\sin \pi/6 = 1/2$ and $\pi/6$ is in the range of Arcsin, we have $\text{Arcsin } 1/2 = \pi/6$. ∎

Work Problem 1 at the side.

Example 2 Find Arccos $-\sqrt{2}/2$

The values for Arccos are in Quadrants I and II, according to the definition given above. Since $-\sqrt{2}/2$ is negative, we are restricted to values in Quadrant II. Let $y = $ Arccos $-\sqrt{2}/2$. Then $\cos y = -\sqrt{2}/2$. We know that in Quadrant II, we have $\cos 3\pi/4 = -\sqrt{2}/2$, so that $y = 3\pi/4$. ∎

2. Find the following.
   (a) Arcsin $-1/2$
   (b) Arctan -1
   (c) Arccos $-1/2$
   (d) Arccos $-\sqrt{3}/2$

Answer:
(a) $-\dfrac{\pi}{6}$
(b) $-\dfrac{\pi}{4}$
(c) $\dfrac{2\pi}{3}$
(d) $\dfrac{5\pi}{6}$

Work Problem 2 at the side.

Example 3 Find $\sin[\text{Arctan}(-3/2)]$ without using tables.

Let $y = $ Arctan$(-3/2)$, so that $\tan y = -3/2$. Since Arctan x is defined only in Quadrants I and IV, and since we have $x = -3/2$, we are interested in Quadrant IV. Sketch y in Quadrant IV and label a triangle as shown in Figure 10.26. The hypotenuse is $\sqrt{13}$, so $\sin y = -3/\sqrt{13}$, or

$$\sin[\text{Arctan}(-3/2)] = \frac{-3}{\sqrt{13}} = \frac{-3\sqrt{13}}{13}. \quad ∎$$

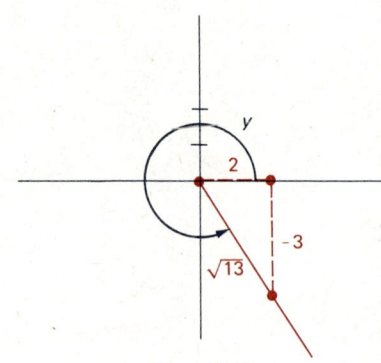

Figure 10.26

3. Find the following.
   (a) cos(Arcsin 3/4)
   (b) tan(Arccos 5/8)

Answer:
(a) $\dfrac{\sqrt{7}}{4}$
(b) $\dfrac{\sqrt{39}}{5}$

Work Problem 3 at the side.

10.4 EXERCISES

Give the value of y in radians for the following. See Examples 1 and 2.

1. $y = $ Arcsin $-\sqrt{3}/2$
2. $y = $ Arccos $\sqrt{3}/2$
3. $y = $ Arctan 1
4. $y = $ Arctan -1
5. $y = $ Arcsin -1
6. $y = $ Arccos -1
7. $y = $ Arccos 1/2
8. $y = $ Arcsin $-\sqrt{2}/2$
9. $y = $ Arccos $-\sqrt{2}/2$
10. $y = $ Arctan $\sqrt{3}/3$
11. $y = $ Arctan $-\sqrt{3}$
12. $y = $ Arccos $-1/2$

Use Table 7 and give the value of the following in degrees.

13. Arcsin −.1392
14. Arccos −.1392
15. Arccos −.8988
16. Arcsin .7880
17. Arccos .9272
18. Arctan 1.7321
19. Arctan 1.111
20. Arcsin .8192
21. Arctan −.9004
22. Arctan −.2867
23. Arcsin .9272
24. Arccos .4384

Give the value of the following without using tables. See Example 3.

25. tan(Arccos 2/3)
26. sin(Arccos 1/4)
27. cos(Arctan −2)
28. sec(Arcsin −1/3)
29. cot(Arcsin −2/5)
30. cos(Arctan 8/5)
31. sec(Arccos 3/5)
32. csc(Arcsin 12/13)
33. sin(Arctan −3)
34. tan(Arccos −4/5)

APPLIED PROBLEMS

Angular measure

A painting 1 meter high and 3 meters from the floor will cut off an angle θ to an observer, where

$$\theta = \text{Arctan} \frac{x}{x^2 + 2}.$$

We assume the observer is x meters from the wall displaying the painting and that the eyes of the observer are 2 meters above the ground. (See the figure.) Find the value of θ for each of the following values of x. Round to the nearest degree.

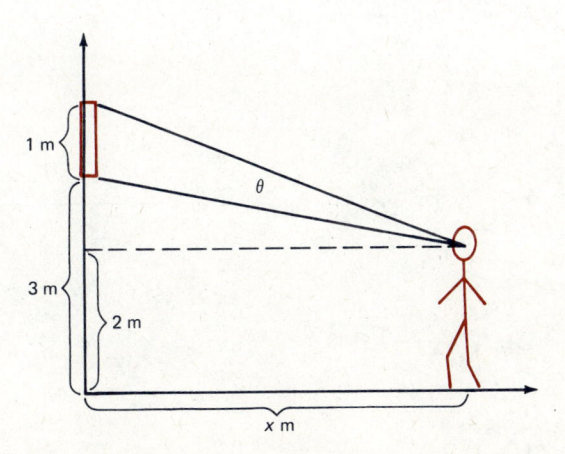

35. 1
36. 2
37. 3
38. 4

10.5 DERIVATIVES OF TRIGONOMETRIC FUNCTIONS

In this section we begin a study of the calculus of the trigonometric functions. In this section we shall find formulas for the derivatives of these functions. All these derivatives can be found from the formula for the derivative of $y = \sin x$. This derivative depends on the fact that

$$\lim_{x \to 0} \frac{\sin x}{x} = 1,$$

one of the key limits in all of mathematics. (We assume throughout the rest of this chapter that all variables represent *radian* measure.)

To see why this limit is reasonable, look at Figure 10.27. The figure shows a circle with radius $OP = 1$, along with points P and S on the circle. Angle POQ has a measure of x radians. By the definition of the trigonometric functions,

$$\sin x = \frac{QP}{OP} = \frac{QP}{1} = QP$$

and
$$\cos x = \frac{OQ}{OP} = \frac{OQ}{1} = OQ.$$

To get from the origin, O, to point P, we go over a length OQ and up a length QP, so that the coordinates of P are (OQ, QP). Since $OQ = \cos x$ and $QP = \sin x$, the coordinates of P can also be written as

$$P = (\cos x, \sin x).$$

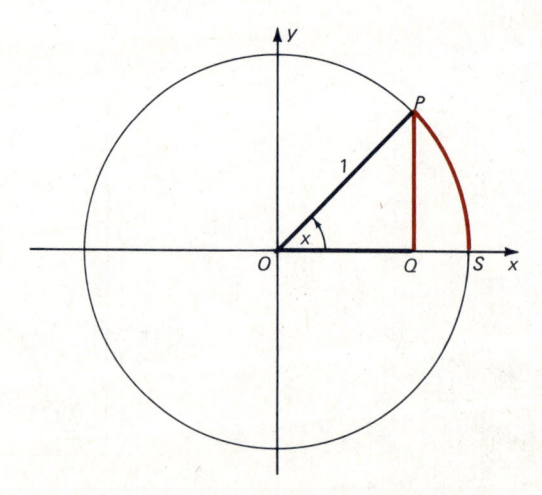

Figure 10.27

10.5 Derivatives of Trigonometric Functions

By the definition of radian measure, the arc of the circle from P to S has a length of x. The segment QP has a length of $\sin x$, as we have seen. The ratio of the length of the segment to the length of the arc is thus

$$\frac{\text{length of } QP}{\text{length of arc } PS} = \frac{\sin x}{x}.$$

As x approaches 0, Figure 10.27 shows that the length of PQ will approach the length of arc PS. Because of this, the ratio of their lengths approaches 1, or

$$\lim_{x \to 0} \frac{\sin x}{x} = 1.$$

The discussion here was really only for positive values of x. A similar argument could be given for negative x by placing triangle OPQ in Quadrant IV.

Example 1 Find $\lim_{h \to 0} \dfrac{\cos h - 1}{h}$.

Use the limit above and some trigonometric identities.

$$\lim_{h \to 0} \frac{\cos h - 1}{h} = \lim_{h \to 0} \frac{(\cos h - 1)}{h} \cdot \frac{(\cos h + 1)}{(\cos h + 1)}$$

$$= \lim_{h \to 0} \frac{\cos^2 h - 1}{h(\cos h + 1)}$$

$$= \lim_{h \to 0} \frac{-\sin^2 h}{h(\cos h + 1)} \qquad (\cos^2 h = 1 - \sin^2 h)$$

$$= \lim_{h \to 0} (-\sin h)\left(\frac{\sin h}{h}\right)\left(\frac{1}{\cos h + 1}\right)$$

$$= 0(1)\left(\frac{1}{1 + 1}\right)$$

$$\lim_{h \to 0} \frac{\cos h - 1}{h} = 0 \quad \blacksquare$$

Now we can find the derivative of $y = \sin x$. First, recall the definition of the derivative of $y = f(x)$:

$$f'(x) = \lim_{h \to 0} \frac{f(x + h) - f(x)}{h}.$$

For $f(x) = \sin x$, we have

$$f'(x) = \lim_{h \to 0} \frac{\sin(x + h) - \sin x}{h}$$

$$= \lim_{h \to 0} \frac{\sin x \cdot \cos h + \cos x \cdot \sin h - \sin x}{h}.$$

436 The Trigonometric Functions

Here we used the identity $\sin(A + B) = \sin A \cos B + \cos A \sin B$.

$$= \lim_{h \to 0} \frac{(\sin x \cdot \cos h - \sin x) + \cos x \cdot \sin h}{h} \quad \text{rearranging terms}$$

$$= \lim_{h \to 0} \frac{\sin x[\cos h - 1] + \cos x \cdot \sin h}{h} \quad \text{factoring}$$

$$= \lim_{h \to 0} \sin x \left[\frac{\cos h - 1}{h} \right] + \lim_{h \to 0} \cos x \left[\frac{\sin h}{h} \right]$$

$$= \sin x(0) + \cos x(1)$$

$$f'(x) = \cos x$$

In summary,

if $f(x) = \sin x$, then $f'(x) = \cos x$.

We can use the chain rule to find derivatives of combinations of the sine functions and other functions, as shown in the following example.

Example 2 Find the derivatives of the following functions.
(a) $y = \sin 6x$
By the chain rule,

$$y' = (\cos 6x) \cdot D_x(6x)*$$
$$= (\cos 6x) \cdot 6$$
$$y' = 6 \cos 6x.$$

(b) $y = 5 \sin(9x^2 + 2)$

$$y' = [5 \cos(9x^2 + 2] \cdot D_x(9x^2 + 2)$$
$$= 18x[5 \cos(9x^2 + 2)]$$
$$y' = 90x \cdot \cos(9x^2 + 2) \quad \blacksquare$$

Work Problem 1 at the side.

Example 3 Find $D_x \sin^4 x$.

We know that $\sin^4 x$ means the same as $(\sin x)^4$. By the generalized power rule,

$$D_x \sin^4 x = 4 \cdot \sin^3 x \cdot D_x \sin x$$
$$= 4 \cdot \sin^3 x \cdot \cos x. \quad \blacksquare$$

Work Problem 2 at the side.

1. Find the derivatives of the following.
(a) $y = -3 \sin(4x)$
(b) $f(x) = 8 \sin(2x^4 + 6x^2)$
(c) $y = -2 \sin(x^3 + 1)$

Answer:
(a) $-12 \cos 4x$
(b) $8(8x^3 + 12x)\cos(2x^4 + 6x^2)$
(c) $-6x^2 \cos(x^3 + 1)$

2. Find the following derivatives.
(a) $D_x \sin^6 x$
(b) $D_x 3 \sin^5 x$

Answer:
(a) $6 \sin^5 x \cos x$
(b) $15 \sin^4 x \cos x$

---

* The symbol $D_x f(x)$ means the same thing as $f'(x)$.

10.5 Derivatives of Trigonometric Functions

We can use the fact that $D_x \sin x = \cos x$ and the trigonometric identities to find $D_x \cos x$. First, use the identity for $\sin(A - B)$ to get

$$\sin\left(\frac{\pi}{2} - x\right) = \sin\frac{\pi}{2} \cdot \cos x - \cos\frac{\pi}{2} \cdot \sin x$$
$$= 1 \cdot \cos x - 0 \cdot \sin x$$
$$= \cos x.$$

In the same way, $\cos\left(\frac{\pi}{2} - x\right) = \sin x$. Therefore,

$$D_x \cos x = D_x \sin\left(\frac{\pi}{2} - x\right).$$

By the chain rule,

$$D_x \sin\left(\frac{\pi}{2} - x\right) = \cos\left(\frac{\pi}{2} - x\right) \cdot D_x\left(\frac{\pi}{2} - x\right)$$
$$= \cos\left(\frac{\pi}{2} - x\right) \cdot (-1) \qquad \pi/2 \text{ is constant}$$
$$= -\cos\left(\frac{\pi}{2} - x\right)$$
$$= -\sin x.$$

In summary,

$$D_x \cos x = -\sin x.$$

Example 4 Find the following derivatives.
(a) $D_x \cos(3x) = -\sin(3x) \cdot D_x(3x) = -3\sin 3x$
(b) $D_x \cos^4 x = 4\cos^3 x \cdot D_x \cos x = 4\cos^3 x (-\sin x)$
$\qquad = -4 \sin x \cdot \cos^3 x$
(c) $D_x 3x \cdot \cos x$
Use the product rule.
$\qquad D_x 3x \cdot \cos x = 3x(-\sin x) + (\cos x)(3)$
$\qquad = -3x \sin x + 3\cos x$ ∎

Work Problem 3 at the side.

Example 5 We know that $\tan x = \sin x / \cos x$. We can find the derivative of $y = \tan x$ by using the quotient rule to find the derivative of $y = \sin x / \cos x$.

---

3. Find the following derivatives.
(a) $D_x(2\cos 5x)$
(b) $D_x[-3\cos(4x^2 + 1)]$
(c) $D_x(2\cos^8 x)$
(d) $D_x[-2x \cdot \cos 3x]$

Answer:
(a) $-10 \sin 5x$
(b) $24x \sin(4x^2 + 1)$
(c) $-16 \sin x \cos^7 x$
(d) $6x \sin 3x - 2\cos 3x$

438 The Trigonometric Functions

$$D_x \tan x = D_x \frac{\sin x}{\cos x} = \frac{\cos x \cdot D_x \sin x - \sin x \cdot D_x \cos x}{\cos^2 x}$$

$$= \frac{\cos x (\cos x) - \sin x (-\sin x)}{\cos^2 x}$$

$$= \frac{\cos^2 x + \sin^2 x}{\cos^2 x}$$

$$= \frac{1}{\cos^2 x}$$

$$D_x \tan x = \sec^2 x$$

In the same way,

$$D_x \cot x = -\csc^2 x. \quad \blacksquare$$

4. (a) Write $\sec x$ as $1/\cos x$. Use the quotient rule on $1/\cos x$ to find $D_x \sec x$.
(b) Find $D_x \csc x$.

Answer:

(a) $\dfrac{\sin x}{\cos^2 x} = \sec x \tan x$

(b) $\dfrac{-\cos x}{\sin^2 x} = -\csc x \cot x$

Work Problem 4 at the side.

Example 6 Find the derivatives of the following functions. In part (b) you need the result of Problem 4(a) at the side.
(a) $D_x \tan 9x = \sec^2 9x \cdot D_x(9x) = 9 \sec^2 9x$
(b) $D_x 4 \sec 6x = 4 \cdot 6 \tan 6x \sec 6x = 24 \tan 6x \sec 6x$
(c) $D_x \cot^6 x = 6 \cot^5 x \cdot D_x \cot x = -6 \cot^5 x \csc^2 x$
(d) $D_x \ln|6 \tan x| = \dfrac{D_x 6 \tan x}{6 \tan x} = \dfrac{6 \sec^2 x}{6 \tan x} = \dfrac{\sec^2 x}{\tan x} \quad \blacksquare$

5. Find the derivatives of the following functions.
(a) $y = -8 \cot x$
(b) $y = \csc^5 x$
(c) $y = \ln|2 \sec x|$

Answer:
(a) $8 \csc^2 x$
(b) $-5 \csc^5 x \cot x$
(c) $\tan x$

Work Problem 5 at the side.

The derivatives of the trigonometric functions given above are themselves trigonometric functions. However, the derivatives of the *inverse* trigonometric functions are algebraic functions:

$$D_x(\text{Arcsin } u) = \frac{1}{\sqrt{1-u^2}} \cdot D_x(u)$$

$$D_x(\text{Arccos } u) = \frac{-1}{\sqrt{1-u^2}} \cdot D_x(u)$$

$$D_x(\text{Arctan } u) = \frac{1}{1+u^2} \cdot D_x(u).$$

Example 7 Find the following derivatives.
(a) $D_x[\text{Arcsin}(5x^2 + 1)] = \dfrac{1}{\sqrt{1-(5x^2+1)^2}} \cdot D_x(5x^2+1)$

$$= \frac{10x}{\sqrt{1-(5x^2+1)^2}}$$

Here we let $u = 5x^2 + 1$.

(b) $D_x[\text{Arctan } 8x^2] = \dfrac{1}{1+(8x^2)^2} \cdot D_x(8x^2)$

$= \dfrac{16x}{1+64x^4}$

(c) $D_x[\text{Arctan } e^{2x}] = \dfrac{1}{1+(e^{2x})^2} \cdot D_x e^{2x}$

$= \dfrac{2e^{2x}}{1+e^{4x}}$ ■

Work Problem 6 at the side.

6. Find the derivatives of the following functions.
   (a) $y = \text{Arcsin}(4x+2)$
   (b) $y = \text{Arccos}\dfrac{1}{x}$
   (c) $y = \text{Arctan } 9x^3$

Answer:

(a) $\dfrac{4}{\sqrt{1-(4x+2)^2}}$

(b) $\dfrac{1}{x\sqrt{x^2-1}}$

(c) $\dfrac{27x^2}{1+81x^6}$

10.5 EXERCISES

Find the derivatives of the following functions. See Examples 2–6.

1. $y = 2\sin 6x$
2. $y = 8\csc 6x^2$
3. $y = 12\tan(9x+1)$
4. $y = -3\cos(8x^2+2)$
5. $y = \cos^4 x$
6. $y = -9\sin^5 x$
7. $y = \tan^5 x$
8. $y = 10\sec^7 x$
9. $y = -5x \cdot \sin 4x$
10. $y = 6x \cdot \cos 3x$
11. $y = \dfrac{\sin x}{x}$
12. $y = \dfrac{\tan x}{2x+4}$
13. $y = \sin e^{5x}$
14. $y = \cos 4e^{2x}$
15. $y = e^{\sin x}$
16. $y = -8e^{\tan x}$
17. $y = \sin(\ln 4x^2)$
18. $y = \cos(\ln 2x^3)$
19. $y = \ln|\sin x^2|$
20. $y = \ln|\tan^2 x|$

Find the derivatives of the following inverse trigonometric functions. See Example 7.

21. $y = \text{Arcsin } 12x$
22. $y = \text{Arccos } 10x$
23. $y = \text{Arctan } 3x$
24. $y = \text{Arcsin } 1/x$
25. $y = \text{Arccos } -2/x$
26. $y = \text{Arccos } \sqrt{1-x^2}$
27. $y = \text{Arctan}(\ln|-2/x|)$
28. $y = \text{Arctan}\left(\ln\left|\dfrac{x+1}{2}\right|\right)$
29. $y = \ln|\text{Arctan}(x+1)|$
30. $y = \ln|\text{Arctan}(3x-5)|$

APPLIED PROBLEMS

Revenue *The revenue received from the sale of electric fans is seasonal, with maximum revenue coming in the summer. Let the rate of revenue received from the sale of x units of fans be approximated by*

$$R(x) = 100 \cos 2\pi x,$$

where x is time in years, measured from July 1.

31. Find $R'(x)$.

32. Find $R'(x)$ for August 1. [Hint: August 1 is 1/12 of a year from July 1. Use a calculator or the "Radians" column of Table 7.]

33. Find $R'(x)$ for January 1.

34. Find $R'(x)$ for June 1.

Velocity *A particle moves along a straight line. The distance of the particle from the origin at time t is given by*

$$s(t) = \sin t + 2 \cos t.$$

Find the following.

35. The velocity when $t = 0$

36. When $t = \pi/2$

37. When $t = \pi$

38. The acceleration when $t = 0$

39. When $t = \pi/2$

40. When $t = \pi$

10.6 ANTIDERIVATIVES

We developed formulas for finding the derivatives of the trigonometric functions in the previous section. Each formula for a derivative leads to a corresponding formula for an antiderivative. In particular, the formulas of the last section lead to the following antiderivatives.

$$\int \sin x \, dx = -\cos x + C$$

$$\int \cos x \, dx = \sin x + C$$

$$\int \sec^2 x \, dx = \tan x + C$$

$$\int \csc^2 x \, dx = -\cot x + C$$

$$\int \tan x \sec x \, dx = \sec x + C$$

10.6 Antiderivatives

$$\int \cot x \csc x \, dx = -\csc x + C$$

$$\int \frac{1}{\sqrt{1-x^2}} dx = \operatorname{Arcsin} x + C \qquad -1 < x < 1$$

$$\int \frac{-1}{\sqrt{1-x^2}} dx = \operatorname{Arccos} x + C \qquad -1 < x < 1$$

$$\int \frac{1}{1+x^2} dx = \operatorname{Arctan} x + C$$

Example 1 Find the following antiderivatives.

(a) $\int \sin 7x \, dx$

Use substitution. Let $u = 7x$, so that $du = 7 \, dx$. Then

$$\int \sin 7x \, dx = \frac{1}{7} \int \sin 7x (7 \, dx)$$

$$= \frac{1}{7} \int \sin u \, du$$

$$= -\frac{1}{7} \cos u + C$$

$$\int \sin 7x \, dx = -\frac{1}{7} \cos 7x + C.$$

(b) $\int \sec^2 12x \, dx = \frac{1}{12} \int \sec^2 12x (12 \, dx) = \frac{1}{12} \tan 12x + C$

(c) $\int \frac{2}{\sqrt{1-x^2}} dx = 2 \int \frac{1}{\sqrt{1-x^2}} dx = 2 \operatorname{Arcsin} x + C$ ∎

Work Problem 1 at the side.

The method of integration by parts discussed in Chapter 7 is often useful for finding the antiderivatives of trigonometric functions.

Example 2 Find $\int 2x \sin x \, dx$.

Let $u = 2x$ and $dv = \sin x \, dx$. Then $du = 2 \, dx$ and $v = -\cos x$. Using integration by parts we have

$$\int u \, dv = uv - \int v \, du$$

$$\int 2x \sin x \, dx = -2x \cos x - \int (-\cos x) 2 \, dx$$

$$= -2x \cos x + 2 \int \cos x \, dx$$

$$= -2x \cos x + 2 \sin x + C.$$ ∎

Check the result by differentiating.

---

1. Find the following antiderivatives.

(a) $6 \int \cos \frac{1}{2} x \, dx$

(b) $\int \csc^2 9x \, dx$

(c) $\int \frac{11}{1+x^2} dx$

Answer:

(a) $12 \sin \frac{1}{2} x + C$

(b) $-\frac{1}{9} \cot 9x + C$

(c) $11 \operatorname{Arctan} x + C$

2. Find $\int 5x \cos x \, dx$.

Answer: $5x \sin x + 5 \cos x + C$

3. Evaluate the following.

(a) $\int_0^{\pi/2} \sin x \, dx$

(b) $\int_0^{\pi/4} \sec^2 x \, dx$

Answer:
(a) 1
(b) 1

4. Find the shaded area.

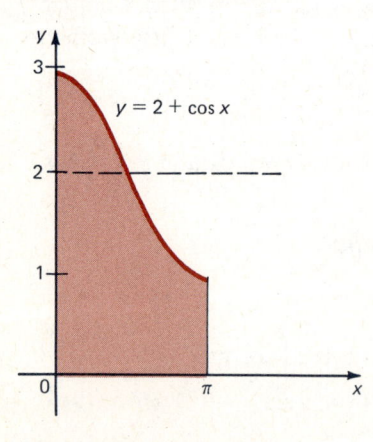

[Graph showing $y = 2 + \cos x$ from 0 to π, with y-axis values 1, 2, 3]

Answer: 2π

Work Problem 2 at the side.

We can also use the antiderivatives of trigonometric functions to find areas under the graphs of trigonometric curves, as the next example shows.

Example 3 Find the shaded area in Figure 10.28.

The shaded area in Figure 10.28 is bounded by $y = \cos x$, $y = 0$, $x = -\pi/2$, and $x = \pi/2$. By the fundamental theorem of calculus, this area is given by

$$\int_{-\pi/2}^{\pi/2} \cos x \, dx = \sin x \Big]_{-\pi/2}^{\pi/2}$$

$$= \sin \frac{\pi}{2} - \sin\left(-\frac{\pi}{2}\right)$$

$$= 1 - (-1)$$

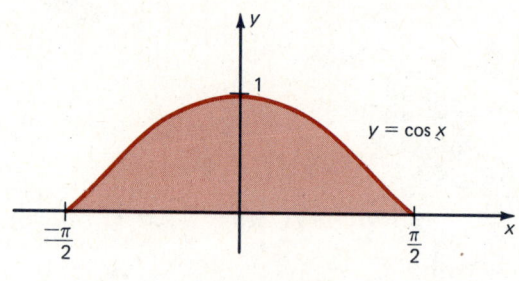

$$= 2. \blacksquare$$

[Figure 10.28: Graph of $y = \cos x$ from $-\pi/2$ to $\pi/2$ with shaded area]

Figure 10.28

Work Problems 3 and 4 at the side.

Tables of Integrals Taking antiderivatives of trigonometric functions can quickly get very complicated. For this reason, tables of integrals are very popular with people who need to find such antiderivatives often. The table of integrals inside the back cover of this book gives antiderivatives of the most common functions. Table 6 in the Appendix is more extensive and includes many antiderivatives involving trigonometric functions. The next few examples show how to use Table 6.

Example 4 Find $\int \tan^2 x \, dx$.

From entry 38 of the table, we have

$$\int \tan^2 x \, dx = \frac{1}{2-1} \tan^1 x - \int \tan^0 x \, dx$$

$$= \tan x - \int dx \qquad \tan^0 x = (\tan x)^0 = 1$$

$$= \tan x - x + C. \blacksquare$$

5. Find the following.
   (a) $\int \tan x \, dx$
   (b) $\int \sec x \, dx$
   (c) $\int \csc x \, dx$

   Answer:
   (a) $\ln|\sec x| + C$
   (b) $\ln|\sec x + \tan x| + C$
   (c) $\ln|\csc x - \cot x| + C$

6. Find the following.
   (a) $\int x \cdot \sin 4x \, dx$
   (b) $\int \sin 2x \cdot \sin x \, dx$

   Answer:
   (a) $\dfrac{\sin 4x}{16} - \dfrac{x \cos 4x}{4} + C$
   (b) $\dfrac{\sin x}{2} - \dfrac{\sin 3x}{6} + C$

7. Find the following.
   (a) $\int \dfrac{\sqrt{x^2 - 25}}{x} \, dx$
   (b) $\int \dfrac{\sqrt{4 - x^2}}{x^2} \, dx$

   Answer:
   (a) $\sqrt{x^2 - 25} - 5 \operatorname{Arccos} \dfrac{5}{x} + C$
   (b) $-\dfrac{1}{x}\sqrt{4 - x^2} - \operatorname{Arcsin}\dfrac{x}{2} + C$

Work Problem 5 at the side.

Example 5 Find $\int 6x \cdot \sin x \, dx$.
From entry 43 of Table 6, we have

$$\int 6x \cdot \sin x \, dx = 6\int x \cdot \sin x \, dx = 6[\sin x - x \cdot \cos x] + C$$
$$= 6\sin x - 6x \cdot \cos x + C. \quad \blacksquare$$

Work Problem 6 at the side.

Example 6 Find $\int \dfrac{9x^2}{\sqrt{16 - x^2}} \, dx$.

This antiderivative is similar to entry 34 in Table 6. Let $a = 4$; we then have

$$\int \dfrac{9x^2}{\sqrt{16 - x^2}} \, dx = 9 \int \dfrac{x^2}{\sqrt{16 - x^2}} \, dx$$
$$= 9\left[-\dfrac{x}{2}\sqrt{16 - x^2} + \dfrac{4^2}{2} \cdot \operatorname{Arcsin}\dfrac{x}{4}\right] + C$$
$$= -\dfrac{9x}{2}\sqrt{16 - x^2} + 72\operatorname{Arcsin}\dfrac{x}{4} + C. \quad \blacksquare$$

Work Problem 7 at the side.

10.6 EXERCISES

Find the following antiderivatives. See Example 1.

1. $\int \cos 5x \, dx$
2. $\int \sin 8x \, dx$
3. $\int (5 \cos x + 2 \sin x) \, dx$
4. $\int (7 \sin x - 8 \cos x) \, dx$
5. $\int 4 \tan 2x \sec 2x \, dx$
6. $\int 8 \cot 6x \csc 6x \, dx$
7. $-\int 6 \sec^2 2x \, dx$
8. $-\int 2 \csc^2 8x \, dx$
9. $\int \sin^7 x (\cos x) \, dx$
10. $\int \sin^6 x (\cos x) \, dx$
11. $\int \sqrt{\sin x}(\cos x) \, dx$
12. $\int \dfrac{\cos x}{\sqrt{\sin x}} \, dx$
13. $\int \dfrac{\sin x}{1 + \cos x} \, dx$
14. $\int \dfrac{\cos x}{1 - \sin x} \, dx$

Use the method of integration by parts to find the following. See Example 2.

15. $\int -6x \cos 5x \, dx$
16. $\int 9x \sin 2x \, dx$
17. $\int 8x \sin x \, dx$
18. $\int -11x \cos x \, dx$
19. $\int -6x \cos 8x \, dx$
20. $\int 10x \sin \tfrac{1}{2}x \, dx$

Evaluate the following definite integrals. See Example 3.

21. $\int_0^{\pi/4} \sin x \, dx$
22. $\int_{-\pi/2}^{0} \cos x \, dx$
23. $\int_0^{\pi/3} \tan x \sec x \, dx$
24. $\int_{-\pi/4}^{\pi/4} \sec^2 x \, dx$
25. $\int_{\pi/2}^{3\pi/2} \cos x \, dx$
26. $\int_{\pi}^{2\pi} \sin x \, dx$

Use Table 6 to find the following. See Examples 4–6.

27. $\int 3 \tan 5x \, dx$
28. $\int \tfrac{1}{2} \csc 2x \, dx$
29. $\int \tan^3(3x - 1) \, dx$
30. $\int 4 \sec^3(8 - 2x) \, dx$
31. $\int \sin^4 x \, dx$
32. $\int \cos^4 x \, dx$
33. $\int 2x \operatorname{Arccos} x \, dx$
34. $\int 5x \operatorname{Arcsin} x \, dx$
35. $\int e^x \cdot \cos 3x \, dx$
36. $\int e^x \cdot \sin 6x \, dx$
37. $\int 6 \sin 5x \cos 4x \, dx$
38. $\int \sin 8x \sin 4x \, dx$
39. $\int \cos 5x \cos x \, dx$
40. $\int \dfrac{5}{x\sqrt{9 - x^2}} \, dx$
41. $\int \sqrt{16 - x^2} \, dx$
42. $\int x^2 \sqrt{49 - x^2} \, dx$

10.7 APPLICATIONS

In this last section of the book we shall look at some applications of trigonometric functions.

Example 1* Homing pigeons are very careful to avoid flying over large bodies of water; they will go around them if at all possible. The reason for this is thought to be the extra energy required to fly over water, due to the fact that air falls over water in daytime.

---

* Example 1 and Exercises 23–36 with diagram of blood vessel from *Introduction to Mathematics for Life Scientists* by Edward Batschelet. Copyright © 1971 Springer-Verlag New York, Inc. Reprinted by permission.

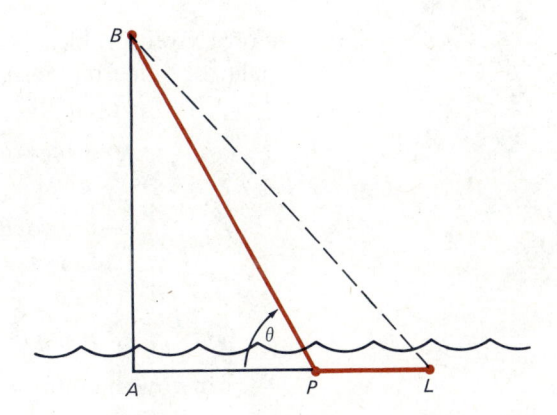

Figure 10.29

Assume that a pigeon is released from a boat at point B in a lake, as shown in Figure 10.29. The pigeon's loft is at point L. The shortest possible path for the pigeon is shown by the dashed line. The pigeon doesn't take the shortest possible path; rather it flies to point P on the shore and then along the shore from P to L.

The problem for the pigeon: where should point P be chosen in order to minimize the energy required for the flight from B to L? In other words, what is the optimum value of angle θ?

Let A be the point on the bank exactly south of B. Let $AB = r$ and let $AL = s$. By the definitions of the trigonometric functions, we have

$$\sin\theta = \frac{r}{BP}, \quad \text{so that} \quad BP = \frac{r}{\sin\theta}.$$

Also,

$$\cot\theta = \frac{AP}{r}, \quad \text{and} \quad AP = r \cdot \cot\theta.$$

Finally,

$$PL = AL - AP = s - r \cdot \cot\theta.$$

Assume that 8 units of pigeon energy are required for flying one unit of length over the lake, with 6 units required on the shore. The energy required to fly from B to P is thus

energy = energy per unit of distance × distance = $8(BP)$.

The energy required to fly from P to L is $6(PL)$. The total energy for the entire trip, E, is thus

$$E = 8(BP) + 6(PL)$$

$$= 8\left(\frac{r}{\sin\theta}\right) + 6(s - r \cdot \cot\theta)$$

$$E = \frac{8r}{\sin\theta} + 6s - 6r \cdot \cot\theta.$$

For a given problem, s and r are constants. We want to find the value of θ that minimizes E. To do this, we first find the derivative E'.

$$E' = \frac{-8r \cdot \cos\theta}{\sin^2\theta} + 6r \cdot \csc^2\theta$$

$$= \frac{-8r \cdot \cos\theta}{\sin^2\theta} + \frac{6r}{\sin^2\theta} \qquad \csc^2\theta = \frac{1}{\sin^2\theta}$$

$$E' = \frac{6r - 8r \cdot \cos\theta}{\sin^2\theta}$$

We can now minimize E by setting E' equal to 0.

$$E' = 0$$

$$\frac{6r - 8r \cdot \cos\theta}{\sin^2\theta} = 0$$

$$6r - 8r \cdot \cos\theta = 0 \qquad \text{Multiply by } \sin^2\theta$$

$$6r = 8r \cdot \cos\theta$$

$$\frac{6r}{8r} = \cos\theta$$

$$\frac{3}{4} = \cos\theta$$

$$.75 = \cos\theta$$

Look through Table 7 for a value of θ such that $\cos\theta = .7500$. To the nearest degree,

$$\theta = 41°.$$

The pigeon will minimize total energy by flying to point P so that $\theta = 41°$. (Experiments have shown that pigeons do indeed fly as this theory shows that they should.) ■

Work Problem 1 at the side.

Example 2 Suppose a light source is hanging at the end of a pendulum, as shown in Figure 10.30. As the pendulum swings back and forth, the light will oscillate on the floor. If a roll of photographic paper is moved at a constant speed under the light, a sine curve will be traced out on the paper. This curve will have an equation of the form

$$y = A\sin(Bt + C)$$

for some constants A, B, and C and time t. Different values for A cause the range of the graph to stretch or contract vertically, while B stretches or contracts the function horizontally, and C shifts the whole graph horizontally. ■

---

1. Rework Exercise 1, assuming that 4 units of energy are required over land and 8 units over water.

Answer: $\theta = 60°$

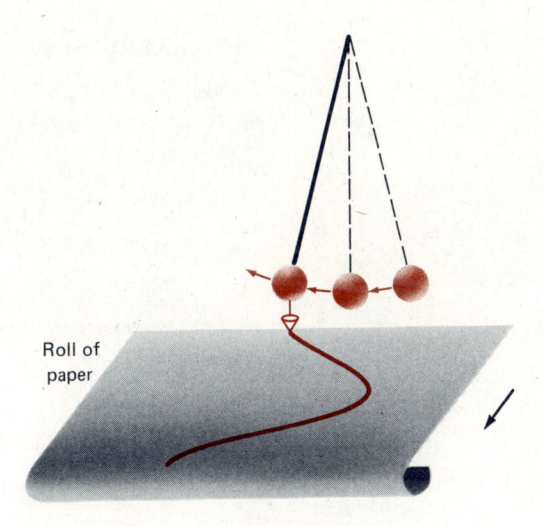

Figure 10.30

Motion of this type is called **simple harmonic motion** and comes up in such diverse areas as sound and wave mechanics. Examples of this are shown in the Exercises.

Work Problem 2 at the side.

2. Find the maximum and minimum values of the function $y = 5\sin(3x + 2)$. Do not use calculus. (Hint: The maximum of $y = \sin\theta$ is 1, while the minimum is -1.)

Answer: maximum is 5, minimum is -5

Example 3 Many biological populations, both plant and animal, experience seasonal growth. For example, an animal population might flourish during the spring and summer and die back in the fall. If $f(t)$ represents the population at time t, then the rate of change of the population with respect to time might be governed by the differential equation

$$\frac{df}{dt} = c \cdot f(t) \cdot \cos t. \qquad \text{\textcolor{red}{c is a positive constant}}$$

Since $\cos t$ ranges from -1 to 1, df/dt will change from negative to positive, so that $f(t)$ is both increasing and decreasing. The solution of the differential equation above is

$$f(t) = f(0) \cdot e^{c \cdot \sin t}, \qquad (1)$$

where $f(0)$ is the size of the population when $t = 0$. ■

3. Equation 1 takes on a maximum value when $\sin t$ is a maximum. (Why?) The maximum possible value of $\sin t$ is $\sin t = 1$. When does this take place?

Answer: at $t = \dfrac{\pi}{2}, \dfrac{5\pi}{2}, \dfrac{9\pi}{2}, \ldots$

Work Problem 3 at the side.

10.7 EXERCISES

In Exercises 1–6, recall that the slope of the tangent to a graph is given by the derivative of the function. Find the slope of the tangent to the given graph at the given point.

1. $y = \sin x$, $x = 0$
2. $y = \sin x$, $x = \pi/4$
3. $y = \cos x$, $x = \pi/2$
4. $y = \cos x$, $x = -\pi/4$
5. $y = \tan x$, $x = 0$
6. $y = \cot x$, $x = \pi/2$

Find the maximum and minimum values of y for the following equations for simple harmonic motion.

7. $y = 2\sin(3x - 1)$
8. $y = 4\sin(8x + 2)$
9. $y = -9\sin(5x + 4)$
10. $y = -3\sin(2x - 7)$
11. $y = \frac{1}{2}\sin(\frac{1}{4}x - 2)$
12. $y = -\frac{2}{3}\sin(\frac{3}{5}x - \frac{1}{2})$

In Example 3, we found that one mathematical model for seasonal growth is given by

$$f(t) = f(0) \cdot e^{c \cdot \sin t}.$$

Suppose that $f(0) = 1000$ and $c = 2$. Find the following. Use Table 7; then use Table 4 to find the power of e closest to the one in the equation. You will need a calculator with trigonometric functions for Exercises 19–22.

13. $f(.2)$
14. $f(.4)$
15. $f(.5)$
16. $f(.8)$
17. $f(1)$
18. $f(1.4)$
19. $f(1.8)$
20. $f(2.3)$
21. $f(3)$
22. $f(3.1)$

APPLIED PROBLEMS

Physiology

The body's system of blood vessels is made up of arteries, arterioles, capillaries, and veins. The transport of blood from the heart through all organs of the body and back to the heart should be as effective as possible. One way this can be done is by having large enough blood vessels to avoid turbulence, with blood cells small enough to minimize viscosity.

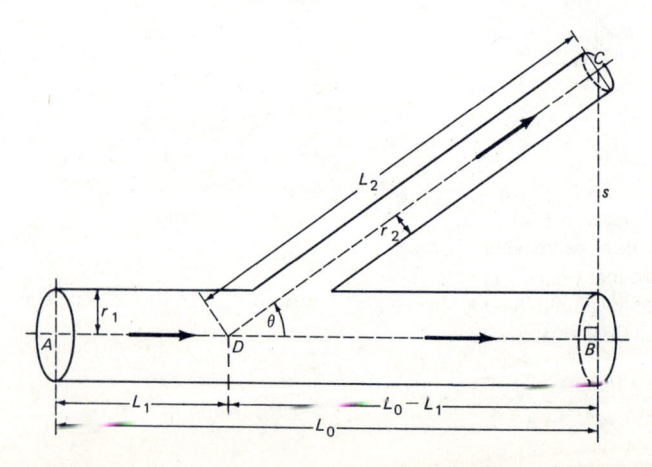

In the rest of the Exercises for this section, we shall find the value of angle θ (see the figure) such that total resistance to the flow of blood is minimized. Assume that a main vessel of radius r_1 runs along the horizontal line from A to B. A side artery, of radius r_2, heads for a point C. Choose point B so that CB is perpendicular to AB. Let $CB = s$ and let D be the point where the axis of the branching vessel cuts the axis of the main vessel.

According to Poiseuille's law, the resistance R in our system is proportional to the length L of the vessel and inversely proportional to the fourth power of the radius r. That is,

$$R = k \cdot \frac{L}{r^4}, \tag{2}$$

where k is a constant determined by the viscosity of the blood. Let $AB = L_0$, $AD = L_1$, and $DC = L_2$.

23. Use right triangle BDC and find $\sin \theta$.

24. Solve the result of Exercise 23 for L_2.

25. Find $\cot \theta$ in terms of s and $L_0 - L_1$.

26. Solve the result of Exercise 25 for L_1.

27. Write an expression similar to equation (2) for the resistance R_1 along AD.

28. Write a formula for the resistance along DC.

29. The total resistance, R, is given by the sum of the resistances along AD and DC. Use your answers to Exercises 27 and 28 to write an expression for R.

30. In your formula for R, replace L_1 with the result of Exercise 26, and L_2 with the result of Exercise 24.

31. Find R'. [Remember: k, L_1, L_0, s, r_1, and r_2 are all constants.]

32. Place R' equal to 0.

33. Multiply through by $(\sin^2 \theta)/s$.

34. Solve for $\cos \theta$.

35. Suppose $r_1 = 1$ cm and $r_2 = 1/4$ cm. Find $\cos \theta$ and then find θ. [Use a calculator or Table 7.]

36. Find θ if $r_1 = 1.4$ cm and $r_2 = .8$ cm.

CHAPTER 10 SUMMARY

Key Words

ray
angle
initial side
vertex

trigonometric functions
Pythagorean theorem
reciprocal identities
identities

450 The Trigonometric Functions

terminal side
degree
minute
acute angle
right angle
obtuse angle
straight angle
standard position
arc
radian
radius vector

equilateral triangle
isosceles triangle
periodic function
period
amplitude
sum and difference identities
double-angle identities
half-angle identities
inverse trigonometric functions
simple harmonic motion

Things to Remember

Fundamental Identities

$$\tan A = \frac{\sin A}{\cos A} \qquad \cot A = \frac{\cos A}{\sin A}$$

$$\cot A = \frac{1}{\tan A} \qquad \csc A = \frac{1}{\sin A} \qquad \sec A = \frac{1}{\cos A}$$

$$\sin^2 A + \cos^2 A = 1$$

$$\sin^2 A = 1 - \cos^2 A \qquad \cos^2 A = 1 - \sin^2 A$$

$$\tan^2 A + 1 = \sec^2 A \qquad 1 + \cot^2 A = \csc^2 A$$

$$\sin(-A) = -\sin A \qquad \cos(-A) = \cos A \qquad \tan(-A) = -\tan A$$

Sum and Difference Identities

$$\cos(A - B) = \cos A \cos B + \sin A \sin B$$
$$\cos(A + B) = \cos A \cos B - \sin A \sin B$$
$$\sin(A - B) = \sin A \cos B - \cos A \sin B$$
$$\sin(A + B) = \sin A \cos B + \cos A \sin B$$

Cofunction Identities

$$\sin\left(\frac{\pi}{2} - A\right) = \cos A \qquad \cos\left(\frac{\pi}{2} - A\right) = \sin A \qquad \tan\left(\frac{\pi}{2} - A\right) = \cot A$$

Double Angle and Half-Angle Identities

$$\cos 2A = \cos^2 A - \sin^2 A$$
$$\cos 2A = 1 - 2\sin^2 A \qquad \cos 2A = 2\cos^2 A - 1$$
$$\sin 2A = 2 \sin A \cos A$$

$$\cos\frac{A}{2} = \pm\sqrt{\frac{1 + \cos A}{2}} \qquad \sin\frac{A}{2} = \pm\sqrt{\frac{1 - \cos A}{2}} \qquad \tan\frac{A}{2} = \pm\sqrt{\frac{1 - \cos A}{1 + \cos A}}$$

CHAPTER 10 TEST

[10.1] *Convert the following degree measures to radians.*

1. 90°
2. 150°

Convert the following radian measures to degrees.

3. 7π
4. $9\pi/20$

[10.2] *Find the following.*

5. $\sin 60°$
6. $\tan 120°$
7. $\cos -30°$
8. $\cos -45°$
9. $\tan 47°$
10. $\cos 38°$
11. $\sin 1.4661$
12. $\cos .3142$

[10.3] *Use trigonometric identities to find the following.*

13. $\sin^2 29°10' + \cos^2 29°10'$
14. $\sin 15°$
15. $\cos 22\frac{1}{2}°$
16. $\sin 10° \cos 50° + \cos 10° \sin 50°$

[10.4] *For each of the following, give the value of y in radians.*

17. $y = \text{Arcsin} -\frac{1}{2}$
18. $y = \text{Arctan} -1$

Find the following without using tables.

19. $\sin(\text{Arccos} \frac{1}{2})$
20. $\tan\left(\text{Arccos} \frac{\sqrt{3}}{6}\right)$

[10.5] *Find the derivatives of the following functions.*

21. $y = -4 \sin 7x$
22. $y = 6 \tan 2x$
23. $y = 3 \cos^6 x$
24. $y = \dfrac{x-2}{\sin x}$
25. $y = \text{Arcsin}(-3x)$

[10.6] *Find the following antiderivatives.*

26. $\int 2 \tan x \sec x \, dx$
27. $\int \sqrt{\cos x} \cdot \sin x \cdot dx$
28. $\int \tan 9x \, dx$
29. $\int 6 \sin^2 x \, dx$
30. $\int \dfrac{1}{\sqrt{25 - x^2}} dx$
31. $\int \dfrac{1}{x\sqrt{x^2 - 100}} dx$

32. Find $\int_0^{\pi/2} \cos x \, dx$.

CASE 11   THE MATHEMATICS OF A HONEYCOMB*

The bee's cell is a regular hexagonal prism with one open end and one trihedral apex (Figure 1(a)). We may construct the surface by starting with a regular hexagonal base $abcdef$ with side s (Figure 1(b)). Over the base we raise a right prism of height h and top $ABCDEF$. The corners B, D, F are cut off by planes through the lines AC, CE, EA, meeting in a point V on the axis VN of the prism, and intersecting Bb, Dd, Ff in X, Y, Z. The three cut-off pieces are the tetrahedrons $ABCX$, $CDEY$, $EFAZ$. We put these pieces on top of the remaining solid such that X, Y, and Z coincide with V. Hereby, the lines AC, CE, EA act as "hinges". The faces $AXCV$, $CYEV$, $EZAV$ are rhombuses, that is, quadrilaterals with equal sides. The new body is the bee's cell and has the same volume as the original prism. The hexagonal base $abcdef$ is the open end.

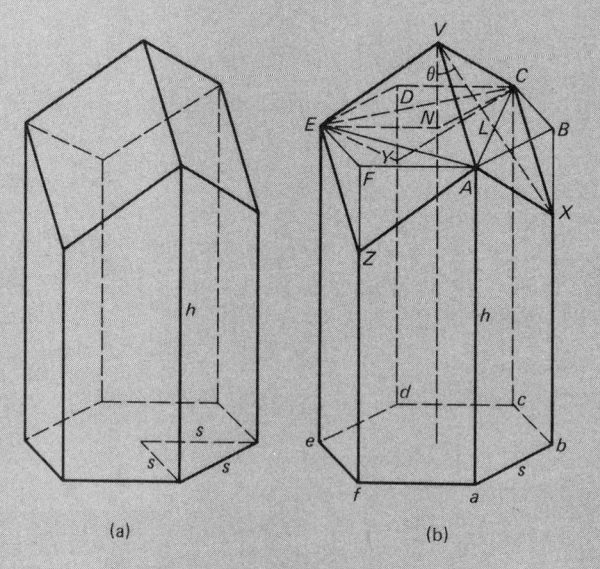

Figure 1   The bee's cell.[†]

The bees form the faces by using wax. When the volume is given, it is economic to spare wax and, therefore, to choose the angle of inclination, $\theta = NVX$, in such a way that the *surface of the bee's cell is minimized*.

The problem can be solved mathematically as follows. Let L be the intersection of CA and VX. Then L bisects the segment NB and, hence, $NL = s/2$. The segment CL is the height of the equilateral triangle BCN. Therefore,

$$CL = \frac{s}{2}\sqrt{3}. \tag{1}$$

---

* From *Introduction to Mathematics for Life Scientists* by Edward Batschelet. Copyright © 1971 Springer-Verlag, New York, Inc. Reprinted by permission.

[†] "The bee's cell," from *On Growth and Form* by d'Arcy W. Thompson, Cambridge University Press, as adapted in *Introduction to Mathematics for Life Scientists* by Edward Batschelet, Springer-Verlag, New York, Inc. Reprinted by permission.

In the triangle NLV we have the relationship

$$VL = \frac{s}{2\sin\theta}. \qquad (2)$$

The rhombus $AXCV$ has its center in L and consists of four congruent right triangles with legs equal to CL and VL. Therefore, from (1) and (2) we get

$$\text{area } AXCV = 4 \cdot \frac{1}{2} \cdot \frac{s}{2}\sqrt{3} \cdot \frac{s}{2\sin\theta} = \frac{s^2\sqrt{3}}{2\sin\theta}.$$

The surface of the bee's cell contains three such areas.

The six lateral faces of the bee's cell, such as $abXA$, are congruent trapezoids. Since $BX = VN$, we obtain from triangle VNL

$$BX = \frac{s}{2}\cot\theta.$$

Hence,

$$\text{area } abXA = \frac{s}{2}(aA + bX) = \frac{s}{2}(h + h - BX) = hs - \frac{s^2}{4}\cot\theta.$$

The total area made of wax amounts to

$$6hs - \frac{3}{2}s^2\cot\theta + \frac{3}{2}\frac{s^2\sqrt{3}}{\sin\theta}.$$

This area is a function of the variable angle θ and, thus, we denote it by $f(\theta)$. We may rewrite $f(\theta)$ in the form

$$f(\theta) = 6hs + \frac{3}{2}s^2\left(-\cot\theta + \frac{\sqrt{3}}{\sin\theta}\right).$$

Only the expression in the parentheses contains the variable θ. Some numerical values rounded off to two decimals are given in the following table.

θ	$-\cot\theta + \dfrac{\sqrt{3}}{\sin\theta}$
10°	4.30
20°	2.32
30°	1.73
40°	1.50
50°	1.42
60°	1.42
70°	1.48
80°	1.58
90°	1.73

The minimum of $f(\theta)$ is reached somewhere between $\theta = 50°$ and $\theta = 60°$. To get the optimal angle, say θ_0, we differentiate $f(\theta)$.

$$f'(\theta) = \frac{3}{2}s^2\left(\frac{1}{\sin^2\theta} - \frac{\sqrt{3}\cos\theta}{\sin^2\theta}\right) \qquad (3)$$

The derivative vanishes if, and only if,

$$1 = \sqrt{3}\cos\theta. \tag{4}$$

Hence, $\cos\theta_0 = 1/\sqrt{3} = 0.57735$ and $\theta_0 = 54.7°$. Notice that the optimal angle θ_0 is independent of the choice of s and h.

It is worth comparing the result with the actual angle chosen by the bees. It is difficult to measure this angle. However, the average of all measurements does not differ significantly from the theoretical value $\theta_0 = 54.7°$. Therefore, the bees strongly prefer the optimal angle. It is unlikely that the result is due to chance. We may rather suppose that selection pressure had an effect on the angle θ.

EXERCISES

1. Show how equation (3) above was found.
2. Show how the result given in equation (4) was obtained.

Appendix

Table 1   Selected Powers of Numbers

Table 2   Squares and Square Roots

Table 3   Common Logarithms

Table 4   Powers of *e*

Table 5   Natural Logarithms

Table 6   Integrals Involving Trigonometric Functions

Table 7   Trigonometric Functions

List of Symbols

Answers to Selected Exercises

Bibliography

Glossary

Index

Tables

TABLE 1   SELECTED POWERS OF NUMBERS

n	n^2	n^3	n^4	n^5	n^6
2	4	8	16	32	64
3	9	27	81	243	729
4	16	64	256	1024	4096
5	25	125	625	3125	
6	36	216	1296		
7	49	343	2401		
8	64	512	4096		
9	81	729	6561		
10	100	1000	10,000		

TABLE 2 SQUARES AND SQUARE ROOTS

n	n²	√n	√10n	n	n²	√n	√10n
1	1	1.000	3.162	51	2601	7.141	22.583
2	4	1.414	4.472	52	2704	7.211	22.804
3	9	1.732	5.477	53	2809	7.280	23.022
4	16	2.000	6.325	54	2916	7.348	23.238
5	25	2.236	7.071	55	3025	7.416	23.452
6	36	2.449	7.746	56	3136	7.483	23.664
7	49	2.646	8.367	57	3249	7.550	23.875
8	64	2.828	8.944	58	3364	7.616	24.083
9	81	3.000	9.487	59	3481	7.681	24.290
10	100	3.162	10.000	60	3600	7.746	24.495
11	121	3.317	10.488	61	3721	7.810	24.698
12	144	3.464	10.954	62	3844	7.874	24.900
13	169	3.606	11.402	63	3969	7.937	25.100
14	196	3.742	11.832	64	4096	8.000	25.298
15	225	3.873	12.247	65	4225	8.062	25.495
16	256	4.000	12.649	66	4356	8.124	25.690
17	289	4.123	13.038	67	4489	8.185	25.884
18	324	4.243	13.416	68	4624	8.246	26.077
19	361	4.359	13.784	69	4761	8.307	26.268
20	400	4.472	14.142	70	4900	8.367	26.458
21	441	4.583	14.491	71	5041	8.426	26.646
22	484	4.690	14.832	72	5184	8.485	26.833
23	529	4.796	15.166	73	5329	8.544	27.019
24	576	4.899	15.492	74	5476	8.602	27.203
25	625	5.000	15.811	75	5625	8.660	27.386
26	676	5.099	16.125	76	5776	8.718	27.568
27	729	5.196	16.432	77	5929	8.775	27.749
28	784	5.292	16.733	78	6084	8.832	27.928
29	841	5.385	17.029	79	6241	8.888	28.107
30	900	5.477	17.321	80	6400	8.944	28.284
31	961	5.568	17.607	81	6561	9.000	28.460
32	1024	5.657	17.889	82	6724	9.055	28.636
33	1089	5.745	18.166	83	6889	9.110	28.810
34	1156	5.831	18.439	84	7056	9.165	28.983
35	1225	5.916	18.708	85	7225	9.220	29.155
36	1296	6.000	18.974	86	7396	9.274	29.326
37	1369	6.083	19.235	87	7569	9.327	29.496
38	1444	6.164	19.494	88	7744	9.381	29.665
39	1521	6.245	19.748	89	7921	9.434	29.833
40	1600	6.325	20.000	90	8100	9.487	30.000
41	1681	6.403	20.248	91	8281	9.539	30.166
42	1764	6.481	20.494	92	8464	9.592	30.332
43	1849	6.557	20.736	93	8649	9.644	30.496
44	1936	6.633	20.976	94	8836	9.695	30.659
45	2025	6.708	21.213	95	9025	9.747	30.822
46	2116	6.782	21.448	96	9216	9.798	30.984
47	2209	6.856	21.679	97	9409	9.849	31.145
48	2304	6.928	21.909	98	9604	9.899	31.305
49	2401	7.000	22.136	99	9801	9.950	31.464
50	2500	7.071	22.361	100	10000	10.000	31.623

TABLE 3 COMMON LOGARITHMS

n	0	1	2	3	4	5	6	7	8	9
1.0	.0000	.0043	.0086	.0128	.0170	.0212	.0253	.0294	.0334	.0374
1.1	.0414	.0453	.0492	.0531	.0569	.0607	.0645	.0682	.0719	.0755
1.2	.0792	.0828	.0864	.0899	.0934	.0969	.1004	.1038	.1072	.1106
1.3	.1139	.1173	.1206	.1239	.1271	.1303	.1335	.1367	.1399	.1430
1.4	.1461	.1492	.1523	.1553	.1584	.1614	.1644	.1673	.1703	.1732
1.5	.1761	.1790	.1818	.1847	.1875	.1903	.1931	.1959	.1987	.2014
1.6	.2041	.2068	.2095	.2122	.2148	.2175	.2201	.2227	.2253	.2279
1.7	.2304	.2330	.2355	.2380	.2405	.2430	.2455	.2480	.2504	.2529
1.8	.2553	.2577	.2601	.2625	.2648	.2672	.2695	.2718	.2742	.2765
1.9	.2788	.2810	.2833	.2856	.2878	.2900	.2923	.2945	.2967	.2989
2.0	.3010	.3032	.3054	.3075	.3096	.3118	.3139	.3160	.3181	.3201
2.1	.3222	.3243	.3263	.3284	.3304	.3324	.3345	.3365	.3385	.3404
2.2	.3424	.3444	.3464	.3483	.3502	.3522	.3541	.3560	.3579	.3598
2.3	.3617	.3636	.3655	.3674	.3692	.3711	.3729	.3747	.3766	.3784
2.4	.3802	.3820	.3838	.3856	.3874	.3892	.3909	.3927	.3945	.3962
2.5	.3979	.3997	.4014	.4031	.4048	.4065	.4082	.4099	.4116	.4133
2.6	.4150	.4166	.4183	.4200	.4216	.4232	.4249	.4265	.4281	.4298
2.7	.4314	.4330	.4346	.4362	.4378	.4393	.4409	.4425	.4440	.4456
2.8	.4472	.4487	.4502	.4518	.4533	.4548	.4564	.4579	.4594	.4609
2.9	.4624	.4639	.4654	.4669	.4683	.4698	.4713	.4728	.4742	.4757
3.0	.4771	.4786	.4800	.4814	.4829	.4843	.4857	.4871	.4886	.4900
3.1	.4914	.4928	.4942	.4955	.4969	.4983	.4997	.5011	.5024	.5038
3.2	.5051	.5065	.5079	.5092	.5105	.5119	.5132	.5145	.5159	.5172
3.3	.5185	.5198	.5211	.5224	.5237	.5250	.5263	.5276	.5289	.5302
3.4	.5315	.5328	.5340	.5353	.5366	.5378	.5391	.5403	.5416	.5428
3.5	.5441	.5453	.5465	.5478	.5490	.5502	.5514	.5527	.5539	.5551
3.6	.5563	.5575	.5587	.5599	.5611	.5623	.5635	.5647	.5658	.5670
3.7	.5682	.5694	.5705	.5717	.5729	.5740	.5752	.5763	.5775	.5786
3.8	.5798	.5809	.5821	.5832	.5843	.5855	.5866	.5877	.5888	.5899
3.9	.5911	.5922	.5933	.5944	.5955	.5966	.5977	.5988	.5999	.6010
4.0	.6021	.6031	.6042	.6053	.6064	.6075	.6085	.6096	.6107	.6117
4.1	.6128	.6138	.6149	.6160	.6170	.6180	.6191	.6201	.6212	.6222
4.2	.6232	.6243	.6253	.6263	.6274	.6284	.6294	.6304	.6314	.6325
4.3	.6335	.6345	.6355	.6365	.6375	.6385	.6395	.6405	.6415	.6425
4.4	.6435	.6444	.6454	.6464	.6474	.6484	.6493	.6503	.6513	.6522
4.5	.6532	.6542	.6551	.6561	.6571	.6580	.6590	.6599	.6609	.6618
4.6	.6628	.6637	.6646	.6656	.6665	.6675	.6684	.6693	.6702	.6712
4.7	.6721	.6730	.6739	.6749	.6758	.6767	.6776	.6785	.6794	.6803
4.8	.6812	.6821	.6830	.6839	.6848	.6857	.6866	.6875	.6884	.6893
4.9	.6902	.6911	.6920	.6928	.6937	.6946	.6955	.6964	.6972	.6981
5.0	.6990	.6998	.7007	.7016	.7024	.7033	.7042	.7050	.7059	.7067
5.1	.7076	.7084	.7093	.7101	.7110	.7118	.7126	.7135	.7143	.7152
5.2	.7160	.7168	.7177	.7185	.7193	.7202	.7210	.7218	.7226	.7235
5.3	.7243	.7251	.7259	.7267	.7275	.7284	.7292	.7300	.7308	.7316
5.4	.7324	.7332	.7340	.7348	.7356	.7364	.7372	.7380	.7388	.7396
n	0	1	2	3	4	5	6	7	8	9

n	0	1	2	3	4	5	6	7	8	9
5.5	.7404	.7412	.7419	.7427	.7435	.7443	.7451	.7459	.7466	.7474
5.6	.7482	.7490	.7497	.7505	.7513	.7520	.7528	.7536	.7543	.7551
5.7	.7559	.7566	.7574	.7582	.7589	.7597	.7604	.7612	.7619	.7627
5.8	.7634	.7642	.7649	.7657	.7664	.7672	.7679	.7686	.7694	.7701
5.9	.7709	.7716	.7723	.7731	.7738	.7745	.7752	.7760	.7767	.7774
6.0	.7782	.7789	.7796	.7803	.7810	.7818	.7825	.7832	.7839	.7846
6.1	.7853	.7860	.7868	.7875	.7882	.7889	.7896	.7903	.7910	.7917
6.2	.7924	.7931	.7938	.7945	.7952	.7959	.7966	.7973	.7980	.7987
6.3	.7993	.8000	.8007	.8014	.8021	.8028	.8035	.8041	.8048	.8055
6.4	.8062	.8069	.8075	.8082	.8089	.8096	.8102	.8109	.8116	.8122
6.5	.8129	.8136	.8142	.8149	.8156	.8162	.8169	.8176	.8182	.8189
6.6	.8195	.8202	.8209	.8215	.8222	.8228	.8235	.8241	.8248	.8254
6.7	.8261	.8267	.8274	.8280	.8287	.8293	.8299	.8306	.8312	.8319
6.8	.8325	.8331	.8338	.8344	.8351	.8357	.8363	.8370	.8376	.8382
6.9	.8388	.8395	.8401	.8407	.8414	.8420	.8426	.8432	.8439	.8445
7.0	.8451	.8457	.8463	.8470	.8476	.8482	.8488	.8494	.8500	.8506
7.1	.8513	.8519	.8525	.8531	.8537	.8543	.8549	.8555	.8561	.8567
7.2	.8573	.8579	.8585	.8591	.8597	.8603	.8609	.8615	.8621	.8627
7.3	.8633	.8639	.8645	.8651	.8657	.8663	.8669	.8675	.8681	.8686
7.4	.8692	.8698	.8704	.8710	.8716	.8722	.8727	.8733	.8739	.8745
7.5	.8751	.8756	.8762	.8768	.8774	.8779	.8785	.8791	.8797	.8802
7.6	.8808	.8814	.8820	.8825	.8831	.8837	.8842	.8848	.8854	.8859
7.7	.8865	.8871	.8876	.8882	.8887	.8893	.8899	.8904	.8910	.8915
7.8	.8921	.8927	.8932	.8938	.8943	.8949	.8954	.8960	.8965	.8971
7.9	.8976	.8982	.8987	.8993	.8998	.9004	.9009	.9015	.9020	.9025
8.0	.9031	.9036	.9042	.9047	.9053	.9058	.9063	.9069	.9074	.9079
8.1	.9085	.9090	.9096	.9101	.9106	.9112	.9117	.9122	.9128	.9133
8.2	.9138	.9143	.9149	.9154	.9159	.9165	.9170	.9175	.9180	.9186
8.3	.9191	.9196	.9201	.9206	.9212	.9217	.9222	.9227	.9232	.9238
8.4	.9243	.9248	.9253	.9258	.9263	.9269	.9274	.9279	.9284	.9289
8.5	.9294	.9299	.9304	.9309	.9315	.9320	.9325	.9330	.9335	.9340
8.6	.9345	.9350	.9355	.9360	.9365	.9370	.9375	.9380	.9385	.9390
8.7	.9395	.9400	.9405	.9410	.9415	.9420	.9425	.9430	.9435	.9440
8.8	.9445	.9450	.9455	.9460	.9465	.9469	.9474	.9479	.9484	.9489
8.9	.9494	.9499	.9504	.9509	.9513	.9518	.9523	.9528	.9533	.9538
9.0	.9542	.9547	.9552	.9557	.9562	.9566	.9571	.9576	.9581	.9586
9.1	.9590	.9595	.9600	.9605	.9609	.9614	.9619	.9624	.9628	.9633
9.2	.9638	.9643	.9647	.9652	.9657	.9661	.9666	.9671	.9675	.9680
9.3	.9685	.9689	.9694	.9699	.9703	.9708	.9713	.9717	.9722	.9727
9.4	.9731	.9736	.9741	.9745	.9750	.9754	.9759	.9763	.9768	.9773
9.5	.9777	.9782	.9786	.9791	.9795	.9800	.9805	.9809	.9814	.9818
9.6	.9823	.9827	.9832	.9836	.9841	.9845	.9850	.9854	.9859	.9863
9.7	.9868	.9872	.9877	.9881	.9886	.9890	.9894	.9899	.9903	.9908
9.8	.9912	.9917	.9921	.9926	.9930	.9934	.9939	.9943	.9948	.9952
9.9	.9956	.9961	.9965	.9969	.9974	.9978	.9983	.9987	.9991	.9996
n	0	1	2	3	4	5	6	7	8	9

TABLE 4   POWERS OF e

x	e^x	e^{-x}	x	e^x	e^{-x}
0.00	1.00000	1.00000			
0.01	1.01005	0.99004	1.60	4.95302	0.20189
0.02	1.02020	0.98019	1.70	5.47394	0.18268
0.03	1.03045	0.97044	1.80	6.04964	0.16529
0.04	1.04081	0.96078	1.90	6.68589	0.14956
0.05	1.05127	0.95122	2.00	7.38905	0.13533
0.06	1.06183	0.94176			
0.07	1.07250	0.93239	2.10	8.16616	0.12245
0.08	1.08328	0.92311	2.20	9.02500	0.11080
0.09	1.09417	0.91393	2.30	9.97417	0.10025
0.10	1.10517	0.90483	2.40	11.02316	0.09071
			2.50	12.18248	0.08208
0.11	1.11628	0.89583	2.60	13.46372	0.07427
0.12	1.12750	0.88692	2.70	14.87971	0.06720
0.13	1.13883	0.87810	2.80	16.44463	0.06081
0.14	1.15027	0.86936	2.90	18.17412	0.05502
0.15	1.16183	0.86071	3.00	20.08551	0.04978
0.16	1.17351	0.85214			
0.17	1.18530	0.84366	3.50	33.11545	0.03020
0.18	1.19722	0.83527	4.00	54.59815	0.01832
0.19	1.20925	0.82696	4.50	90.01713	0.01111
0.20	1.22140	0.81873	5.00	148.41316	0.00674
0.30	1.34985	0.74081	5.50	224.69193	0.00409
0.40	1.49182	0.67032			
0.50	1.64872	0.60653	6.00	403.42879	0.00248
0.60	1.82211	0.54881	6.50	665.14163	0.00150
0.70	2.01375	0.49658	7.00	1096.63316	0.00091
0.80	2.22554	0.44932	7.50	1808.04241	0.00055
0.90	2.45960	0.40656			
1.00	2.71828	0.36787	8.00	2980.95799	0.00034
			8.50	4914.76884	0.00020
1.10	3.00416	0.33287			
1.20	3.32011	0.30119	9.00	8103.08393	0.00012
1.30	3.66929	0.27253	9.50	13359.72683	0.00007
1.40	4.05519	0.24659			
1.50	4.48168	0.22313	10.00	22026.46579	0.00005

TABLE 5   NATURAL LOGARITHMS

x	ln x	x	ln x	x	ln x
		4.5	1.5041	9.0	2.1972
0.1	7.6974 − 10	4.6	1.5261	9.1	2.2083
0.2	8.3906 − 10	4.7	1.5476	9.2	2.2192
0.3	8.7960 − 10	4.8	1.5686	9.3	2.2300
0.4	9.0837 − 10	4.9	1.5892	9.4	2.2407
0.5	9.3069 − 10	5.0	1.6094	9.5	2.2513
0.6	9.4892 − 10	5.1	1.6292	9.6	2.2618
0.7	9.6433 − 10	5.2	1.6487	9.7	2.2721
0.8	9.7769 − 10	5.3	1.6677	9.8	2.2824
0.9	9.8946 − 10	5.4	1.6864	9.9	2.2925
1.0	0.0000	5.5	1.7047	10	2.3026
1.1	0.0953	5.6	1.7228	11	2.3979
1.2	0.1823	5.7	1.7405	12	2.4849
1.3	0.2624	5.8	1.7579	13	2.5649
1.4	0.3365	5.9	1.7750	14	2.6391
1.5	0.4055	6.0	1.7918	15	2.7081
1.6	0.4700	6.1	1.8083	16	2.7726
1.7	0.5306	6.2	1.8245	17	2.8332
1.8	0.5878	6.3	1.8405	18	2.8904
1.9	0.6419	6.4	1.8563	19	2.9444
2.0	0.6931	6.5	1.8718	20	2.9957
2.1	0.7419	6.6	1.8871		
2.2	0.7885	6.7	1.9021	25	3.2189
2.3	0.8329	6.8	1.9169	30	3.4012
2.4	0.8755	6.9	1.9315	35	3.5553
				40	3.6889
2.5	0.9163	7.0	1.9459		
2.6	0.9555	7.1	1.9601	45	3.8067
2.7	0.9933	7.2	1.9741	50	3.9120
2.8	1.0296	7.3	1.9879		
2.9	1.0647	7.4	2.0015	55	4.0073
				60	4.0943
3.0	1.0986	7.5	2.0149	65	4.1744
3.1	1.1314	7.6	2.0281		
3.2	1.1632	7.7	2.0412	70	4.2485
3.3	1.1939	7.8	2.0541	75	4.3175
3.4	1.2238	7.9	2.0669	80	4.3820
				85	4.4427
3.5	1.2528	8.0	2.0794	90	4.4998
3.6	1.2809	8.1	2.0919		
3.7	1.3083	8.2	2.1041	95	4.5539
3.8	1.3350	8.3	2.1163	100	4.6052
3.9	1.3610	8.4	2.1281		
4.0	1.3863	8.5	2.1401		
4.1	1.4110	8.6	2.1518		
4.2	1.4351	8.7	2.1633		
4.3	1.4586	8.8	2.1748		
4.4	1.4816	8.9	2.1861		

TABLE 6 INTEGRALS INVOLVING TRIGONOMETRIC FUNCTIONS*

18. $\displaystyle\int \sin u \, du = -\cos u + C$

19. $\displaystyle\int \cos u \, du = \sin u + C$

20. $\displaystyle\int \sec^2 u \, du = \tan u + C$

21. $\displaystyle\int \csc^2 u \, du = -\cot u + C$

22. $\displaystyle\int \sec u \tan u \, du = \sec u + C$

23. $\displaystyle\int \csc u \cot u \, du = -\csc u + C$

24. $\displaystyle\int \tan u \, du = \ln |\sec u| + C$

25. $\displaystyle\int \cot u \, du = \ln |\sin u| + C$

26. $\displaystyle\int \sec u \, du = \ln |\sec u + \tan u| + C$

27. $\displaystyle\int \csc u \, du = \ln |\csc u - \cot u| + C$

28. $\displaystyle\int \frac{du}{\sqrt{a^2 - u^2}} = \text{Arcsin}\,\frac{u}{a} + C$

29. $\displaystyle\int \frac{du}{a^2 + u^2} = \frac{1}{a}\,\text{Arctan}\,\frac{u}{a} + C$

30. $\displaystyle\int \frac{du}{u\sqrt{u^2 - a^2}} = \frac{1}{a}\,\text{Arcsec}\,\frac{u}{a} + C$

31. $\displaystyle\int \sqrt{a^2 - u^2}\, du = \frac{u}{2}\sqrt{a^2 - u^2} + \frac{a^2}{2}\,\text{Arcsin}\,\frac{u}{a} + C$

32. $\displaystyle\int u^2\sqrt{a^2 - u^2}\, du = \frac{u}{8}(2u^2 - a^2)\sqrt{a^2 - u^2} + \frac{a^4}{8}\,\text{Arcsin}\,\frac{u}{a} + C$

33. $\displaystyle\int \frac{\sqrt{a^2 - u^2}}{u^2}\, du = -\frac{1}{u}\sqrt{a^2 - u^2} - \text{Arcsin}\,\frac{u}{a} + C$

34. $\displaystyle\int \frac{u^2\, du}{\sqrt{a^2 - u^2}} = -\frac{u}{2}\sqrt{a^2 - u^2} + \frac{a^2}{2}\,\text{Arcsin}\,\frac{u}{a} + C$

35. $\displaystyle\int \frac{\sqrt{u^2 - a^2}}{u}\, du = \sqrt{u^2 - a^2} - a\,\text{Arccos}\,\frac{a}{u} + C$

---

*From *Calculus with Analytic Geometry* by Earl W. Swokowski. Prindle, Weber & Schmidt. Copyright © 1979.

36. $\displaystyle\int \sin^n u \, du = -\frac{1}{n} \sin^{n-1} u \cos u + \frac{n-1}{n} \int \sin^{n-2} u \, du$

37. $\displaystyle\int \cos^n u \, du = \frac{1}{n} \cos^{n-1} u \sin u + \frac{n-1}{n} \int \cos^{n-2} u \, du$

38. $\displaystyle\int \tan^n u \, du = \frac{1}{n-1} \tan^{n-1} u - \int \tan^{n-2} u \, du$

39. $\displaystyle\int \sec^n u \, du = \frac{1}{n-1} \tan u \sec^{n-2} u + \frac{n-2}{n-1} \int \sec^{n-2} u \, du$

40. $\displaystyle\int \sin au \sin bu \, du = \frac{\sin(a-b)u}{2(a-b)} - \frac{\sin(a+b)u}{2(a+b)} + C$

41. $\displaystyle\int \cos au \cos bu \, du = \frac{\sin(a-b)u}{2(a-b)} + \frac{\sin(a+b)u}{2(a+b)} + C$

42. $\displaystyle\int \sin au \cos bu \, du = -\frac{\cos(a-b)u}{2(a-b)} - \frac{\cos(a+b)u}{2(a+b)} + C$

43. $\displaystyle\int u \sin u \, du = \sin u - u \cos u + C$

44. $\displaystyle\int u^n \sin u \, du = -u^n \cos u + n \int u^{n-1} \cos u \, du$

45. $\displaystyle\int u \operatorname{Arccos} u \, du = \frac{2u^2 - 1}{4} \operatorname{Arccos} u - \frac{u\sqrt{1-u^2}}{4} + C$

46. $\displaystyle\int u \operatorname{Arcsin} u \, du = \frac{2u^2 - 1}{4} \operatorname{Arcsin} u + \frac{u\sqrt{1-u^2}}{4} + C$

47. $\displaystyle\int e^{au} \sin bu \, du = \frac{e^{au}}{a^2 + b^2}(a \sin bu - b \cos bu) + C$

48. $\displaystyle\int e^{au} \cos bu \, du = \frac{e^{au}}{a^2 + b^2}(a \cos bu + b \sin bu) + C$

TABLE 7 TRIGONOMETRIC FUNCTIONS

Deg	Rad	Sin	Cos	Tan	Deg	Rad	Sin	Cos	Tan
0	0.0000	0.0000	1.0000	0.0000	45	0.7854	0.7071	0.7071	1.0000
1	0.0175	0.0175	0.9998	0.0175	46	0.8029	0.7193	0.6947	1.0355
2	0.0349	0.0349	0.9994	0.0349	47	0.8203	0.7314	0.6820	1.0724
3	0.0524	0.0523	0.9986	0.0524	48	0.8378	0.7431	0.6691	1.1106
4	0.0698	0.0698	0.9976	0.0699	49	0.8552	0.7547	0.6561	1.1504
5	0.0873	0.0872	0.9962	0.0875	50	0.8727	0.7660	0.6428	1.1918
6	0.1047	0.1045	0.9945	0.1051	51	0.8901	0.7771	0.6293	1.2349
7	0.1222	0.1219	0.9925	0.1228	52	0.9076	0.7880	0.6157	1.2799
8	0.1396	0.1392	0.9903	0.1405	53	0.9250	0.7986	0.6018	1.3270
9	0.1571	0.1564	0.9877	0.1584	54	0.9425	0.8090	0.5878	1.3764
10	0.1745	0.1736	0.9848	0.1763	55	0.9599	0.8192	0.5736	1.4281
11	0.1920	0.1908	0.9816	0.1944	56	0.9774	0.8290	0.5592	1.4826
12	0.2094	0.2079	0.9781	0.2126	57	0.9948	0.8387	0.5446	1.5399
13	0.2269	0.2250	0.9744	0.2309	58	1.0123	0.8480	0.5299	1.6003
14	0.2443	0.2419	0.9703	0.2493	59	1.0297	0.8572	0.5150	1.6643
15	0.2618	0.2588	0.9659	0.2679	60	1.0472	0.8660	0.5000	1.7321
16	0.2793	0.2756	0.9613	0.2867	61	1.0647	0.8746	0.4848	1.8040
17	0.2967	0.2924	0.9563	0.3057	62	1.0821	0.8829	0.4695	1.8807
18	0.3142	0.3090	0.9511	0.3249	63	1.0996	0.8910	0.4540	1.9626
19	0.3316	0.3256	0.9455	0.3443	64	1.1170	0.8988	0.4384	2.0503
20	0.3491	0.3420	0.9397	0.3640	65	1.1345	0.9063	0.4226	2.1445
21	0.3665	0.3584	0.9336	0.3839	66	1.1519	0.9135	0.4067	2.2460
22	0.3840	0.3746	0.9272	0.4040	67	1.1694	0.9205	0.3907	2.3559
23	0.4014	0.3907	0.9205	0.4245	68	1.1868	0.9272	0.3746	2.4751
24	0.4189	0.4067	0.9135	0.4452	69	1.2043	0.9336	0.3584	2.6051
25	0.4363	0.4226	0.9063	0.4663	70	1.2217	0.9397	0.3420	2.7475
26	0.4538	0.4384	0.8988	0.4877	71	1.2392	0.9455	0.3256	2.9042
27	0.4712	0.4540	0.8910	0.5095	72	1.2566	0.9511	0.3090	3.0777
28	0.4887	0.4695	0.8829	0.5317	73	1.2741	0.9563	0.2924	3.2709
29	0.5061	0.4848	0.8746	0.5543	74	1.2915	0.9613	0.2756	3.4874
30	0.5236	0.5000	0.8660	0.5774	75	1.3090	0.9659	0.2588	3.7321
31	0.5411	0.5150	0.8572	0.6009	76	1.3265	0.9703	0.2419	4.0108
32	0.5585	0.5299	0.8480	0.6249	77	1.3439	0.9744	0.2250	4.3315
33	0.5760	0.5446	0.8387	0.6494	78	1.3614	0.9781	0.2079	4.7046
34	0.5934	0.5592	0.8290	0.6745	79	1.3788	0.9816	0.1908	5.1446
35	.6109	0.5736	0.8192	0.7002	80	1.3963	0.9848	0.1736	5.6713
36	0.6283	0.5878	0.8090	0.7265	81	1.4137	0.9877	0.1564	6.3138
37	0.6458	0.6018	0.7986	0.7536	82	1.4312	0.9903	0.1392	7.1154
38	0.6632	0.6157	0.7880	0.7813	83	1.4486	0.9925	0.1219	8.1442
39	0.6807	0.6293	0.7771	0.8098	84	1.4661	0.9945	0.1045	9.5144
40	0.6981	0.6428	0.7660	0.8391	85	1.4835	0.9962	0.0872	11.4301
41	0.7156	0.6561	0.7547	0.8693	86	1.5010	0.9976	0.0698	14.3007
42	0.7330	0.6691	0.7431	0.9004	87	1.5184	0.9986	0.0523	19.0811
43	0.7505	0.6820	0.7314	0.9325	88	1.5359	0.9994	0.0349	28.6363
44	0.7679	0.6947	0.7193	0.9657	89	1.5533	0.9998	0.0175	57.2900
					90	1.5708	1.0000	0.0000	

*From *Calculus for the Life Sciences* by Rodolfo De Sapio. W. H. Freeman and Company. Copyright © 1978.

List of Symbols

The number of the page where the symbol first appears is on the left.

2	$\{\ \}$	set braces
2	$\approx$	approximately equal to
3	$<$	less than
3	$\leq$	less than or equal to
3	$>$	greater than
3	$\geq$	greater than or equal to
5	$\|x\|$	absolute value of x
32	a^n	$a \cdot a \cdot a \cdots a$; a appears n times
33	a^0	1, if $a \neq 0$
36	$\sqrt{x}$	square root of x
44	(a, b)	ordered pair
45	$f(x)$	value of the function f for the number x
59	x_1	x-sub-one
59	$\triangle x$	change in x ($\triangle$ is the Greek letter delta)
61	m	slope of a line
94	$\lim_{x \to a} f(x)$	limit of f as x approaches a
98	$x \to \infty$	x gets larger and larger without bound
109	(a, b)	open interval, $a < x < b$
109	$[a, b]$	closed interval, $a \leq x \leq b$
109	$(-\infty, a)$	open interval, $x < a$
109	(b, ∞)	open interval, $x > b$
122	$f'(x)$	derivative of $f(x)$
131	y'	derivative of $y = f(x)$
170	$f''(x)$	second derivative of $f(x)$
171	y''	second derivative of $y = f(x)$
206	dy	differential of y
221	e	an irrational number; $e \approx 2.7182818$
226	$\log_a x$	logarithm of x to the base a
228	$\log x$	common logarithm of x
229	$\ln x$	natural logarithm of x
255	$F(x)$	antiderivative of $f(x)$
264	$\int_a^b f(x)\,dx$	definite integral
265	Σ	Greek letter sigma, represents summation
270	$F(x)\Big]_a^b$	$F(b) - F(a)$

List of Symbols

270	$\int f(x)\,dx$	indefinite integral, antiderivative
282	$\int_a^b [g(x) - f(x)]dx$	area between two curves
314	$\int_a^b \pi[f(x)]^2 dx$	volume
319	$\int_a^\infty f(x)dx$	improper integral
324	$P(t = a)$	probability that a random variable has value a
328	μ	Greek letter mu, represents the mean
328	σ	Greek letter sigma, represents the standard deviation
338	C_1	arbitrary constant
363	$z = f(x, y)$	function of two variables
364	$f(x_1, x_2, x_3, \cdots, x_n)$	function of several variables
366	(x, y, z)	ordered triple
373	f_x	partial derivative
373	$\partial f/\partial x$	partial derivative
375	f_{xx}	second partial derivative
375	$\partial^2 z/\partial x^2$	second partial derivative
388	λ	Greek letter lambda, used in setting up a Lagrange function
404	$1°$	one degree
404	$1'$	one minute
405	θ	Greek letter theta
408	α	Greek letter alpha
408	β	Greek letter beta

Answers To Selected Exercises

Chapter 1

Section 1.1 (page 5)

1. Counting number, whole number, integer, rational number, real number   3. Integer, rational number, real number   5. Rational number, real number   7. Irrational number, real number   9. Irrational number, real number   11. True   13. True   15. False   17. True
19. [number line with points at -4, -2, 0, 2, 4]
21. [number line 0 to 3]
23. [number line 1 to 4]
25. [number line ending at 4]
27. [number line at 6]
29. [number line at -2]
31. [number line from -5 to -3]
33. [number line from -3 to 6]
35. [number line from 1 to 6]
37. 8   39. -4   41. 2
43. -4   45. 17   47. 4
49. -19   51. =   53. <
55. =   57. =   59. =
61. =   63. =   65. Yes
67. Let x represent the percent of the money paid to the distributor. Then $x \geq 35$.   69. $x \geq 12{,}600$
71. Let x be the manufacturing cost of a product, in percent of the retail price. Then $x \geq 15$.   73. 100; 0
75. 67; 11

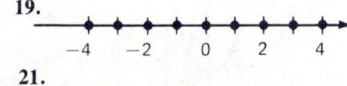

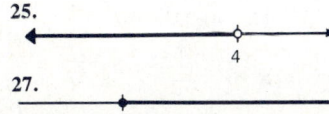

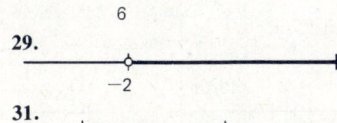

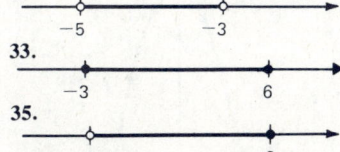

Section 1.2 (page 13)

1. 3   3. -9   5. 2   7. 3
9. 4   11. 7   13. -1   15. 3
17. -3   19. 5   21. $x \leq -3$
23. $p > -6$ 25. $a > 0$
27. $x \leq 4$ 29. $k < 1$
31. $m > -1$ 33. $y \leq 1$
35. 68°F   37. 15°C
39. 37.8°C   41. 104°F
43. 13%   45. $432
47. $1500   49. $205.41
51. $66.50   53. $C = 1.2x$
55. Profit $= 5x - 1.2x = 3.8x$
57. Avis is better if $14(7) < 54 + .07x$, where x is the number of miles driven in a week. They must drive 629 miles for Avis to be a better deal.

Section 1.3 (page 19)

1. $14m + 6$ 3. $-9k + 5$
5. $-x^2 + x + 9$ 7. $-6y^2 + 3y + 10$
9. $-10x^2 + 4x - 2$
11. $6p^2 - 15p$ 13. $-18m^3 - 27m^2 + 9m$ 15. $12k^2 - 20k + 3$
17. $6y^2 + 7y - 5$ 19. $36y^2 - 4$
21. $5(5k + 6)$ 23. $4(z + 1)$
25. $2(4x + 3y + 2z)$
27. $m(m^2 - 9m + 6)$
29. $8a(a^2 - 2a + 3)$ 31. $(m + 7)(m + 2)$ 33. $(x + 5)(x - 1)$
35. $(z + 5)(z + 4)$ 37. $(b - 7)(b - 1)$ 39. Cannot be factored
41. $(y - 7)(y + 3)$ 43. $6(a - 10)(a + 2)$ 45. $3m(m + 3)(m + 1)$
47. $(2x + 1)(x - 3)$
49. $(3a + 7)(a + 1)$ 51. $(5y + 2)(3y - 1)$ 53. Cannot be factored
55. $(5a + 3)(a - 2)$
57. $2a^2(4a - 1)(3a + 2)$
59. $4z^3(8z + 3)(z - 1)$
61. $(x + 8)(x - 8)$ 63. $(3m + 5)(3m - 5)$ 65. Cannot be factored
67. $(z + 7)^2$ 69. $(m - 3)^2$
71. $(3p - 4)^2$ 73. $(a - 6)(a^2 + 6a + 36)$ 75. $(2r - 3)(4r^2 + 6r + 9)$

Section 1.4 (page 26)

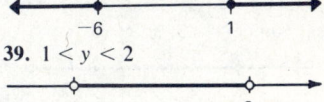

1. 5; -4   3. -2; -3   5. 6; -1
7. -8; 3   9. 4   11. 5/2; -2
13. 4/3; -1/2   15. 5; 2   17. 0; 1
19. $(5 + \sqrt{13})/6 \approx 1.434$; $(5 - \sqrt{13})/6 \approx .232$ 21. $(1 + \sqrt{33})/4 \approx 1.686$; $(1 - \sqrt{33})/4 \approx -1.186$
23. $(10 + \sqrt{20})/2 \approx 7.236$; $(10 - \sqrt{20})/2 \approx 2.764$
25. $(-12 + \sqrt{104})/4 \approx -.450$; $(-12 - \sqrt{104})/4 \approx -5.550$
27. 4/3; 1/2   29. -5; 2   31. No real number solutions   33. 0; -1
35. $-2 < m < 4$
[number line from -2 to 4]
37. $t \leq -6$ or $t \geq 1$
[number line showing -6 and 1]
39. $1 < y < 2$
[number line from 1 to 2]
41. $k < -4$ or $k > 1/2$
[number line showing -4 and 1/2]
43. $-3 \leq y \leq 1/2$
[number line from -3 to 1/2]
45. $-5 \leq x \leq 5$
[number line from -5 to 5]

468 Answers

47. $54   **49.** 5/4 months and 6 months

Section 1.5 (page 31)

1. $m/4$ **3.** $z/2$ **5.** $8/9$
7. $2(x+2)/x$ **9.** $(m-2)/(m+3)$
11. $(x+4)/(x+1)$ **13.** $3k/5$
15. $25p^2/9$ **17.** $6/(5p)$ **19.** 2
21. $2/9$ **23.** $2(a+4)/(a-3)$
25. $(k+2)/(k+3)$ **27.** $(m+6)/(m+3)$
29. $(m-3)/(2m-3)$ **31.** $14/r$
33. $19/(6k)$ **35.** $5/(12y)$ **37.** 1
39. $(6+p)/(2p)$ **41.** $(8-y)/(4y)$
43. $137/(30m)$ **45.** $(r-12)/[r(r-2)]$
47. $14/[3(a-1)]$ **49.** $23/[20(k-2)]$
51. $(x+1)/(x-1)$ **53.** $-1/(x+1)$

Section 1.6 (page 37)

1. 343 **3.** $1/8$ **5.** $1/8$
7. $27/64$ **9.** 8 **11.** 9
13. 3 **15.** 4 **17.** 100
19. 4 **21.** -25 **23.** $2/3$
25. $1/32$ **27.** $4/3$ **29.** $3/4$
31. 3^6 **33.** $1/3^6$ **35.** $1/6^2$
37. 4^3 **39.** $1/7^7$ **41.** 8^5
43. $1/10^8$ **45.** 5 **47.** 2^2
49. $27^{1/3}$ or 3 **51.** 4^2
53. $(\sqrt{7})^3$ **55.** $1/(\sqrt[3]{60})^2$
57. $12/(\sqrt{x})^3$ **59.** $1/[\sqrt[3]{(3r)}]^2$
61. $64   **63.** $64,000,000
65. About 86 miles   **67.** About 211 miles   **69.** 29 **71.** 177
73. 11.7

Chapter 1 Test (page 41)

1. Integer, rational number, real number   **2.** Irrational number, real number   **3.** $x \geq -3$

4. $-4 < x \leq 6$

5. -6 **6.** 4 **7.** 2 **8.** $2/3$
9. 7 **10.** $k \geq -4$ **11.** $m < 14$
12. $84   **13.** $8(y+2)$
14. $(p-7)(p-2)$ **15.** $(k+9) \cdot (k-5)$ **16.** $(2a-3)(a+5)$
17. $(3m+7)(m-5)$
18. $(4p+3)^2$ **19.** $-6; 3$

20. $-2/3; 5/2$ **21.** $(4+\sqrt{8})/2 \approx 3.414; (4-\sqrt{8})/2 \approx .586$
22. $(6+\sqrt{108})/18 \approx .911; (6-\sqrt{108})/18 \approx -.244$
23. $-3 \leq x \leq 1$ **24.** $y < -1/3$ or $y > 2$ **25.** $8p^2/5$ **26.** $2r/3$
27. 4 **28.** $17/(2r)$
29. $119/(72y)$ **30.** $(5r+3)/[r(r-1)]$
31. $1/16$ **32.** 36 **33.** $1/8$
34. 512 **35.** $1/243$ **36.** $7/12$

Chapter 2

Section 2.1 (page 49)

1. (a) 21 (b) 0 (c) 9 (d) $3a+9$
3. (a) -12 (b) 2 (c) -4 (d) $-2a-4$
5. (a) 48 (b) 6 (c) 0 (d) $2a^2+4a$
7. (a) 5 (b) -23 (c) 1
(d) $-a^2+5a+1$ **9.** (a) 18 (b) 4
(c) -2 (d) $(a-1)(a+2)$ **11.** -3
13. $2a-3$ **15.** $2m+3$
17. -1 **19.** $(-2, 8), (-1, 7), (0, 6), (1, 5), (2, 4), (3, 3)$

$(-2, 8)$
$(-1, 7)$
$(0, 6)$
$(1, 5)$
$(2, 4)$
$(3, 3)$

$y = -x + 6$

21. $(-2, 4), (-1, 3), (0, 2), (1, 1), (2, 0), (3, 1)$

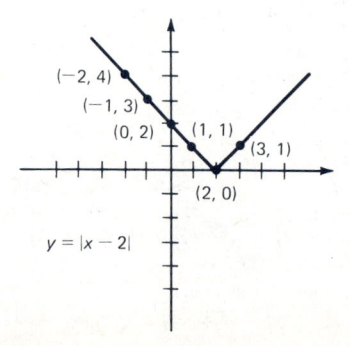

$(-2, 4)$
$(-1, 3)$
$(0, 2)$ $(1, 1)$
$(3, 1)$
$(2, 0)$

$y = |x - 2|$

23. $(-2, -2), (-1, -1), (0, 0), (1, -1), (2, -2), (3, -3)$

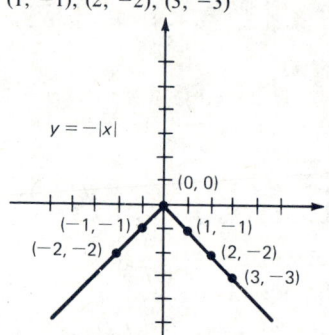

$y = -|x|$
$(0, 0)$
$(-1, -1)$ $(1, -1)$
$(-2, -2)$ $(2, -2)$
$(3, -3)$

25. $(-2, 6), (-1, 3), (0, 2), (1, 3), (2, 6), (3, 11)$

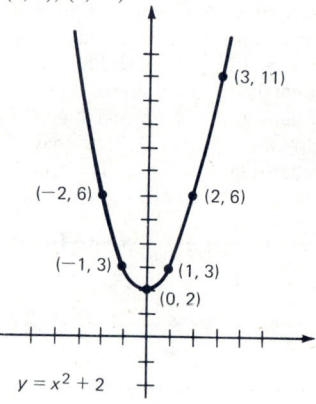

$(3, 11)$
$(-2, 6)$ $(2, 6)$
$(-1, 3)$ $(1, 3)$
$(0, 2)$

$y = x^2 + 2$

27. $(-2, -4), (-1, -1), (0, 0), (1, -1), (2, -4), (3, -9)$

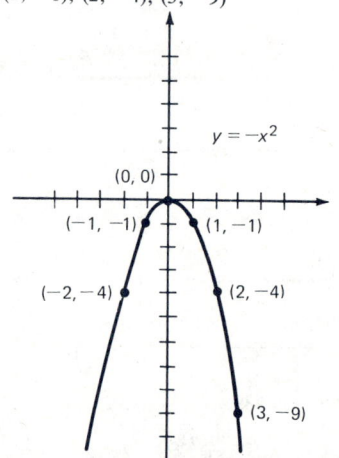

$y = -x^2$
$(0, 0)$
$(-1, -1)$ $(1, -1)$
$(-2, -4)$ $(2, -4)$
$(3, -9)$

29. Function   **31.** Not a function   **33.** $1,050,000
35. $1,200,000   **37.** $11
39. $18   **41.** $32   **43.** $39
45. $58   **47.** $58   **49.** $94

Section 2.2 (page 57)

1. $y = 2x + 1$

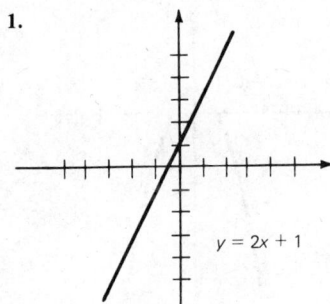

3. $y = 4x$

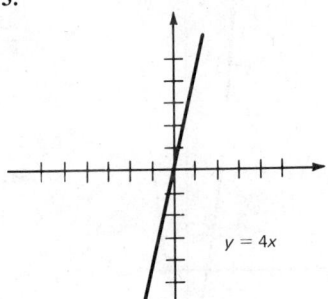

5. $3y + 4x = 12$

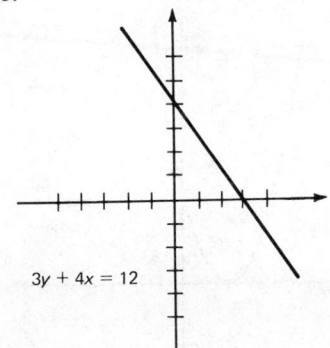

7. Horizontal; $y = -2$

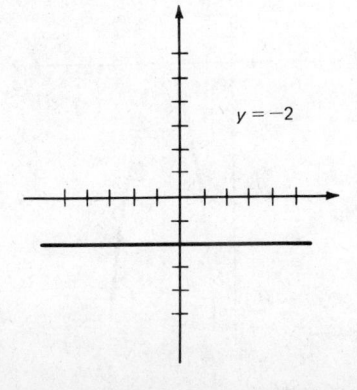

9. $6x + y = 12$

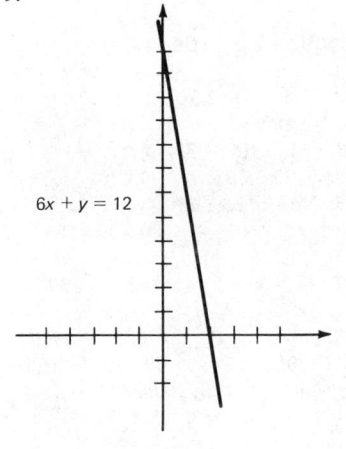

11. $x - 5y = 4$

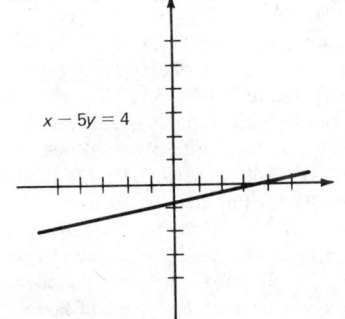

13. Vertical; $x + 5 = 0$

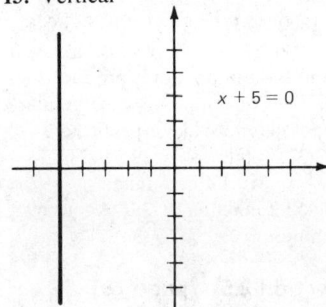

15. $5y - 3x = 12$

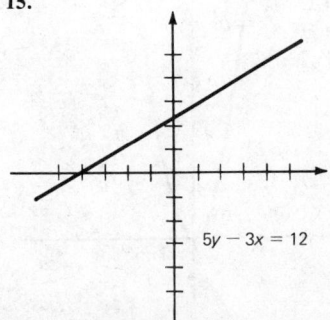

17. $8x + 3y = 10$

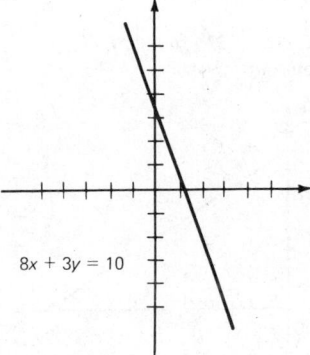

19. $y = 2x$

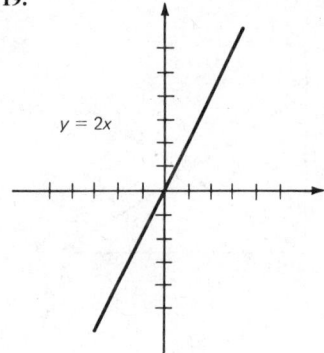

21. $y = -4x$

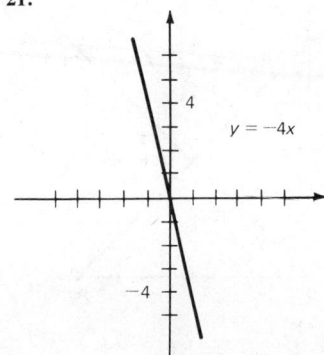

23. $x + 4y = 0$

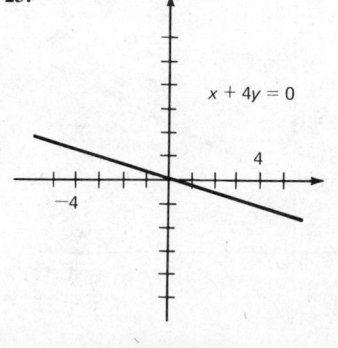

470 Answers

25. $16   **27.** $6   **29.** 24/5 thousands or 4800

31.

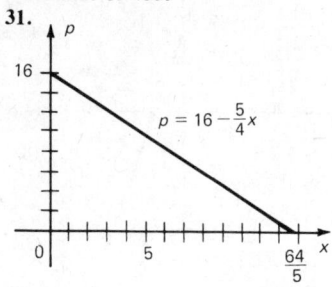

[Graph: $p = 16 - \frac{5}{4}x$, from (0,16) to (64/5, 0)]

33. 8

35.

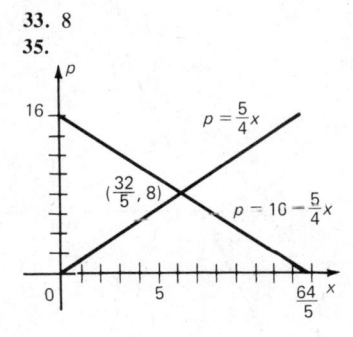

[Graph showing two lines: $p = \frac{5}{4}x$ and $p = 10 - \frac{5}{4}x$, intersecting at $(\frac{32}{5}, 8)$]

37. $8   **39.** 100/3 or 33 1/3

41.

[Graph showing two lines: $p = 100 - \frac{2}{5}x$ and $p = \frac{2}{5}x$, intersecting at (125, 50)]

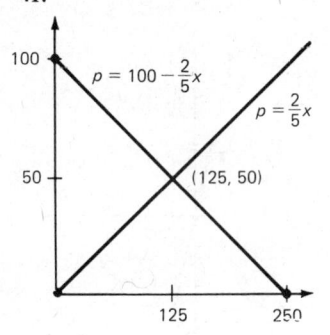

43. $50   **45.** 1.12
47. $14.40   **49.** $43.20
51. $135   **53.** $275
55.

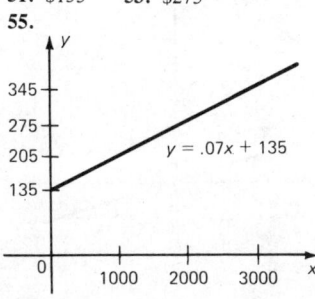

[Graph of $y = .07x + 135$ with y-values 135, 205, 275, 345 at x = 0, 1000, 2000, 3000]

Section 2.3 (page 63)

1. $-1/5$ **3.** 2/3   **5.** $-3/2$
7. No slope   **9.** 0   **11.** 3
13. -4 **15.** $-3/4$ **17.** -3
19. $-2/5$ **21.** 0   **23.** 0
25. No slope   **27.** $4y = -3x + 16$
29. $2y = -x - 4$ **31.** $4y = 6x + 5$
33. $y = 2x + 9$ **35.** $y = -3x + 3$
37. $4y = x + 5$ **39.** $3y = 4x + 7$
41. $3y = -2x$ **43.** $x = -8$
45. $y = 3$ **47.** $y = 640x + 1100$; $12,620; $18,380   **49.** $y = -1000x + 40,000$; 33,000; 26,000
51. $y = 2.5x - 70$; 55%; 72.5%

Section 2.4 (page 69)

1. (a) 2600 (b) 2900 (c) 3200 (d) 2000 (e) 300   **3.** (a) 100 thousand (b) 70 thousand (c) 0 thousand (d) -5 thousand; the number is decreasing   **5.** (a) $y = 6500x + 85,000$ (b) $137,000 (c) $169,500   **7.** (a) 480 (b) 360 (c) 120 (d) June 30 (e) -20 **9.** If $C(x)$ is the cost of renting a saw for x hours, then $C(x) = 12 + x$.   **11.** If $P(x)$ is the cost in cents for parking for x half hours, then $P(x) = 35 + 30x$.   **13.** $C(x) = 30x + 100$ **15.** $C(x) = 25x + 1000$ **17.** $24   **19.** $48.048
21. 500; $30,000   **23.** Break-even point is 45 units; don't produce
25. Break-even point is -50 units; impossible to make a profit here
27. 32.5 minutes   **29.** 145 minutes   **31.** 13.95 minutes
33. 52.2 minutes   **35.** About 69 minutes

Section 2.5 (page 78)

1.

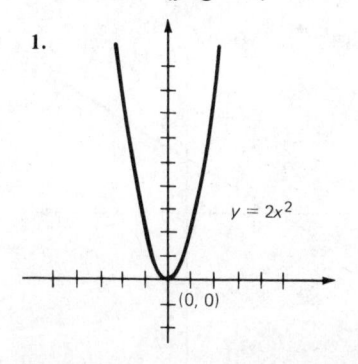

[Graph of $y = 2x^2$, vertex at (0, 0)]

3.

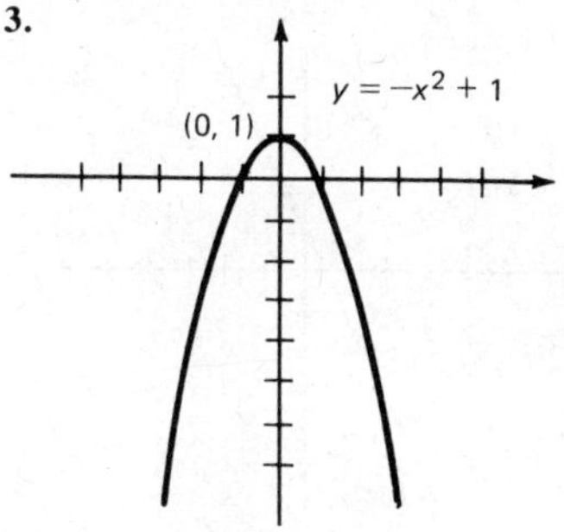

[Graph of $y = -x^2 + 1$, vertex at (0, 1)]

5.

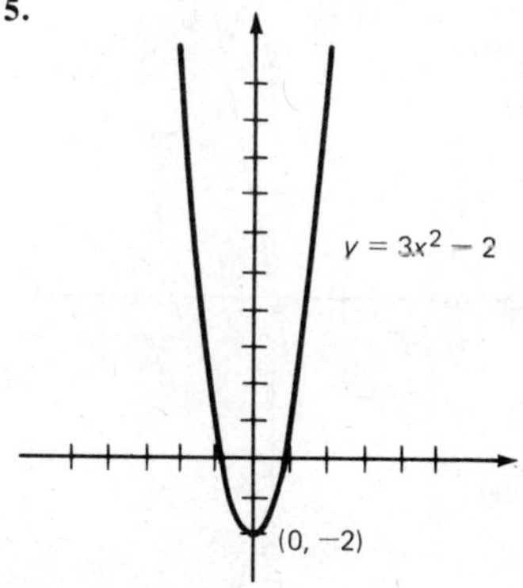

[Graph of $y = 3x^2 - 2$, vertex at (0, -2)]

7.

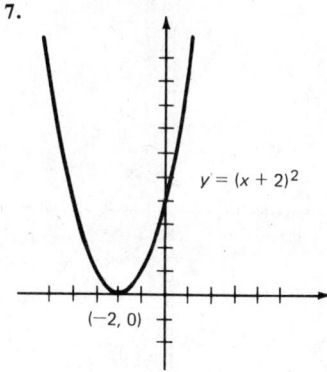

[Graph of $y = (x + 2)^2$, vertex at (-2, 0)]

9.

[Graph of $y = -2(x - 3)^2$, vertex at (3, 0)]

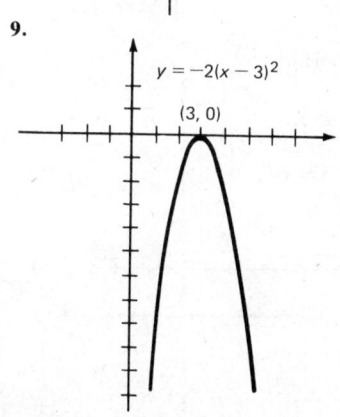

Answers 471

11.

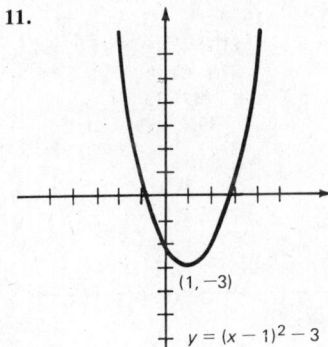

[Graph: $y = (x-1)^2 - 3$, vertex $(1, -3)$]

13. (a)

[Graph: parabola with vertex $(5, 15)$]

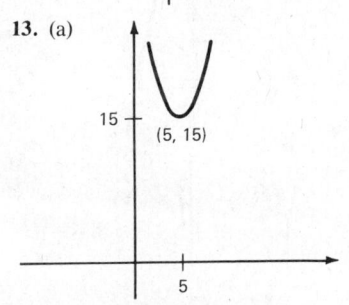

(b) 5 (c) $15   **15.** (a) 10 milliseconds (b) 140 impulses
17. (a) 640 (b) 515 (c) 140
(d)

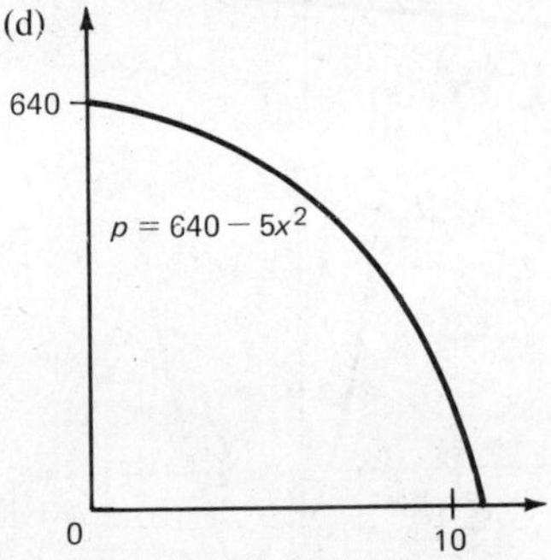

[Graph: $p = 640 - 5x^2$]

19. (a) About 12 (b) 10 (c) about 7
(d) 0
(e)

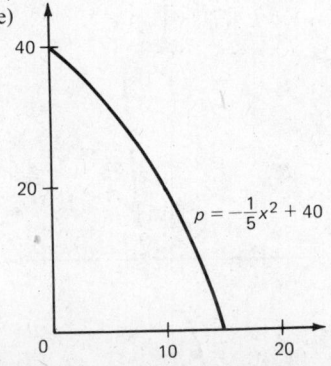

[Graph: $p = -\frac{1}{5}x^2 + 40$]

21. 10
23.

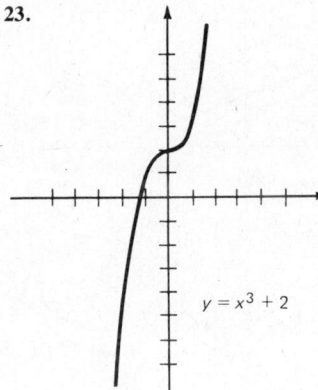

[Graph: $y = x^3 + 2$]

25.

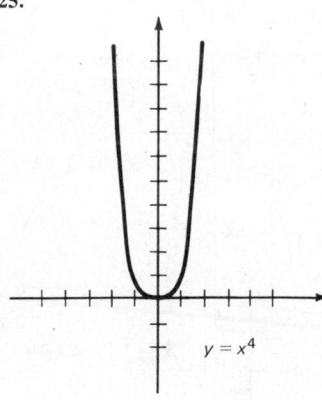

[Graph: $y = x^4$]

27.

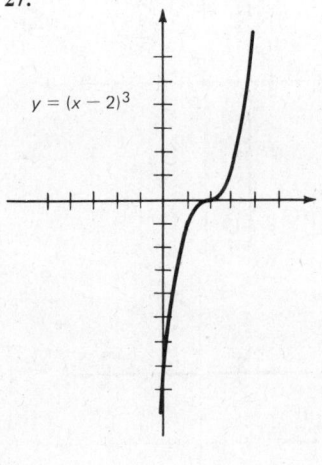

[Graph: $y = (x-2)^3$]

29.

[Graph: $y = x^3 - 2x^2 - 5x + 6$]

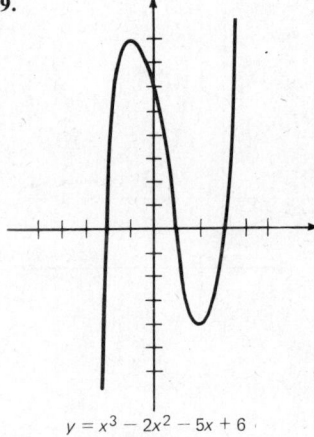

31. 0   **33.** About 2.0 or .20%
35. About 3.2 or .32%   **37.** About 1.0 or .1%   **39.** Between 4 and 5 hours, closer to 5 hours

Section 2.6 (page 86)

1.

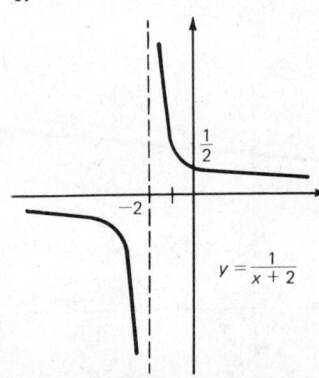

[Graph: $y = \frac{1}{x+2}$, vertical asymptote at -2, passes through $\frac{1}{2}$]

3.

[Graph: $y = 2/x$]

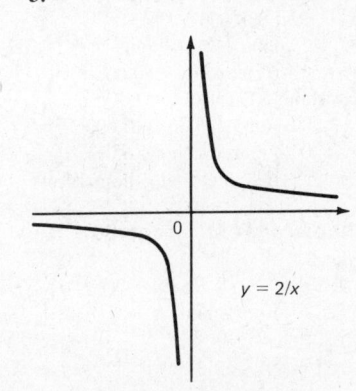

5.

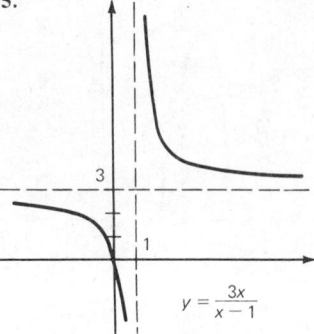

[Graph of $y = \dfrac{3x}{x-1}$]

7.

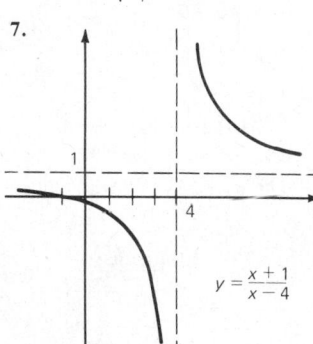

[Graph of $y = \dfrac{x+1}{x-4}$]

9. 37.5¢   **11.** 18.75¢
13.

[Graph of $C(x) = \dfrac{1500}{x+30}$]

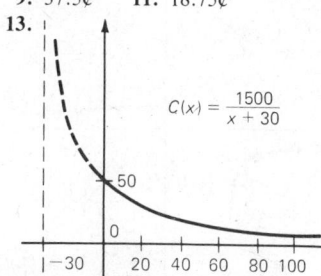

15. $440   **17.** About $232
19. $0   **21.** About $24,000
23. About $88,000
25. $325,000   **27.** About $16,000   **29.** About $60,000
31. About $127,000   **33.** About $663,000   **35.** About $10,000
37. 5000 hundred gallons of gasoline; 1000 hundred gallons of oil

Chapter 2 Test   (page 88)

1. $(-2, 10), (-1, 8), (0, 6), (1, 4), (2, 2), (3, 0)$; domain: $\{-2, -1, 0, 1, 2, 3\}$; range: $\{10, 8, 6, 4, 2, 0\}$
2. -13 **3.** 2 **4.** $-2p^2 + 3p + 1$

5.

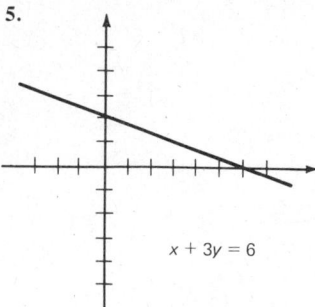

[Graph of $x + 3y = 6$]

6.

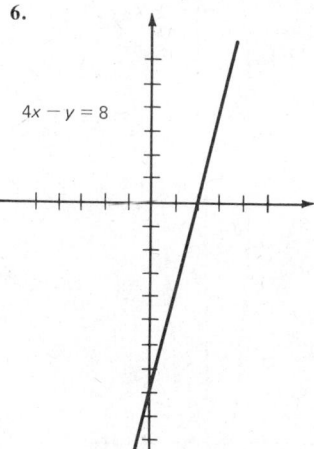

[Graph of $4x - y = 8$]

7.

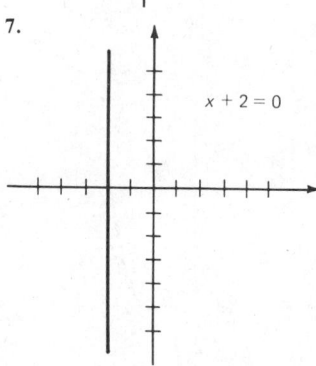

[Graph of $x + 2 = 0$]

8.

[Graph of $y = 3$]

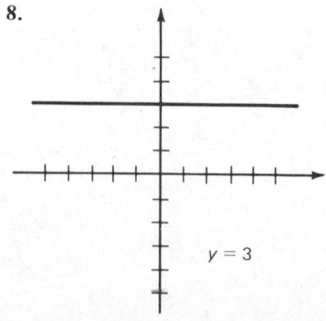

9. 1/3   **10.** 8/3   **11.** 0   **12.** $2y = x + 8$ **13.** $5y = x - 13$
14. $x = 3$ **15.** $C(x) = 30x + 60$; $60   **16.** 2   **17.** 15
18. 5 units   **19.** $200   **20.** 7
21. 10
22.

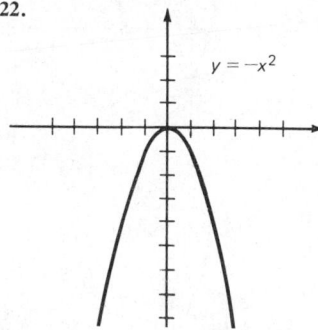

[Graph of $y = -x^2$]

23.

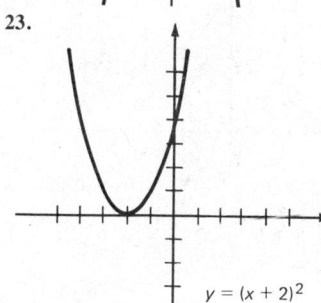

[Graph of $y = (x + 2)^2$]

24.

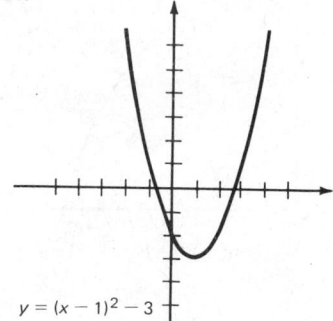

[Graph of $y = (x - 1)^2 - 3$]

25.

[Graph of $y = x^3 - 2$]

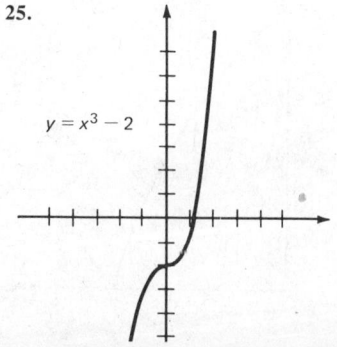

Answers

26.

[graph of $y = \dfrac{1}{x-3}$]

27.

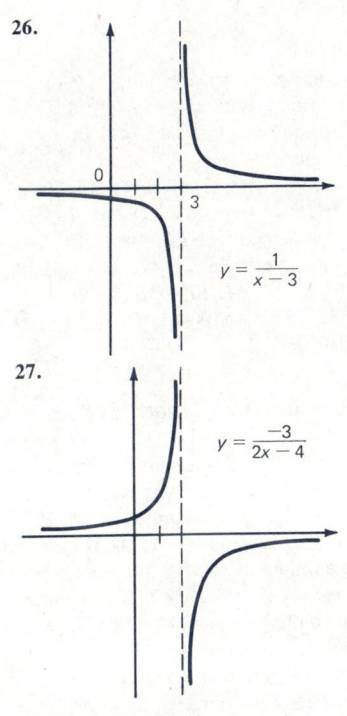

[graph of $y = \dfrac{-3}{2x-4}$]

Case 1 (page 90)

1. .1A + 200   **2.** .0001A − .3
3. 8000   **4.** 8000   **5.** 800
6. 800(10,000) = $8,000,000

Case 2 (page 92)

1. 4.8 million units
2.

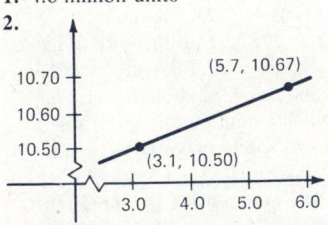

[graph with points (3.1, 10.50) and (5.7, 10.67)]

3. In the interval under discussion (3.1 to 5.7 million units) the marginal cost always exceeds the selling price.   **4.** (a) 9.88; 10.22
(b)

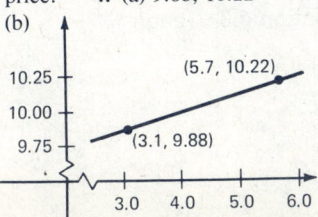

[graph with points (3.1, 9.88) and (5.7, 10.22)]

(c) .83 million units, which is not in the interval under discussion

Chapter 3

Section 3.1 (page 101)

1. 3   **3.** Does not exist
5. 2   **7.** 0   **9.** −11   **11.** 9
13. No limit   **15.** 10   **17.** No limit   **19.** 12   **21.** 29
23. 41   **25.** 9/7   **27.** −1
29. 6   **31.** −5   **33.** −1
35. $\sqrt{5}$ **37.** 1/10   **39.** 500
41. $1500   **43.** 3/5   **45.** 1/2
47. 1/2   **49.** 0   **51.** No limit
53. .57   **55.** .5

Section 3.2 (page 110)

1. −1   **3.** −1   **5.** 0; 2
7. 1   **9.** Yes; no   **11.** No; no; yes   **13.** Yes; no; no   **15.** Yes; no; yes   **17.** Yes; no; yes
19. Yes; yes; yes   **21.** $96
23.

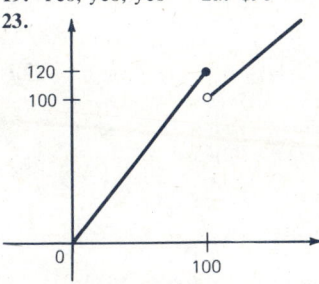

[graph]

25. $260   **27.** $525
29.

[graph]

31. m   **33.** [−9, 15]   **35.** (10, ∞)
37. [−8, 0]   **39.** (12, 20)
41. (−∞, 3)   **43.** (−10, ∞)
45. (−∞, −6); (−6, 0); (0, 4); (4, ∞)
47. (−∞, 0); (0, ∞)

Section 3.3 (page 118)

1. 1   **3.** 6/10 or 3/5   **5.** 2 million   **7.** −.8 million   **9.** −2.2 million   **11.** 3   **13.** 1
15. −2   **17.** −3   **19.** About $5 million   **21.** About $35 million
23. About −$150/7 ≈ −$21 million   **25.** 5   **27.** 8   **29.** 1/3
31. 6   **33.** −1/3   **35.** 7
37. 3.02; 3.002; 3.0002   **39.** 3
41. 15   **43.** 13.01; 13.001; 13.0001   **45.** 17

Section 3.4 (page 128)

1. $f'(x) = 4x$; 8; 0; −12   **3.** $f'(x) = -8x + 11$; −5; 11; 35   **5.** $f'(x) = -9$; −9; −9; −9   **7.** $f'(x) = 8$; 8; 8; 8   **9.** $f'(x) = 3x^2 + 3$; 15; 3; 30   **11.** $f'(x) = 2/x^2$; 1/2; does not exist; 2/9   **13.** $f'(x) = -4/(x-1)^2$; −4; −4; −1/4   **15.** $f'(x) = 1/(2\sqrt{x})$; $1/(2\sqrt{2})$; does not exist; does not exist   **17.** −4   **19.** −6
21. has a derivative everywhere
23. 0   **25.** −2   **27.** −1; 2
29. −4; 0   **31.** −8   **33.** 0; no   **35.** −16; no   **37.** 30
39. 10   **41.** −10

Section 3.5 (page 135)

1. $y' = 10x$ **3.** $y' = -30x^4$
5. $y' = 30x$ **7.** $y' = 18x - 8$
9. $y' = 6x - 4$ **11.** $y' = -12x^2 + 4x$ **13.** $y' = 30x^2 - 18x + 6$
15. $y' = 4x^3 - 15x^2 + 18x$
17. $y' = 9x^{.5}$ **19.** $y' = -48x^{2.2}$
21. $y' = -27x^{1/2}$ **23.** $y' = 16x^{-1/2}$ or $16/x^{1/2}$ **25.** $y' = 75x^{1/2} - 3x^{-1/2}$ or $75x^{1/2} - 3/x^{1/2}$ **27.** $y' = 4x^{-1/2} + 6$ or $4/x^{1/2} + 6$ **29.** $y' = -30x^{-6}$ or $-30/x^6$ **31.** $y' = 8x^{-3} - 3x^{-2}$ or $8/x^3 - 3/x^2$ **33.** $y' = -20x^{-3} - 12x^{-5} - 6$ or $-20/x^3 - 12/x^5 - 6$
35. $y' = -6x^{-2} + 16x^{-3}$ or $-6/x^2 + 16/x^3$ **37.** $y' = -6x^{-3/2}$ or $-6/x^{3/2}$ **39.** $y' = 5x^{-3/2} - 12x^{-5/2}$ or $5/x^{3/2} - 12/x^{5/2}$ **41.** $x = 5/3$
43. $x = 2/3$; $x = 1$ **45.** $x = 5/2$; $x = -1$ **47.** $x = \sqrt[3]{4}$ **49.** 10; increasing   **51.** 1   **53.** 1%
55. 450   **57.** The blood sugar level is decreasing at a rate of 4 points per unit of insulin.   **59.** $dV/dr = 160\pi r$

61. 960π cubic mm   **63.** $15
65. $207   **67.** $8   **69.** $428
71. 4   **73.** 2 2/3 or 8/3
75. (a) $v(t) = 6$ (b) 6; 6; 6
77. (a) $v(t) = 22t + 4$ (b) 4; 114; 224   **79.** (a) $v(t) = 12t^2 + 16t$ (b) 0; 380; 1360

Section 3.6 (page 141)

1. $y' = 12x - 15$ **3.** $y' = 4x + 3$
5. $y' = 48x + 66$ **7.** $y' = 18x^2 - 6x + 4$ **9.** $y' = 9x^2 - 4x - 5$
11. $y' = 36x^3 + 21x^2 - 18x - 7$
13. $y' = 4(2x - 5)$ or $8x - 20$
15. $y' = 4x(x^2 - 1)$ or $4x^3 - 4x$
17. $y' = 3x^{1/2}/2 + x^{-1/2}/2 + 2$ or $3\sqrt{x}/2 + 1/(2\sqrt{x}) + 2$
19. $y' = 10 + 3x^{-1/2}/2$ or $10 + 3/(2\sqrt{x})$
21. $y' = -3/(2x - 1)^2$ **23.** $y' = 53/(3x + 8)^2$ **25.** $y' = -6/(3x - 5)^2$
27. $y' = -17/(4 + x)^2$
29. $y' = (x^2 - 2x - 1)/(x - 1)^2$
31. $y' = (-x^2 + 4x + 1)/(x^2 + 1)^2$
33. $y' = (-2x^2 - 6x - 1)/(2x^2 - 1)^2$
35. $y' = (x^2 + 6x - 14)/(x + 3)^2$
37. $y' = [-\sqrt{x}/2 - 1/(2\sqrt{x})]/(x - 1)^2$ or $(-x - 1)/[2\sqrt{x}(x - 1)^2]$
39. $y' = (5\sqrt{x}/2 - 3/\sqrt{x})/x$ or $(5x - 6)/(2x\sqrt{x})$ **41.** 86.8
43. $A(x) = 9x - 4 + 8/x$
45. $72.75   **47.** $A(x) = 10x - 5 - 18/x$ **49.** $G'(20) = -1/200$; go faster   **51.** $N'(t) = 6t^2 - 80t + 200$
53. 46   **55.** $s'(x) = m/(m + nx)^2$
57. $5000/3600 \approx 1.39$

Section 3.7 (page 148)

1. $y' = 4(2x + 9)$ **3.** $y' = 90(5x - 1)^2$
5. $y' = -144x(12x^2 + 4)^2$
7. $y' = 36(2x + 5)(x^2 + 5x)^3$
9. $y' = 36(2x + 5)^{1/2}$ **11.** $y' = -21(8x + 9)(4x^2 + 9x)^{1/2}/2$
13. $y' = 16(4x + 7)^{-1/2}$ or $16/\sqrt{4x + 7}$ **15.** $y' = -(2x + 4) \cdot (x^2 + 4x)^{-1/2}$ or $-(2x + 4)/\sqrt{x^2 + 4x}$
17. $y' = 16x(2x + 3) + 4(2x + 3)^2$ or $12(x + 3)(2x + 1)$
19. $y' = 2(x + 2)(x - 1) + (x - 1)^2 = 3(x - 1)(x + 1)$ **21.** $y' = 10(x + 3)^2(x - 1) + 10(x - 1)^2 \cdot (x + 3)$ or $10(x + 3)(x - 1)(2x + 2)$ or $20(x + 3)(x - 1)(x + 1)$
23. $y' = (x + 1)^2 \cdot x^{-1/2}/2 + 2x^{1/2}(x + 1)$ or $x^{-1/2}(x + 1)(5x + 1)/2$

25. $y' = -2(x - 4)^{-3}$ or $-2/(x - 4)^3$
27. $y' = (x^2 - 2x - 15)/(x - 1)^2$
29. $y' = 8/(x + 2)^3$ **31.** -3.2
33. -2.4 **35.** $A(x) = 1000/x - 4 + x/250$ **37.** 6   **39.** $138/7 \approx 19.71$
41. $-1/2 = -.5$ **43.** $-.011$
45. $R'(Q) = -Q/[6\sqrt{C - Q/3}] + \sqrt{C - Q/3}$

Chapter 3 Test (page 150)

1. 17/3   **2.** 8   **3.** -13
4. 1/6   **5.** 1/2   **6.** 0
7. none   **8.** $-4; 2$ **9.** $[-4, 2]$
10. $(4, \infty)$ **11.** 30   **12.** -60
13. 9/77   **14.** $y' = 4$
15. $y' = 10x + 6$ **16.** 315
17. 1015   **18.** 1515   **19.** 55/3 = 18 1/3   **20.** 65/4 = 16 1/4
21. 15   **22.** (a) $v(t) = 60$ (b) 60; 60; 60   **23.** (a) $v(t) = 24t^2 - 8t$ (b) 0; 1872; 9440   **24.** $y' = 3x^2 - 8x$
25. $y' = 21x^{3/4}/2$ **26.** $y' = 36x + 27$
27. $y' = 2(x + 2)/\sqrt{x} + 4\sqrt{x}$ or $6x^{1/2} + 4x^{-1/2}$ **28.** $y' = 16/(2x + 1)^2$
29. $y' = (x^2 - 2x)/(x - 1)^2$
30. $y' = 60x(x^2 + 2)^4$
31. $y' = 3(6x - 11)^{-1/2}$ or $3/\sqrt{6x - 11}$
32. $y' = 45x(x + 1)^4 + 9(x + 1)^5$ or $9(6x + 1)(x + 1)^4$ **33.** $y' = (-2x + 10)/(x + 5)^3$

Chapter 4

Section 4.1 (page 161)

1. Minimum of -4 at 1
3. Maximum of 3 at -2
5. Maximum of 3 at -4; minimum of 1 at -2 **7.** Maximum of 3 at -4; minimum of -2 at -7 and -2 **9.** Maximum of 2 at -4 and 5 at 2; minimum of -3 at -2
11. Minimum of -44 at -6
13. Minimum of -46 at 7
15. Maximum of 22 at -4
17. Maximum of 17 at -3
19. Maximum of -8 at -3; minimum of -12 at -1
21. None   **23.** Maximum of 57 at 2; minimum of 30 at 5
25. Minimum of $-377/6$ at -5; maximum of 827/96 at $-1/4$
27. Maximum of -4 at 0; minimum of 85 at 3 and -3 **29.** $(-6, -42)$

31. $(2, -10)$ **33.** $(-1, 2)$
35. $(4/3, -4/3)$ **37.** Increasing on $(-6, \infty)$; decreasing on $(-\infty, -6)$
39. Increasing on $(-\infty, 4/3)$; decreasing on $(4/3, \infty)$ **41.** Increasing on $(-\infty, -3)$ and $(4, \infty)$; decreasing on $(-3, 4)$ **43.** Increasing on $(-\infty, -3/2)$ and $(4, \infty)$; decreasing on $(-3/2, 4)$ **45.** Always decreasing   **47.** 30; 1080
49. 12°   **51.** $(0, 7/3); (5, \infty)$
53. (a) 6 (b) $1604   **55.** (a) 10 (b) 500

Section 4.2 (page 168)

1. Absolute maximum at x_3; no absolute minimum   **3.** Neither
5. Absolute minimum at x_1; no absolute maximum   **7.** Absolute maximum at x_1; absolute minimum at x_2 **9.** Absolute maximum at x_2; absolute minimum at x_3 **11.** Absolute maximum at 0; absolute minimum at -3 **13.** Absolute maximum at -3; absolute minimum at $-3/4$ **15.** Absolute maximum at -1; absolute minimum at -5 **17.** Absolute maximum at -2; absolute minimum at 4
19. Absolute maximum at -2; absolute minimum at 3   **21.** Absolute maximum at 4; absolute minimum at -1 **23.** Absolute maximum at 6; absolute minimum at -4 and 4   **25.** About 34.4; 60   **27.** Use all the wire to make a circle   **29.** Absolute maximum is about 3.4, at about $x = 1.2$ or 1.3; absolute minimum is -7 at $x = 0$
31. Absolute maximum on $[-5, -4]$ is about 60.4 at -4.5 or -4.4; absolute minimum is 58, at -5; absolute maximum on $[0, 1]$ is about 3 at 0; absolute minimum is about 1.6 at about .4 or .5

Section 4.3 (page 177)

1. $f''(x) = 6x + 8$; 8; 20; -10
3. $f''(x) = -12x^2 + 12x - 2$; -2; -26; -146 **5.** $f''(x) = 16$; 16; 16; 16   **7.** $f''(x) = 6(x - 2)$; -12; 0; -30 **9.** $f''(x) = 4/(x - 1)^3$; -4; 4; $-1/16$ **11.** $f'''(x) = 48x - 18$;

$f^{(4)}(x) = 48$ **13.** $f'''(x) = 180x^2 - 24x + 12; f^{(4)}(x) = 360x - 24$
15. $f'''(x) = -6x^{-4} = -6/x^4; f^{(4)}(x) = 24x^{-5} = 24/x^5$ **17.** $f'''(x) = 24(2x + 1)^{-4} = 24/(2x + 1)^4; f^{(4)}(x) = -192(2x + 1)^{-5} = -192/(2x + 1)^5$
19. Relative minimum at 6

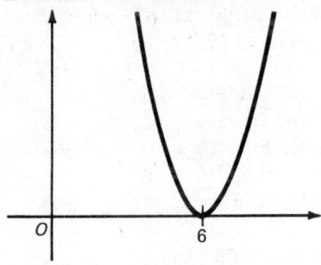

21. Relative minimum at 1

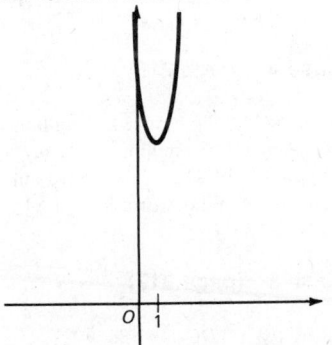

23. Relative maximum at 0; relative minimum at 4/3

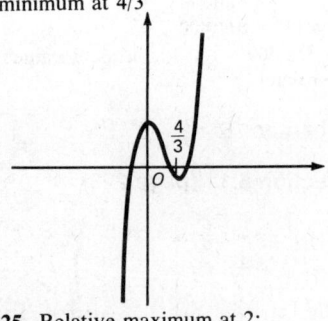

25. Relative maximum at 2; relative minimum at -5

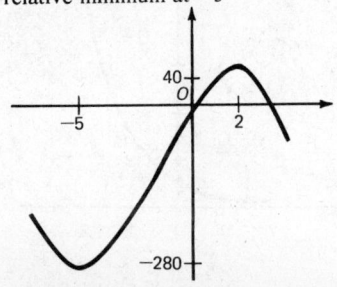

27. Relative maximum at -1; relative minimum at 6

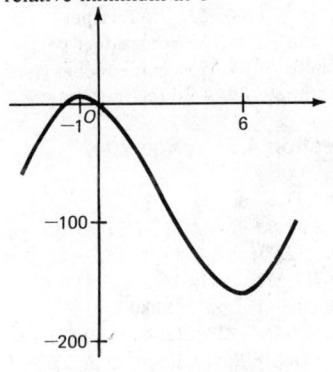

29. Critical value at 0, but neither a maximum nor a minimum there

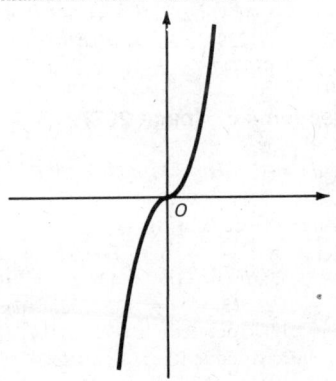

31. Relative maximum at 0; relative minimum at -2 and 2

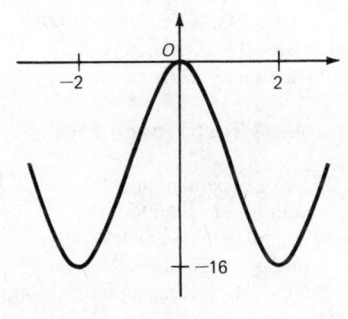

33. Relative maximum at -2; relative minimum at 2

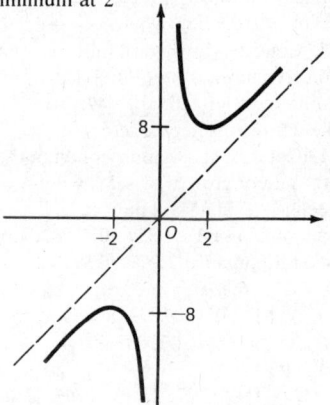

35. Relative maximum at -2; relative minimum at 2

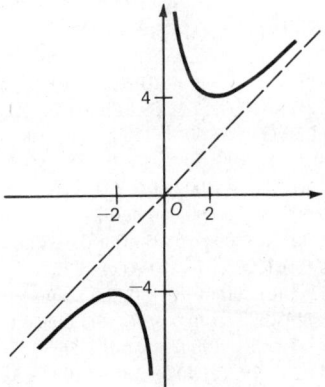

37. No critical values; no maximums or minimums

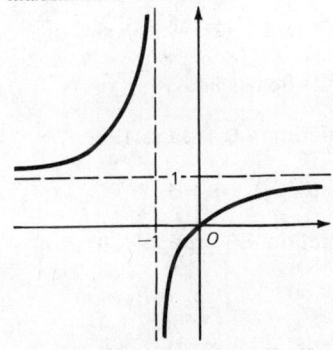

39. Always concave up; no points of inflection **41.** Always concave downward; no points of inflection
43. Concave downward on $(-\infty, 1/2)$; concave upward on $(1/2, \infty)$; point of inflection when $x = 1/2$

476 Answers

45. Concave downward on $(-4, \infty)$; concave upward on $(-\infty, -4)$; point of inflection where $x = -4$
47. Concave downward on $(-1, \infty)$; concave upward on $(-\infty, -1)$; no point of inflection   **49.** To the left of trash compactors
51. Color tv's, room air conditioners; rate of growth of sales will now decline   **53.** Maximum at 8; minimum at 4/3   **55.** 50   **57.** (a) After 3 hours (b) 2/9%   **59.** $v(t) = -6t - 6$; $a(t) = -6$; -6; -30; -6; -6 **61.** $v(t) = 9t^2 - 8t + 8$; $a(t) = 18t - 8$; 8; 120; -8; 64
63. $v(t) = -(t + 3)^{-2}$; $a(t) = 2(t + 3)^{-3}$; $-1/9$; $-1/49$; $2/27$; $2/343$ **65.** -96 feet per second   **67.** -256 feet per second   **69.** 260   **71.** 32

Section 4.4 (page 185)

1. (a) $y = 100 - x$ (b) $P = x(100 - x)$ (c) $P' = 100 - 2x$; $x = 50$ (d) 50 and 50 (e) $50 \cdot 50 = 2500$ **3.** 100; 100   **5.** 100; 50   **7.** 5; -5
9. (a) $R(x) = 100{,}000x - 100x^2$ (b) 500 (c) 25,000,000 cents
11. (a) $\sqrt{3200} \approx 56.6$ mph (b) $22.63
13. (a) $1200 - 2x$ (b) $A(x) = 1200x - 2x^2$ (c) 300 m (d) 180,000 sq m
15. 405,000 sq m   **17.** 200 feet on the $3 sides; 100 feet on the $6 sides   **19.** (a) $40 - 2x$ (b) $100 + 5x$ (c) $R(x) = 4000 - 10x^2$ (d) pick now (e) $40 per tree   **21.** (a) 90 (b) $405   **23.** 4 by 4 by 2   **25.** 3 by 6 by 2   **27.** 49 mph gives the minimum cost of $490.   **29.** 8 miles from point A

Section 4.5 (page 193)

1. 10   **3.** 60   **5.** 5

Section 4.6 (page 198)

1. $-4x/(3y)$ **3.** $-y/(y + x)$
5. $2/y$ **7.** $-3y^2/(6xy - 4)$
9. $(-y - x)/x$ **11.** $(-6x - 4y)/(4x + y)$ **13.** $3x^2/(2y)$
15. $-2y/x$ **17.** $-y/x$
19. $-2xy/(x^2 + 3y^2)$ **21.** $-y^{1/2}/x^{1/2}$
23. $-y^{1/2}x^{-1/2}/(x^{1/2}y^{-1/2} + 2) = -y/(x + 2x^{1/2}y^{1/2})$
25. $4y = 3x + 25$ **27.** $y = x + 2$
29. $y = -24x + 57$ **31.** $4y = -x + 12$ **33.** 7/6 feet per minute   **35.** 6π square feet per minute   **37.** 50π cubic inches per minute   **39.** 1/16 feet per minute

Section 4.7 (page 204)

1. 2.73   **3.** .33   **5.** .67
7. 2.33   **9.** -2.20 **11.** 3.10
13. -2.00; .67; 3.50   **15.** -1; 3.67   **17.** 1.414   **19.** 3.317
21. 15.811   **23.** 2.080
25. 4.642   **27.** Relative maximum at -1.6; relative minimum at 3.6
29. Relative minimum at $-.7$ and 1.8; relative maximum at 1.2   **31.** $f'(i) = [-1 + ni(1 + i)^{-n-1} + (1 + i)^{-n}]/i^2$
33. .01371229   **35.** $i_2 = .02075485$; $i_3 = .02075742$

Section 4.8 (page 207)

1. $dy = 12x\, dx$ **3.** $dy = x^{-1/2}dx$
5. $dy = (14x - 9)\, dx$
7. $dy = [-22/(x - 3)^2]dx$
9. $\Delta y = .6$; $dy = .6$ **11.** $\Delta y = -1.52$; $dy = -1.6$ **13.** $\Delta y = .109$; $dy = .1$ **15.** $\Delta y = .125$ (rounded); $dy = .130$ (rounded)   **17.** $\Delta y = -.024$ (rounded); $dy = -.023$ (rounded)   **19.** 3 1/6   **21.** 3 7/8
23. 11 1/11   **25.** 1 11/12
27. 4 1/48   **29.** 2 1/32
31. 2.005   **33.** 1 119/120 $\approx$ 1.992
35. 12.8π cubic centimeters
37. $.48\pi$ square miles

Chapter 4 Test (page 210)

1. Relative maximum at 2; maximum is -4 **2.** Relative maximum of $-151/54$ at 1/3; relative minimum of $-27/8$ at $-1/2$
3. None   **4.** Increasing on $(5/2, \infty)$; decreasing on $(-\infty, 5/2)$ **5.** Increasing on $(-4, 2/3)$; decreasing on $(-\infty, -4)$ and $(2/3, \infty)$
6. Absolute maximum of 29/4 at 5/2; absolute minimum of 5 at 1 and 4   **7.** Absolute maximum of 39 at -3; absolute minimum of $-319/27$ at 5/3   **8.** $v(t) = 18t - 7$; $a(t) = 18$; -7; 47; 18; 18   **9.** $v(t) = -6t^2 + 8t - 6$; $a(t) = -12t + 8$; -6; -36; 8; -28 **10.** $f''(x) = 36x - 18$; -18; -198 **11.** $f''(x) = -68(2x + 3)^{-3}$ or $-68/(2x + 3)^3$; $-68/27$; 68/343   **12.** $f'''(x) = 54$; $f^{(4)}(x) = 0$ **13.** $f'''(x) = 12(x - 4)^{-4} = 12/(x - 4)^4$; $f^{(4)}(x) = -48(x - 4)^{-5} = -48/(x - 4)^5$ **14.** 6 hundred boxes
15. $178   **16.** 225 by 450
17. 40   **18.** $(-4y - 2xy^3)/(3x^2y^2 + 4x)$
19. $(y - 3y^2)/(4y^2 + x)$
20. $16y = -23x + 94$ **21.** 4.73
22. 5.196   **23.** $dy = (24x^2 - 4x)\, dx$
24. $dy = 24x(x^2 - 1)^2 dx$
25. 12 23/26

Case 3 (page 212)

1. $r_P = 4r_0/5$ **2.** $r_P = 2r_0/3$

Case 4 (page 215)

1. (a) $520 million, or $4.33 million per plane (b) $80 million   **2.** (a) 200 planes (b) 80   **3.** (a) $310 million (b) -4 million, or a loss of $4 million

Case 5 (page 217)

1. $-C_1/m^2 + DC_3/2$ **2.** $m = \sqrt{2C_1/(DC_3)}$ **3.** About 3.33
4. $m^+ = 4$ and $m^- = 3$ **5.** $Z(m^+) = Z(4) = $11{,}400$; $Z(m^-) = Z(3) = $11{,}300$ **6.** 3 months; 9 trainees per batch

Chapter 5

Section 5.1 (page 223)

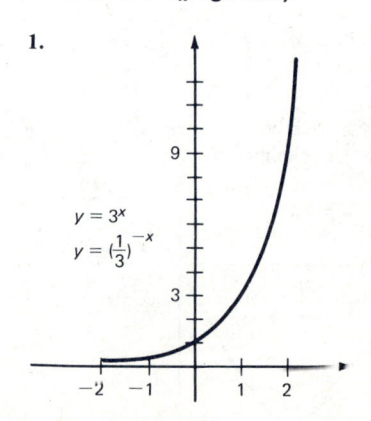

1.

3. [graph of $y = 3^{-x}$]

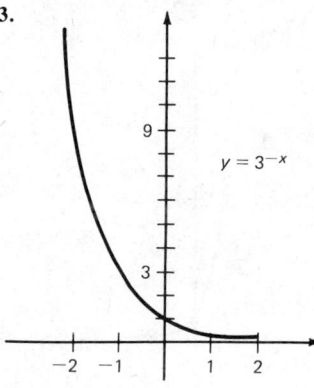

5. [graph of $y = (\frac{1}{4})^x$]

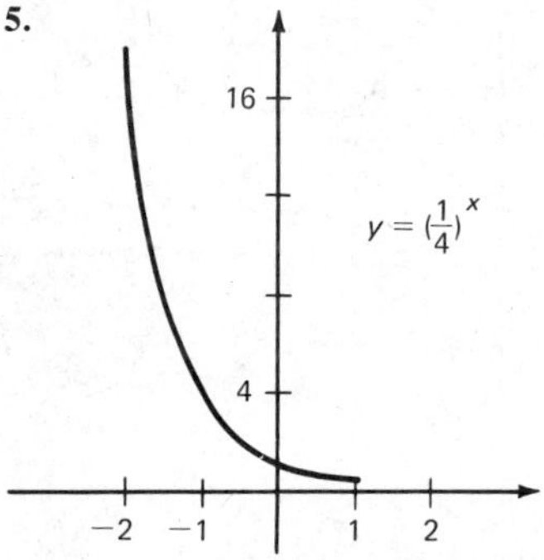

7. [graph of $y = 3^x$, $y = (\frac{1}{3})^{-x}$]

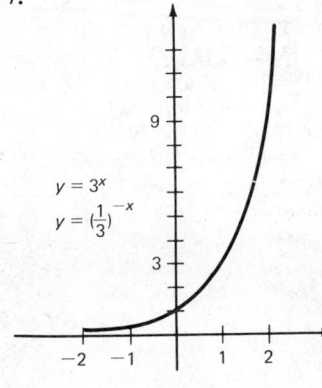

9. [graph of $y = 4^{-|x|}$]

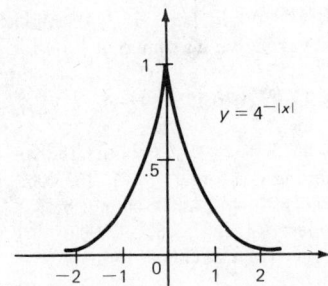

11. [graph of $y = 3^{-x^2}$]

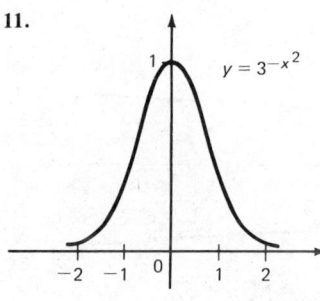

13. [graph of $y = 2^{1-x}$]

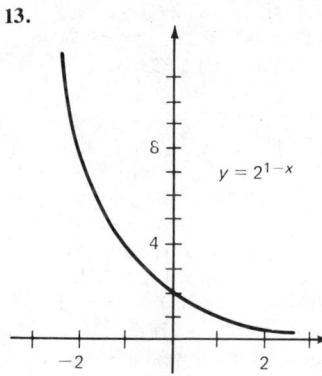

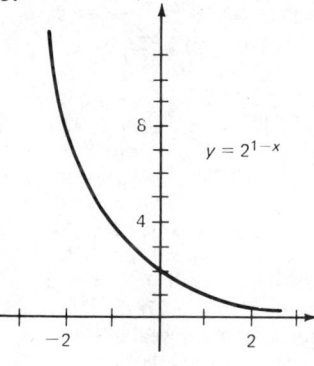

15. [graph of $y = e^{x+1}$]

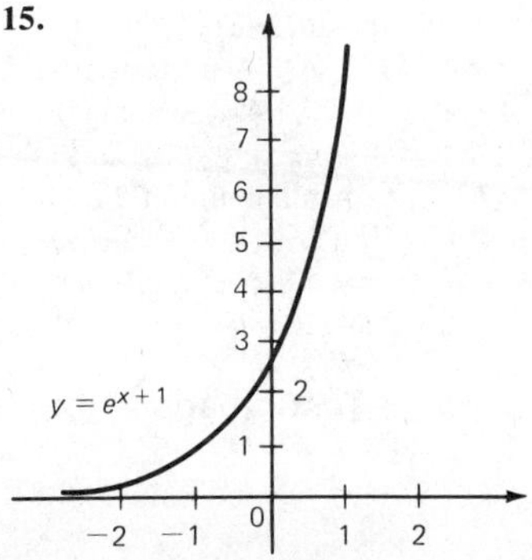

17. [graph of $y = e^{-3x}$]

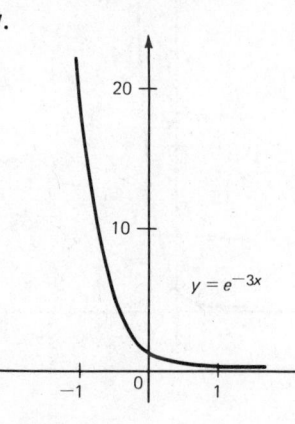

19. 2 **21.** −3 **23.** 3/2
25. −3 **27.** 1 **29.** 1/4
31. 1.06; 1.12; 1.19; 1.26; 1.42; 1.50; 1.59; 1.69 **33.** .92; .85; .78; .72; .61; .56; .51; .47 **35.** About $116,000 **37.** $1338.23
39. $52,632.03 **41.** $8818.60
43. 12; 24; 48; 96; 384; 768; 1536; 3072 **45.** About 1,120,000
47. About 1,410,000
49. 2,000,000 **51.** 8,000,000
53. About 40 **55.** About 11
57. About 3 **59.** About 9900

Section 5.2 (page 231)

1. $\log_2 8 = 3$ **3.** $\log_3 81 = 4$
5. $\log_{1/3} 9 = -2$ **7.** $2^3 = 8$
9. $10^2 = 100$ **11.** $10^5 = 100{,}000$
13. 4 **15.** 3 **17.** −2
19. $\log_4 16 = 2$ **21.** $\log_7 15/11$
23. $\log_{10} 30^{0.2}$ or $\log 30^{0.2}$
25. $\ln e = 1$ **27.** $\ln (5^{.3})/(6^{.4})$
29. 2.9957 **31.** 4.0943
33. 6.6847 **35.** 6.2767
37. 6.6439 **39.** 10.9770
41. 3; 2; 1; 0; −1; −2; −3

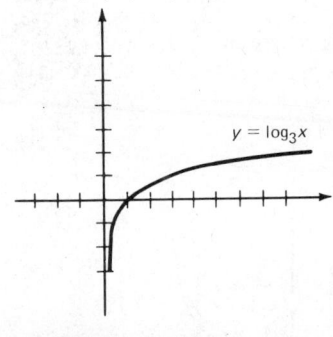

[graph of $y = \log_3 x$]

43. 2; 1; 0; −1; −2; −3

[graph]

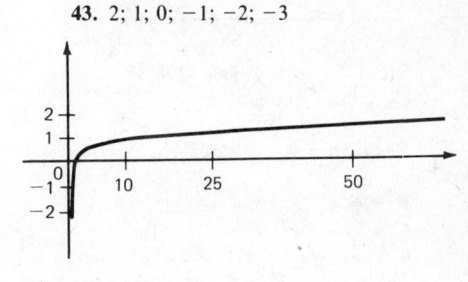

45. $-3; -2; -1; 0; 1; 2; 3$

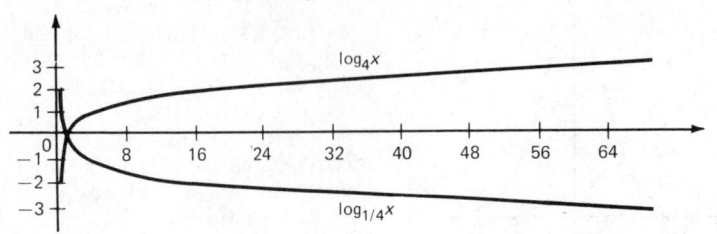

47. (a) $0 (b) about $9000 (c) about $16,100 (d) about $19,600

Section 5.3 (page 238)

1. $\ln 6/(2 \ln 5) \approx .56$ **3.** $(-\ln 5)/.3 \approx -5.4$ **5.** $(\ln 10)/.02 \approx 115$
7. $(\ln 4) - 2 \approx -.61$
9. $(5 \ln 8)/2 \approx 5.2$ **11.** 1,000,000
13. About 1,080,000 **15.** 500
17. About 335 **19.** 25,000
21. About 37,300 **23.** 50,000
25. About 40,900 **27.** 1000
29. About 4500 **31.** About 5000 **33.** 0 **35.** About 430
37. About 500 **39.** 527
41. 6020 **43.** 1800 years
45. 18,600 years **47.** $1051.27
49. $27,598.67 **51.** About 13.9 years **53.** 1000 **55.** 3700/404 or about 9
57.

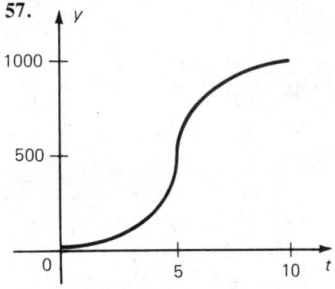

59. .125 **61.** About 12.9
63. About 20.0

Section 5.4 (page 245)

1. $y' = 4e^{4x}$ **3.** $y' = 12e^{-2x}$
5. $y' = -16e^{2x}$ **7.** $y' = -16e^{x+1}$
9. $y' = 2xe^{x^2}$ **11.** $y' = 12xe^{2x^2}$
13. $y' = 16xe^{2x^2-4}$ **15.** $y' = xe^x + e^x = e^x(x+1)$ **17.** $y' = 2(x-3) \cdot (x-2)e^{2x}$ **19.** $y' = -1/(3-x)$ or $1/(x-3)$ **21.** $y' = (4x-7)/(2x^2-7x)$
23. $y' = 1/[2(x+5)]$
25. $y' = (2xe^x - x^2e^x)/e^{2x} = x(2-x)/e^x$ **27.** $y' = x + 2x \ln |x| = x(1 + 2 \ln |x|)$ **29.** $y' = (4x + 7 - 4x \ln |x|)/[x(4x+7)^2]$
31. $y' = [x(\ln |x|)e^x - e^x]/[x(\ln |x|)^2]$ or $e^x[x \ln |x| - 1]/[x(\ln |x|)^2]$
33. $y' = [x(e^x - e^{-x}) - (e^x + e^{-x})]/x^2$
35. $y' = -20,000e^{.4x}/(1 + 10e^{.4x})^2$
37. $y' = 8000e^{-.2x}/(9 + 4e^{-.2x})^2$
39. $y' = 1/(x \ln |x|)$ **41.** $200e^{.4} \approx 298; 200e^{1.6} \approx 991$ **43.** (a) $e^{-.02} \approx .98$ (b) $e^{-.2} \approx .82$ (c) $e^{-2} \approx .14$ (d) $-.02e^{-2} \approx -.0027$; the rate of change in the proportion wearable when $x = 100$ **45.** (a) .005 (b) .0007 (c) .00001 (d) $-.022$ (e) $-.0029$ (f) $-.000054$ **47.** A maximum of $1/e$ at $x = -1$ **49.** A minimum of 0 at $x = 0$; a maximum of $4/e^2$ at $x = 2$ **51.** A minimum of 2 at $x = 0$
53. $C'(t) = [k/(b-a)](-ae^{-at} + be^{-bt})$
55. $b/a = e^{-at}/e^{-bt}$ **57.** $(b-a)t = \ln b/a$ **59.** .04

Chapter 5 Test (page 249)

1. 4 **2.** -1 **3.** $\ln 7/\ln 2 \approx 2.81$
4.

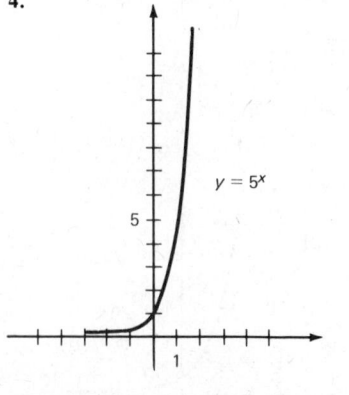

5.

6.

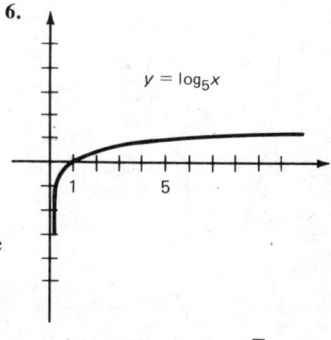

7. $\log_2 64 = 6$ **8.** $\log_3 \sqrt{3} = 1/2$
9. $\log_{1000} .001 = -1$ **10.** $2^5 = 32$ **11.** $10^2 = 100$ **12.** $27^{1/3} = 3$ **13.** 2 **14.** 1/2
15. $\log_5 18$ **16.** $\log_3 4$ **17.** $\log_4 2^6 = \log_4 64 = 3$ **18.** 1.8245
19. 6.5511 **20.** 6.1799 **21.** (a) 100,000 (b) about 149,000 (c) about 272,000 **22.** (a) 5000 (b) about 3350 (c) $-(\ln .5)/.2 \approx 3.5$ days
23. (a) 0 (b) about 55 (c) about 98 (d) about 100 (e) about 100 **24.** $y' = 12xe^{6x^2}$ **25.** $y' = 3/(3x-2)$
26. $y' = -1/[2(3-x)]$ or $1/[2(x-3)]$ **27.** $y' = e^{2x}(2x-1)/(5x^2)$ **28.** $y' = e^x(2/x + \ln x^2)$ or $e^x(2 + x \ln x^2)/x$
29. $y' = (6x \ln |x| - 6x - 4)/[x(\ln |x|)^2]$
30. A relative maximum at $-1/10$

Case 6 (page 253)

1. .0315 **2.** λy_0 is about 118,000; painting is a forgery **3.** 163,000; forgery **4.** 24,000; cannot be modern forgery **5.** 142,000; forgery **6.** 134,000; forgery

Answers

Chapter 6

Section 6.1 (page 259)

1. $5x^3/3 + C$ 3. $6x + C$
5. $x^2 + 3x + C$ 7. $x^3/3 - 2x^2 + 5x + C$ 9. $x^4 + x^3 + x^2 - 6x + C$
11. $2x^{3/2}/3 + 2x^{5/2}/5 + C$ 13. $2x^{3/2}/3$
15. $4x^{5/2} - 4x^{7/2} + C$ 17. $6x^{3/2} + 4x + C$
19. $-x^{-1} + C$ or $-1/x + C$
21. $-x^{-2}/2 - 2x^{1/2} + C$ or $-1/(2x^2) - 2x^{1/2} + C$ 23. $-4e^x + C$
25. $30e^x + 5x^3/3 + C$ 27. $2 \ln |x| + C$ 29. $3 \ln |x| + e^x + C$
31. $C(x) = 2x^2 - 5x + 8$ 33. $C(x) = .07x^3 + 10$ 35. $C(x) = 2x^{3/2}/3 - 8/3$ 37. $C(x) = x^3/3 - x^2 + 3x + 6$ 39. $P(x) = -x^2 + 20x - 50$
41. $y = 2x^3 - 2x^2 + 3x + 1$
43. $v(t) = t^3/3 + t + 6$
45. $s(t) = -16t^2 + 6400$; 20 seconds

Section 6.2 (page 266)

1. 32; 38 3. 15; 31/2 5. 20; 30 7. 16; 14 9. 12.8; 27.2
11. 2.67; 2.08
13. (a) 15.656 (b) 18.258 (c) 21
15. About 10,000,000 17. We get about 20 19. A rough answer is about 2900 feet 21. 30 23. 65
25. 357 27. 20

Section 6.3 (page 272)

1. 35/2 3. 44 5. 24
7. 42 9. 12 11. 124
13. 46 15. 36 17. 768
19. 14/3 21. $3(\ln 5 - \ln 2) \approx 2.75$ 23. $e^2 - e \approx 4.67$
25. $e^3 - e + 4 \approx 21.37$ 27. $3/4 + \ln 4 \approx 2.14$ 29. 161 31. 21
33. 81/4 35. $\ln 2 \approx .69$

Section 6.4 (page 277)

1. $280; $11.67 3. $31; $216
5. No 7. 220 9. 49.5 days
11. 850 13. 8700 15. 6960
17. 1.37 19. (a) 75 (b) 100
21. (a) 14.26 (b) 3.55

Section 6.5 (page 286)

1. 15 3. 4 5. 40
7. 1735/6 9. 41/24 11. (a) 8 years (b) about 148 (c) about 771
13. 19.6 days 15. $807.52 hundreds or $80,752 17. 5733.33
19. 83.33
21. (a)

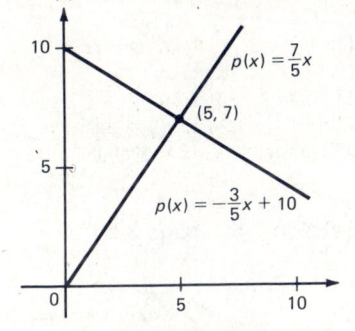

(b) $x^* = 5$ (c) 7.50 (d) 17.50
23. $I(.1) = .019$; the lower 10% of the income producers earn 1.9% of the total income of the population. 25. $I(.6) = .384$; the lower 60% of the income producers earn 38.4% of the total income of the population.
27.

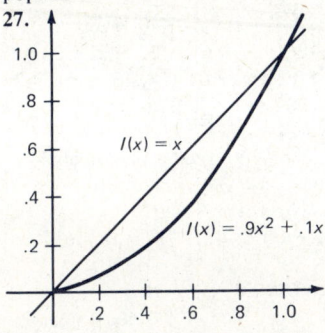

Section 6.6 (page 293)

1. $2(2x + 3)^5/5 + C$ 3. $-2(x - 2)^{-2} + C$ or $-2/(x - 2)^2 + C$
5. $2(x - 4)^{3/2}/3 + C$ 7. $(x^2 + 1)^4/4 + C$ 9. $2(x^2 + 12x)^{3/2}/3 + C$
11. $e^{2x}/2 + C$ 13. $-2e^{2x} + C$
15. $e^{2x^3}/2 + C$ 17. $-8 \ln |1 + x| + C$ 19. $(\ln |2x + 1|)/2 + C$
21. $\ln |x^2 + 4x| + C$

23. $-(3x^2 + 2)^{-3}/3 + C$ or $-1/[3(3x^2 + 2)^3] + C$ 25. $-(2x^2 - 4x)^{-1}/4 + C$ or $-1/[4(2x^2 - 4x)] + C$
27. $[(1/x) + x]^2/2 + C$
29. 14/3 31. 15/4
33. 1024/3 35. 1 37. $(e^2 - 1)/2 \approx 3.19$ 39. $5(e^{-.4} - e^{-.8}) \approx 1.105$
41. About $1350 43. $c(t) = 1.2e^{.04t}$
45. 14.75 47. .5 49. 15 years

Section 6.7 (page 298)

1. $x \ln 4 + x(\ln |x| - 1) + C$
3. $-4 \ln |(x + \sqrt{x^2 + 36})/6| + C$
5. $\ln |(x - 3)/(x + 3)| + C$, $(x^2 > 9)$
7. $(4/3)\ln |[3 + \sqrt{9 - x^2}]/x| + C$, $(0 < x < 3)$ 9. $-2x/3 + 2 \ln |3x + 1|/9 + C$
11. $(-2/15) \ln |x/(3x - 5)| + C$
13. $\ln |(2x - 1)/(2x + 1)| + C$
15. $-3 \ln |(1 + \sqrt{1 - 9x^2})/(3x)| + C$
17. $x^5(\ln |x|/5 - 1/25) + C$
19. $(1/x)(-\ln |x| - 1) + C$
21. $-xe^{-2x}/2 - e^{-2x}/4 + C$
23. $-x^2 e^{-2x} - xe^{-2x} - e^{-2x}/2 + C$
25. $x^3 e^x - 3x^2 e^x + 6xe^x - 6e^x + C$

Chapter 6 Test (page 299)

1. $x^4/2 + C$ 2. $x^3/3 + 5x^2/2 + C$
3. $2x^{5/2}/5 + C$ 4. $-x^{-3}/3 + C$ or $-1/(3x^3) + C$ 5. $3e^x + C$
6. $5 \ln |x| + C$ 7. 24
8. 28 9. 24 10. $\ln 3 \approx 1.10$
11. $-2(e^4 - 1) \approx -107$
12. 36,000 13. 1000 14. 64/3
15. 2.5 years; about $99,000
16. $(5x^2 + 6)^{3/2}/15 + C$
17. $(x^2 - 5x)^5/5 + C$
18. $-e^{-4x}/4 + C$ 19. $12 \cdot \ln |x^2 + 9x + 1| + C$ 20. 243
21. $\ln |(x + \sqrt{x^2 - 64})/8| + C$
22. $-\ln |(5 + \sqrt{25 + x^2})/x| + C$
23. $-(1/2) \ln |x/(3x + 4)| + C$
24. $x\sqrt{x^2 + 49}/2 + 49 \ln |(x + \sqrt{x^2 + 49})/2| + C$
25. $3x^3[(\ln |x|)/3 - 1/9] + C$ or $3x^3[3 \ln |x| - 1]/9$

Case 7 (page 303)

1. About 102 years 2. About 98 years 3. About 45½ years
4. About 90 years

Chapter 7

Section 7.1 (page 307)

1. $x(x+1)^6/6 - (x+1)^7/42 + C$
3. $2x^2(1+x^2)^{5/4}/5 - 8(1+x^2)^{9/4}/45 + C$
5. $2x(x-1)^{1/2} - 4(x-1)^{3/2}/3 + C$
7. $-2x(x+5)^{-5}/5 - (x+5)^{-4}/10 + C$
9. $-x^2(3-x^2)^{1/2} - 2(3-x^2)^{3/2}/3 + C$
11. $xe^x - e^x + C$
13. $-(5x-9)e^{-3x}/3 - 5e^{-3x}/9 + C$
15. $(x^2+1)e^x - 2xe^x + 2e^x + C$
17. $(1-x^2)e^{2x}/2 + xe^{2x}/2 - e^{2x}/4 + C$
19. $x \ln|x| - x + C$ 21. $(x^2 \ln|x|)/2 - x^2/4 + C$ 23. $(x^2-x) \cdot \ln|3x| - x^2/2 + x + C$ 25. $e^4 + e^2 \approx 61.9872$

Section 7.2 (page 311)

1. (a) 2.75 (b) 2.67 3. (a) 6.76 (b) 6.79 5. (a) 16 (b) 14.67
7. (a) .94 (b) .84 9. (a) .10 (b) .10
11. (a)

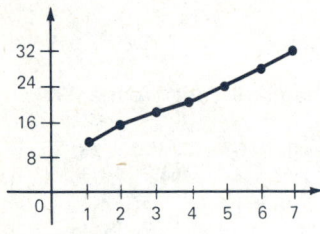

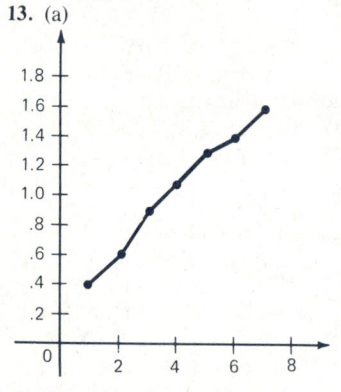

(b) 128 (c) 128
13. (a)

(b) 6.3 (c) 6.27 15. About 34
17. About 9

Section 7.3 (page 316)

1. $8\pi/3$ 3. $364\pi/3$
5. $386\pi/27$ 7. $3\pi/2$ 9. 18π
11. $\pi(e^4-1)/2 \approx 84.2$ 13. $\pi \ln 4 \approx 4.36$ 15. $3124\pi/5$
17. $16\pi/15$ 19. $4\pi/3$
21. $32\pi/3$ 23. $4\pi r^3/3$
25. $2\pi k \int_0^R r(R^2 - r^2)dr$

Section 7.4 (page 322)

1. 1/2 3. Divergent 5. -1
7. 1000 9. 1 11. 3/5
13. 1 15. 4 17. Divergent
19. 1 21. Divergent
23. $(2 \ln 2.5)/21 \approx .087$
25. $-4(1/2 + \ln 1/2)/9 \approx .086$
27. Divergent 29. $750,000
31. 1250

Section 7.5 (page 329)

1. Yes 3. Yes 5. No; $\int_0^3 4x^3 dx \neq 1$ 7. No; $\int_{-2}^{2}(x^2/16)dx \neq 1$ 9. No; $2x \leq 0$ in the interval $-2 \leq x \leq 0$ 11. 3/14
13. 3/125 15. 2/9 17. 1/12
19. 3/26 21. (a) .46 (b) .17 (c) .37 23. (a) 3/5 (b) 1/5 (c) 3/5
25. .38 27. .45 29. .12

Chapter 7 Test (page 333)

1. $2x(8+x)^{5/2}/5 - 4(8+x)^{7/2}/35 + C$
2. $xe^x - e^x + C$ 3. $-(x+2)e^{-3x}/3 - e^{-3x}/9 + C$ 4. $(3+x^2)e^{2x}/2 - xe^{2x}/2 + e^{2x}/4 + C$
5. .414 6. .387 7. $98\pi/3$
8. 32π 9. 1/2 10. Divergent 11. 3/4 12. 1/10
13. No; $\int_1^4 (2x+4)dx \neq 1$
14. $-1/4$ 15. .91 16. .632

Chapter 8

Section 8.1 (page 341)

1. $y = x^3/3 + C$ 3. $y = -x^2 + x^3 + C$ 5. $y = -4x + 3x^4/4 + C$
7. $y = e^x + C$ 9. $y = -2e^{-x} + C$ 11. $y = -4x^3/9 + C$ 13. $y = x^2/8 + C$ 15. $y = 2x^3 - 2x^2 + 4x + C$ 17. $y = -4x^3/3 + C_1x + C_2$
19. $y = 5x^2/2 - 2x^3/3 + C_1x + C_2$ 21. $y = x^4/3 - x^3/3 + C_1x + C_2$ 23. $y = -5x^3/6 + 3x^2/2 + C_1x + C_2$ 25. $y = e^x + C_1x + C_2$ 27. $y = x^3 - x^2 + 2$
29. $y = 5x^2/2 + 2x - 3$
$= x^3 - 2x^2 + 2x + 8$
$+ x^2/2 - 5x/6 + 2$
$+ x^2/2 - ex + 1$
$- x^2 + x + 4$
$+ 250$ (b) 12 days
31. $y = x^3/3$ 33. $y = x^3/3$ 35. $y = e^x$ 37. $y = x^4/4$ 39. (a) $y = 5t^2$

Section 8.2 (page 346)

1. $y^2 = x^2 + C$ 3. $y^2 = 2x^3/3 - 2x + C$ 5. $y = ke^{x^2}$ 7. $y = ke^{(x^3 - x^2)}$ 9. $y = 2x^3/3 + C_1x + C_2$ 11. $y = x^4/8 + C_1x + C_2$
13. $y^2 = 2x^3/3 + 9$ 15. $y = e^{(x^2 + 3x)}$ 17. $y = x^2 - x^3/6 - 29x/6 + 2$ 19. (a) $y = ke^{-.05x}$ (b) $y = 90e^{-.05x}$ (c) about 55 grams

Section 8.3 (page 352)

1. (a) 156 (b) 198 (c) 206 (d) 6
3. (a) 1080 thousand (b) 1125 thousand (c) 1000 thousand (d) 625 thousand 5. About 13.4 grams 7. About 332,000 9. $y = e^{.06x}$; about 12 years 11. $y = 6e^{-.03t}$; about 4.4 cc 13. (a) $y = 4000e^{.23t}$ (approximately) (b) about 400,000 15. $n = Ce^{-kD}$ 17. (a) About 243 (b) about 388 (c) about 1368 19. $dT/dt = -kce^{-kt}$; since $ce^{-kt} = T - T_F$, we have $dT/dt = -k(T - T_F)$ 21. About 75°F
23. $T = 28e^{-.8t} + 22$

Chapter 8 Test (page 356)

1. $y = -3x + x^3 + x^5 + C$ 2. $y = -2e^{-2x} + C$ 3. $y = x^4/4 + e^x + C_1x + C_2$ 4. $y + y^2/2 = x^3/3 + C$
5. $y = x^2/2 - 2x^{3/2}/3 + 2$
6. $y = -e^{2x^2}$ 7. $1085 8. About 9050 9. (a) $dy/dx = -kx$ (b) equation is $y = 200e^{-.17x}$; about 36.5 grams
10. (a) $y = 60,000/(60 + 940e^{-x})$ (b) about 148

Case 9 (page 359)

1. $x = 100,000e^{140.71t}$; $x \approx 9.999 \times 10^{99}$; increasing 2. $x = 100,000e^{-3.29t}$; $x \approx 3725$; decreasing

Chapter 9

Section 9.1 (page 365)

1. 6 3. -20 5. 92
7. 47 9. $\sqrt{43}$ 11. $\sqrt{19}$
13. 1300 15. 1996 17. 136
19. 304 21. 7.85 23. 18.9
25. 17.3

Section 9.2 (page 372)

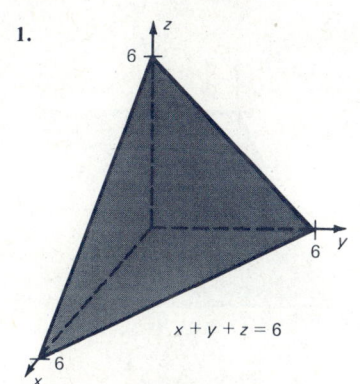

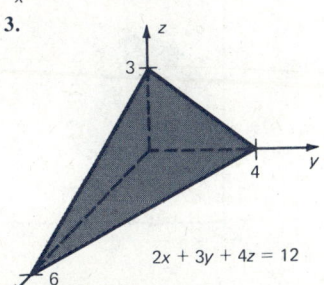

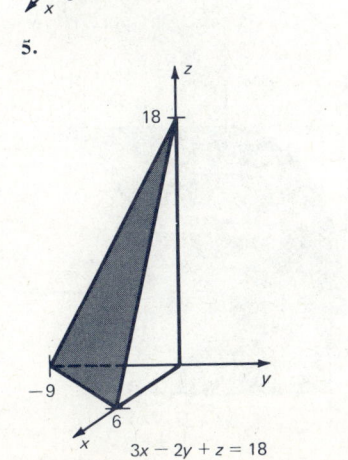

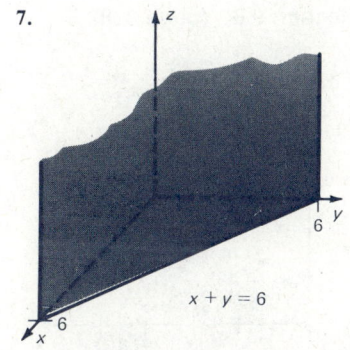

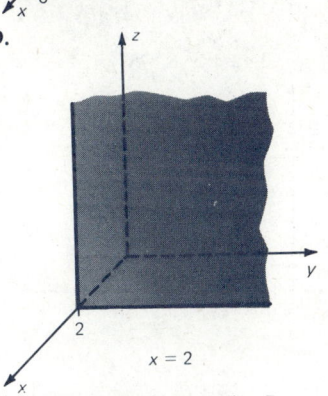

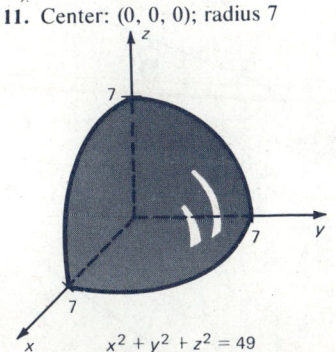

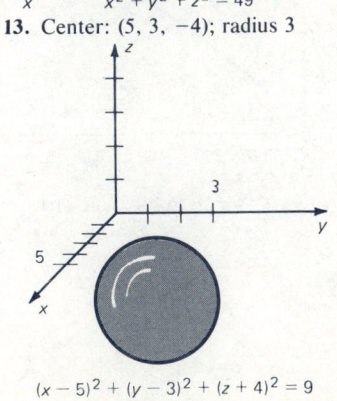

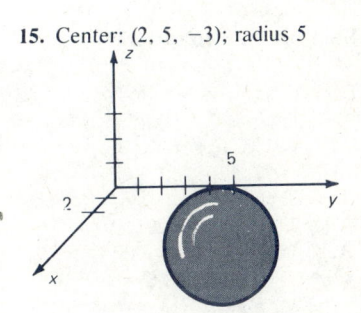

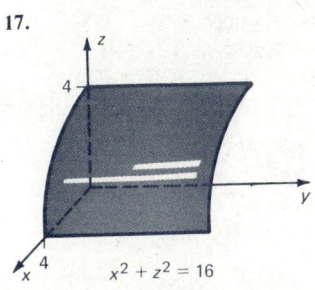

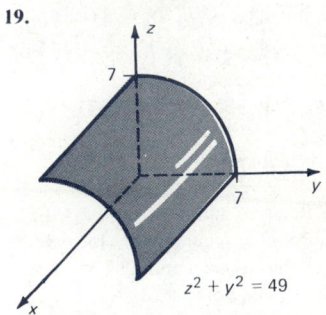

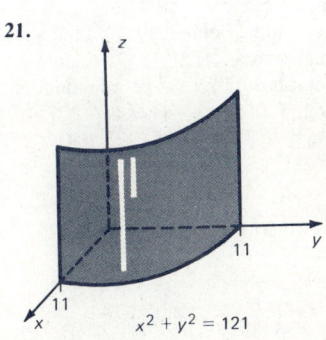

1. $x + y + z = 6$

3. $2x + 3y + 4z = 12$

5. $3x - 2y + z = 18$

7. $x + y = 6$

9. $x = 2$

11. Center: $(0, 0, 0)$; radius 7
    $x^2 + y^2 + z^2 = 49$

13. Center: $(5, 3, -4)$; radius 3
    $(x - 5)^2 + (y - 3)^2 + (z + 4)^2 = 9$

15. Center: $(2, 5, -3)$; radius 5
    $(x - 2)^2 + (y - 5)^2 + (z + 3)^2 = 25$

17. $x^2 + z^2 = 16$

19. $z^2 + y^2 = 49$

21. $x^2 + y^2 = 121$

23. $\sqrt{3}$ 25. $\sqrt{99} = 3\sqrt{11}$
27. $\sqrt{59}$

482 Answers

Section 9.3 (page 377)

1. $f_x = 24x; f_y = 8y;$ 48; 24   3. $f_x = -2y; f_y = -2x + 18y^2;$ 2; 170
5. $f_x = 9x^2y^2; f_y = 6x^3y;$ 36; -1152 7. $f_x = e^{x+y}; f_y = e^{x+y};$ e^1 or $e; e^{-1}$ 9. $f_x = -15e^{3x-4y}; f_y = 20e^{3x-4y}; -15e^{10}; 20e^{-24}$
11. $f_{xx} = 36xy; f_{yy} = -18; f_{xy} = f_{yx} = 18x^2$ 13. $R_{xx} = 8 + 24y^2; R_{yy} = -30xy + 24x^2; R_{xy} = R_{yx} = -15y^2 + 48yx$ 15. $r_{xx} = -8y/(x+y)^3;$ $r_{yy} = 8x/(x+y)^3; r_{xy} = r_{yx} = (4x-4y)/(x+y)^3$ 17. $z_{xx} = 0;$ $z_{yy} = 4xe^y; z_{xy} = z_{yx} = 4e^y$
19. $r_{xx} = -1/(x+y)^2; r_{yy} = -1/(x+y)^2; r_{xy} = r_{yx} = -1/(x+y)^2$
21. 80   23. 110   25. 168
27. 96   29. Increase of $70
31. $\partial z/\partial x = 2.1x^{1.1}y^{-.3}; \partial z/\partial y = .3x^{2.1}y^{-.7}$ 33. $2.5w^{-.67}$
35. $-1.675(T-F)w^{-1.67}$
37. $-2.5w^{-.67}$
39. $.8585W^{-.575}H^{.725}$
41. $1.4645W^{.425}H^{-.275}$ 43. About 5.71   45. $1/(a-v)$

Section 9.4 (page 386)

1. Saddle point at $(1, -1)$
3. Relative minimum at $(-1, -1/2)$
5. Relative minimum at $(-2, -2)$
7. Relative minimum at $(15, -8)$
9. Relative maximum at $(2/3, 4/3)$
11. Saddle point at $(2, -2)$
13. Saddle point at $(1, 2)$
15. Saddle point at $(0, 0)$; relative minimum at $(4, 8)$ 17. Saddle point at $(0, 0)$; relative minimum at $(9/2, 3/2)$ 19. $P(12, 40) = 2744$
21. $x = 12; y = 72$ 23. $C(12, 25) = 2237$

Section 9.5 (page 391)

1. $f(6, 6) = 72$ 3. $f(4/3, 4/3) = 64/27$ 5. $f(5,3) = 28$
7. $f(20, 2) = 360$ 9. 10, 10
11. $x = 6$ and $y = 12$ 13. 30, 30, 30   15. 50 by 50   17. 150 by 300   19. $r = 5, h = 10$ 21. 12.91 m by 12.91 m by 6.46 m   23. Make 10 large, no small   25. $x = 200, y = 300$

Section 9.6 (page 398)

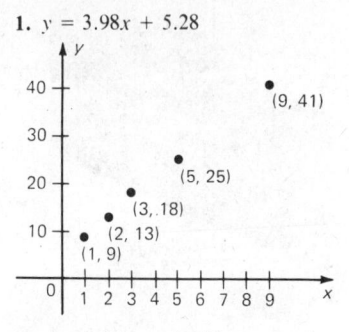

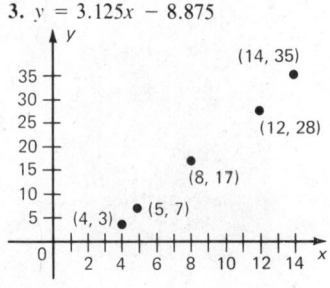

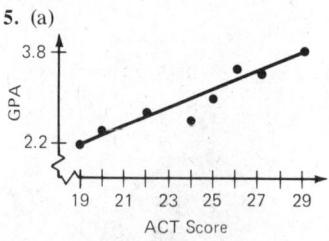

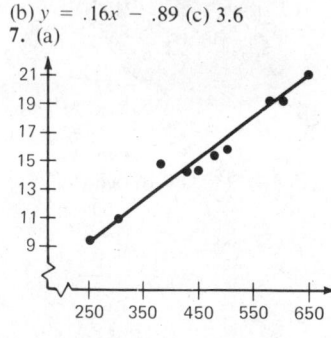

1. $y = 3.98x + 5.28$

3. $y = 3.125x - 8.875$

5. (a)

(b) $y = .16x - .89$ (c) 3.6
7. (a)

(b) $y = .03x + 1.56$ (c) 22.6; 24.1

Chapter 9 Test (page 401)

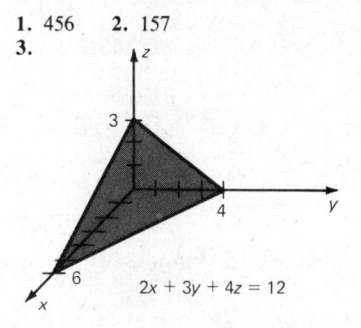

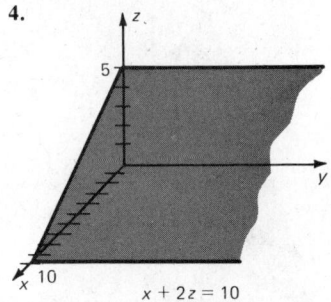

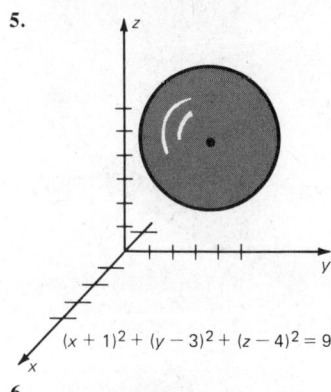

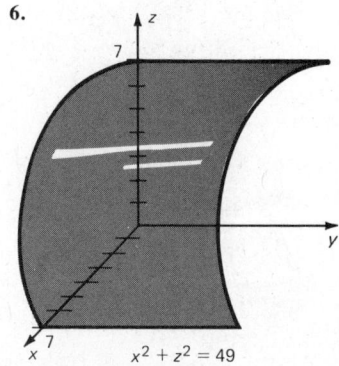

1. 456   2. 157
3.

$2x + 3y + 4z = 12$

4.

$x + 2z = 10$

5.

$(x+1)^2 + (y-3)^2 + (z-4)^2 = 9$

6.

$x^2 + z^2 = 49$

7. $f_x = 27x^2 - 4y$; $f_y = 4y - 4x$; 88; 12 **8.** $f_x = 1/(y+1)$; $f_y = -(x+1)/(y+1)^2$; 1/6; $-4/49$ **9.** $f_{xx} = 2y^3$; $f_{yy} = 6x^2y$; $f_{xy} = f_{yx} = 6xy^2 + 8$
10. $f_{xx} = 2y/(1+x)^3$; $f_{yy} = 0$; $f_{xy} = f_{yx} = -1/(1+x)^2$ **11.** Saddle point at (0, 2) **12.** Saddle point at (0, 0); relative maximum at $(-3/4, -9/32)$ **13.** $x = 11$; $y = 12$
14. $C(11, 12) = 431$ **15.** 100 by 100 **16.** $y = .97x + 31.9$
17. About 216

Chapter 10

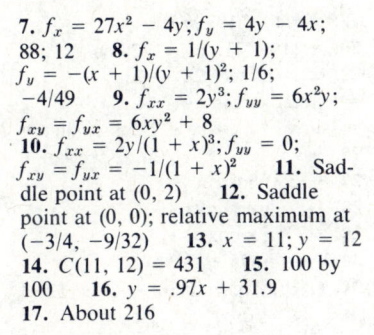

Section 10.1 (page 410)

1. $\pi/3$ **3.** $5\pi/6$ **5.** $7\pi/6$
7. $\pi/6$ **9.** 315° **11.** 330°
13. 288° **15.** 48°
In Exercises 17–23 we give the answers in the order sine, cosine, tangent, cotangent, secant, and cosecant.
17. 4/5; −3/5; −4/3; −3/4; −5/3; 5/4 **19.** 4/5; 3/5; 4/3; 3/4; 5/3; 5/4 **21.** 1; 0; undefined; 0; undefined; 1 **23.** $-1/2$; $-\sqrt{3}/2$; $\sqrt{3}/3$; $\sqrt{3}$; $-2\sqrt{3}/3$; -2 **25.** -3
27. 5 **29.** 1 **31.** -1
33. 1 **35.** 1/3 **37.** $\sqrt{5}/5$
39. +; +; +; +; +; + **41.** −; −; +; +; −; − **43.** II **45.** III
47. IV **49.** I or II **51.** I or III

Section 10.2 (page 420)

In Exercises 1–11 we give the answers in the order sine, cosine, tangent, cotangent, secant, and cosecant.
1. $\sqrt{3}/2$; $-1/2$; $-\sqrt{3}$; $-\sqrt{3}/3$; -2; $2\sqrt{3}/3$ **3.** $1/2$; $-\sqrt{3}/2$; $-\sqrt{3}/3$; $-\sqrt{3}$; $-2\sqrt{3}/3$; 2 **5.** $-\sqrt{3}/2$; $-1/2$; $\sqrt{3}$; $\sqrt{3}/3$; -2; $-2\sqrt{3}/3$
7. $-1/2$; $\sqrt{3}/2$; $-\sqrt{3}/3$; $-\sqrt{3}$; $2\sqrt{3}/3$; -2 **9.** $\sqrt{3}/2$; $1/2$; $\sqrt{3}$; $\sqrt{3}/3$; 2; $2\sqrt{3}/3$ **11.** $1/2$; $-\sqrt{3}/2$; $-\sqrt{3}/3$; $-\sqrt{3}$; $-2\sqrt{3}/3$; 2
13. $\sqrt{3}/3$; $\sqrt{3}$; 2 **15.** $\sqrt{3}/2$; $\sqrt{3}/3$; $2\sqrt{3}/3$ **17.** -1; -1
19. $-\sqrt{3}/2$; $-2\sqrt{3}/3$
21. .6293 **23.** 7.1154
25. .6018 **27.** .8693

29. .3907 **31.** .1564
33.

$y = 2 \sin x$

35.

$y = \cos 2x$

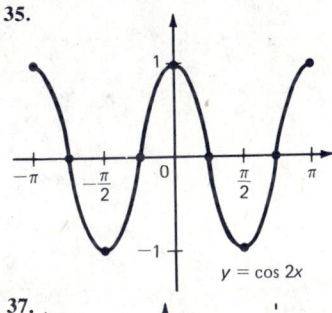

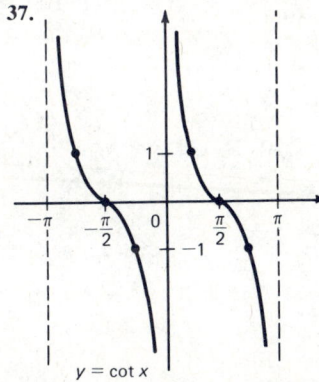

37.

$y = \cot x$

39. 1; 240° **41.** 80° **43.** 15°
45. Down **47.** 44° **49.** 84°
51. 100 **53.** 122 **55.** About 1°C **57.** 25°C **59.** 37

Section 10.3 (page 427)

1. (b) **3.** (e) **5.** (a)
7. (a) **9.** (b) **11.** True
13. False **15.** True
17. False **19.** $(\sqrt{6} + \sqrt{2})/4$
21. $(\sqrt{2} - \sqrt{6})/4$ **23.** $\sqrt{2}/2$
25. $\sqrt{2}/2$ **27.** $(\sqrt{6} + \sqrt{2})/4$
29. $\sqrt{2}/2$ **31.** $-\sqrt{3 - 2\sqrt{2}}$ **33.** $+\sqrt{3 - 2\sqrt{2}}$
35. $\sqrt{2} - \sqrt{2}/2$; $\sqrt{2} + \sqrt{2}/2$;

37. $-\sqrt{2} - \sqrt{3}/2$; $-\sqrt{2} + \sqrt{3}/2$; $\sqrt{7 - 4\sqrt{3}}$
39. $-\sqrt{2 - \sqrt{2}}/2$; $\sqrt{2 + \sqrt{2}}/2$; $-\sqrt{3 - 2\sqrt{2}}$ **41.** 84° **43.** 60°
45. 3.9

Section 10.4 (page 432)

1. $-\pi/3$ **3.** $\pi/4$ **5.** $-\pi/2$
7. $\pi/3$ **9.** $3\pi/4$ **11.** $-\pi/3$
13. $-8°$ **15.** 154° **17.** 22°
19. 48° **21.** $-42°$ **23.** 68°
25. $\sqrt{5}/2$ **27.** $\sqrt{5}/5$
29. $-\sqrt{21}/2$ **31.** 5/3
33. $-3\sqrt{10}/10$ **35.** 18° **37.** 15°

Section 10.5 (page 439)

1. $y' = 12 \cos x$ **3.** $y' = 108 \cdot \sec^2(9x + 1)$ **5.** $y' = -4 \cos^3 x \cdot \sin x$ **7.** $y' = 5 \tan^4 x \sec^2 x$
9. $y' = -5(4x \cos 4x + \sin 4x)$
11. $y' = (x \cos x - \sin x)/x^2$
13. $y' = 5e^{5x} \cos e^{5x}$ **15.** $y' = (\cos x)e^{\sin x}$ **17.** $y' = (2/x) \cdot \cos(\ln 4x^2)$ **19.** $y' = (2x \cos x^2)/\sin x^2$ or $2x \cot x^2$ **21.** $y' = 12/\sqrt{1 - 144 x^2}$
23. $y' = 3/(1 + 9x^2)$ **25.** $y' = -2/(x^2\sqrt{1 - 4/x^2})$ or $-2/(x\sqrt{x^2 - 4})$
27. $y' = -1/[x(1 + [\ln|-2/x|]^2)]$
29. $y' = 1/[(x^2 + 2x + 2)\text{Arctan}(x + 1)]$
31. $y' = -200\pi \sin 2\pi x$ **33.** 0
35. 1 **37.** -1 **39.** -1

Section 10.6 (page 443)

1. $(1/5) \sin 5x + C$ **3.** $5 \sin x - 2 \cos x + C$ **5.** $2 \sec 2x + C$ **7.** $-3 \tan 2x + C$ **9.** $(1/8) \sin^8 x + C$ **11.** $(2/3) \cdot \sin^{3/2} x + C$ **13.** $-\ln|1 + \cos x| + C$
15. $(-6/5) x \sin 5x - (6/25) \cos 5x + C$
17. $-8x \cos x + 8 \sin x + C$
19. $(-3/4) x \sin 8x - (3/32) \cos 8x + C$ **21.** $1 - \sqrt{2}/2$ **23.** 1 **25.** -2
27. $(3/5) \ln|\sec 5x| + C$ **29.** $(1/6) \cdot \tan^2(3x - 1) - (1/3) \ln|\sec(3x - 1)| + C$ **31.** $(-1/4) \sin^3 x \cos x + 3x/8 - (3/8) \sin x \cos x + C$
33. $[(2x^2 - 1)\text{Arccos } x - x\sqrt{1 - x^2}]/2 + C$
35. $e^x(\cos 3x + 3 \sin 3x)/10 + C$
37. $-3 \cos x - (\cos 9x)/3 + C$
39. $(\sin 4x)/8 + (\sin 6x)/12 + C$
41. $x\sqrt{16 - x^2}/2 + 8 \text{ Arcsin } x/4 + C$

Section 10.7 (page 448)

1. 1 **3.** −1 **5.** 1 **7.** −2; 2
9. 9; −9 **11.** 1/2; −1/2 **13.** About 1490 **15.** About 2720 **17.** About 5470 **19.** About 7000 **21.** About 1330 **23.** $\sin \theta = s/L_2$ **25.** $\cot \theta = (L_0 - L_1)/s$ **27.** $k \cdot L_1/r_1^4$
29. $R = k \cdot L_1/r_1^4 + k \cdot L_2/r_2^4$ **31.** $R' = k(s \csc^2 \theta)/r_1^4 + k(-s \cos \theta/\sin^2 \theta)/r_2^4$
33. $0 = k/r_1^4 - k \cos \theta/r_2^4$
35. $\cos \theta = 1/256 \approx .0039$; $\theta \approx 90°$

Chapter 10 Test (page 451)

1. $\pi/2$ **2.** $5\pi/6$ **3.** 1260°
4. 81° **5.** $\sqrt{3}/2$ **6.** $-\sqrt{3}$
7. $\sqrt{3}/2$ **8.** $\sqrt{2}$ **9.** 1.0724
10. .7880 **11.** .9945
12. .9511 **13.** 1
14. $(\sqrt{6} - \sqrt{2})/4$ **15.** $\sqrt{2 + \sqrt{2}}/2$
16. $\sqrt{3}/2$ **17.** $-\pi/6$
18. $-\pi/4$ **19.** $\sqrt{3}/2$
20. $\sqrt{11}$ **21.** $y' = -28 \cos 7x$
22. $y' = 12 \sec^2 2x$ **23.** $y' = -18 \cos^5 x \sin x$ **24.** $y' = [\sin x - x \cos x + 2 \cos x]/\sin^2 x$
25. $-3/\sqrt{1 - 9x^2}$ **26.** $2 \sec x + C$
27. $(-2/3) \cos^{3/2} x + C$
28. $(1/9) \ln |\sec 9x| + C$
29. $-3 \sin x \cos x + 3x + C$
30. $\operatorname{Arcsin} x/5 + C$
31. $(1/10)(\operatorname{Arcsec} x/10) + C$ **32.** 1

Bibliography

Algebra Review

If Chapter 1 of this text does not provide enough algebra review for you, look in any standard intermediate algebra or college algebra text for further algebra.

Lial, Margaret, and Charles Miller. *College Algebra*. 2nd ed. Glenview, Ill.: Scott, Foresman, 1977.

Miller, Charles, and Margaret Lial. *Intermediate Algebra: Text/Workbook*. Glenview, Ill.: Scott, Foresman, 1980. This second book has a set of audio tapes available that make self study easy.

Trigonometry Review

For a review of trigonometry (for those students who will study Chapter 10 of this book) see any standard trigonometry text.

Lial, Margaret, and Charles Miller. *Trigonometry*. Glenview, Ill.: Scott, Foresman, 1977.

Calculus Books for Reference

Ayres, Frank. *Schaum's Outline of Theory and Problems of Calculus*. 2nd ed. New York: McGraw-Hill, 1964. The emphasis of this book in the commonly available Schaum's Outline Series is on engineering and physics. The format of these books makes them difficult to learn from, but the many worked examples that they provide make them reasonable for review.

Oakley, Cletus O. *Calculus*. rev. ed. New York: Barnes & Noble. 1957. Oakley doesn't have as many problems as the Schaum's book, but is probably easier to understand.

Calculus Books for Proofs

The books listed on the next page provide proofs of just about everything that we do not prove. They also show numerous applications of calculus to physics and engineering.

De Sapio, Rodolfo. *Calculus for the Life Sciences*. San Francisco: W. H. Freeman, 1978. An excellent book with applications of calculus to the life sciences, at a higher level than this text.

Leithold, Louis. *Calculus*. 3rd ed. New York: Harper & Row, 1976.

Riddle, Douglas F. *Calculus*. 3rd ed. Belmont, Calif.: Wadsworth, 1979. Offers standard treatments of most topics.

Salas, Saturnino L., and Einar Hille. *Calculus*. 3rd ed. New York: John Wiley & Sons, 1978.

Swokowski, Earl W. *Calculus*. 2nd ed. Boston: Prindle, Weber & Schmidt, 1979. A very good book, with clear explanations.

Thomas, George B., Jr., and Ross L. Finney. *Calculus*. 5th ed. Reading, Mass.: Addison-Wesley, 1979. A classic book in the field.

Glossary

Absolute maximum The highest value a function ever takes on.

Absolute minimum The lowest value a function ever takes on.

Absolute value The distance on the number line from a number to 0. The absolute value of the number x is written $|x|$.

Acceleration The rate of change of velocity.

Acceleration of revenue The second derivative of the revenue function.

Amplitude The maximum distance of the graph $y = \sin x$ or $y = \cos x$ from the x-axis.

Angle An angle is formed by rotating a ray about its end point.

Antiderivative A function whose derivative is some given function.

Asymptote A line which a graph approaches very closely but never reaches.

Average rate of change The difference in two y-values divided by the difference in the two corresponding x-values.

Boundary condition A condition which permits the value of an arbitrary constant in a differential equation to be determined.

Break-even point The point at which costs are recovered. From the break-even point on, profits will be earned.

Carbon-14 dating A process used for dating objects; approximately valid for objects not over 10,000 years old.

Chain rule A rule for finding the derivative of a function of a function. See Section 3.7.

Closed interval Two points on the number line together with all the points between the two points.

Coefficient A number used to multiply a variable.

Common logarithm A logarithm to base 10.

Coefficient of inequality A number which shows how the income distribution in a particular nation differs from a truly theoretical income distribution.

Concave downward A function is concave downward on an interval if the second derivative is negative for all values on the interval. See Figure 4.18.

Concave upward A function is concave upward on an interval if its second derivative is positive for all values in the interval. See Figure 4.18.

Constraint A restriction on a variable.

Consumer surplus The amount of money saved by consumers when they are able to get an item at a price lower than they would be willing to pay.

Continuous compounding Interest compounded every instant.

Continuous function A function with no breaks, gaps, or jumps.

Convergent integral An improper integral which exists.

Coordinate A number which is associated with a point on a graph.

Cost-benefit model A mathematical formulation showing cost as a function of various levels of achievement, for example, the percent of pollutant removed from the environment.

Counting number Any number from the set $\{1, 2, 3, 4, \ldots\}$.

Critical point A value for which a derivative is 0, or a value in the domain of the function for which the derivative does not exist.

Decreasing function A function whose graph goes down as we move from left to right along the x-axis.

Definite integral The limit of a certain kind of sum; it represents the area between a curve and the x-axis and two vertical lines.

Degree A unit for measuring angles; the central angle of a circle has 360°.

Demand curve The curve which shows how the demand for an item is affected by the price of the item.

Derivative A very useful function, found from a given functon by using either the definition of a derivative or theorems. See Chapter 3.

Differential The differential of y is defined as the first derivative of y multiplied by the change in x. See Section 4.8.

Differential equation An equation involving derivatives.

Discontinuity A point where a function has a break, gap, or jump.

Divergent integral An improper integral which does not exist.

Domain of a function The set of all possible values of the independent variable in a function.

e A very important number in applications of mathematics; e is approximately 2.7182818.

Economic lot size The number of items that should be ordered at one time to minimize total order and carrying costs. See Section 4.5.

Element An object belonging to a set.

Equation A statement that two expressions are equal.

Equilibrium demand The demand at which supply and demand of an item are equal.

Equilibrium price The price at which the supply and demand of an item are equal.

Equilibrium supply The supply at which supply and demand of an item are equal.

Explicit function A function of the form $y = f(x)$ in which one variable is expressed in terms of the other.

Exponent A number which tells you how many times a number is used in a product.

Exponential distribution A probability density function of the form ae^{-ax}.

Exponential function A function of the form $y = a^x$ where a is a positive real number and a is not equal to 1.

First derivative test A test using the first derivative to tell whether a critical point leads to a maximum or a minimum.

Fixed cost The cost of getting ready to manufacture an item: building factories, designing equipment, hiring workers, etc.

Function A rule of correspondence where for each value of x, there is exactly one value of y.

Function of several variables A function having more than one independent variable.

Fundamental theorem of calculus The key theorem connecting differential calculus and integral calculus. See Section 6.3.

General solution A solution to a differential equation involving an arbitrary constant.

Half-life The time it takes for half of a radioactive sample to decay.

Identity An equation which is true for all meaningful values of the variable.

Implicit differentiation The method of finding the derivative of an implicit function.

Implicit function A function in which the defining equation is not solved for the dependent variable.

Improper integral An integral where one of the limits of integration is infinity.

Increasing function A function whose graph moves up as we move along the x-axis from left to right.

Indeterminate form A form such as 0/0.

Infinity The symbol $x \to \infty$ is used to express the fact that x gets larger and larger without bound. The symbol ∞ does not represent any real number.

Initial condition A condition which permits the value of an arbitrary constant in a differential equation to be determined.

Instantaneous rate of change The limit of the average rate of change as the differences in the x-values approach 0.

Integer Any number from the set $\{\ldots, -3, -2, -1, 0, 1, 2, 3, \ldots\}$.

Integration by parts A method of integrating certain more complicated functions. See Section 7.1.

Intercept The point where a graph crosses the x-axis or the y-axis.

Irrational number A real number which is not rational. Examples of irrational numbers include $\sqrt{3}, \sqrt{5}, \sqrt{12}$, and π.

Lagrange multipliers A method for solving problems of maximums and minimums involving constraints.

Learning curve A curve which shows how rapidly the learning of various skills progresses.

Least squares regression equation The equation which gives the best possible straight line through a group of data points.

Limit A value approached by the dependent variable as the values of the independent variable approach some fixed number.

Limited growth model A model for growth in which the population approaches some fixed upper limit.

Linear equation An equation of the form $ax + by = c$.

Logarithmic function A function of the form $y = \log_a x$ where a is positive and a is not equal to 1.

Marginal cost The cost of making one more item; approximated by the derivative of the cost function.

Mathematical model A description of a real-world situation using mathematical equations and inequalities.

M-test A test for relative maximums and minimums of functions of two variables. See Section 9.4.

Natural logarithm A logarithm to base e.

Natural number Any number from the set $\{1, 2, 3, 4, \ldots\}$.

Newton's method A method for finding approximate solutions to equations.
Normal distribution The statistical distribution whose graph is a bell-shaped curve.
Number line A line with numbers marked on it, used to illustrate sets of numbers.
Numerical integration A method of approximating a definite integral by numerical methods.
Octant One-eighth of the space in a three-dimensional coordinate system.
Open interval All points on a number line up to a given point or all points on a number line beyond a given point or all points on a number line between two given points.
Ordered pair Two numbers, written inside parentheses and separated by commas, in which the order is important.
Ordered triple Three numbers, written inside parentheses and separated by commas, in which the order is important.
Parabola The graph of a quadratic function.
Partial derivative A derivative of a function of several variables taken with respect to one variable at a time.
Particular solution A solution to a differential equation with no arbitrary constant.
Point of inflection A point where a function changes from being concave downward to concave upward, or vice versa. The point of inflection is usually found where the second derivative is 0.
Probability density function A function which is used to describe a relative frequency curve.
Producers surplus Extra money earned by producers of an item, who can produce the item for less than the going price.
Product exchange function A function giving the relationship between quantities of two items that can be produced by the same machine or factory.
Production function A function which gives the quantity of an item produced as a function of two or more other variables.
Product rule A rule for finding the derivative of a product of functions. See Section 3.6.
Quadrant One fourth of the space in a two-dimensional coordinate system.
Quadratic equation An equation of the form $ax^2 + bx + c = 0$ where a does not equal 0.
Quotient rule A rule for finding the derivative of a quotient of two functions. See Section 3.6.
Radian A unit of measure for angles; one radian is the measure of an angle whose vertex is at the center of a circle and which cuts an arc on the circle equal in length to the radius of the circle.
Range The set of all possible values of the dependent variable in a function.
Rational expression A quotient of two algebraic expressions (with nonzero denominator).
Rational number A number which can be expressed as the quotient of two integers (with nonzero denominator).
Ray A portion of a line that starts at a given point and continues in one direction.
Real number A number which corresponds to a point on a number line.
Relative frequency curve The graph of a probability density function.
Relative maximum The highest value of a function within some neighborhood.
Relative minimum The lowest value of a function within some neighborhood.
Saddle point A point on a graph which is a maximum from one direction and a minimum from another direction.
Scatter diagram A diagram of all the data points in a given distribution.
Secant line A line cutting a graph in at least two points.
Second derivative The derivative of a derivative.
Second derivative test A method used to tell whether a given critical value leads to a maximum or a minimum.
Separation of variables A method of solving differential equations. See Section 8.2.
Set A collection of objects. In mathematics the objects are usually numbers.
Set braces The symbols { } which are used to enclose the elements of a set.
Sigma notation Use of the Greek capital letter sigma (Σ) to indicate summation.
Simpson's rule A method of numerical integration.
Slope A numerical measure of the steepness of a straight line.
Solid of revolution A solid figure which results from rotating a region about a line.
Standard normal distribution A normal distribution having mean zero and standard deviation one.
Supply curve A curve which shows how the supply of an item is affected by the price of the item.
Table of integrals A list of antiderivatives for various useful functions.
Tangent line A line which just touches a graph.
Trapezoidal rule A method of numerical integration.

True annual interest The interest rate which must be stated, according to government regulations, in consumer finance contracts.

Uniform distribution A probability density function in which the probability of every outcome is the same.

Unlimited growth model A model of growth in which there are no limits on the growth built into the model.

Variable A letter which stands for a number.

Velocity The speed and direction of a moving object.

Velocity of revenue The derivative of a revenue function.

Vertex The highest or lowest point on a parabola.

Whole number Any number from the set {0, 1, 2, 3, 4, . . . }.

Index

Absolute maximum, 164
Absolute minimum, 164
Absolute value, 5
   function, 46
Acceleration, 175
Acute angle, 404
Addition property of
   equality, 9
Addition property of
   inequality, 11
Amplitude, 417
Angle, 403
Antiderivative, 254
   of trigonometric
      functions, 440–41
Approximations, 205
Arc, 405
Arcsin function, 430
Area, 261
   between two curves,
      279–86
Asymptote, 81
Average cost, 140
Average rate of change,
   113
Average status, 8
Average velocity, 117
Avis, 57

Base, 32
Bee's cell, 452
Boeing Company, 212
Booz, Allen and
   Hamilton, 91
Boundary condition, 339
Break-even point, 68

Calculus, fundamental
   theorem of, 269
Carbon-14 dating, 236
Celsius, 13
Chain rule, 147
Change in x, 59
Closed interval, 109, 153
Coefficient, 15
Common denominator, 10
Common logarithm, 228
Compound interest, 220,
   239
Concave, upward or
   downward, 172
Constant rule, 131
Constraint, 387
Consumer's surplus, 286

Continuity, 105–9
Continuous
   at a point, 107
   on an interval, 109
Continuous compounding,
   239
Convergent integral, 319
Cooling, Newton's law of,
   355
Cosecant, 407
Cosine, 407
Cost
   average, 140
   fixed, 68
   marginal, 134
Cost function, 67
Cost-benefit model, 85
Cotangent, 407
Counting numbers, 2
Critical point, 155
Cube root, 35
Curve sketching, 170, 177
Cylinder, 370

Decreasing function, 158
Definite integral, 264, 265
Degree, 404
Delta Airlines, 39
Delta x (Δx), 59
Demand, 55
Denominator, common, 10
Dependent variable, 44,
   363
Depletion of minerals, 301
Derivative, 121, 123
   chain rule, 147
   definition of, 123
   of exponential, 243
   of inverse trigonometric
      functions, 438
   of natural logarithm, 243
   of a power, 131
   of a product, 138
   of a quotient, 139
   of trigonometric
      functions, 434–39
   partial, 372
   second, 170
   second partial, 375
   short cuts for finding,
      130
Difference of two cubes,
   19
   of two squares, 19

Differential, 206
Differential equations,
   337
   applications of, 347
   general solution of, 338
   particular solution of,
      339
Differentiation, implicit,
   195
Discontinuous, 105
Distribution
   exponential, 328
   normal, 328
   standard normal, 329
   uniform, 327
Distributive property, 9
Divergent integral, 319
Domain, 46
Double angle identities,
   427
Drug dosage, 67

e, 221
Economic lot size, 190
Electoral college, 39
Equality
   addition property of, 9
   multiplication property
      of, 9
Equation, 8
   linear, 8
   quadratic, 21
Equilateral triangle, 412
Equilibrium
   demand, 56
   price, 56
   supply, 56
Explicit function, 194
Exponent, 15, 32
   properties of, 34, 36
Exponential
   derivative of, 243
   distribution, 328
   form, 226
   functions, 218

Factor, greatest common,
   17
Factoring, 16
Fahrenheit, 13
Fick's law, 260
First derivative test, 160
First octant, 367
Fixed cost, 68

Forgetting curve, 240
Function, 44
   absolute value, 46
   cost, 67
   decreasing, 158
   definition of, 44
   explicit, 194
   exponential, 218
   implicit, 194
   increasing, 158
   inverse, 227
   inverse trigonometric,
      430
   Lagrange, 388
   linear, 45, 51
   logarithmic, 226
   of two variables, 362
   periodic, 416
   polynomial, 77
   probability density, 325
   product exchange, 87
   quadratic, 45, 72
   rational, 80
   trigonometric, 403, 407
Fundamental theorem of
   calculus, 269

General solution, 338
Generalized power rule,
   144
Graph, 2
Greatest common factor,
   17

Half-angle identities, 427
Hares, 360
Harmonic motion, 447
Honeycombs, 452
Horizontal asymptote, 81

Identity, 410, 423–27
   double-angle, 426
   half-angle, 427
   reciprocal, 423
Implicit differentiation,
   194
Implicit function, 194
Improper integral, 318
Increasing function, 158
Indefinite integral, 270
Independent variable, 44,
   363
Indeterminate form, 98
Inequality, 11

Infinity, limits to, 99
Inflection point, 172
Initial condition, 338
Initial side, 403
Instantaneous rate of change, 117
Instantaneous velocity, 117
Insulin, 136
Integers, 2
Integral,
  definite, 264, 265
  improper, 318
  indefinite, 270
  sign, 264
  table of, 295
Integration
  by parts, 304, 305
  by substitution, 289
  numerical, 308
Interest
  compound, 220
  continuous compounding, 239
  true annual, 14, 205
  unearned, 14
Interval notation, 108
Inverse functions, 227
Inverse trigonometric functions, 430
Irrational numbers, 2
Isosceles triangle, 414

Just-noticeable-difference (JND), 70

Lagrange function, 388
Lagrange multipliers, 387
Law of cooling, Newton's, 355
Learning curve, 173, 235
Least squares regression line, 393
Like terms, 16
Limit, 93–101
  definition of, 94
  lower, 264
  of integration, 265
  to infinity, 99
  upper, 264
Limited growth model, 349
Linear equation, 8
Linear function, 45, 51
Logarithmic function, 226
Logarithms
  common, 228
  natural, 229
  properties of, 228
Logistic curve, 240
Lot size, economic, 190

Lower limit, 264
Lynx, 360

M-test, 382
Marginal
  cost, 91, 120, 134
  profit, 134
  revenue, 134, 145
Mathematical model, 1
Maximum
  absolute, 164
  relative, 153, 379
Mineral depletion, 301
Minimum
  absolute, 164
  relative, 153, 380
Minute, 404
Model
  cost-benefit, 85
  limited growth, 349
  mathematical, 1
  training program, 215
  unlimited growth, 348
Multiplication property of equality, 9
Multiplication property of inequality, 11
Multipliers, Lagrange, 387

Natural logarithm, 229
  derivative of, 243
Natural number, 2
Newton's
  method, 201
  law of cooling, 355
Normal distribution, 328
nth root, 35
Number line, 1
Numbers
  counting, 2
  integers, 2
  irrational, 2
  natural, 2
  rational, 2
  real, 2
  whole, 2
Numerical integration, 308

Obtuse angle, 404
Octant, 367
Open interval, 109, 153
Ordered pair, 44
Ordered triple, 366
Oscilloscope, 419

Parabola, 73
Partial derivative, 372
Particular solution, 339
Parts, integration by, 304
Perfect square, 19
Periodic function, 416

Plane, 367
Point of inflection, 172
Point-slope form, 62
Poiseuille's law, 212, 449
Polynomial, 15
  function, 77
Power rule, 131
  generalized, 144
Price, 55
Probability, 324
Probability density function, 323–29
Producer's surplus, 286
Product exchange function, 87
Product life cycle, 178
Product rule, 138
Properties of exponents, 34, 36
  of rational expressions, 27–28
Pythagorean theorem, 408

Quadrant, 46
Quadratic
  equation, 21
  formula, 22
  function, 45, 72
  inequalities, 24
Quotient rule, 139

Radian, 405
Radical, 36
Radius vector, 403
Random variable, 323
Range, 46
Rate of change, 117
  average, 113
Rational expression, 27
Rational function, 80
Rational numbers, 2
Ray, 403
Real numbers, 2
Reciprocal identities, 410, 423
Regression
  equation, least squares, 395
  line, 393–98
Relative maximum, 153, 379
Relative minimum, 153, 380
Revenue, marginal, 135
Revolution, solid of, 313
Right angle, 404
Right circular cylinder, 370
Rule of 78, 14

Saddle point, 380

Sales analysis, 66
Scatter diagram, 393
Secant function, 407
Secant lines, 122
Second derivative, 170
  test, 174
Second partial derivative, 375
Separation of variables, 343
Set, 1
  braces, 1
Seventy-eight, rule of, 14
Shaklee Products, 7
Sigma notation, 265, 396
Sign graph, 25
Simple harmonic motion, 446
Simpson's rule, 310
Sine, 407
Sine waves, 419
Slope, 60
Slope-intercept form, 61
Slope of tangent line, 123
Solid of revolution, 313
Solution, 8
  general, 338
  particular, 339
Sphere, 369
Square, perfect, 19
Square root, 35
Standard normal distribution, 329
Standard position, 404
Status incongruity, 8
Straight angle, 404
Substitution, integration by, 289
Sum of two cubes, 19
Supply, 55
Surplus
  consumers, 286
  producers, 286

Table of integrals, 295
Tangent line, 122
Tangent function, 407
Term, 15
Terminal side, 403
Total value expressed by definite integral, 273
Training program model, 215
Triangle
  equilateral, 412
  isosceles, 414
Trigonometric functions, 403, 407
  inverse, 430–32
True annual interest, 14, 205

Uniform distribution, 327
Unlike terms, 16
Unlimited growth model, 348
Upjohn Company, 90
Upper limit, 264

Van Meegeren, 251

Variable, 8, 15
  dependent, 44, 363
  independent, 44, 363
Variables, separation of, 343
Velocity, 117
Vertex
  of an angle, 403
  of a parabola, 73
Vertical asymptote, 81
Vertical line test, 48

Whole numbers, 2

x-axis, 46
x-coordinate, 46
x-intercept, 51

y-axis, 46
y-coordinate, 46
y-intercept, 51

Zero-factor property, 21

What do you think of *Essential Calculus with Applications?*

We would appreciate it if you would take a few minutes to answer these questions. Then cut the page out, fold it, seal it, and mail it. No postage is required.

Which chapters did you cover?
    (circle)   1   2   3   4   5   6   7   8   9   10   All

Which helped most?
    Explanations _____   Examples _____   Exercises _____
    All three _____

Does the book have enough worked-out examples? Yes _____   No _____

Does the book have enough exercises? Yes _____   No _____

Were the problems in the margin helpful? Yes _____   No _____

Were the answers in the back of the book helpful? Yes _____   No _____

Did you have trouble getting the form of the answer in the back of the book?
    Often _____   Sometimes _____   Never _____

Were the cases in the text helpful? Yes _____   No _____

Did your instructor cover the cases in class? Yes _____   No _____

Did you use the *Study Guide with Computer Problems?*
    Yes _____   No _____   Did not know of it _____

If yes, was the *Study Guide* helpful?
    Yes _____   For some topics _____   No _____

For you, was the course elective _____
    required by _____

Do you plan to take more mathematics courses? Yes _____   No _____

If yes, which ones?
    Statistics _____   College Algebra _____
    Analytic geometry _____   Trigonometry _____
    Calculus (engineering and physics) _____   Other _____

How much algebra did you have before this course?
    Years in high school   (circle)   0   ½   1   1½   2   more
    Courses in college   0   1   2   3

If you had algebra before, how long ago?
    Last 2 years _____   3-5 years _____   5 years or more _____

What is your major or your career goal? _____   Your age? _____

We would appreciate knowing of any errors you found in the book. (Please supply page numbers.)

What did you like most about the book?

-- FOLD HERE --

What did you like least about the book?

College _____ State _____

-- FOLD HERE --

BUSINESS REPLY MAIL
FIRST CLASS   PERMIT NO. 31   GLENVIEW, IL

Postage will be paid by
SCOTT, FORESMAN AND COMPANY
College Division   Attn: Lial/Miller
1900 East Lake Avenue   LM-3
Glenview, Illinois 60025

NO POSTAGE
NECESSARY
IF MAILED
IN THE
UNITED STATES